SIXTH EDITION

INTRODUCTION TO FIRE PROTECTION AND EMERGENCY SERVICES

Robert Klinoff, AA, BS, EFO

Adjunct Instructor, National Fire Academy
Adjunct Instructor, Bakersfield College
Master Instructor, California State Fire Training
Safety Officer Type 1, National Wildfire Coordinating Group
FIRESCOPE Safety Specialist Group
Hazardous Materials Specialist
Emergency Medical Technician

JONES & BARTLETT
LEARNING

World Headquarters
Jones & Bartlett Learning
5 Wall Street
Burlington, MA 01803
978-443-5000
info@jblearning.com
www.psglearning.com

Jones & Bartlett Learning books and products are available through most bookstores and online booksellers. To contact Jones & Bartlett Learning Public Safety Group directly, call 800-832-0034, fax 978-443-8000, or visit our website, www.psglearning.com.

Substantial discounts on bulk quantities of Jones & Bartlett Learning publications are available to corporations, professional associations, and other qualified organizations. For details and specific discount information, contact the special sales department at Jones & Bartlett Learning via the above contact information or send an email to specialsales@jblearning.com.

19339-8

Production Credits

General Manager and Executive Publisher: Kimberly Brophy
VP, Product Development: Christine Emerton
Executive Editor: William Larkin
Senior Managing Editor: Donna Gridley
Editorial Assistant: Alexander Belloli
VP, Sales, Public Safety Group: Phil Charland
Project Specialist: John Fuller
Digital Project Specialist: Angela Dooley
Director of Marketing Operations: Brian Rooney
Production Services Manager: Colleen Lamy

VP, Manufacturing and Inventory Control: Therese Connell
Composition: codeMantra U.S. LLC
Project Management: codeMantra U.S. LLC
Cover Design: Kristin E. Parker
Text Design: Scott Moden
Rights Specialist: John Rusk
Senior Media Development Editor: Troy Liston
Cover image: Courtesy of Robert Klinoff
Printing and Binding: LSC Communications
Cover Printing: LSC Communications

Library of Congress Cataloging-in-Publication Data

Library of Congress Control Number: 2019914450
9781284180152

6048

Printed in the United States of America
25 24 23 22 21   10 9 8 7 6 5 4 3 2

Courtesy of Robert Klinoff

Brief Contents

Contents

Courtesy of Robert Klinoff

CHAPTER 3
Public Fire Protection 47

CHAPTER 4
Chemistry and Physics of Fire 77

CHAPTER **5**
Public and Private Support Organizations   95

CHAPTER **8**
Support Functions   **181**

CHAPTER **9**
Training   **199**

About the Author

Courtesy of Robert Klinoff

Robert W. Klinoff is a California Certified Chief Officer, National Fire Academy Executive Fire Officer, and retired Chief Deputy Fire Chief with the Kern County Fire Department. He is a 35-year fire service veteran. His background includes experience as a fire fighter while a student at Columbia College, as an Emergency Medical Technician, as a fire fighter with the U.S. Forest Service, as a fire fighter with the City of San Gabriel, California, and 29 years with the Kern County Fire Department. He has served the Kern County Fire Department in Operations and as a Training Officer and Fire Marshal.

His education includes an associate of science degree in Fire Science from Columbia College and a bachelor's degree in Occupational Studies—Vocational Arts from California State University Long Beach. He is a California State Fire Service Training and Education System–certified Chief Officer, Master Instructor, Movie and Television Fire Safety Officer, and Hazardous Materials Specialist. As an instructor for the State of California, he teaches both Company Officer- and Chief Officer-level courses. He also serves as adjunct faculty for the National Fire Academy, teaching Infection Control for Emergency Response Personnel, Managing Company Tactical Operations, and Chemistry of Hazardous Materials.

Robert has further served his community as the president of the Kern SAFE Coalition, an agency dedicated to childhood injury prevention. He is a National Fire Protection Association *Risk Watch* Champion and a National Highway Traffic Safety Administration–certified Child Passenger Safety Technician/Instructor.

Under the National Interagency Incident Qualifications System, he is certified as an Incident Commander, Safety Officer, Division Supervisor, and Strike Team/Task Force Leader. Robert served as a Safety Officer on California Interagency Management Teams. He and the team responded to the Pentagon incident on September 11, 2001, as well as numerous other major incidents throughout the country. As a result of the team's efforts at the Pentagon, they received the Group Honor Award for Excellence from the U.S. Department of Agriculture.

Robert continues to serve as a mentor for firefighters and firefighter candidates. He also remains a student of fire history, the future of the fire and rescue service, and fire department operations and administration.

CHAPTER **13**
Emergency Incident Management

CHAPTER **14**
Emergency Operations

Courtesy of Robert Klinoff

Acknowledgments

Reviewers

Robert Jay Alley
Gerton Fire and Rescue
Gerton, NC

Christopher L. Baker
Fire Science Instructor
North Central Fire Protection District
Kerman, CA

Michael Cromer
Captain, Adjunct Faculty
College of the Sequoias
Visalia, CA

Robert J. Healey
University of New Haven
West Haven, CT

Bradley Iverson, MS, CEM
College of Southern Nevada
Las Vegas, NV

Matt P. Johnson
Captain, Fall River Fire Department
Fall River, MA

Keith Kawamoto
Professor
College of the Canyons
Santa Clarita, CA

Kenneth Krebbs
Fire Marshal, Copper Canyon Fire and Medical District
Verde Fire Training Center
Cottonwood, AZ

Chad Landis
Training/Safety Officer
Rapides Parish Fire District 3/Alpine FD

Albert Lannone
Ret. Chief
West Sacramento, CA

Carl J. Mack
Lorain County Community College
Elyria, OH

Kenneth Milligan
Quinsigamond Community College
Westborough Fire Department
Westborough, MA

Dan Reid
Program Director, Fire Protection Technology & Emergency Management
Cape Fire Community College—North Campus Castle Hayne, NC

John E. Rzycki
Fire Protection Technology Instructor
Fayetteville Technical Community College
Fayetteville, NC

Kenneth Staelgraeve
Professor of Fire Science
Macomb Community College
Warren, MI

Scott Ventura
Ret. Fire Chief, Palm Springs Assistant Professor, Fire Technology
College of the Desert
Palm Desert, CA

Beverley E. Walker
East Georgia State College
Swainsboro, GA

Kevin S. Walker
Associate Professor of Business
Eastern Oregon University
Chair, Fire Services Administration
Chair, Emergency Medical Services Administration
Verde Fire Training Center
Cottonwood, AZ

Norcliff W. Wiley
Ret. Deputy Fire Chief/Fire Marshal
Salinas Fire Department
Fire Technology Instructor
Cabrillo College
Aptos, CA

Author's Acknowledgments

Courtesy of Robert Klinoff

I would like to personally thank Jack Amundsen, the retired chief of the Columbia College Fire Department. He is the person who first hired me as a fire fighter in 1971 and showed me what a rewarding career it can be. He always encouraged me to be the best that I could be. Many other instructors and co-workers have helped me to develop myself as a professional and increased my base of knowledge. They have done much to elevate the professional standing of all fire fighters in the eyes of the public.

I would also like to thank my wife, Helen, who has stood by me during my career with long periods away from home due to incidents and course work. She has always been supportive.

Many people have contributed their time and effort in assisting me with gathering the information to write this text. Their interest in developing the fire service should be recognized.

I would also like to thank the staff at Jones & Bartlett Learning for encouraging me to undertake the writing of this revision and for their help in its preparation.

CHAPTER 1

Fire Science/Technology Education and the Fire Fighter Selection Process

OBJECTIVES

After studying this chapter, you should be able to:

- Explain the difference between training and education.
- Describe the kinds of college fire science/technology programs that are available.
- Discuss the value of a background in public administration and other specialized studies.
- Discuss how to assess your career potential.
- Discuss the role of human resources and work ethics in the fire service.
- Identify preservice training programs.
- Identify different types of personnel development programs.
- List and describe the steps in the selection process.
- Describe how to set and meet career goals.

Case Study

A young man graduates high school without distinction and has no clear idea what he wishes to do with his life. He has grown up in a town with a volunteer fire department, but since he has not seen them around, he does not consider firefighting a career. He goes to college and, finding himself in need of financial support, gets a job as a fire fighter in the school fire department. Once there he develops a love for the job and starts taking courses to complete an associate's degree in fire science/technology.

After a few years, including a stint as a seasonal fire fighter with the U.S. Forest Service and working for an ambulance company, he attains a job in a fire department. Once there he wishes to be promoted and, seeing little opportunity where he is (two stations and thirty-three personnel), he seeks a job in another department. While working in the new department, he continues his education to the point of a bachelor's degree and certifications as Company Officer and Chief Officer. He rises through the ranks to the number-two spot in the organization (with 46 stations and 500 personnel). During all of this, he becomes a professor of fire science/technology and writes this textbook. When he retires, he is healthy, happy, and living comfortably.

1. What do you think the chances are that things would work out this way again in today's competitive process of seeking out a job as a fire fighter?
2. How could this person have better prepared himself to compete in today's fire fighter selection process?
3. How could goal setting be applied to this person's pursuit of a career in the fire service?

 Access Navigate for more resources.

Introduction

In the minds of some people, being a fire fighter is the highest calling. This is summed up by the former chief of the FDNY (1899–1911) Edward F. Croker*: "When a man becomes a fireman his greatest act of bravery has been accomplished. What he does after that is all in the line of work." In modern times this would be changed to "When men and women become fire fighters…" and rightfully so; women are taking an ever-increasing role as fire personnel in all areas of the fire and rescue services. This chapter will examine the components of career preparation and goal setting and the importance of higher education to the professionalization of the fire service. Many students confuse training with education. **Training** is the process of skills development; **manipulative training** is the use of tools and equipment. Conversely, **education** includes memorizing specific pieces of information, developing an understanding of concepts or philosophies, and developing the powers of reasoning and judgment. Taking a course in psychology, which would better prepare a fire fighter to deal with the diverse workforce and community he or she serves, would be considered education. A complete fire service–oriented **curriculum** will cover both education and training. An additional benefit to pursuing education, as well as training, before and during a fire service career is that it aids in developing **critical thinking** skills and the ability to communicate clearly and concisely both verbally and in writing. Both of these skills are sought in the fire fighter selection process and on promotional examinations.

Many fire departments are adding the requirement of either a bachelor's degree or a master's degree to high-ranking positions. As a company or chief officer, you may be required to perform budgeting, personnel administration, public contact, policy creation and revision, and other tasks that require a higher order of thinking and communication ability in both writing and speaking. College-level curriculum requires that the student demonstrate competency in critical thinking, decision making, and other skills highly desirable in a fire officer. Having completed an advanced college degree indicates that you have performed to a higher level in at least some of these areas. It is recommended that students get started early and just keep going instead of being in a position of not being eligible for promotion later because of the lack of a higher education degree.

Another critical distinction that must be made in preparing for a career in the fire service, and pursuing promotion once employed, is certification versus a college degree. A level of certification is based primarily on training. To be certified as a Fire Fighter I, emergency

* Croker, Edward. n.d. "Wise Old Sayings." Accessed September 5, 2018. http://www.wiseoldsayings.com/firefighter-quotes/

medical technician (EMT), paramedic, etc., a person must have successfully completed a course of training. This usually involves classroom training, field training, and testing—both in knowledge and in practical skills. Certification varies by state and organization. Some organizations require the prerequisite training and experience combined with the completion of a task book, referred to as performance-based certification, which requires the skills to be demonstrated in a workplace and/or classroom environment to prove competency.

College Fire Science/ Technology Programs

In some colleges, fire science/technology curricula allow the student to earn a certificate without completing all of the requirements of a degree in the program. The certificate program requires the completion of a set number of accredited core courses and additional specified courses in the area of general education. Although the certification program is not a degree in and of itself, the certificate attests to the accumulation of a body of knowledge in the fire science/technology subject area. Some fire agencies will pay fire fighters more for earning a certificate or degree.

> **Tip**
>
> Research the college's catalog and meet with a college counselor to devise a plan to achieve the desired goal.

The completion of an associate's degree in fire science/technology requires more general education units than a certificate does. Some of the courses may be transferable to a 4-year college; others are not.

The core courses may or may not be transferable. They may serve as prerequisites for acceptance and count as credit in an upper division program. Each school differs in its requirements. It benefits the student to research the college's catalog, ensure that it is an accredited institution, and meet with a college counselor to create a plan that achieves the desired educational goal.

Prior to the year 2000, there existed no nationally recognized fire science/technology curriculum for students. The colleges providing fire science/technology degrees developed courses they thought best based on state or local guidelines. In 2000, a conference was held at the National Fire Academy (NFA) in Maryland. At this conference, fire service leaders throughout the country and state directors of fire service training came together to establish recommendations for fire service–related training curricula. This conference created the Fire and Emergency Services Higher Education (FESHE) model curriculum for the associate's degree program. A result of the conference was the development of the model fire associate's curriculum.

In 2001, the National Fire Science Programs Committee (NFSPC) was formed. The plan the NFSPC produced is called the *Fire and Emergency Services Higher Education (FESHE) Model Curriculum: Transforming to a National System* (U.S. Fire Administration 2013a). The NFSPC was formed to develop a model curriculum including standard titles, descriptions, outcomes, and outlines for the curriculum. The curriculum was developed to create a theoretical core of courses, which would allow problem-free student transfers between schools. Fire service–related textbook publishers also agreed to publish textbooks for some, if not all, of the courses. It is now common practice for fire science/technology textbook publishers to design their texts to meet the FESHE guidelines of the model curriculum. This textbook is written to meet the FESHE guidelines.

The committee continues to meet on an annual basis to update and refine the model curriculum. At the 2002 conference, the 2002 FESHE IV, an experience-based model that recommends an efficient path for fire service professional development, was produced. This model addresses how education and training should be integrated. The national professional development model illustrates the relationship between education and training in a professional development matrix **FIGURE 1-1**. As of 2018, the upper division courses are as follows.

> **Tip**
>
> Do not let these long lists of courses overwhelm or discourage you. You should begin by focusing on the core associate level courses and then go on from there as you progress in your career.

The core six-course curriculum for the associate's degree in fire science/technology is as follows:

- Principles of Emergency Services
- Building Construction for Fire Protection
- Fire Behavior and Combustion
- Principles of Fire and Emergency Services Safety and Survival
- Fire Prevention
- Fire Protection Systems

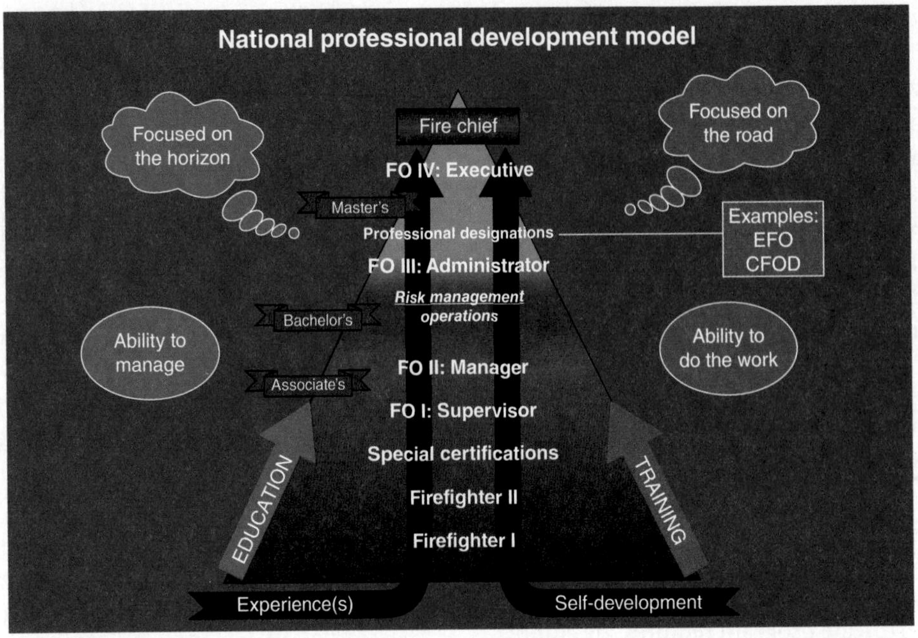

National professional development model

Focused on the horizon

Focused on the road

Fire chief

FO IV: Executive

Master's

Professional designations

Examples:
EFO
CFOD

FO III: Administrator

Risk management operations

Bachelor's

Ability to manage

Associate's

Ability to do the work

FO II: Manager

FO I: Supervisor

Special certifications

Firefighter II

Firefighter I

EDUCATION

TRAINING

Experience(s)

Self-development

FIGURE 1-1 National professional development model.
Courtesy of FEMA

Non-core courses such as the following may also be offered:

- Principles of Fire and Emergency Services Administration
- Fire Investigation I
- Fire Investigation II
- Fire Protection Hydraulics and Water Supply
- Hazardous Materials Chemistry
- Legal Aspects of Emergency Services
- Occupational Safety and Health for the Emergency Services
- Strategy and Tactics
- Hazardous Materials Chemistry

In addition, there are five non-core associate courses in aircraft rescue and firefighting management (ARFF):

- Airport Fire Fighter
- Aircraft Mishaps
- Aircraft Related Mass Casualty Incidents
- Aviation Terrorism Response
- Airport Operations for the Emergency Responder

In 2000, the NFA also released its FESHE 13-course, upper-level Degrees at a Distance Program (DDP) curriculum to accredited baccalaureate degree programs. This program consists of six core courses and seven non-core courses. The DDP is a way for students to take college courses online to earn a bachelor's degree with a concentration in fire administration or fire prevention technology. The DDP is offered through a national network of 4-year colleges and universities that adopt the NFA-developed curriculum to form their own individual degree programs. The core courses are:

- Political and Legal Foundations of Fire Protection
- Applications of Fire Research
- Fire Prevention Organization and Management
- Personnel Management for the Fire Service
- Fire and Emergency Services Administration
- Community Risk Reduction for the Fire and Emergency Services

The eight non-core courses include:

- Fire Dynamics
- Fire Related Human Behavior
- Analytical Approaches to Public Fire Protection
- Managerial Issues in Hazardous Materials
- Fire Investigation and Analysis
- Fire Protection Structures and System Design
- Disaster Planning and Control
- Fire Service Ethics

In addition, FESHE has added bachelor-level course specializations in Emergency Medical Services (EMS), Fire Prevention, and Fire Protection Engineering.

The six core EMS courses are:

- EMS Risk Management and Safety
- Management of Emergency Medical Services
- Legal, Political, and Regulatory Environment of EMS
- EMS Quality Management
- Foundations of EMS Systems
- Community Risk Reduction in EMS

The six bachelor's EMS non-core courses are:

- EMS Education
- EMS Communications Management
- Finance of EMS Systems
- EMS Public Information and Community Relations
- Management of Transport Services
- Analytical Approaches to EMS

The bachelor's Fire Prevention courses are:

- Fire and Life Safety Education
- Principles of Code Enforcement
- Fire Plans Review

The bachelor's Fire Protection Engineering courses include:

- Performance-Based Design Fire protection
- Advanced Concepts in Structural Fire Protection

There are two additional courses offered in both associate and bachelor's level aimed at reducing line-of-duty deaths:

- Principles of Fire and Emergency Services Safety and Survival
- Advanced Principles of Fire and Emergency Services Safety and Survival

Several 4-year colleges in the United States offer on-site bachelor's degree programs in fire-related fields. These degree programs include fire protection administration, fire protection and safety technology, and fire protection engineering. The on-site, 4-year colleges offer, along with course instruction, summer internships for students, giving them actual experience of working in fire departments and industry. The Federal Emergency Management Agency (FEMA 2018), through the Emergency Management Institute (EMI), has a list of schools that offer higher education programs from the associate to the doctoral level in emergency management and homeland security.

Online Programs

Fire fighter training and education are becoming more and more available through online programs. These programs allow fire fighters and prospective fire fighters to earn degrees in fire science/technology and related training without having to attend in-person courses at a traditional brick-and-mortar facility. These programs, such as the DDP, are available in numerous community colleges and schools of higher learning throughout the country and take the form of associate's through master's degrees. One advantage of online programs is that the student does not need to live close to the school to complete the courses. Another is that fire fighters with a set duty schedule may not be off duty on the days that the course is taught in a traditional classroom setting, and online programs offer some leeway in scheduling. Previously, personnel would have to arrange duty trades or other time off to attend the courses, but now they can participate in the course over the Internet and complete the work as time permits within the course requirements, such as completing the work for week two during the second week of the course. Primary examples of these types of courses, from a training standpoint, are the National Incident Management System (NIMS) courses offered by FEMA through EMI in an online format, which include the following:

- IS-100.a—Introduction to the Incident Command System (ICS)
- IS-200.a—ICS for Single Resources and Initial Action Incidents
- IS-700.a—National Incident Management Systems (NIMS), An Introduction

Some schools, such as the International Association of Fire Fighters (IAFF) Virtual Academy, offer distance learning or extended university programs. These programs work much the same way as the open learning program, allowing students to complete their course work away from the college campus. These schools are just a sample of those offering these types of degree programs.

Pursuing a higher education in fire service–related courses can make you a more effective member of the fire service community. It may also help you get promotions after you gain employment. An education may have a direct dollar value in the workplace. Many fire departments offer, as a part of their compensation package, a pay incentive for a fire science/technology certificate or degree or for other specified types of training certification. This usually ranges from 2.5 to 10 percent. Calculated over a 30-year career and carried over into retirement, this amounts to quite a bit of money.

In some departments the completion of certain courses is a condition for completion of the probationary period, which is described later in this chapter.

Other College Programs

Another popular course of study for fire professionals is public administration. Most fire departments operate as public agencies governed by local or state government; this makes a public administration educational background vital to the fire executive. Another reason for this program's popularity is the availability of a master's degree in public administration.

There are other courses of specialized study beneficial to the fire professional in the fields of emergency management, risk management, industrial hygiene, law, emergency medicine, and chemistry. These fields are directly applicable to the fire service. Today's fire fighters respond to all manner of incidents. Advanced knowledge in these areas can better prepare a fire fighter to be safer, more effective, and more efficient in the performance of his or her duties.

An alternate and complementary course of study to the fire science/technology technical education is the fire fighter certification. Based on National Fire Protection Association (NFPA) 1001: *Standard for Fire Fighter Professional Qualifications*, this course of study is primarily manipulative in nature, with technical instruction where necessary. The course of study includes instruction about fire behavior, fire extinguishers, self-contained breathing apparatus, ropes and knots, forcible entry, rescue, water supply, fire streams, ventilation, salvage and overhaul, fire cause determination, fire suppression techniques, automatic sprinklers, and fire prevention inspection. As expanded on later in this chapter, preparation in this area can make you a more viable candidate and assist you in securing a fire fighter position in a fire department. It can also assist you in doing your job safely and efficiently once you are employed as a fire fighter.

Career Potential Assessment

Becoming a fire fighter requires the highest moral and ethical character. You represent one of the proudest professions there is. When you pin on a fire fighter's badge, you represent hundreds of years of tradition of selfless service and sacrifice. Your actions affect the reputation of the fire service as a whole. You are expected to act at great personal risk to save the lives and property of others without seeking recognition or acclaim. Fire fighters must be prepared to act honestly and ethically. They must follow a code of personal and organizational conduct that is beyond reproach. They are in people's neighborhoods and homes when they are away or excluded due to evacuation. They will have to take tests for completion of the academy probation and promotion. In the words of David Batchelor, a fire service instructor, "fire fighters do not lie, cheat, or steal nor tolerate those who do." Fire fighters must have genuine compassion for others. They will deal with victims at quite possibly the worst time of the victim's life: Someone close to them is very sick, or has been injured or died. The victim(s) have lost their possessions, been displaced, or both. Fire fighters must be prepared to show compassion to others no matter what the victim's social status, race, background, or sexual preference is. Fire fighters must do their best to save lives and property every time they are called upon to do so. What does not appear to have much value to the fire fighter may be someone else's most valued possession. In the famous words of former Phoenix Fire Chief and nationally known author and speaker Alan Brunacini, "be nice". This means all the time to all customers, inside and outside your organization. This includes people within your organization and from other fire agencies. An acronym for all of this, as posted by T. Correia on LinkedIn, is that fire fighters have to have HEART, meaning they must be Honor(able), Ethical, Accountable, Responsible, and Trust(worthy).

Fire fighters must be ready and willing to follow rules and policies. They do not necessarily have to agree with them, but they must follow them as the situation dictates. If they do not like the rules or policies, they must be able to work constructively to change them. They are not to just sit around and complain. They may not know the history of why the rule or policy is in place.

Fire fighters must respond and be prepared to act at all times of the day and night, and carry on the work until it is finished or they are relieved by others. This may require hours or days. They may be pressed into service with their days off cancelled even up to a month at a time. Some incidents can take short periods of time, such as a car fire, and other incidents can go on for months, such as a large wildfire.

That does not mean the job is without its level of satisfaction. In an article written for the Monster online job search website, contributing writer Catherine Conlan quoted statistics from Pay Scale's "Most & Least Meaningful Jobs" survey. The survey found that 93 percent of fire fighters reported high levels of meaning from their work. This was exceeded only by clergy, with 97 percent. When watching the news on television, you may have noticed that no matter what the disaster is, the fire department is usually there. Fire department personnel are usually in the background quietly performing their jobs, whether rescuing people from flood or fire, providing medical attention to victims of crimes or accidents, delivering babies, or preventing the spread of hazardous materials. The public does not often think much about the fire department until they need it, and they expect a high level of professionalism and competence. Political candidates do not run on fire protection platforms as they do with law and order. Fire departments may not get much press, but they still give the public much more than their money's worth.

Some people want to become fire fighters because they see it as their chance to become a hero. Anyone who seeks this career for the singular purpose of becoming a hero is misguided. There are few opportunities to become a hero; even heroic actions can go unnoticed. Firefighting is not about being a hero; it is about doing the best you can to save lives and property every time you are given the opportunity. The respected fire fighter is quietly competent and lets his or her actions speak for himself or herself.

Firefighting entails a certain amount of danger and excitement. Approximately 100 fire fighters each year make the ultimate sacrifice by giving their lives in the line of duty (NFPA 2018). This number has declined somewhat in the last few years, with only 60 on-duty fire fighter deaths in 2017, but is subject to increasing. Firefighting is dangerous, but it is not a profession for those who disregard their own safety or the safety of others. With proper training and care, you should be able to make it through your career and enjoy your retirement in relatively good health.

Being a fire fighter is not all about doing the exciting work of responding to incidents. There is also much work to do in the area of prevention. By the time the fire department responds to a fire, damage is already being done. The best case scenario is always that no fires occur. That would save the most lives and property, and be of the greatest service to the community. Looking at the FESHE course models, it is obvious that modern prevention efforts are also spreading into the areas of EMS and Community Risk Reduction as a whole.

Firefighting involves many hours of preparation through physical fitness, education, and training. The job requires drill and study to master the myriad tasks you are expected to perform, often in extremely stressful situations. It has been proven that people do what they were trained to do when things go badly and danger surrounds them. You must be willing to perform the preparation before you can perform the job. The constantly changing demands of the workplace ensure that training never truly ends. When new chemicals are developed, new industrial processes are invented, building construction techniques change, new subdivisions of homes are built in more remote locations, or changes in the public's behavior occur, the fire department must take notice and stay involved.

> **Tip**
>
> You must be willing to perform the preparation before you can perform the job.

Workplace Stress

Fire fighters suffer the same problems as the rest of society. The stress level of the job is high, leading some to divorce, alcoholism, and drug abuse. To many fire fighters, their co-workers are their extended family, and what affects one affects all. Fire fighters spend long periods of time together **on duty** and see each other under the worst of conditions. Responding to other people's tragedies has a way of drawing fire fighters closer together and forming strong bonds among them.

Fire fighters are required to show compassion and become skilled in dealing with people at the worst times in their lives—at accident scenes when loved ones have been killed or seriously injured and at fires and other incidents of devastating loss. Fire fighters must be able to deal with the injury and death of people of all ages, from infants to elders, under horrible conditions. As fire fighters become more involved in the delivery of medical aid, they see more instances of child abuse, elder abuse, and other tragedies.

The modern fire service has started to address the problem of stress in the workplace. Many agencies have employee assistance programs that allow fire fighters to talk confidentially with a counselor when they are having stress problems at work or home. There are also provisions for alcohol and drug abuse treatment programs. No one can be a fire fighter and be totally unaffected by what he or she sees in the line of duty. **Critical incident stress management (CISM)** programs utilizing **critical incident stress debriefings (CISD)** have been introduced to help fire fighters cope with particularly difficult or emotional incidents resulting in **post-traumatic stress disorder (PTSD)**. Another area that is being studied as to how it affects fire fighters is **recurring exposure to trauma (RET)**. RET is the effects of responding numerous times in one's career to traumatic events and their effect on the responder. Imagine responding to the crash of an airliner with the expectation of rescuing people. The crash is such that all 300 persons aboard are killed. Fire fighters still search the wreckage to see if there is anyone left alive, the same as they did after the bombing of the Federal Building in Oklahoma City in 1995 and the attack on the World Trade Center in New York City in 2001. Another possible scenario that fire fighters face is responding to a vehicle accident where a family member, personal friend, or co-worker has been killed or seriously injured. In such situations it is not uncommon to ask yourself, "What could I have done to save him?" Another common question when a co-worker is killed or injured is, "What if it had been me?" After experiencing enough of these kinds of incidents, they start to build up and manifest themselves in home or work-related problems. You must be willing to ask for help to cope when necessary.

Teamwork

Fire fighters must be team members. They must be willing to give up personal desires to benefit the team. When one member succeeds, the *whole* team succeeds.

When the team performs extraordinarily well and praise is theirs, it should be shared equally. When a baby was saved from a burning house, one fire fighter carried the baby out and got his picture in the paper—but without the help and backup of the others at the scene, the rescue would not have happened. No one can perform the job alone. Whether at emergencies or on routine work assignments, the whole team needs to pitch in and help until the work is done.

Washing the fire apparatus, maintaining equipment, preparing meals, and doing dishes are not glamorous or fun, but they are a part of the station routine **FIGURE 1-2**. Working together as a group makes the job go by much more quickly.

Physical Health and Fitness

A career-long commitment to physical fitness is part of being a fire fighter. In 2016, 45 percent of fire fighter line-of-duty deaths (LODD) were attributed to sudden cardiac events (USFA 2016). Fire fighters are often compared to athletes because of the requirements of their job. Your fellow fire fighters expect you to be able to carry your share of the load in a physically demanding profession. Letting yourself get out of shape endangers not only you but also your

FIGURE 1-2 Fire fighter performing routine station duties.
© Jones & Bartlett Learning

FIGURE 1-3 Fire fighter maintaining level of physical fitness.
© Jones & Bartlett Learning

co-workers and the public. They have every right to expect you to stay fit **FIGURE 1-3**. If one of them goes down, it may be up to you alone to get him or her out.

> **Tip**
>
> A career-long commitment to physical fitness is part of being a fire fighter.

Changing Demands of the Fire Service

The role of the fire service has changed and will continue to do so. The job of a fire fighter has changed to meet the demands of the new fire service. It used to be that a fire fighter's main job was to control and extinguish hostile fires. The fire service has now taken responsibility for providing EMS, fire prevention, hazardous materials response, search and rescue, homeland security, and other emergency services. The modern fire fighter is expected to be an educator, a professional, and a highly skilled technician. As public employees, fire fighters are expected to serve the public by providing the services it requires. If you think that all you are going to do is prepare for and extinguish fires, you are wrong. Instead of saying, "We are doing enough already," the contemporary fire chief is asking, "What more can we do, and in what ways can we better serve the community?"

As a member of the fire service you may very well be asked to participate in community programs during your time off. These programs are beneficial because they promote the image of the fire service and fire fighters as a whole. If you were to ask fire service professionals for the most important things the fire department must have, the items at the top of the list would be the trust and support of the public. For the fire department to exist and receive the budget it needs to function, the political support of the public is essential. As a fire fighter, you are not solely a community employee, but a part of the community. You must always keep in mind who ultimately pays your wages.

You may ask yourself, "Am I ready and willing to proceed into a burning building when everyone else is running out?" If you can meet all of these criteria and successfully complete the selection process, you could pursue one of the most exciting and personally satisfying careers there is.

Human Relations and Work Ethics

The standards fire fighters are held to, on and off duty, are well above the traditional standards of many professions. During your career in the fire service, you will not only work with other fire fighters, you will also often live with them, sometimes up to 24 hours a day (with 48-hour shifts becoming more popular) in the fire station during your tour of duty (shift). The ability to get along with others is very important in the often-cramped confines of the fire station. In some departments, the situation is much like a military barracks. You must also perform such tasks as station and vehicle maintenance, training, and meal preparation.

In the area of human relations, the ability to work within a diverse environment is critical. The Equal Employment Opportunity Act prohibits discrimination against any person in the classified service or any person seeking admission thereto because of race, national origin, sex, age, physical disability,

color, medical condition, marital status, ancestry, or union activity. Discrimination on the basis of age, sex, or physical disability is prohibited except where age, sex, or physical requirements constitute a bona fide occupational requirement. Physical disability is one of the few reasons a person cannot be hired by the fire department as a fire fighter. The Americans with Disabilities Act (ADA) specifies that there must be a clear reason why the disability excludes the person from being hired (the Fire Protection Career Opportunities chapter further discusses jobs in the fire service, other than fire fighter, that a person with a disability can hold).

One of the quickest ways to lose your job as a fire fighter is to engage in harassment. There is no tolerance for this type of activity in the modern fire department. Sensitivity must be shown to all groups—at the station and on the scene of an emergency. One careless remark can cost you your job and tarnish the image of all fire fighters. As a fire fighter, you will respond to assignments requiring you to serve people of all backgrounds and social status, from the affluent to the homeless. It does not matter how much money they have or how they live; all of your customers deserve to be treated with dignity and respect, and to be provided with the full benefit of your best efforts.

Tip

One of the quickest ways to lose your job as a fire fighter is to harass someone. This is regardless of whether or not you are on duty.

Many attributes make a fire fighter a valued member of the fire department community. Loyalty in the fire service is hard to define, but it is essentially sticking by your fellow fire fighters through good times and bad. It involves not criticizing your superiors, subordinates, or co-workers. Much of the time, you will not be responding to emergencies. During this time, it is very easy to fall into the habit of griping and gossiping. Doing so only makes others think that you will talk about them when they are not around. If other fire fighters do not feel that they can trust you, they will not want to work with you.

Dedication to duty is how you approach your job. As a new fire fighter, you are expected to be the first one to volunteer for the dirty jobs, such as crawling under a house to retrieve a lost kitten. It does not stop there, however. Your dedication to duty should last your entire career. It is not enough to apply yourself for the first couple of years until you find yourself a comfortable station and crew that you can call home. You must constantly strive to be the best fire fighter you can be.

A fire fighter must be able to accept hardship without complaint. When the alarm sounds, the fire fighter goes to work. It does not matter that you are tired from the last incident or that you have not had a chance to eat, even though the food is on your plate. Long hours under stressful and often extremely tough conditions are considered part of the job.

In some departments, or station assignments in larger departments, there are few calls for service and members must constantly train to maintain their skill proficiency. One of the greatest dangers to fire fighters is complacency, or showing the attitude that you do not need to practice because your skills will not be necessary.

The ability to follow orders is part of the team effort required of all fire fighters **FIGURE 1-4**. The fire ground is no place to argue with your officer or to freelance. Only through the total directed effort of all of the people on the scene can the problems be overcome. Around the station, many orders are stated as requests; this does not mean they can be ignored. It is just a polite way of getting things accomplished and establishing a more relaxed, stress-free atmosphere.

You must be able and willing to learn. The field of fire suppression and prevention is dynamic and ever changing. The fire service itself has changed drastically in the last few years. The fire department has an ever-expanding role in providing service to the community. Major changes have come about in the types of hazards encountered and how they are handled. The equipment available has also changed at a rapid rate. Fire fighters must be ready and willing to accept

and adapt to these changes or become dinosaurs—and we all know what happened to the dinosaurs.

You must be willing to accept personal responsibility for your actions. When given a job, it is your responsibility to complete it, not someone else's. You should not have to be closely supervised once you are trained to perform the required job. Emergency activity can and does disrupt the other work that fire fighters perform on a regular basis. If this happens, it is your responsibility to return to the previously assigned work once the equipment is placed back in service after an incident. If you are unable to complete the assigned task by yourself, seek help or advise your supervisor.

Last but not least, every fire fighter must have a positive safety attitude. Firefighting is an inherently dangerous profession. Numerous fire fighters are disabled, and others lose their lives every year in the line of duty. With a proper safety attitude, you can do your best to avoid serious injury and death while still performing your job aggressively.

SAFETY TIP

Every fire fighter must have a positive safety attitude.

The following is a Code of Ethics developed by the National Society of Executive Fire Fighters (USFA. FEMA.gov):

Firefighter Code of Ethics

I understand that I have the responsibility to conduct myself in a manner that reflects proper ethical behavior and integrity. In so doing, I will help foster a continuing positive public perception of the fire service. Therefore, I pledge the following…

- Always conduct myself, on and off duty, in a manner that reflects positively on myself, my department, and the fire service in general.

- Accept responsibility for my actions and for the consequences of my actions.

- Support the concept of fairness and the value of diverse thoughts and opinions.

- Avoid situations that would adversely affect the credibility or public perception of the fire service profession.

- Be truthful and honest at all times and report instances of cheating or other dishonest acts that compromise the integrity of the fire service.

FIGURE 1-4 Fire fighters attacking a structure fire.
Courtesy of Kern County Fire Department

- Conduct my personal affairs in a manner that does not improperly influence the performance of my duties, or bring discredit to my organization.

- Be respectful and conscious of each member's safety and welfare.

- Recognize that I serve in a position of public trust that requires stewardship in the honest and efficient use of publicly owned resources, including uniforms, facilities, vehicles, and equipment, and that these are protected from misuse and theft.

- Exercise professionalism, competence, respect, and loyalty in the performance of my duties and use information, confidential or otherwise, gained by virtue of my position, only to benefit those I am entrusted to serve.

- Avoid financial investments, outside employment, outside business interests, or activities that conflict with or are enhanced by my official position or have the potential to create the perception of impropriety.

- Never propose or accept personal rewards, special privileges, benefits, advancement, honors, or gifts that may create a conflict of interest, or the appearance thereof.

- Never engage in activities involving alcohol or other substance use or abuse that can impair my mental state or the performance of my duties and compromise safety.

- Never discriminate on the basis of race, religion, color, creed, age, marital status, national origin, ancestry, gender, sexual preference, medical condition, or handicap.

- Never harass, intimidate, or threaten fellow members of the service or the public, and stop or report the actions of other fire fighters who engage in such behaviors.

- Responsibly use social networking, electronic communications, or other media technology opportunities in a manner that does not discredit, dishonor, or embarrass my organization, the fire service, and the public. I also understand that failure to resolve or report inappropriate use of this media equates to condoning this behavior.

Training Programs

Preservice training programs in manipulative skills are available through Explorers, volunteer firefighting programs, reserve/cadet programs, the National

FIGURE 1-5 Explorer program fire fighters attending a hose stream training.
Courtesy of Kern County Fire Department

Junior Firefighter Program from the National Volunteer Fire Council, colleges, and training associations **FIGURE 1-5**. The goal of these programs is to teach the actual skills necessary for a fire fighter to perform on the fire ground. These programs start with an academy that teaches skills in handling ladders, fire extinguishers, salvage equipment, hand and power tools, self-contained breathing apparatus (SCBA), and hose lays.

Sometimes having medical training, such as EMT or paramedic training, can be of benefit when seeking employment. This varies by department and the level of medical service it provides. As discussed in the Fire Protection Career Opportunities chapter, this may even be a prerequisite to employment.

Some programs are sponsored by individual fire departments, associations, or professional groups, with college credit issued. Generally, you are required to attend on your own time at your expense. Often incentives are offered in the way of special treatment in the hiring process. They may even be used to establish a direct hiring pool for the department(s) involved. The completion of the fire science/technology program on the college level, along with a certificate from a certified preservice academy, can assist you in competing for the job of fire fighter.

In-service training programs have been developed to train active fire fighters. These programs start with the academy and move to the station, battalion, department, area, state, and national levels. Training programs are sponsored by the departments themselves, state and local training officer's associations, the offices of state fire marshals, and colleges. The courses run the gamut of subject matter from hose lays to specialty courses in hazardous materials and heavy rescue and often require

FIGURE 1-6 Fire fighters performing physical fitness training during training academy.
© Jones & Bartlett Learning

FIGURE 1-7 Sample career ladder showing training certifications.
© Jones & Bartlett Learning

department-sponsored attendance due to **worker's compensation** coverage.

The first level of training for newly hired fire fighters is the academy. The new fire fighters report to the school instead of to the fire station. The purpose of the academy is to train the new fire fighters in department equipment and methods, in courses that are required by law (such as emergency medical training and hazardous materials first responder operations), and to observe the new fire fighter's physical and mental performance **FIGURE 1-6**. While attending the academy, the new fire fighters are evaluated on their performance on written tests and during **drills**. If for some reason new fire fighters do not measure up to department standards, they are dismissed. This time can be very stressful for new fire fighters. They are being watched very closely for the slightest infraction. A great amount of homework and studying is required to perform well on the written exams.

Numerous schools around the country offer technical training programs. Two of the best known are the NFA and the EMI (both located at the National Emergency Training Center), which offer courses year-round at the facility in Emmitsburg, Maryland. Instructors from all over the country are employed to present the widest viewpoint and to give the courses national appeal. The instructors are of the highest caliber and are recognized as experts in their field. There is no tuition charged to attend the NFA. For students to attend, they must be sponsored by their department and be accepted after filing an application. When the student is accepted, his or her travel is reimbursed and lodging is provided. All the student pays for while attending the course is food.

One state's model fire training program follows the Fire Service Career Ladder **FIGURE 1-7**. The completion of the courses on the ladder leads to certification at different levels. Most of the instructors are fire service professionals and relate the material to the fire fighter's needs very well. Several of the courses parallel those offered by the NFA and in some cases may be substituted.

Throughout the United States, there are colleges that sponsor fire department–related courses at their facilities. A program that covers a wide variety of fire-related subjects is offered at Texas A&M and across the country, through its extension service (TEEX). Specialized training in various types of firefighting is offered at other locations as well.

Personnel Development Programs

Personnel development programs are becoming more widespread in the fire service. In these programs, a fire fighter is trained as high as two ranks above the one currently held. The purpose of these programs is to develop an understanding of how the department works and to prepare the leaders of tomorrow. In some instances, students are assigned a **mentor** to aid in goal setting, to help monitor their progress, and to assist them as necessary.

Modern fire fighters must be **generalists**, cross-trained to be able to perform many firefighting and rescue functions. They may also want to become **specialists** in one or more areas of fire department operations. The modern fire service has so many responsibilities that no one can know everything or be an expert in every aspect of the job to be performed. It is the duty of every fire fighter to seek training on all levels.

Selection Process

The selection process for the fire department contains a number of steps. Different departments use various combinations of the steps presented here. It is important to research the department you are applying to in order to determine which of the steps are used and to prepare accordingly. The steps presented here, for the purpose of illustrating the process, are recruitment, application, written examination, skills test, oral interview, physical ability/agility, medical examination, psychological examination, background check, final oral examination, and probationary period. A representative selection process, from start to finish, is presented here.

Tip

It is important to research the department you are applying to in order to determine which of the steps in the hiring process are used and to prepare accordingly.

Recruitment

The selection process starts during recruitment. Fire departments are looking for the most qualified applicants they can find. When you apply for a fire department job, you will be competing for a limited number of openings against others who have prepared themselves to varying degrees.

Students who are currently in or have completed fire science/technology programs have already shown an interest in a fire department career and a willingness to invest time and money in pursuing an education in this field. They have also shown the drive and ability to study and learn in a classroom environment.

Most fire departments have prerequisites for application. The most basic prerequisites are a valid driver's license, a high school diploma or general equivalency diploma (GED), and no felony convictions. The exclusion of an applicant because of a felony conviction is now being questioned in some departments. The Fire Department of New York (FDNY), for example, is accepting applicants with a certificate of good conduct from the state parole board, and such applicants will receive a special review (Edelman 2013). This is not to be considered a widespread practice in the fire service.

Another prerequisite often used by smaller departments that do not have the money or the staffing to place a newly hired fire fighter in an academy is a Fire Fighter I certification. Smaller departments are usually funded to have only the minimum staffing required to keep their apparatus responding; they need new personnel on the equipment and ready to go the first day they report for work. When new personnel are hired and placed in an academy, they are getting paid and the department does not receive any direct benefit from their employment. Some departments promise employment at the completion of an academy program of their choosing, but applicants must attend on their own time. With most academies utilizing instructors from the local fire department(s), it also allows the instructors an additional opportunity to evaluate applicants before they are hired. Another common prerequisite may be emergency medical technician (EMT) or paramedic certification.

Application Process

Nowadays departments are accepting applications one of two ways, either a physical form that is submitted to the required address, or an online application. Administering examinations to large numbers of applicants is time consuming and also expensive for public agencies. Some charge a fee to apply. There are several ways of limiting applications. One way is to limit the number of applications given out. It does not make much sense to give out 600 applications when the department's anticipated need is for fewer than 10 new fire fighters. Another method is to give out applications for 1 day only. Limited advertising is also used by publishing the job announcement only in the local newspaper. In some departments that have reserve/cadet fire fighter programs, those in the program get a guaranteed application out of an already limited number. With the rise in healthcare costs and as a long-term cost-saving measure, many departments have gone to a no tobacco policy. Applicants agree not to use tobacco in any form for the term of their employment. Other departments are even more restrictive in that they require applicants to sign an affidavit at the time of applying that they have not used tobacco for a year prior to being hired. Violation of the no tobacco agreement can result in disciplinary action, up to and including termination.

A limit affecting applicants is sometimes imposed as a result of court action. In some cases, the fire department has been found not to represent the demographics of the community. In these cases, the court may impose an order requiring an affirmative action program to be put into place that would require the department to hire specific groups, such as women and minorities, over other candidates.

Some jurisdictions also have a residency requirement. A part of the reason for these requirements is political—to hire local residents for public jobs. Another reason is to have at least some of the force available on a **call-back** basis in case of a major disaster or large fire. Some of these residency requirements require the applicants to reside in the jurisdiction prior to their application being accepted. Others require the new fire fighter to reside in the jurisdiction or within a certain radius for a specified period of time after being hired. Fire departments, like any other public agency or business, are trying to attract only the best applicants and at the same time keep costs down in the hiring process.

The application process consists of first finding out when the application will be available and, second, finding out when it must be completed and returned. One way to avoid the problems with learning about application dates is to subscribe to an application notification service. These services will, for an annual fee, send you a postcard (or post online) with information on upcoming fire fighter examinations and application filing periods along with prerequisites such as having a Fire Fighter I, certificate of successful completion of the completed Physical Ability/Agility test (page 17) from a regional testing site, EMT, or paramedic certification. Some departments are now posting job openings online and allowing the application to be filed online. Another way to receive notification of application filing dates is to visit the personnel department of the jurisdiction of your choice (or their website) and fill out a job interest card **FIGURE 1-8**. This card will include your name and address and is left with the personnel department. When the application filing date is announced, the department will notify you that they are accepting applications. Most personnel departments will not mail out applications—they must be picked up in person. Note: Because some

departments limit the number of applications distributed, they also may only accept so many applications. They will leave the submission website open only until the predetermined number of applications are submitted. Sometimes, this can be only a matter of minutes. If you are submitting an online application with application limiting in place, it is very important to be prepared and submit your online application as soon as the site opens.

Tip

Most personnel departments will not mail out applications—they must be picked up in person.

After you have picked up your printed application, make a copy. Use the copy for practice and return the original. The application should come with a job announcement (flyer) **FIGURE 1-9**. The job announcement is very important because it contains the information you need to fill out the application. Most applications are a standard form for any job in that jurisdiction. Always ask if there are any supplemental materials with the application. There may be additional materials in the form of a pretest guide and/or study booklet. There may also be a preparation class or recruitment seminar offered to assist you in performing well on the exam.

Before leaving the personnel department and filling out the application, read the job announcement carefully. Make sure you have the right job announcement for the position for which are applying. They are usually all the same color. You do not want to drive home and then return because you have the job announcement for a painter. Be sure to pay particular

FIGURE 1-8 Online job interest card.

Courtesy of Clark County

Job Announcement: FireFighter I, August 20- August 22, 2013 Test Dates

Apply Online

Job Announcement: FireFighter I, August 20- August 22, 2013 Test Dates
COBB COUNTY GOVERNMENT
Human Resources Department
Employment Center

Cobb County is an equal opportunity employer.
Cobb County Government does not discriminate on the basis of race, color, national origin, sex, religion, age or disability in employment or the provision of services.

Requisition Details
Requisition Title/Job Title: FIREFIGHTER I
Department: Department Of Public Safety
Position information: This position reports to the Fire Department. The entry-level written test for this position will be given on August 20 - August 22, 2013.
Position Type: Full-Time
Hours worked per week: 24 hours on duty/ 48 hours off duty
Issue Date: 07-01-2013
Final Filing Date: 07-26-2013

Essential Job Functions:
Performs work in controlling and extinguishing fires.

Performs first aid and emergency rescue and medical treatment at the emergency scene.

Responds to and performs duties as assigned, at accidents, disasters, emergency rescues, searches, and any other emergency situations.

Performs other related duties as assigned.

Minimum Qualifications:
High School Diploma or GED

Age 18 or older on or before test date

Valid Driver's License

Prior experience as a certified firefighter or certified medical provider preferred

College coursework in Fire Science is desirable

Must successfully complete the following:
Written entry level test
Physical ability test
In-depth background investigation
Polygraph examination
Psychological examination
Medical examination and drug screen

Special Instructions to Applicants:

Important Information - Please Read Thoroughly

The entry-level written test for this position will be given on August 20 - August 22, 2013.

There will be representatives from Fire and Emergency Services, the Department of Public Safety and Human Resources at the test to explain the hiring process, and to answer questions you have about the position.

Grade: 50

Status: Non-Exempt

Physical Requirements:

While performing the essential functions of this job, depending on area of assignment, the incumbent is regularly required to lift, move or carry objects over 200 pounds; ascend or descend ladders, stairs, scaffolding, ramps, poles; utilize feet and legs or hands and arms; maintain body equilibrium to prevent falling when walking, standing, crouching, or navigating narrow, slippery, or erratically moving surfaces; bend body downward and forward by bending spine and legs; move about on hands and knees or hands and feet; use hands handle or feel objects; and to reach with hands and arms.

While performing the essential functions of this position, depending on area of assignment, the employee is frequently exposed to fire, fumes or airborne particles, toxic or caustic substances, excessive noise, temperature extremes, and dampness/humidity. While performing the essential functions of this job, the incumbent is regularly exposed to possible bodily injury from falling from high, exposed places; and moving mechanical parts of equipment, tools, and machinery.

FIGURE 1-9 Job announcement flyer for fire fighter position. The job announcement is very important because it contains much of the information you need to fill out the application.

Courtesy of Cobb County Fire and Emergency Services and Cobb County Government

attention to the date the application must be returned. Applications turned in late are not accepted.

A lot of information on the job announcement will assist you in filling out the application. The first item is the test number and title. This is entered on the application for personnel department sorting purposes. The job announcement also contains information about pay, working hours, and locations, and also specifies the minimum requirements of the position for applying.

When filling out the portion of the application that asks for job history and duties, use the job flyer as a guide. Specify how your current and past jobs meet the duties of the new job for which you are applying.

If you have an expanded job history or description that will not fit on the application, attached sheets are allowed. For these to be accepted, they must be turned in at the same time as the application. Plan to bring a résumé to your oral exam and attach a copy to your application as well. Attach a copy of your high school diploma or GED, and if you have a college certificate or degree or any other special job-related training certificates, attach them as well. If the jurisdiction awards veterans' points, you must submit acceptable proof of military service along with the application.

A hard copy job application should be typed. The application must, at the minimum, be printed neatly in ink. This has changed somewhat because many departments now require applications be filed online. Your first impression on the oral panel may be your application. The members of the panel usually have your application before them as they conduct the interview. They may also review it after you are interviewed. Spelling and grammar are important in your answers. Be careful and take your time on applications, whether online or in a printed format. Answering questions incorrectly or incompletely may lead to their rejection during the application process. If you have any questions, be sure to contact the appropriate department and seek the answer prior to returning the application.

Written Examination

The written examination is designed to test a candidate's ability to learn firefighting procedures and techniques. The test evaluates mechanical aptitude, general intelligence, mathematical ability, behavioral reactions to given situations or events, mental alertness, adaptability to the work of firefighting, and the ability to understand orders and written material (reading comprehension). Because firefighting experience is not a prerequisite for employment, questions on the written examination are geared to the inexperienced candidate. This does not mean that there cannot be a reading comprehension question with a firefighting situation and related questions. It just means that the answer will be found in the reading material. Questions are multiple choice and computer scored. Written examinations are weighted differently by different departments. In some, the written examination will be worth 40–50 percent of your overall score. In others, it is scored pass/fail.

Several resources are available in preparing for the written exam. One of these is to go to the local library and check out fire fighter exam preparation texts and videos. These preparation manuals have exercises in mechanical aptitude, reading comprehension, mathematics, and so forth. If you have not done any long division or have not been taking multiple-choice tests for a while, the experience you will gain by practicing with these texts is worthwhile.

Skills Test

The skills test portion of the exam is not used by all departments. When it is used, the top performers from the written exam are invited to participate. The skills test simulates real-life occurrences likely to be encountered on the job. The applicants' mental ability to deal with these situations is graded. A few examples of this sort of test are as follows:

- Applicants are given a written procedure to study for a specified period of time, and then they are asked questions about the material.
- Applicants are shown a video of operations at a fire scene and then are asked questions about the video, such as "How many fire fighters were on the roof?" or "Where was the engine parked?"
- Applicants listen to an audio recording of a dispatcher giving directions and then are asked questions, such as "What street were you on after the third turn?"

Oral Examination/Interview

The next step is an interview with the oral interview panel FIGURE 1-10. There is usually a representative from the personnel department in attendance, and the interview is usually recorded (video or audio). The purpose of this is twofold: to ensure consistency in the way the interviews are conducted should they be challenged, and to avoid disagreement on what the answers were. This also establishes a record if the applicant claims education or experience that is untrue. Oral panels commonly consist of three interviewers, usually of fire officer's rank. They sit on one side of a table and the applicant sits on the other.

The oral exam is designed to evaluate education and work experience and to measure the personal attributes required of firefighting personnel. It may evaluate the following characteristics: ability to act under stress, ability to accept authority, ability to get along with fellow fire fighters, ability to deal with the public, and motivation to be a fire fighter. This

FIGURE 1-10 Candidate being interviewed by oral examination panel.
© Jones & Bartlett Learning

interview may also include situational judgment questions, which are usually presented as a scenario followed by the applicant explaining what he or she would do in that situation. For example, "You are in a burning building searching for a victim and your partner becomes injured. What would you do?" There is no real right or wrong answer; the interviewers are looking for your ability to think through the scenario before giving your answer and your ability to defend the answer you provide.

The panel of interviewers will attempt to put you at ease in discussing your qualifications as a potential fire fighter. This portion of the examination is competitive and may be rated as much as 100 percent of your final score when a pass/fail written exam is used solely as a means to qualify you for the oral exam.

Punctuality and appearance are important. Neat grooming and attire will give a good first impression of you to the panel. The old saying, "Look in the mirror; would you hire this person?" applies here.

Project an image of self-confidence by thinking you are the best person for the job. Be courteous to the interviewers; speak clearly and loudly enough to be heard by all members of the panel. When responding to a question, look directly at the questioner. Avoid a harsh or hasty answer. If necessary, take time to think out your answer before responding. In a positive way, stress the value of your knowledge, skills, and abilities and how you can contribute to the fire service. You may not have any firefighting experience, but a training certificate, fire science/technology degree, or having worked in a team environment with a diverse workforce are definite pluses.

Avoid distracting mannerisms that might draw the interviewer's attention away from your statements.

Maintain good habits, posture, and poise during the entire interview process. If you are seated in a chair that rocks, sit still. Place your hands in your lap and try to relax.

You can prepare for the oral examination by first assessing your delivery. Write down likely questions and then answer them while looking into a mirror. You may want to record your answers to hear how you sound. A very common request is, "Tell us about yourself," or a close variation. This response is an excellent one to practice. Another method is to contact the local fire department and see if you can set up a mock oral exam. In a mock oral exam, the room is set up much as it would be in a real oral exam, and fire fighters ask you questions just as an oral exam panel would. Sources for video recordings and written material on preparing for oral examinations can be found in fire service magazines and online.

The oral interview exam is the final step in competing for placement on the fire department list of certified eligible candidates. Your score on the oral exam is determined by averaging the scores given to you by the individual panel members. A minimum average score of 70 percent is required to pass the oral exam. Applicants receiving less than 70 percent will be disqualified.

In situations where both the written and oral examinations are weighted, the average of your two scores will determine your ranking on the list of eligible candidates. In an examination process where the two scores are weighted 50/50, a written score of 80 and an oral of 90 will give you an overall score of 85. This score will be compared to all the other candidates' scores and you will be ranked 1, 2, 3, etc., accordingly. Being eligible for veterans' points can raise your score, so read your job announcement carefully or ask the personnel department for information on how to receive credit.

Another method of ranking candidates is called **banding**. In this process, candidates with scores between specified percentages are placed in groups. When positions become available, the candidates in the first band are offered the job. If the number in the group exceeds the number of available positions, a lottery is conducted to offer a position to a certain number of candidates in the band being used.

Physical Ability/Agility

Physical ability/agility tests are administered to judge the candidates' overall physical conditioning and ability to perform firefighting-related tasks **FIGURE 1-11**. This test is expensive and time consuming to administer, and usually only a select number of the

FIGURE 1-11 Physical ability/agility testing.
Courtesy of Kern County Fire Department

top-performing candidates are asked to participate. The test consists primarily of climbing, hoisting, carrying, lifting, and dragging. The emphasis is on overall body strength and endurance. There is no set national standard for this type of test as there are other physical ability/agility tests in current use. The basic events are much the same in most tests because they are required to be firefighting related.

As with the written and oral tests, preparation is important. The people administering the test will be your future employers if you are hired, and they may very well remember your performance. It is a good idea to contact someone already on the fire department and have them help you practice the basic types of activities. Some agencies and colleges offer preparatory physical fitness courses for this type of test. Take advantage of them if they are available.

Just like any other physical endeavor, technique is involved in many of the test activities. Dragging a hose around several obstacles may appear very simple, but until you have done it several times you are not likely to be very good at it.

Many departments with ladder trucks will require you to ascend and descend a raised ladder. This event can be very frightening the first time you perform it, but you can become more comfortable with practice **FIGURE 1-12**. When you are taking the physical agility test and being timed is not the occasion to find out you have not prepared properly.

The physical ability/agility test may or may not be competitive and part of your final score. Most are pass/fail. It is good to know this and prepare accordingly.

Candidate Physical Ability Test

An example of candidate physical ability/agility testing is the Candidate Physical Ability Test (CPAT).

FIGURE 1-12 Candidate climbing aerial ladder.
Courtesy of Kern County Fire Department

It was developed as a joint effort of the International Association of Fire Fighters (IAFF), the International Association of Fire Chiefs (IAFC), 10 professional fire departments, and their union locals from across the United States and Canada. The fire departments that participated in the development of the CPAT were Austin, Texas; Calgary, Alberta, Canada; Charlotte, North Carolina; Fairfax County, Virginia; Indianapolis, Indiana; Los Angeles County, California; Miami Dade, Florida; New York City, New York; Phoenix, Arizona; and Seattle, Washington.

The test was designed to simulate activities performed by fire fighters at incidents. The CPAT was developed to allow a fire department to obtain a pool of trainable potential candidates for employment who are physically able to perform essential job tasks at fire scenes.

The CPAT consists of eight separate sequential events that require the candidate to progress along a predetermined path from event to event in a continuous manner. This is a pass/fail test based on a validated maximum total time of 10 minutes and 20 seconds.

Throughout all events, the candidate must wear long pants, a hard hat with chin-strap, work gloves, and footwear with no open heel or toe. Watches and loose or restrictive jewelry are not permitted.

In these events, the candidate wears a 50-pound vest to simulate the weight of a self-contained breathing apparatus and fire fighter protective clothing. An additional 25 pounds, using two 12.5-pound weights that simulate a high-rise pack (hose bundles), is added for the stair climb event.

All props are designed to obtain the necessary information regarding the candidate's physical ability.

The tools and equipment are chosen to provide the highest level of consistency, safety, and validity in measuring the candidate's physical abilities.

The props are placed in a sequence that best simulates their use in a fire scene while requiring an 85-ft walk between events. To ensure the highest level of safety and to prevent candidates from exhaustion, no running is allowed between events. This walk allows the candidate approximately 20 seconds to recover and regroup before each event.

The CPAT includes eight sequential events as follows:

1. Stair climb
2. Hose drag
3. Equipment carry
4. Ladder raise and extension
5. Forcible entry
6. Search
7. Rescue
8. Ceiling breach and pull

The CPAT is not only a physical ability testing process. It also includes information on recruiting and mentoring personnel to prepare for the testing; on diversity and its importance in the fire service; and on recruiting methods through community outreach, community organizations, the Internet, the media, brochures, public announcements, and colleges and high schools.

The program has a preparation guide for candidates and information as to why a fire department must provide a preparation guide. Included in the preparation guide is information on who should receive it and what other activities fire departments can provide to prepare applicants. There are sources of information about this type of testing and the required physical preparation available on the Internet, such as Firefighter Functional Fitness, Fit to Fight Fire, and 555 Fitness. For further information on the CPAT, contact an IAFF union local or check the IAFF website.

Note: With the CPAT being so closely regulated and expensive to conduct, it is often only provided at regional sites and may require a fee be paid to participate. Some fire departments provide a scholarship to cover the fee. It is cost saving to ask if this is available.

Work Capacity Test for Wildland Fire Fighters

Federal wildland fire fighters are required to pass the work capacity test (pack test) at the arduous fitness level (National Wildfire Coordinating Group 2003). Some fire departments in areas with wildland firefighting responsibilities require this test as well. The test used in the Wildland Fire Qualification Subsystem for positions requiring an arduous fitness level is as follows: The person being tested is to carry a 45-pound pack a distance of 3 mi in 45 minutes or less (equivalent to a pace of 4 mph). Altitude corrections are applied for elevations over 4000 ft **TABLE 1-1**.

Psychological Test

The psychological test for fire fighters is usually performed as a series of written true or false questions, answers to questions read off a computer screen, and answers to interview questions. The three components of the test are occupational, personality, and polygraph. The occupational test is performed to match the interests, knowledge, and abilities of the candidate with those of fire fighters already in the profession. Some feel that those who become fire fighters have a "rescue personality," meaning that they are action oriented, are easily bored, enjoy being needed, are highly dedicated, are inner directed, like control of situations and themselves, are obsessed with high standards of performance, and are socially conservative and traditional (Spadafora 2011).

TABLE 1-1 Pack Test Elevation Corrections for Elevations over 4000 ft

Altitude (ft)	Correction Added to Time Allowed (seconds)
4000–5000	30
5000–6000	45
6000–7000	60
7000–8000	75
8000–9000	90

Data from National Wildfire Coordinating Group. 2003. *Work Capacity Test Administrator's Guide NFES 1109*. Boise, ID: National Wildfire Coordinating Group. http://www.nwcg.gov/var/products/work-capacity-test-administrators-guide

Background Investigation

One of the components of the hiring process that is becoming more prevalent is the background investigation. Before offering candidates a job, many departments conduct a comprehensive check of the prospective employee's background. The candidate signs a disclosure agreement that lets the department's investigators look at all of his or her records, with the exception of medical files.

Components of the background investigation may include a review of the application for errors and omissions, a personal information check, a fingerprint check, and a polygraph examination. Department investigators may also visit social networking sites to see if the candidate has posted any information or photos that may compromise his or her application for employment. Keep in mind that nothing posted on the Internet is really private, nor do things sent electronically ever go away. Tweets, emails, blogs, and other submissions may be discovered. The personal information component consists of a questionnaire to be filled out by the applicant. One of these questionnaires can easily extend to 20 or 30 pages. It may include any or all of the following:

- All residences for the past 10 years
- Names and addresses of all relatives and several references
- All educational experience, including high school
- Any experience or employment, including voluntary and temporary, for the past 10 years
- Military information, including all locations and supervisors
- Legal information regarding perjury, convictions for felonies or misdemeanors, lawsuits, and traffic violations
- Ownership, carriage, and use of firearms, including concealed weapons permit
- Prejudice against any groups and any discriminatory actions
- Drug use or any involvement with drugs directly or indirectly

Note: Even in states where recreational marijuana use is legal the fire department does not have to allow it due to health and safety concerns.

- Motor vehicle operation, including driver's license and insurance information
- Financial information, including accounts, mortgages, credit cards and loans, late payments, judgments, child support, and bankruptcy

There may also be questions regarding any physical altercations or domestic abuse incidents that you may have been involved in. The final question may be worded as: "Did you in any way cheat, lie, or commit fraud during the application or evaluation process or during any of the background processes?"

All relatives, employers, roommates, current and former spouses, and references may be sent a personal, confidential inquiry about you. Typical questions are:

- Does the applicant have good communications skills?
- Do you feel the applicant can make logical decisions and use common sense?
- Do you feel the applicant retains information?
- Do you feel the applicant has the willingness to confront problems?
- Is the applicant dependable and well-motivated?
- Is the applicant generous and willing to help others even at his or her inconvenience?
- Do you think the applicant has the desire for self-improvement?
- Please describe the applicant's demeanor, grooming, and personal care.
- Do you feel the applicant is of strong moral character and is honest?
- Do you feel the applicant is physically able to perform the duties of a fire fighter?
- How do you feel about the applicant working with the public?

These are subjective questions and not all responses are going to be weighted the same. You may be asked to explain why someone said a particular thing about you in his or her response.

Once the packet is reviewed by the employer and all of the references that reply have been reviewed, a polygraph examination may be administered. Its purpose is to verify that all of the information presented on the personal information questionnaire is correct. It is considered the best practice not to misrepresent any of the information. If there is something in your background, such as recreational drug use, possession of alcohol, or theft, be honest and forthcoming. Everything is subject to review, and it may be considered juvenile, versus adult, behavior. If you are doing something now that will ruin your chances of being selected, stop it. The duration of time between your last bad behavior and now may also be considered in the process. Every question needs an answer, and one of the worst things that can be done is to leave something out or lie about it. Doing so is grounds for failure of the background portion of the examination. In addition, you should keep records of the information that the employer is likely to request so you can truthfully and completely fill out the background investigation packet.

Final Interview

The last interview is with the fire chief or designated representative(s). Conducted in a manner similar to the oral examination interview, the primary purpose is to discuss various phases of the fire service career. This interview takes place when your name appears within the top positions on the eligible list.

The interviewer(s) on this panel can change your ranking among the people they are interviewing. More than one person may be interviewed for a vacancy.

Stay alert and pay attention. This interview may make the difference in whether you get hired. You could walk into the interview as the top-rated candidate on the list and walk out without a job.

Medical Examination

A step in your evaluation process prior to appointment as a probationary fire fighter is a complete medical examination, which may include drug screening (NFPA). It is given by a physician appointed by the department at the department's expense. Each applicant must be in good general physical condition, free from disease or defects that would interfere with the satisfactory performance of the duties of the position. The results of the pre-employment medical examination are used to determine whether applicants possess the prescribed standards of physical health and physique required for the position of fire fighter. Applicants who fail this examination have their names removed from the eligible list.

Probationary Period

The probationary period is the last step in the selection process. The academy program can be considered a part of the probationary period. At the end of the academy program there may be a final examination that covers all of the skills and technical material presented. The fire fighter then goes to an assignment at a fire station. Probationary fire fighters are expected to perform independent study to learn the required departmental policies and procedures. The probationary period may last 12–18 months because a fire fighter working a 24-hour shift schedule works only an average of 10 days a month. Twelve months is often considered to be enough time to observe how the new

fire fighter has adjusted to fire department life and performs at emergencies and regular work assignments.

During the probationary period, new fire fighters may not enjoy full civil service protection of their employment. This allows the department to remove fire fighters who cannot adjust or who are found to be unable to perform their duties. Removals from service are not done without cause. By this time the department has invested a lot of money in the selection process and training.

Many departments have a comprehensive combination written/manipulative test at the end of the probationary period. The purpose of this test is to assess the knowledge and skill of the fire fighter. The candidate is expected to perform correctly the skills taught in the academy as well as those learned during the probationary period. The written test consists of knowledge of policies and procedures and technical information about the department and equipment. If, at this point, the fire fighter cannot pass these tests, dismissal may occur.

Goal Setting

To achieve your goal of becoming a skilled professional fire fighter, you must first clearly define what your goals are. You must create a road map to success with a well-defined destination. If you wander aimlessly, no matter how fast you go you will not arrive at the desired destination. Planning and working hard are required to achieve your goal. Without both you will not succeed.

The time to start preparing and planning is now. No one is going to wait for you. The job you desire is going to come open, and unless you are truly prepared you are not going to be ready to get it. Fire fighter entrance examinations are competitive. Not all who apply are going to be accepted. With few exceptions, those who are well prepared are going to perform best. In civil service hiring, lists are established and those at the top of the list get first consideration for vacancies.

Goal-Setting Process

When you develop your goals, keep the acronym SMART in mind. Goals should be simple, measurable, accountable, realistic, and timely.

The first step is to visualize what your goals are and write them down. Keeping a list in your head is ineffective. The interference from everyday events will soon have you losing sight of your goals unless they are clearly written and easily referred to on a regular basis. The process of writing down your goals assists you in determining whether they are truly realistic.

By writing your goals down, you are making a contract with yourself. This personalizes the goals to you and no one else. Goal accomplishment requires commitment. Establishing a contract with yourself helps to create that commitment.

Long-term planning is based on current decisions. What you do now will have consequences—positive or negative—later. The background investigation requires you to stay out of trouble and to avoid doing certain things. If you choose to start or continue doing things that will preclude you from becoming a fire fighter in the present, they are going to cost you in the future. Look at the news and count the number of people who derail themselves through current actions. They were on the road to success, and some decision they made caused them to lose the ability to accomplish their career goals. It is the same with physical fitness. You will be required to be in good physical condition to pass the physical ability test. You cannot get in shape overnight. It takes months to prepare for a physical ability test such as the CPAT. The CPAT involves both physical ability and skills at performing operations. Both of these can and should be practiced ahead of time.

Preparation is important, but so is taking action to achieve your goals. Preparation paralysis can set in, and you may never really take the action required to achieve the goal. Setting a time line on certain goals may assist you in preparation. Projecting goals into the future with no completion date lets you take forever to achieve them.

Look at the job requirements, and once you meet the minimum, fill out the application and take the entrance exam. You can continue to prepare as you continue to take entrance exams. Do not wait until some unspecified time in the future to start taking entrance exams. The more often you take them, the better you can prepare to take them in the future. Testing can be

practiced just like other skills. You may not do well at first because you are not used to the testing process. Do not be discouraged or give up. You were not able to run when you first learned to walk. The intimidation factor of that first oral interview panel can be overcome with practice. If at first you lack poise, work on it.

You need not do this all by yourself. Do not be afraid to ask for help. Many of the instructors in fire science/technology programs are working or retired fire fighters. Ask for assistance. For the most part they are involved in this type of instruction because they wish to give something back to the service they love. They can do this through you. If they are unable to assist you, they probably know someone who can. Do not be hesitant about asking. Fire fighters tend to be can-do people, and they respect this in others.

Your goals have been established; now what if something happens that causes you to be unable to achieve one or more of them? This does not mean that they are forever unattainable. You just have to make a course correction and move on. If you did not do well on the exam in your hometown, take other exams. Focus on the positive. Everyone has a bad day now and then. Life has a way of taking strange twists and turns that adaptable people use to their advantage. To adapt, overcome, and continue is the way successful people react to setbacks and obstacles.

Celebrate success. Take some time to reflect when you have succeeded in achieving a personal goal—the completion of your degree, Fire Fighter I, or EMT certification, for example. A feeling of success on a regular basis will help energize you to accomplish the goals that are left.

Goals are accomplished in three steps. The first is to visualize the goals. Second, clearly define the goals and write them down. Finally, take the actions required to achieve them. The absence of one of these three steps does not guarantee failure, but it will hinder accomplishment.

Wrap-Up

CHAPTER SUMMARY

- Training is the pursuit of a specific skill; education includes memorization of specific pieces of information and developing an understanding of concepts or philosophies.
- Fire science/technology education options include certificate programs, associate's degrees, bachelor's degrees, and master's degrees.
- Advantages of online programs are that the student does not need to live close to the school to complete the courses and that there is some leeway in scheduling, allowing fire fighters with set duty schedules to complete the assignments when it is convenient for them.

- Most fire departments operate as public agencies governed by local or state government; this makes a public administration educational background vital to the fire executive.
- When you pin on a fire fighter's badge, you represent hundreds of years of tradition of selfless service and sacrifice. Your actions affect the reputation of the fire service as a whole. You must be ready and willing to follow a Firefighter Code of Ethics both on and off duty.
- Responding to other people's tragedies has a way of drawing fire fighters closer together and forming strong bonds among them.
- Fire fighters must be team members. They must be willing to give up personal desires to benefit the team.
- Letting yourself get out of shape endangers not only you but also your co-workers and the public. They have every right to expect you to stay fit.
- Fire fighters must watch out for signs and symptoms of PTSD and/or RET in themselves and their co-workers and know what to do if it appears.
- The main job of a fire fighter used to be to control and extinguish hostile fires. The fire service has now taken responsibility for providing EMS, fire prevention, hazardous materials response, search and rescue, homeland security, and other emergency services.
- The Equal Employment Opportunity Act prohibits discrimination against any person in the classified service or any person seeking admission thereto because of race, national origin, sex, age, physical disability, color, medical condition, marital status, ancestry, or union activity.
- Sometimes having medical training such as EMT or paramedic training can be of benefit when seeking employment. This varies by department and the level of medical service it provides.
- In a personnel development program, a fire fighter is trained as high as two ranks above the one currently held.
- The steps in the selection process may include the following but will vary by department: recruitment, application, written examination, skills test, oral interview, physical ability/agility, medical examination, psychological examination, background check, final oral examination, and probationary period.
- Most fire departments have prerequisites for application. The most basic prerequisites are a valid driver's license, a high school diploma or general equivalency diploma (GED), and no felony convictions.
- The application process consists of first finding out when the application will be available and second, finding out when it must be completed and submitted. There may also be measures in place to limit the number of applicants.
- The written examination is designed to test a candidate's ability to learn firefighting procedures and techniques.
- The skills test portion of the exam is not used by all departments. When it is used, the top performers from the written exam are invited to participate.
- There is usually a representative from the personnel department in attendance at the oral examination/interview, and the interview is recorded.
- Physical ability/agility tests are administered to judge the candidates' overall physical conditioning and ability to perform firefighting-related tasks.
- Federal wildland fire fighters are required to pass the work capacity test (pack test) at the arduous fitness level. Some fire departments in areas with wildland firefighting responsibilities require this test as well.
- The psychological test for fire fighters is usually performed as a series of written true or false questions, answers to questions read off a computer screen, and answers to interview questions.
- One of the components of the hiring process that is becoming more prevalent is the background investigation.
- The final interview is with the fire chief or designated representative(s). Conducted in a manner similar to the oral examination interview, the primary purpose is to discuss various phases of the fire service career.
- A step in your evaluation process prior to appointment as a probationary fire fighter is a complete medical examination, which may include drug screening.

- Probationary fire fighters are expected to perform independent study to learn the required departmental policies and procedures. Twelve months is often considered to be enough time to observe how the new fire fighter has adjusted to fire department life and performs at emergencies and regular work assignments.

- To achieve your goal of becoming a skilled professional fire fighter, you must first clearly define what your goals are. You must create a road map to success with a well-defined destination.

- When you develop your goals, keep the acronym SMART in mind. Goals should be simple, measurable, accountable, realistic, and timely.

KEY TERMS

Banding A civil service selection process tool in which candidates who score between certain points on the scale (such as between 90 and 100 percent) are grouped, and the employer may choose candidates for employment from within the group.

Call-back A recall of personnel to on-duty status, usually because of an emergency situation.

Certification A document that specifies that a student has successfully completed the prerequisite education and training to perform a job function, such as Fire Fighter I or emergency medical technician.

Critical incident stress debriefing (CISD) A discussion in which personnel are encouraged to express their feelings after responding to and operating at particularly stressful incidents that result in high loss of life, loss of life by a co-worker, or other significant conditions. These are conducted after an incident to assist personnel to better deal with their emotions.

Critical incident stress management (CISM) An intervention protocol developed specifically for dealing with traumatic events. It is a formal, highly structured, and professionally recognized process for helping those involved in a critical incident to share their experiences, vent emotions, learn about stress reactions and symptoms and given referral for further help, if required. It is not psychotherapy. It is a confidential, voluntary, and educative process, sometimes called "psychological first aid." This is performed through a peer process utilizing Critical incident stress debriefings.

Critical thinking The process of examining, analyzing, questioning, and challenging situations, issues, and information of all kinds.

Curriculum A particular course of study.

Demographics The statistical characteristics (e.g., age, race, gender, income) of the population of an area.

Drills Tasks and jobs being practiced to improve performance.

Education Memorization of specific pieces of information and development of an understanding of concepts or philosophies.

Eligible list A certified list of persons who have successfully completed the testing process.

Emergency medical technician (EMT) A specified level of medical training that usually consists of approximately 100 hours of classroom and practical training and the completion of a national registry examination.

Explorers A program of the Boy Scouts of America for persons 15–21 years of age. The Explorers work in conjunction with a professional organization such as the fire or police department to learn the operation and job requirements. Females are now allowed in the Boy Scouts and its programs.

Freelance The act of performing operations without a coordinated effort or the knowledge of one's superior officer.

Generalist A person with general knowledge of varying levels of depth in many subject areas.

Manipulative training Training in the operation of tools and equipment.

Mechanical aptitude The ability to figure out the operation and construction of equipment from drawings.

Mentor A person who guides and directs another person toward a goal.

Model curriculum A series of courses meeting standardized criteria, including titles, descriptions, outcomes, and outlines.

On duty The time fire fighters spend performing their jobs.

Oral interview panel An interview technique in which the interviewers ask questions and evaluate the answers given by job candidates. They assign a score to the candidate's responses for ranking purposes during the selection process.

Paramedic A person with an advanced level of medical training. Paramedics can perform invasive procedures on the patient, such as starting intravenous lines.

Performance-based certification A training program in which the student must meet prerequisites of education and experience to take a training course. Once the course is completed, the student then completes a task book for the position, requiring classroom and/or incident experience. Once the task book is completed, the person receives certification for the new position.

Post-Traumatic Stress Disorder (PTSD) A disorder that develops in some people who have experienced a shocking, scary, or dangerous event.

Probationary fire fighter A person hired by the fire department who has not been granted permanent status.

Recurring Exposure to Trauma (RET) The effects of responding numerous times in one's career to traumatic events and their effect on the responder.

Reserve/cadet programs Organized programs sponsored by paid fire departments that provide training in return for personnel volunteering their time.

Résumé A listing of a person's areas of experience and education.

Specialists People with extensive training in one or more areas of operations or information.

Task book Log used to verify competency in particular skills. A trainer must certify that the skill was performed in a satisfactory manner in a field and/or classroom environment.

Training The pursuit of a particular skill.

Upper division College-level courses that are applicable to a degree program for a bachelor's degree or higher.

Veterans' points Points added to a person's final score on a competitive examination process; given to persons who have satisfactorily performed military service.

Volunteer firefighting programs Performing firefighting services without pay. In some areas a variation of this is the Paid Call Fire Fighter program. Under this program, fire fighters are paid a specified sum when they respond to incidents or attend training.

Worker's compensation Money paid to persons who have been injured in the course of their employment and are unable to work either temporarily or permanently.

CASE STUDY

A young woman seeks a career in the fire service. She makes an attempt to complete the hiring process but is unsuccessful. She has the training, education, and experience to do the job and do it well. Her failure is during the physical ability/agility step because she lacks the upper body strength to complete the events in the time required.

Focusing her efforts on the area in which she does not meet the standards, she sets the goal of completing the physical ability/agility within prescribed time limits. She seeks training and guidance, specifically addressing the areas in which she is deficient. The next time she attempts the physical ability/agility she is successful and continues on in the selection process, ultimately ending up with a job offer with the fire department.

1. Of the steps in the employment process, where is this young woman deficient?

   A. Oral examination
   B. Physical examination
   C. Written test
   D. Physical ability/agility

2. In terms of goal setting, what is the first step in the process she should address?

   A. Time certain
   B. Specific
   C. Measurable
   D. Realistic

3. To address her upper body strength issue, lacking the funds to hire a personal trainer, she should seek help from:

   A. fire fighters
   B. friends
   C. Internet sources
   D. college professors

4. A readily-available source of information on how to address her particular issue would be from:

   A. the library
   B. a newspaper
   C. a magazine
   D. the Internet

REVIEW QUESTIONS

1. The fire science/technology curriculum is aimed at providing the student with what types of skills?

2. Before enrolling in classes, the student should meet with which college official?

3. What is the basic college degree in fire science/technology?

4. List several advantages of attending online courses over traditional course settings.

5. List two ways to attend fire academies.

6. What manipulative certification should you have before applying for a position with the fire department?

7. Training programs are offered at various levels, such as state and local. List two others.

8. List two preservice opportunities for gaining firefighting experience.

9. List two basic prerequisites in applying for the fire fighter exam.

10. What is the first step in the selection process?

11. List two ways to find out when application periods are open for fire fighter examinations.

12. List a source of material used to prepare for the written examination.

13. List two ways to prepare for the physical ability/agility test.

14. List two ways to prepare for the oral examination.

15. What is the purpose of the probationary period?

16. What are the steps in preparing SMART goals for yourself?

DISCUSSION QUESTIONS

1. What are some ways you can prepare yourself to perform well in the fire fighter selection process?

2. Why is hiring for diversity (e.g., women and minorities) an important issue in public agencies, such as the fire service?

3. Prepare answers to the following questions you may be required to answer in an oral interview:

   - Why are you the best person for the job?

   - What have you done to prepare yourself to be a fire fighter?

   - What is your goal in the fire service?

   - What is your definition of customer service; give us an example of customer service you have provided to someone else.

4. Why is the national professional development model based on education *and* training?

REFERENCES AND ADDITIONAL RESOURCES

Conlan, Catherine. 2014. "Clergy, Firefighters among Most 'Meaningful' Jobs." *The Bakersfield Californian*, E1, August 24, 2014.

Croker, Edward. n.d. "Wise Old Sayings." Accessed September 5, 2018. http://www.wiseoldsayings.com/firefighter-quotes/

Edelman, Susan. 2013. "FDNY: Felons Welcome." *New York Post*, January 27, 2013. http://www.nypost.com/p/news/local/fdny_felons_welcome_URda7GQscBVj N65N05bL1M

Fahy, Rita F., Paul R. LeBlanc, and Joseph L. Molis. June, 2018. *Firefighter Fatalities in the United States—2017*. Quincy, MA: National Fire Protection Association. Accessed July 23, 2018. https://www.nfpa.org/News-and-Research/Data-research-and-tools/Emergency-Responders/Firefighter-fatalities-in-the-United-States

Federal Emergency Management Agency, Emergency Management Institute. 2013. "The College List." Emmitsburg, MD:

FEMA EMI. Modified May 20, 2013. Accessed July 16, 2013. http://www.training.fema.gov/EMIWeb/edu/collegelist/

International Association of Fire Fighters. 2013. *Candidate Physical Ability Test Program Summary*. Washington, DC: International Association of Fire Fighters. http://www.iaff.org/HS/CPAT/cpat_index.html

Jahnke, Sara Anne, Walker S.C. Poston, Christopher Keith Haddock, and Beth Murphy. 2016. "Firefighting and Mental Health: Experiences of Repeated Exposure to Trauma." *Work* 53, no. 4: 737–44. doi:10.3233/wor-162255. Accessed July 23, 2018. U.S. Fire Administration https://www.usfa.fema.gov/current_events/011718.html

National Fire Protection Association. *NFPA 1001: Standard for Fire Fighter Professional Qualifications*. Quincy, MA: National Fire Protection Association.

National Fire Protection Association. *NFPA 1582: Standard on Comprehensive Occupational Medical Program for Fire Departments*. Quincy, MA: National Fire Protection Association.

National Fire Protection Association. 2018. *The U.S. Fire Problem*. Quincy, MA: National Fire Protection Association.

National Wildfire Coordinating Group. 2003. *Work Capacity Test Administrator's Guide NFES 1109*. Boise, ID: National Wildfire Coordinating Group. http://www.nwcg.gov/var/products/work-capacity-test-administrators-guide

Spadafora, Ronald R. 2011. "Becoming a Firefighter: Psychological Tests for Firefighters." Accessed April 22, 2019. www.education.com/reference/article/psychological-tests-firefighters/

U.S. Fire Administration. December, 2017. *Firefighter Fatalities in the United States in 2016*. Federal Emergency Management Agency. Accessed May 1, 2019. https://www.usfa.fema.gov/downloads/pdf/publications/ff_fat16.pdf

U.S. Fire Administration. 2018a. *Fire and Emergency Services Higher Education Model Course Outlines*. Accessed May 1, 2019. https://www.usfa.fema.gov/training/prodev/model_courses.html

U.S. Fire Administration. 2018b. *Firefighter Code of Ethics*. Accessed April 23, 2019. https://www.usfa.fema.gov/downloads/pdf/code_of_ethics.pdf

CHAPTER 2

Fire Protection Career Opportunities

OBJECTIVES

After studying this chapter, you should be able to:

- Identify fire protection careers in the public fire service.
- Identify civilian positions in the fire service.
- Identify fire protection careers in the private fire service.

Case Study

Three students meet in the fire science program at their local college. They all wish to be fire fighters but, coming from differing backgrounds, seek different things from the job.

Student 1 seeks a job in the urban/suburban environment with a focus on medical aid incidents and structural firefighting. She does not mind the paramilitary organization and prefers the more regular work schedule of either 24- or 48-hour shifts. She likes the idea of having a prescribed work schedule that is planned out on an annual calendar so that she even knows whether she will be on shift for each of the holidays over the coming year.

Student 2 is an outdoorsman and has no desire to "work in town." He seeks wide open spaces and a more relaxed work atmosphere that will take him around the country going to wildland fires. He does not mind that his career will most likely begin as seasonal and then transition to full-time, all-year employment in the future.

Student 3 seeks seasonal employment for now, so he can pursue his hobbies of snow skiing and other adventure travel in the winter off-season. He will then consider the urban/suburban path of employment in the future.

1. What would be the best training, experience, and education for Student 1 to pursue? What type of fire agencies should she apply to for a job as a fire fighter?
2. What would be the best training, experience, and education for Student 2 to pursue? What type of fire agencies should he apply to for a job as a fire fighter?
3. What would be the best training, experience, and education for Student 3 to pursue? What type of fire agencies should he apply to for a job as a fire fighter?

 *Access Navigate for more resources.*

Introduction

In this chapter, the wide variety of firefighting and other fire service–related careers is examined. The modern fire service offers numerous career paths in both the public and private sectors. With the rise of emergency management as a profession, some will choose to pursue a career in emergency management instead of a career in firefighting. Some fire service positions are filled with firefighting personnel, and others are filled with civilian (non-firefighting) personnel.

As the fire service becomes broader in mission, some personnel tend to focus their training on specific areas. Specialization within the firefighting ranks is available with a focus on urban search and rescue (USAR), swift water rescue, emergency medicine, high-rise firefighting, wildland firefighting, fire prevention, public education, incident management, and many other areas. These specialized services are largely dependent on the needs of the organization and its location in the country.

The process of pursuing a career in the fire service frequently begins with training to become a fire fighter, completing probation, and working with a firefighting force. A diverse organization will expect its own firefighting personnel to play many specialized roles and to perform a variety of functions. In addition, with resources now more commonly being utilized on a regional, rather than a local, basis, the size of the organization in which a fire service professional is employed is less of a limiting factor in choosing a specialization.

Public Fire Protection Careers

The first selection of jobs described directly involves firefighting. These job descriptions are representative of those in municipal and rural fire departments.

Fire Fighter Recruit–Fire Department

In cases in which the fire department employs new fire fighters under specific trainee programs, the job title fire fighter recruit or trainee–fire department would be representative. A person working in this capacity typically receives less pay and fewer benefits than a fire fighter who has completed the training program.

Under close supervision and in a learning capacity, the fire fighter recruit–fire department assists in the various phases of fire suppression and prevention with the goal of learning the functions carried on by the department. Typical tasks include performing a variety of work assignments:

- Responding to fire alarms and other emergency incidents to protect life and property

- Assisting fire personnel by performing selected duties of significant learning value
- Participating in continuing training and instruction programs by individual study of technical material and attendance at scheduled drills and classes
- Driving and operating fire engines and similar equipment

Employment standards include knowledge of modern fire prevention and suppression methods or the ability to acquire such knowledge; physical endurance and agility; mechanical aptitude; and the ability to learn technical firefighting techniques and principles of hydraulics applied to fire suppression, to understand and follow oral directions, to establish and maintain cooperative relationships with fellow employees and the public, and to keep simple records and prepare reports (Kern County Human Resources Department 2018). The emphasis of this role is on close supervision in a learning capacity. These positions are often used to target specific groups for entry into the fire department because of affirmative action requirements.

After a specified period of time and satisfactory completion of the program, the person is promoted to fire fighter–probationary. This program would not have a prerequisite of Fire Fighter I certification but would instead lead to this certification.

Fire Fighter–Fire Department

This is the standard entry-level position for the fire department and may or may not require the completion of a Fire Fighter I academy or certification to apply. Some departments also require a certification as an Emergency Medical Technician (EMT). Under supervision, the fire fighter–fire department will respond to fire alarms and other emergency incidents to protect life and property and will do related work as required. Tasks include:

- Responding to alarms and assisting in suppression of fires
- Cleaning up and performing salvage operations after fires
- Assisting in maintaining and caring for fire apparatus, equipment, fire station, and grounds
- Responding to emergency incidents
- Operating oxygen administration equipment and automated external defibrillator and administering first aid
- Making residential and business inspections to discover and eliminate potential fire hazards and to educate the public in fire prevention

FIGURE 2-1 Fire fighters performing extrication of a trapped victim.
© Rick McClure/AP Images

- Participating in continuing training and instruction programs by individual study of technical material and attendance at scheduled drills and classes
- Driving and operating fire engines and similar equipment
- Training or assisting in training auxiliary fire fighters
- Acting as a reliever for a driver/operator or company officer **FIGURE 2-1**.

The employment standards are the same as for fire fighter recruit–fire department.

In departments with several ranks, the fire fighter may become eligible to test for the position of driver/operator or company officer after a specified period of time in rank and completion of any department-specified prerequisites. The time in rank and prerequisites vary depending on the department.

In departments where the driver/operator position is of a higher rank than fire fighter, the position usually has more responsibility to ensure that company functions, such as program work, are completed. When it is a promotional position, there is an increase in pay.

Fire Fighter–Fire Department Federal

The federal government has numerous positions for fire fighters at federal installations, mostly located at military bases. The job descriptions, requirements, and promotional opportunities are much the same as for municipal fire fighters.

Fire Fighter Paramedic

As the traditional role of the fire department has changed, the position of fire fighter paramedic has

FIGURE 2-2 Fire fighter paramedic and other personnel checking victim for injuries.
© Jones & Bartlett Learning

FIGURE 2-3 Fire heavy equipment specialist constructing fire line.
Courtesy of Kern County Fire Department

become much more prevalent and has increased responsibility over that of fire fighters and requires the completion of advanced medical training.

Depending on the department, the fire fighter paramedic may respond as part of an engine crew and provide medical aid as needed as an adjunct to fire-fighting duties. In other departments, the fire fighter paramedic responds in either a special squad vehicle or ambulance. If responding in a squad vehicle, patient transportation is handled by others. If responding in an ambulance, the fire fighter paramedic treats victims at the scene and transports them to the hospital **FIGURE 2-2**.

In addition to the required standards for a fire fighter, the fire fighter paramedic must possess or be capable of acquiring a paramedic certification valid for the jurisdiction served. Professional opportunities are the same as for fire fighters. As with any other preparation for promotion, a willingness to accept additional responsibility is considered an asset.

Advantages of becoming a fire fighter paramedic are that persons with these qualifications are in great demand and lateral transfer to another department may be available. Another advantage is that along with the increased responsibility and ability to assist the public in medical emergencies, there is usually a pay incentive.

Fire Heavy Equipment Specialist

The fire heavy equipment specialist position is mostly limited to departments that provide fire protection for wildland areas **FIGURE 2-3**. Under direction, the fire heavy equipment specialist operates heavy motorized equipment in fire control work and constructs and maintains fire breaks and roads, as well as completing

other tasks as required. Typical tasks include the following:

- Operating a bulldozer in wildland fire areas over steep, rough terrain to establish fire control lines or in fire hazard reduction and conservation work
- Operating motor graders, heavy duty transports, fire trucks, and other types of equipment used in fire suppression and maintenance work
- Servicing and assisting in making mechanical repairs to equipment, including welding and limited body repair
- Working with or supervising others on fire line, road construction, conservation, and in-camp work projects or fire incidents
- Maintaining records of work accomplished and equipment serviced and submitting reports
- Responding to alarms and assisting in suppression of fires
- Cleaning up and performing salvage operations after fires
- Assisting in maintaining and caring for fire apparatus, equipment, fire station, and grounds
- Responding to emergency incidents, operating oxygen administration equipment, and administering first aid
- Continuing training and instruction programs by individual study of technical material and being in attendance at scheduled drills and classes
- Assisting in roadside burning and in building or clearing fire breaks or fire roads

Employment standards include graduation from high school or equivalent and experience in operating heavy motorized equipment, including bulldozers and heavy transport trucks, some of which has been in rugged terrain.

Safety Section Retirement

In some states, firefighting positions, along with law
enforcement, are classified as safety section positions
and receive a higher level of retirement compensation than those employees classified as general (non-
safety) members. For example, consider Employee A
operates heavy equipment for the roads and bridges
department for the jurisdiction. Upon retirement
after 30 years of service, at 55 years of age, his percentage of final compensation is 44 percent. Also consider Employee B operates heavy equipment as a fire
heavy equipment specialist for the same jurisdiction.
Upon retirement after 30 years, at 55 years of age, his
percentage of final compensation is 78 percent. This
works out to Employee B's pension being 1¾ times
as much as that of Employee A. This increased pension amount is based on the fact that firefighting is an
extremely dangerous and stressful occupation.

Another difference under this law is that safety section members are eligible for service retirement after
20 years, regardless of age, versus 30 years, regardless
of age, for general members (State of California 2018).

The next group of positions includes positions
in the wildland fire agencies, including the U.S. Forest Service, Bureau of Indian Affairs, National Park
Service, Bureau of Land Management, and state
departments of forestry, which tend to be seasonal
employment for entry-level personnel. It may take
several summers as a seasonal employee to secure a
permanent position. Several items of note are that the
federal wildfire agencies have a mandatory retirement
age of 57, and their employees can be furloughed or
work without pay during government shutdowns.

Fire Fighter (Forestry Aid) Wildland General Schedule 3

The fire fighter wildland General Schedule 3 (GS 3)
position is that of an advanced trainee on a wildland fire suppression engine or hand crew and/or a
fuels management crew **FIGURE 2-4**. The work is
performed in a forest environment in steep terrain,

FIGURE 2-4 Forestry aids constructing fire line with
assistance of water-dropping helicopter.
Courtesy of Casey Christie/ Kern County Fire Department

where surfaces may be extremely uneven, rocky,
covered with thick, tangled vegetation, and so forth.
Temperatures are frequently extreme, both from the
weather and from the fire. Smoke and dust conditions
are frequently severe. This position is usually considered to be entry level, and the training required may
be provided on the job or from a local community college or high school program prior to the fire fighter
being allowed to engage in firefighting operations.
Tasks may include the following:

- Performing assignments to develop knowledge
  of fuels management and fire suppression
  techniques and practices, such as fire line
  construction, use of pumps and engines,
  hose lays, foam and retardant, working
  around aircraft, safety rules, and fire and fuels
  terminology
- Searching out and extinguishing burning
  materials by moving dirt, applying water by
  hose or backpack pump, and other methods

- Chopping brush or felling small trees to build a fire line using various tools such as an axe, a shovel, a **Pulaski**, a **McLeod**, and power saws to control spreading wildland fire and to prepare lines prior to **prescribed burning**
- Chopping, carrying, and piling **logging slash**
- Patrolling the fire line to locate and extinguish sparks, flare-ups, and hot spot fires that may threaten developed fire lines
- Cleaning, reconditioning, and storing fire tools and equipment
- Performing other resource management activities such as recreation, timber, or **reforestation** when not performing fire suppression or fuels management duties

The purpose of the work is to carry out assigned tasks in the reduction of ground fuels and the suppression of wildland fires. The scope varies from a small prescribed fire that may be managed by the crew to a large fire involving several thousand people. The aim of the work performed is to manage fire under controlled conditions to accomplish work or to minimize resource loss under wildfire conditions.

Employment standards include knowledge of forestry practices and techniques, including accepted fire suppression and prescribed burning methods to be used in various types of fuels and under a variety of weather and terrain. This position also requires working knowledge of fire behavior and fire control techniques to carry out assigned fuels and wildland fire tasks. Skills in the use of hand tools, such as an axe, a shovel, a Pulaski, a McLeod, portable pumps, and chain saws to build fire lines, extend hose lays, and extinguish burning materials are also needed, as is knowledge of safety practices to prevent injury and loss of life.

The work requires strenuous physical exertion for extended periods, including walking, climbing, shoveling, chopping, throwing, lifting, and frequently carrying objects weighing 50 pounds or more. The duties of the position require that the candidate meet prescribed physical requirements as measured by the Work Capacity Test for Wildland Firefighters (discussed in the Fire Science Education and the Fire Fighter Selection Process chapter) (U.S. Forest Service 2018).

With satisfactory performance and available positions, the fire fighter wildland GS 3 may be promoted to a higher pay position, fire fighter, or transfer into a position requiring greater knowledge and proven ability, such as **helitack** crew person, **smoke jumper**, fire prevention technician, or fire engine operator trainee.

Many college students fill these positions because of their seasonal nature. It gives them the ability to make money during the summer for educational and living expenses while gaining experience in fire suppression. A person trained in wildland fire suppression is a great asset to fire departments that have wildland responsibilities. This experience can be a valuable addition to the résumé when trying to secure a firefighting job. Considering the work ethic and human relations elements of fire service work, it is obvious that experience of this type is directly applicable. More information on becoming a federal wildland fire fighter, including information on how to become a wildland fire fighter, can be found at the National Wildfire Coordinating Group website. For state-level positions, contact your state department of forestry.

Civilian Positions in the Fire Service

There are numerous civilian (non-safety) positions in the public fire service. These jobs fall into several areas and are an integral part of the delivery of a fire protection system. These civilian jobs do not have the rigorous physical requirements of fire fighter, although many are stressful and mentally challenging in their own right.

Fire Prevention Specialist

Under direct supervision, a fire prevention specialist performs a variety of routine fire prevention, training, and hazardous materials disclosure work. Typical tasks include the following:

- Performing routine field checks of fire prevention systems, including, but not limited to, hydrant flow test, fire sprinkler system check, and access requirements **FIGURE 2-5**
- Assisting in hazard reduction/weed abatement inspections to ensure compliance with federal, state, and local standards

FIGURE 2-5 Fire prevention specialist inspecting fire sprinkler system.

- Participating in fire safety, emergency medical services, and hazardous materials disclosure, training, and education programs
- Researching federal, state, and local policies, procedures, codes, and ordinances
- Gathering and correlating hydrant information, such as location, size, type, water pressure, and flow rates, and updating water maps as directed
- Providing fire prevention and hazardous materials disclosure information to the public, department personnel, and other related agencies
- Assisting in developing procedures for, and coordination of, a hazardous materials disclosure program
- Evaluating information provided on disclosure forms and assigning fees
- Assisting in developing, producing, coordinating, and maintaining hazardous material data on facilities in jurisdiction area
- Assisting in gathering and correlating statistical information
- Writing reports based on field notes
- Assisting in filing and record keeping
- Assisting fire prevention officers and inspectors with office coverage

Employment standards are determined by the employing jurisdiction. Upon obtaining experience and proficiency, the person may be promoted to Fire Prevention Specialist II. People have also used this position to gain a foothold in the fire department and have used the work experience, background, and knowledge gained to perform well on the fire fighter exam.

Fire Hazardous Materials Program Specialist

Under direction, the fire hazardous materials program specialist evaluates potential hazards of unusual chemicals and materials and determines which chemicals are subject to the hazardous materials disclosure sections of the health and safety code. This person serves as the technical advisor to the hazardous materials control unit of the fire department and does related work as required **FIGURE 2-6**. Typical tasks include the following:

- Evaluating potential hazards of unusual chemicals and materials
- Determining which chemicals are subject to the hazardous materials disclosure sections of the health and safety code

FIGURE 2-6 Hazardous materials specialist performing business inspection.
© Jones & Bartlett Learning

- Developing, maintaining, and updating lists of hazardous materials
- Analyzing industrial manufacturing processes to identify those potential chemical usages requiring regulations
- Providing technical support to fire department staff regarding regulations
- Reviewing and analyzing requirements for chemical storage facilities, hazardous material inventory statements, and business emergency plans
- Assisting in the development, implementation, and review of the area plan
- Reviewing completed disclosure forms and emergency plans involving the most complex substances
- Interpreting sections of the laws pertaining to hazardous materials and providing information to emergency personnel, health officials, elected officials, businesses, and the general public
- Coordinating hazardous materials disclosure to the fire code
- Reviewing and commenting on hazardous material preparedness programs
- Assisting in technical evaluation of materials at emergency scenes as required
- Assisting other agencies as required regarding the effects of hazardous materials on the environment
- Providing information and assistance to industries, other agencies, and the public on compliance with the hazardous materials disclosure regulations

- Working with other local, state, and federal agencies involved in determining the effects of hazardous materials on the environment
- Providing technical support to the fire department personnel conducting onsite inspections and participating in inspections of facilities involving complex and volatile chemicals
- Writing reports based on field notes
- Preparing and reviewing technical documents, reports, correspondence, documentation, and evidentiary material

Employment standards are determined by the jurisdiction offering the position. The following standards are offered as an example: Bachelor's degree in a physical or biological science, environmental engineering, industrial hygiene, or related field (fire protection technology, fire protection engineering), and the requisite years of increasingly responsible experience in hazardous materials control, industrial hygiene, industrial processes, or related areas; or a master's degree in industrial hygiene, chemistry, environmental engineering, industrial engineering or related field; and 2 years of experience in hazardous materials control, industrial hygiene, industrial processes, or related areas.

Additional requirements could be knowledge of the principles of physical, organic, and inorganic chemistry, including qualitative and quantitative analysis; knowledge of chemistry and laboratory techniques, including toxicology, hazardous materials identification, and biology; knowledge of effects of hazardous materials and their interaction with the environment; knowledge of local, state, and federal regulations and laws relating to hazardous materials; ability to perform research work on technical problems; ability to analyze manufacturing processes; ability to analyze and interpret data; ability to critically review chemical reports and documents; ability to interpret local, state, and federal laws and regulations relating to hazardous materials; ability to prepare clear, accurate, and concise written technical reports; and ability to present evidence in court (Kern County Human Resources Department 2018).

Fire Department Training Specialist

Under supervision, the fire department training specialist plans, develops, and produces training, informational, and educational materials. He or she also supervises or participates in the preparation of

FIGURE 2-7 Training specialist shooting video for training.
Courtesy of Kern County Fire Department

multimedia instructional and informational materials **FIGURE 2-7**. Typical tasks include the following:

- Planning and developing materials for use in training programs
- Developing training materials for the fire department and the public
- Presenting training programs to fire department personnel and the public
- Assessing training needs through testing and evaluation of personnel
- Evaluating effectiveness of current and future training programs
- Researching training needs in regard to department needs and state and federal laws
- Implementing training programs
- Developing and maintaining a record-keeping system to ensure and verify compliance with legal requirements
- Scheduling in-service training programs to maintain company efficiency
- Reviewing standard operating procedures for accuracy and consistency with fire department operations
- Ensuring department members' ability to utilize new equipment
- Evaluating and recommending use of outside-developed training programs and materials
- Preparing training budgets
- Researching and recommending new procedures

Employment standards include college-level study in instruction and evaluation. This person must also be able to communicate effectively, both orally and in writing, and organize and schedule required training.

Public Fire Safety/Education Specialist

The public fire safety/education specialist provides public education within the community in all aspects of life/fire safety, including providing instruction in fire prevention, fire escape planning, fire extinguishment methods, and other related aspects. This specialist coordinates department-sponsored community awareness; community relations; and community improvement programs, services, and activities related to public fire and burn safety education. He or she also assists public educators with fire safety education **FIGURE 2-8**. Typical tasks include the following:

- Planning, conducting, and developing safety, fire, and burn prevention programs
- Investigating and evaluating various community needs and implementing programs to address these needs
- Presenting programs to homeowners' groups, public and private schools, large groups, civic organizations, industry, businesses, and other segments of the community
- Providing instruction for identifying and correcting potential fire and burn hazards in the home
- Reviewing data to determine areas of the community requiring an emphasis on fire prevention instruction
- Coordinating engine company activities in implementing fire safety programs
- Providing instruction to citizens regarding escape planning in the event of fire and other disasters

FIGURE 2-8 Fire agency mascots participating in a county fair fire safety booth. Left to right are Seymour Antelope, Smokey Bear, Sparky the Fire Dog, and Woodsy Owl.
Courtesy of Captain Brandon Smith

- Describing function and benefits of various types of smoke and fire detectors
- Instructing citizens regarding extinguishing minor fires pending the arrival of the fire department
- Coordinating a juvenile fire-setters program
- Preparing press releases and public service announcements in coordination with community programs
- Performing a variety of administrative and research assignments in response to requests from management staff
- Conducting special studies of organizational policies, procedures, and practices relating to departmental and state-mandated policies
- Compiling and preparing oral and written reports
- Compiling data and preparing grant proposals
- Organizing fund-raising activities
- Assisting in budget preparation
- Attending continuing education classes
- Performing other duties as required

Employment standards include a college degree with studies emphasizing fire science, public relations, communications, or education; and work experience, or an equivalent combination of training and experience that provides the capabilities to perform the desired duties. The public fire safety/education specialist should be able to communicate effectively both orally and in writing; to establish and maintain effective working relationships with staff, public, and private representatives; to compile, analyze, and summarize statistical and technical data; to prepare and present reports and programs related to public fire and burn safety education; to create brochures and informational packets; and to organize, coordinate, and schedule work projects efficiently.

Dispatcher/Telecommunicator

In some departments, the position of dispatcher is held by fire fighters. Under direction of the shift supervisor, the dispatcher/telecommunicator receives emergency and nonemergency telephone and radio calls and dispatches resources. This person should be able to operate a computer-aided dispatch system and to do related work as required **FIGURE 2-9**. Typical tasks include the following:

- Receiving and acting upon emergency and nonemergency incidents in accordance with established policies and procedures

FIGURE 2-9 Fire department telecommunicator (dispatcher) at console utilizing computer-aided dispatch software and equipment.
© Jones & Bartlett Learning

- Operating communications equipment
- Instructing the public on proper medical techniques under emergency situations
- Securing and recording incident information
- Referring inquiries to appropriate public and private agencies
- Keeping fire control officers advised on situations and dispatching additional personnel or equipment when so advised by the incident commander
- Continually monitoring the status of fire units
- Logging all departmental emergency activity
- Operating other emergency service radios
- Operating computer systems, including video display terminals
- Operating telephone equipment
- Sending and receiving telephone messages
- Preparing and typing reports
- Keeping necessary records
- Performing related work as required

Employment standards include a high school diploma or successful completion of GED examination; a minimum typing proficiency of required net words per minute; and knowledge of basic principles and techniques of communications equipment and computer terminal operations. The telecommunicator must also know the local geography, communities, and location of streets and highways. He or she must be able to learn the operation of emergency communication equipment and computer-aided dispatch systems; have the ability to remain calm, act quickly, and exercise good judgment in emergency situations; speak clearly and concisely in a well-modulated voice; perform basic mathematics; follow oral and written instructions; develop reports; and keep records (City of Bakersfield Human Resources Department 2018). This position also requires the positive conclusion of a background check, explained in Chapter 1.

Emergency Services Planner/Manager

A field that is becoming more established and professionalized across the country is emergency management. With the establishment of homeland security funding after the terrorist attacks of September 11, 2001, and increased focus on disaster management, more and more personnel are being recruited and seeking careers in this field. These are primarily civilian (non-safety section) positions.

This position is responsible for developing plans for meeting emergencies; leading the development of operational concepts and orders; and directing coordination with heads of departments, agencies, and cities. This position requires a high degree of initiative, resourcefulness, diplomacy, and organizational ability. Typical tasks include the following:

- Formulating plans for the organization and implementation of a jurisdiction's emergency services program
- Planning for mobilization of personnel and resources
- Acting as an assistant to the director of emergency services during times of declared emergencies
- Integrating and coordinating activities and programs with emergency officials and local and area emergency offices
- Completing studies and inventories of resources in the jurisdiction relating to emergency services
- Developing and preparing public service announcements, press releases, and other media-oriented reports
- Supervising public informational broadcasts during emergencies
- Conducting tests to determine adequacy of preparation
- Performing other tasks, such as:
  - Preparing correspondence and comprehensive reports
  - Maintaining necessary records

- Supervising the coordination of the county-wide shelter program
- Supervising training programs to acquaint workers with their emergency roles
- Performing other tasks as required for the management of the jurisdiction's emergency services

Employment standards include a degree in emergency management, business administration, public administration, or closely related field from an accredited 4-year college and 2 years of increasingly responsible administrative experience in emergency service or other civil defense activities. Applicants must possess the physical capacity to perform all essential tasks.

Other requirements are knowledge of the Federal Disaster Relief and Emergency Act and the applicable state emergency services act; knowledge of fiscal disaster assistance programs of local, state, and federal governments; and knowledge of policies related to emergency action.

Also required are an ability to analyze emergency situations accurately and to adopt effective courses of action; ability to develop and manage interagency emergency service programs; ability to speak effectively before public gatherings; ability to work with others and gain their respect and cooperation; ability to organize and coordinate training programs; ability to prepare informational and directive materials and formulate correspondence; and ability to supervise others engaged in emergency services activities.

Private Fire Protection Careers

Fire Fighter

There are opportunities in the private sector as well for the position of fire fighter. An example of a large corporation that hires its own fire fighters for plant protection is the Northrop Grumman Corporation. The Northrop Grumman B-2 Division Fire Department is wholly subsidized by the Northrop Grumman Corporation and provides plant protection for its aircraft production facilities in Palmdale and Pico Rivera, California. Another private fire department is the Reedy Creek Fire Department operated by the Walt Disney Corporation at Disney World Properties in Orlando, Florida. Two of the more well-known private firefighting companies are Red Adair Wild Well Control and Fire Fighting and Boots and Coots Fire Fighting, both from Texas. The Boots and Coots Company was very much involved in oilfield firefighting in Kuwait after

the Gulf War. Another company that hires firefighting and security personnel on a worldwide basis is Wackenhut Services.

Some companies have their employees form fire brigades for plant protection. The personnel are not employed primarily as fire fighters. They receive training in self-contained breathing apparatus (SCBA) and basic firefighting methods and perform firefighting duties when required.

There are also private wildland firefighting companies. Some of these companies contract out to public agencies on large wildland fires. Contracting with the federal government through the National Interagency Fire Center in Boise, Idaho, these contractors provide engines and other wildland firefighting apparatus along with the crews to operate them. Some of these companies are branching out into structure protection from wildland fires, disaster response, and hazardous materials response as well to give them a year-round operation. One example is a company named Mt. Adams Wildfire. This company has personnel respond to wildfire areas in advance of the fire front. Their job is to "wrap" structures with foam or fire retardant in the path of the advancing fire. They also place covers over openings (i.e., vents and windows) and trim back brush and other vegetation to reduce the heat and ember exposure of the structure. Another company preinstalls under-eave foam systems that are automatically activated when the fire is within a half mile of the structure. These systems rely on onsite water and generator-driven electrically powered pumps to dispense the foam.

Insurance Companies

Insurance companies require people with fire science backgrounds in the area of loss prevention, inspection, emergency plan development, claims adjustment, and investigation, especially in arson-related cases.

Industry

As well as employing fire fighters, industry utilizes fire science graduates as loss prevention specialists and safety consultants. These positions inspect properties for fire and other hazards and develop and present employee training programs.

Fire Protection Systems Engineer

In any building or structure with an installed fire protection system, the system had to be engineered. Engineers design the specifications and plan the installation of systems **FIGURE 2-10**. As manufacturing processes become

FIGURE 2-10 Fire protection systems engineer performing a plans check of a water supply system.
© Jones & Bartlett Learning. Photographed by Glen E. Ellman

FIGURE 2-11 Fire protection equipment company that employs fire extinguisher and fire protection system maintenance specialists.
© Jones & Bartlett Learning

more exotic and higher in dollar value, special systems need to be developed to prevent damage from system operation as well as fire. These jobs require advanced degrees, usually a bachelor's or master's degree.

Fire Protection System Installation and Maintenance Specialist

Numerous contractors across the country sell, install, and maintain fire protection systems as their primary business. Where required by law, they service fire extinguishers annually. They also install, inspect, and maintain fixed fire protection systems and equipment, including fire sprinklers and special systems to protect computer rooms and other special applications **FIGURE 2-11**.

Invention and Innovation

Over the years, many fire fighters have contributed their inventions to the fire service. As they were actually doing the job, they saw the need for tools and techniques to make operations more efficient. The following are several examples of the ingenuity displayed by fire fighters to improve their ability to do their jobs.

Around 1800, George Smith, a New York fire fighter, invented the fire hydrant. In 1913, Edward Pulaski of the U.S. Forest Service invented the grub hoe and axe combination tool that carries his name and is still used by wildland fire fighters. The FIRE-SCOPE incident command system, the basis of the National Incident Management System (NIMS) used by firefighting agencies across the country, was developed by fire fighters as well.

As the business of firefighting becomes more complex, there is always a need for new and innovative equipment and techniques. A background in fire science can aid a person in assessing these needs and assisting in the development of new firefighting equipment and methods.

Fire Marks

The following story is an example of a fire fighter—Captain Scott Park of the Kern County Fire Department—recognizing a need and using his background and experience to invent a device to fill the need.

On October 17, 1989, at 5:03 PM, a magnitude 6.8 earthquake hit the San Francisco Bay Area. Later named the Loma Prieta earthquake for the location of its epicenter, this earthquake cost more than 60 people their lives. The earthquake hit just as game three of the World Series between the San Francisco Giants and the Oakland Athletics was about to begin at Candlestick Park in San Francisco.

People all over the world had tuned in to watch the pregame show when disaster struck.

The extent of the disaster was not immediately known. The Goodyear Blimp was overhead for the game and was broadcasting live shots of the numerous large fires in the area as the sun was setting. At first the focus was on the fires because they were the most evident and communications lines were disrupted. As reports finally got through, it became evident that there was widespread structural collapse throughout the area **FIGURE 2-12**. The collapses were not only homes and other businesses, but also a section of the Bay Bridge and Interstate 880, the Cypress Freeway.

FIGURE 2-12 Structure collapse due to earthquake.
Courtesy of FEMA

As the hours went by, it became evident that the worst rescue problems involved the freeway collapse. The freeway was built as a double deck with post and lintel construction, which left a void space between the underside of the upper deck and the lower roadbed. An undetermined number of cars and their occupants were trapped in this void space.

A massive search effort was immediately launched to rescue these victims, but it was difficult to locate those trapped and to assess their condition. The operation was severely impeded by the falling darkness and sheer mass of the freeway sections.

Where there was a sufficient opening, brave rescuers crawled into the void space to search for trapped victims. In other areas there was not enough space for a person to squeeze between the lower part of the upper deck and the wall on the lower deck. To compound the problem, rescuers were faced with numerous aftershocks and the very real possibility of further collapse of the weakened structure.

A way needed to be found, and quickly, to locate the trapped vehicles and their occupants. To remove the top layer of freeway from the whole collapsed structure would take too long. Metal detectors would not work to locate the cars because of the thickness of the concrete and the presence of steel reinforcing rod.

The decision was made to dig through from the top, which entailed cutting concrete and reinforcing rod and removing material as they went. There was no way to know if the holes were anywhere near a trapped vehicle. The danger of further structural collapse had to be evaluated before heavy

equipment could be brought in on the top deck to dig through. These holes were needed to check out areas that were not accessible in any other way. In the end, the last person to be found alive was extricated after 4 days. Captain Park watched what was going on and realized the need for a faster way to check void spaces, which occur in almost any structural collapse. To make a man-sized hole in reinforced concrete 6 in. thick takes up to 2 hours. It often requires the use of heavy equipment that may not be available or feasible to use.

To save lives in any structural collapse situation, speed is essential. The first thing that must be done is to locate the victims. Assessing where to dig and the mechanism of entrapment is necessary to make a decision as to the best way to proceed. The use of heavy equipment is often not possible because of the problem of further endangering the victims.

Captain Park was already an active member of the Kern County Fire Department's USAR Team and had a long-time interest in this sort of work and its related problems. Captain Park's intention was to invent a device that could gain audio/visual access to a void space in a minimum amount of time. After much thought he came up with the idea of the Searchcam **FIGURE 2-13**.

The Searchcam is a compact unit with headphones and a video screen worn at waist level. It has a hand-held probe with a miniature video camera, lights, and microphone on the end. The probe is inserted into a hole drilled through the concrete or other surface to check out the area on the other side.

FIGURE 2-13 Searchcam in use at a structure collapse.
Courtesy of Savox Communications

(continues)

Fire Marks   (continued)

Drilling the probe-sized hole and inserting the probe are much faster and easier than trying to cut a man-sized hole. It also requires no special or heavy equipment. A hole big enough to insert the probe can be drilled through 6-in. concrete in only 2–3 minutes with a hand-held drill. By properly determining where the victims are and what is holding them, precision digging can take place, speeding up the extrication process tremendously.

Using the video/lighting capability, the presence or absence of trapped victims can rapidly be determined. The audio capability helps to assess whether the person is still alive. Overall, the device aids in the live rescue of trapped victims.

These kinds of inventions do not come into production easily. It took 2½ years from the concept to the first prototype. It took another year from prototype to production. Captain Park used his education, experience, and contacts as part of the USAR Team to aid in the development of the Searchcam. This device has been very successful, with each of the active Federal Emergency Management Agency (FEMA) Urban Search Task Forces having at least one. There have been sales to other countries as well, and the Searchcam is now being used worldwide.

The Searchcam was utilized on both of the following incidents. In the early morning hours of January 17, 1993, a magnitude 6.6 earthquake hit the Northridge area of the San Fernando Valley in Los Angeles, California. Considered the most expensive disaster in the history of the United States, the earthquake caused an estimated $20 billion in damage. This quake, much like the one in San Francisco, toppled freeway overpasses, caused widespread structural collapse, and ignited numerous fires. Miraculously, there were only 33 deaths attributed to the quake. Part of the reason for this is it occurred early in the morning before there was heavy traffic on the freeways. A year later, almost to the day, a devastating magnitude 7.2 quake struck the city of Kobe, Japan. This quake occurred in a densely populated area, causing almost 5000 deaths; it virtually destroyed the city.

These disasters occurred in two of the most technologically advanced and best prepared countries in the world. As the population density of the major cities continues to increase, disasters of this level are bound to happen, illustrating the need for continued efforts to design new and better equipment to deal with emergency situations.

Wrap-Up

CHAPTER SUMMARY

- The modern fire service offers numerous career paths in both the public and private sectors.
- In cases in which the fire department employs new fire fighters under specific trainee programs, the job title fire fighter recruit–fire department would be representative.
- Under supervision, the fire fighter–fire department will respond to fire alarms and other emergency incidents to protect life and property and will do related work as required.
- The federal government has numerous positions for fire fighters at federal installations, mostly located at military bases.
- The position of fire fighter paramedic has increased responsibility over that of fire fighter and requires the completion of advanced medical training.
- The fire heavy equipment specialist position is mostly limited to departments that provide fire protection for wildland areas.
- In some states, firefighting positions, along with law enforcement, are classified as safety section and receive a higher level of retirement compensation than those employees classified as general (non-safety) members.
- The fire fighter wildland GS 3 position is that of an advanced trainee on a wildland fire suppression engine or hand crew and/or a fuels management crew.
- Private companies are becoming more prevalent in wildfire prone areas. They contract with state and federal agencies and homeowners to provide various types of firefighting and structure protection services.

- Civilian positions in the fire service do not have the rigorous physical requirements of fire fighter, although many are stressful and mentally challenging in their own right.
- Under direct supervision, a fire prevention specialist performs a variety of routine fire prevention, training, and hazardous materials disclosure work.
- Under direction, the fire hazardous materials program specialist evaluates potential hazards of unusual chemicals and materials and determines which chemicals are subject to the hazardous materials disclosure sections of the health and safety code.
- The fire department training specialist participates in or supervises the planning, development, and production of training, informational, and educational materials.
- The public fire safety/education specialist provides public education within the community in all aspects of life/fire safety, including providing instruction in fire prevention, fire escape planning, fire extinguishment methods, and other related aspects.
- Under direction of the shift supervisor, the dispatcher/telecommunicator receives emergency and nonemergency telephone and radio calls.
- With the increased homeland security funding and focus on disaster management, more and more personnel are being recruited and seeking careers in emergency management.
- Some companies have their employees form fire brigades for plant protection. The personnel are not employed primarily as fire fighters. They receive training in SCBA and basic firefighting methods and perform fire-fighting duties when required.
- Insurance companies require people with fire science backgrounds in the area of loss prevention, inspection, emergency plan development, claims adjustment, and investigation, especially in arson-related cases.
- As well as employing fire fighters, industry utilizes fire science graduates as loss prevention specialists and safety consultants.
- Fire protection system engineers design the specifications and plan the installation of fire protection systems.
- Fire protection system installation and maintenance specialists across the country sell, install, inspect, and maintain fire protection systems as their primary business.
- Over the years, many fire fighters have contributed their inventions to the fire service. As they were actually doing the job, they saw the need for tools and techniques to make operations more efficient.

KEY TERMS

Company officer The first-line supervisor in the fire department. Depending on the jurisdiction involved, this position may be identified by various titles, such as captain, lieutenant, sergeant, station manager, module leader, or unit manager.

Control lines An area where fuel has been removed, water or other extinguishing agent has been applied, or natural barriers exist to stop a wildland fire from spreading.

Driver/operator The position responsible for operating the pumping or aerial apparatus assigned to the fire department. Depending on the jurisdiction involved, this position may be identified by various titles, such as engineer, chauffeur, or truck operator.

Foam The finished product of water combined with certain agents that aid in the water's ability to extinguish fires.

Fuels management A program where naturally growing fuels, such as brush, are reduced to lessen fire intensity or to open up areas for wildlife and cattle.

Helitack Personnel whose primary means of transportation to fires is by helicopter. They also assist in helicopter operations when the helicopter is being used for water drops or for crew and equipment transportation.

Hose lays The method of laying out hose at a fire scene.

Hydraulics The required pressure to be applied to water to overcome the effects of pressure loss because of friction in piping and fire hose.

Lateral transfer A change of jobs from one fire department to another without moving up or down in rank.

Logging slash The remnants of logging operations, including limbs trimmed from downed trees and broken tree trunks.

McLeod A tool with a scraping blade on one side of the head and a rake on the other; used for fighting wildland fires.

Prescribed burning Planned application of fire under specified conditions in a predetermined area to achieve management objectives; includes removal or modification of fuels, clearing paths through brush, and killing unwanted plant growth.

Pulaski A tool for fighting wildland fires with an axe on one side of the head and a grub hoe on the other.

Reforestation The planting of seedling trees in areas destroyed by fire or logging.

Retardant A material spread on fuels that inhibits their burning.

Safety section A body of law that sets the retirement benefit rate for certain professions, primarily those that are high hazard and deal with public safety, namely fire and police.

Salvage A firefighting procedure for protecting building contents from damage due to water or falling debris.

Smoke jumper Highly trained personnel who parachute in to suppress fires in remote areas.

Toxicology The science of materials that are poisonous to living things, especially humans and animals.

Wildland Open land in its natural state.

CASE STUDY

There are many different types of jobs in the fire service. Some are the physically and mentally challenging, high-risk job of firefighting, be it wildland or structural (and in some cases both). Others are focused on prevention, public education, emergency management, or engineering. Within each of these areas of employment are specialties. Each individual must decide for himself or herself where their interests lie and what type of career they will pursue.

In a large fire department, many specialized jobs may be available to firefighting personnel. A fire fighter, while still retaining rank and privileges, can specialize in performing tasks such as fire and injury prevention, medical treatment of the injured, dispatching, or public education.

1. A student is challenged with a physical disability and will not be able to pass the physical ability/agility test for fire fighter. Which of the following jobs in the fire service might they qualify for?

   A. Fire heavy equipment specialist
   B. Paramedic fire fighter
   C. Dispatcher
   D. Fire fighter trainee

2. A student has a desire to assist others in need of medical treatment in emergency situations.

Which of the following jobs in the fire service might they be most interested in?

   A. Fire prevention specialist
   B. Dispatcher
   C. Fire heavy equipment specialist
   D. Paramedic fire fighter

3. A student finds that he gets the most satisfaction from assisting others through teaching them to assist themselves. Which of the following jobs in the fire service might he be most interested in?

   A. Public education specialist
   B. Fire fighter trainee
   C. Emergency manager
   D. Paramedic fire fighter

4. A student is employed in a fire department that has a large and diverse geographical area to protect. The area includes urban, suburban, and rural areas and a large wildland area. His interests lie in wildland firefighting. Of the following jobs, the one he should seek is:

   A. Fire prevention specialist
   B. Dispatcher
   C. Fire heavy equipment specialist
   D. Paramedic fire fighter

REVIEW QUESTIONS

1. What is the difference between the fire fighter trainee and fire fighter positions?

2. What is the meaning of *under supervision* as it relates to being a fire fighter?

3. What additional requirements are placed on a fire fighter paramedic over those of a regular fire fighter?

4. What particular skills are required of a fire heavy equipment operator?

5. What are the differences between a safety section retirement and a general retirement?

6. What are the benefits of having a summer job as a fire fighter (forestry aid)?

7. What are some of the services provided by private fire protection companies, both in firefighting and fire protection?

8. What jobs are available in the fire service that do not require actual firefighting to be performed?

9. What are some of the private sector jobs you could have that would help you prepare for the position of fire fighter–fire department?

DISCUSSION QUESTIONS

1. How would you go about identifying the need for a new device or procedure for the fire service?

2. What is the job description for fire fighter in the area in which you reside?

3. How do you fit the requirements for the position you are seeking in the fire service?

REFERENCES AND ADDITIONAL RESOURCES

Brugger, Kelsey. February 26, 2018. *Rise of Privatized Firefighting.* Santa Barbara, CA: Santa Barbara Independent. Accessed August 15, 2018. https://www.independent.com/news/2017/dec/20/rise-privatized-firefighting/

City of Bakersfield Human Resources Department. 2018. *Fire Dispatcher.* Bakersfield, CA: City of Bakersfield Human Resources Department.

Kern County Human Resources Department. 2018a. *Fire Hazardous Materials Program Specialist.* Bakersfield, CA: Kern County Human Resources Department.

Kern County Human Resources Department. 2018b. *Emergency Services Manager.* Bakersfield, CA: Kern County Human Resources Department.

Kern County Human Resources Department. 2018c. *Fire Heavy Equipment Specialist.* Bakersfield, CA: Kern County Human Resources Department.

Kern County Human Resources Department. 2018d. *Firefighter Recruit.* Bakersfield, CA: Kern County Human Resources Department.

National Fire Protection Association. *NFPA 1001: Standard for Fire Fighter Professional Qualifications.* Quincy, MA: National Fire Protection Association.

National Fire Protection Association. *NFPA 1031: Standard for Professional Qualifications for Fire Inspector and Plan Examiner.* Quincy, MA: National Fire Protection Association.

National Fire Protection Association. *NFPA 1035: Standard on Fire and Life Safety Educator, Public Information Officer, Youth Firesetter Intervention Specialist, and Youth Firesetter Program Manager Professional Qualifications.* Quincy, MA: National Fire Protection Association.

National Fire Protection Association. *NFPA 1041: Standard for Fire Service Instructor Professional Qualifications.* Quincy, MA: National Fire Protection Association.

National Wildfire Coordinating Group. 2019. *How to Become a Wildland Firefighter.* www.nwcg.gov/how-to-become-a-wildland-firefighter

State of California. 2018. *County Employees Retirement Law of 1937.* Sacramento, CA: Strategic Local Government Services.

U.S. Forest Service. 2018. *Forestry Aid-Fire Suppression GS-0462-03, Standard Job No. N2011.* Washington, DC: U.S. Department of Agriculture Forest Service. www.usa.gov/jobs

CHAPTER **3**

Public Fire Protection

OBJECTIVES

After studying this chapter, you should be able to:

- Describe the evolution of fire protection in the United States.
- Describe the history of wildland fire in the United States.
- Describe the evolution of modern firefighting equipment.
- Identify and describe fire service symbols.
- Describe the evolution of fire stations.
- Describe how major fire losses have affected the modern fire service.
- Identify statistics of the U.S. fire problem.
- Discuss the purpose and scope of fire agencies.
- Discuss the future of fire protection.

Case Study

The summer of 1871 was very dry in the Great Lakes region of the United States. The city of Chicago was densely packed with wooden buildings. Wood was used for sidewalks, and sawdust was used for dust control on the streets. On Sunday evening, October 8, 1871, a fire broke out at 13 De Koven Street. The common belief is that the fire was sparked by a cow kicking over a lantern in a barn. The city's fire fighters were tired from a large fire the day before and were at first sent to the wrong neighborhood when responding to the fire.

The fire soon spread and burned for 2 days. The fire was finally suppressed when it rained on the morning of October 10, 1871. The fire completely devastated the center of the city. At least 300 people were dead; 100,000 were left homeless; and 17,500 buildings, 73 mi of streets, and $200 million worth of property were destroyed.

1. How would elements of the modern fire service have positively impacted the Great Chicago Fire?
2. In hindsight, what would have been the most critical factor in avoiding a fire on this scale?
3. What was the greatest loss to the city of Chicago in this fire?

Modified from sources: Chicago History Museum, 1999, and Oracle Think Quest, 2003

JONES & BARTLETT LEARNING
NAVIGATE 2   *Access Navigate for more resources.*

Introduction

The only creature in the world that has learned how to initiate and utilize fire is man. Humans are also unique in that they do not immediately flee from flame. This is apparent with campers sitting around the campfire and people sitting in front of fireplaces in their homes. From the beginning, humans had reason to fear fire. It could destroy their homes and food sources. If they were caught in the open in a fire on a grassy plain or in a forest fire, it could overtake and kill them. It is known that fire existed for thousands of years before people learned to use it. Fire initiated by natural phenomena—lightning striking trees and starting fires, flaming meteors falling to earth, volcanoes spewing fire—likely frightened people because they did not understand what fire was or how it could either harm or help them.

We do not know exactly when people learned to use fire, but there is reported evidence that it was between 200,000 and 400,000 years ago. In caves near Beijing, China, the remains of a hearth and charred animal bones were discovered along with the bones of a humanlike creature known as Peking Man. It is not certain whether this individual had learned to make fire or just learned to carry it to the cave, because the charred animal bones only suggest the fire was used for cooking.

Once people learned how to create and control fire, it lost much of its mystique. When people first learned to use fire, their culture and society changed dramatically and fire became humanity's constant companion. Previously unexplored, colder areas of the world could be inhabited, and people gained more control over their environment. As the world population expanded, new land could be explored in the constant hunt for food.

Later, people used fire to make tools, implements, and pottery, which could be used to store food. This important discovery of the use of fire to fashion implements is the very basis of our civilization. Before people learned to farm and store grains, they had to spend most of their time hunting for fresh game and fresh wild food. When they learned how to grow and store food, it meant they could develop permanent settlements. Populations no longer had to be nomadic. The Industrial Revolution of the 18th century was initiated with the development of steam power—steam generated by heat from controlled fire. Steam was used for transportation, manufacturing, and the generation of electricity.

We have become a nation that relies heavily on energy, and much of that energy is a by-product of fire. We still use fire for the same processes that early humans did. We use heat generated by fire to keep us warm and cook our food, to make glass and shape metal, and to propel our means of transportation. Most homes have at least two essential fires—heat sources for cooking and heating. Some houses also have fireplaces or wood stoves for warmth and comfort. As with any tool, fire can be used correctly or incorrectly. As we have seen over the centuries, fire used incorrectly can have tragic and disastrous results.

Fire has also long been a weapon of war. In ancient times, it was common to burn the enemy's crops, stored food, and homes. The British burned Washington D.C. in 1814 during the War of 1812. General Sherman burned Atlanta in 1864 during the Civil War. Firebombing enemy cities was a tactic of choice by most parties during World War II. Napalm bombs and white phosphorus grenades were used during the Vietnam Conflict. Fire is still used by some militaries in the world to destroy enemy supplies and to terrify residents.

When people become too comfortable with fire and lose their respect for it, problems arise. In North America, people who are adversely affected by fires are considered victims, even if the fire is the result of their own negligence. It has not always been this way. Over the centuries, as people have become city dwellers, fire has become more of a destructive force. The fire service, as we will see in this chapter, has evolved to deal with the problems this presents.

Man's use of fire as a tool can be summed up by the inscription on the outside of Union Station in Washington, D.C. (built in 1907). "Fire, greatest of discoveries, enabling man to live in various climates, use many foods, and compel the forces of nature to do his work.*" The inscription was chosen by Dr. Charles W. Elliott and approved by President Woodrow Wilson (President 1913–1921).

Evolution of Fire Protection

Earliest Known Firefighting

The first recognized firefighting force was organized in Rome, composed of slaves carrying primitive fire pumps. These fire fighters were known as the *Familia Publica*. The slaves were slow to respond, in regard for their own safety. After numerous disastrous fires, the Corps of *Vigiles* was formed by the Emperor Augustus. The *Vigiles* were slaves who were promised freedom after 6 years of service in the *Vigiles*. The *Vigiles* comprised 7000 men divided into battalions of 1000 men each. Each battalion was commanded by a man responsible to the emperor himself. The *Vigiles*, equipped with buckets and axes, patrolled the streets and fought fires. They also performed fire prevention duties by warning the citizens to be careful with their fires and to keep water in their homes for fire extinguishment. The cost of maintaining the *Vigiles* was paid out of public funds. Unplanned fires were investigated and people found responsible for setting the fires were subject to corporal punishment. The *Vigiles* gained such acclaim that after a century, free men would join for the prestige of being a member (Kentucky Educational Television 2013).

American Settlements

The first settlement in the United States, Jamestown, was almost a failure due to a fire. Upon landing in the New World, settlers hastily constructed shelters out of readily available materials—brush, sticks, clay, and grasses used as roof thatching. These homes were equipped with large central fireplaces for warmth and cooking. The chimneys were constructed of brush and sticks or wooden planks with a thick coating of mud or clay. After a while, the mud or clay would come loose, and the chimneys, well dried from numerous fires, would burst into flame. Combined with the thatched roofs and the closeness of the structures to each other, the colony was a design for disaster **FIGURE 3-1**.

The first permanent colony was founded in 1607. The first conflagration followed in the winter of 1608. This conflagration destroyed almost all of the colonists' homes and provisions. Because of the severity of the winters in Virginia, many of the colonists died from exposure or starvation (ICMA 2012). To this day, the combination of structures closely spaced, either to each other or to combustible vegetation, still leads to conflagration-scale fires.

Fire protection at the time of the first settlers mostly consisted of creating firebreaks by pulling down structures. The bucket brigade also came into use. The townspeople would gather and form two lines between the water source and the fire **FIGURE 3-2**.

FIGURE 3-1 Thatch roof house with wooden chimney.
© unkas_photo/iStock/Getty Images Plus/Getty Images

Fire Marks

The first recognized firefighting force was organized in Rome and was known as the *Familia Publica*. They were succeeded by the *Vigiles*, created by the Emperor Augustus.

---

* Quote by Dr. Charles W. Elliott

FIGURE 3-2 Bucket brigade.
© Niday Picture Library/Alamy Stock Photo

The men would pass the full buckets toward the fire and the women and children would return the empty buckets to be refilled.

In 1647, the governor of New Amsterdam (later renamed New York), Peter Stuyvesant, took the first steps to save the new city from fire. He drew up a building code prohibiting wood or plaster chimneys. He appointed four volunteer fire wardens, whose primary function was fire prevention by enforcing the law and making sure that chimneys were properly maintained. The wardens were allowed to levy fines when violations were found. The money accrued from the fines was used to purchase fire equipment.

The greatest fire threat was that fires would get a good head start while people slept. New Amsterdam, Boston, and other towns adopted a curfew. At 9:00 PM a bell was rung and all fires were to be extinguished or covered until 4:30 AM. Stuyvesant went a step further and appointed a group of young men to patrol the streets at night carrying wooden noisemakers that were twirled to sound an alarm if a fire was discovered. They became known as the "rattle watch."

In 1666, the Great Fire of London destroyed almost two-thirds of the city. The fire went on for five straight days, destroying the city's primarily wood frame structures. It was fought with the implements available at the time—buckets, hooks, and primitive fire engines. As a result of this tremendous fire, the city passed a code of building regulations. Many new firefighting methods and architectural improvements were proposed but were not enacted. These included wider streets, green spaces, building setbacks, and the use of noncombustible materials. One major effect of this fire was the creation of fire insurance companies. This idea, like so many others, soon spread to the New World.

At this time, the cities in the American colonies had no well-organized firefighting forces of their own. The owners of the fire insurance companies saw the need for, and benefit of, an organized firefighting force. They were soon advertising the availability of

FIGURE 3-3 Fire marks were used to identify which insurance company protected a home or building.
© PURPLE MARBLES YORK 1/Alamy Stock Photo

their own firefighting forces to protect the properties they insured. The insurance company would place a plaque, called a fire mark, on the buildings that they protected **FIGURE 3-3**. Because this was a business and not a public service, the responding fire brigades would fight fires only in the businesses they insured. If they arrived at the scene and it was a business that was uninsured or with another provider, they let it burn.

In 1679, Boston established the first publicly funded, paid fire department in America. Volunteer organizations called mutual fire societies were organized to assist the fire department. Ben Franklin observed the operation of the mutual fire societies and in 1736 organized the Union Volunteer Fire Company in Philadelphia. He is considered to have conceived the idea of the volunteer fire companies that continues today.

In England, the fire insurance companies had their own firefighting forces. In America, the insurance companies performed salvage work, but the firefighting was left up to the volunteers. The insurance

FIGURE 3-4 Section of wire-wrapped wooden water main.
© Jones & Bartlett Learning

companies would pay the first company to arrive to put water on the fire. This, along with the great pride the fire fighters had in their company and its abilities, led to fierce and often violent competition between the companies of volunteers.

City water systems consisted of wooden logs hollowed out to carry water **FIGURE 3-4**. Every half block or so there was a wooden plug placed in the log. This plug was removed to access the water for firefighting, thus giving us the term *fire plug*. There was great competition to see which company could get first water on the fire. Some companies would pay boys to race ahead to the scene of the fire and hide the plug from other companies. These boys were called "plug uglies." Another source of water was cisterns, dug under the streets, whose lids were removed to allow access during fires (Frady 1991).

Volunteer Fire Companies

Membership in a volunteer fire company was a great source of pride for many Americans. There were long waiting lists of adventurous young men who wanted to join. Many famous Americans served on volunteer fire companies, including George Washington, John Hancock, Alexander Hamilton, Samuel Adams, and

Paul Revere. After the Revolutionary War, the concept of volunteer fire companies spread across the nation. Men joined the local volunteer company for social reasons as well as to serve the public and to protect their towns.

The volunteer fire companies continued to develop and grow in strength. Pride in the companies grew to the point that the equipment became very ornate, with the advent of uniforms, fancy painting on the engines, bells, and a constant desire to have the latest and best equipment.

These were rough men and rough times. The competition got to the point that companies arriving on the scene at the same time would fist fight to see who could claim first water. Some companies would send a man with a bucket of water to the scene as fast as possible so he could throw it on the fire and claim first water for his company.

These early fire fighters reveled in the excitement and danger of attacking fires. They were readily prepared to be injured in the line of duty. They often took bad falls, were run over by the equipment when they pulled it to the fire, or were singed and blistered by the heat. They wore their scars and bandages proudly, their injuries proclaiming to all that they were real men. It was common to go out in the dead of winter and have their wet clothes freeze to their bodies as they brought a blaze under control. Not even hazards of a warehouse storing black powder or chemicals, or falling walls and roofs, daunted the fire fighters in the pursuit of their task. The fire was their enemy and they would not surrender.

The competition among the companies hastened their demise in the large cities. The fighting and rowdiness led to city leaders organizing paid fire departments. In 1853, Cincinnati became the first city with a fully paid fire department (City of Cincinnati Fire & EMS 2013).

The Modern Fire Department

Since these early beginnings, the fire department has continued to evolve and expand its role. The time of waiting around the firehouse for the alarm to go off is

Fire Marks

- In 1679, Boston established the first publicly funded, paid fire department in America.
- Benjamin Franklin organized the Union Volunteer Fire Company in Philadelphia in 1736.
- In 1853, Cincinnati became the first city with a fully paid fire department.

gone. The modern fire department is proactive in the community in trying to minimize loss from fire and other disasters. Fire departments now routinely provide such diverse services as fire prevention, public safety education, fire suppression, medical aid, rescue services, and hazardous materials response. The term encompassing these activities is "community risk reduction." Fire fighters are now considered professionals, not just technicians.

The fire service is still full of traditions. New fire fighters are still required to prove themselves on the fire ground. Most fire engines are still painted red, even though other colors are more visible. People are still required to come in at the bottom of the organization and work their way up through the ranks. Tradition and a sense of history help to instill pride in the organization and the people who work there. Many of the traditions, like that of service to the public, are what make the fire department one of the most respected public agencies. After especially large and destructive fires, the public has been known to come out in force and show its appreciation for the job done by the fire fighters. It is a great feeling to watch the evening news and see signs showing the public's appreciation for a tough battle fought and won **FIGURE 3-5**.

Some see tradition as negatively affecting the fire service and impeding progress. A statement sometimes made in reference to the fire service is "100 years of tradition unimpeded by progress." Fire departments and fire fighters must be able to change with the times. The fire service should not lose touch with its long and honorable history, but it must be able to see where change is necessary and adapt for the future.

FIGURE 3-5 Sign erected by citizens to thank fire fighters for saving their homes and businesses.
Courtesy of Kern County Fire Department

Fire-Based Emergency Medical Services

Actual fire suppression activities occupy very little of the fire fighters' time in most departments. The fire service has become more involved in medical aid due to a perceived need from the communities they serve. Doctors stopped making house calls in the 1960s and ambulances were run by the local undertaker in many cases. The Department of Transportation also recognized the need and issued standards for Emergency Medical Technicians (EMT's)

Some departments report that requests for medical aid are 70 percent of their incident volume. Others provide ambulances and paramedic programs to their communities. Television shows such as Emergency!, which aired from 1972 to 1977 about the Los Angeles County Fire Department's paramedic program, have raised the public's expectations of the fire service. The public is no longer content to pay for fire fighters to sit in the station—it wants service for its tax dollars.

The majority of fire departments are providing some level of emergency medical response, from first responder to advanced life support paramedics. Some departments are evolving to use fire department vehicles to transport the sick and injured. This has led to departments generating revenue for these services and returning that revenue back to the community to help cover increased costs. This demand for increased service helped fuel the evolution of the fire service education system. As illustrated in the Fire Science Education and the Fire Fighter Selection Process chapter, the educational system has progressed to the point where many fire fighters are no longer content with time on the job as their only educational opportunity. As in many other professions, national and state standards of qualification have been adopted. Certifications and degrees are available in a wide variety of fire service specialties.

Fire Stations

As paid fire fighters became more common, the fire station needed to have sleeping quarters because the fire fighters were there 24 hours a day. Previously, a shed for the apparatus was all that was needed. A stable for the horses was added as horse-drawn steamers were brought into service. Many old fire stations still have a hole in the ceiling that went up to the hay loft. In multistory fire stations, the sliding pole was introduced to give quick access to the apparatus floor when the sleeping quarters were upstairs. The first poles were wood; then brass was used. In some stations, because of injuries from sliding down the poles, slides were introduced. The design of the modern fire station is presented in the Fire Department Resources chapter.

Personal Protective Equipment

In the early days of the volunteers, the fire fighters fought fire in whatever they happened to be wearing at the time. As pride in the companies grew, uniforms were used as a means of identifying the companies' members **FIGURE 3-6**. The uniforms were mostly worn at parades and gatherings. They were trimmed in white or blue and had large letters or numbers on the front of the shirt. A thick leather belt with the company number on the buckle was also worn. The belt came in handy as a place to hang tools. Standard uniforms were adopted by all of the larger paid departments during the early 1800s. The badge identifying the member's rank and department took the place of the large numbers and letters (Hasenmeier 2008).

A recognized symbol of the firefighting profession is the shape of the fire helmet. The first fire fighters in the United States wore whatever hat they wore on a day to day basis. These provided little protection from embers, sprayed water, and the elements. Jacob Turck of New York City is given credit for inventing the first "fire cap" around 1740. The fire cap was round, with a high crown, and narrow brim and was made of leather. The name of the fire company or insurance company was painted on the front of the cap. The primary purpose of the fire cap at the time of its introduction was more to identify which company a fire fighter was a member of, rather than to provide head protection. Mathew DuBois sewed an iron wire into the rim to add strength, help the fire cap retain its shape when wet, and resist warping. In the early 1820s a New York luggage maker and volunteer fireman, Henry Gratacap, is credited with designing the fire helmet shape that is revered by fire fighters today.

The crown of the first fire helmets was made of four pieces of leather sewn together forming ribs. The ribs aided in adding strength to the crown and in deflecting falling material away from the fire fighter's head. The helmet designed by Gratacap is composed of specially treated ¼ inch thick leather with eight pieces sewn together to make the crown; then the brim with the long tail is added. The crown still has four main ribs with four lesser ribs joining the other pieces of leather in the crown. The design of the helmet has several functions. The ribs formed where the crown pieces are sewn together add strength to protect the head from fallen debris. The long tail of the brim aids in keeping water, hot embers, melted roofing tar, and other materials from going down the back of the fire fighter's neck. The helmet may also be worn backward to shield the fire fighter's face from heat. Prior to windshields and enclosed cabs the "tillerman" (the person driving the rear section of a tractor drawn ladder truck) would wear the helmet in this fashion when it was raining or snowing to protect their face. Prior to radios for communication the helmet thrown from a window was a signal that the fire fighter inside the building required assistance. The brass eagle was added to the front of the helmet after an unknown sculptor created a figure on a volunteer fire fighter's grave at the Trinity Churchyard in Manhattan, New York in the early 1800s. This figure showed a person emerging from the flames, one hand holding a trumpet and the other holding a sleeping child; an eagle was on his helmet. Because of this, despite new studies that show the brass eagle hinders fire fighters by entanglement and that the eagle can get knocked off or dented, fire fighters continue to wear the eagle on their helmets. In Canada the eagle is often replaced with a beaver. The metal eagle attached to the helmet could also be used as a striking tool when breaking window glass for ventilation or rescue.

The Cairns Brothers, contemporaries of Gratacap, had a metal badge and insignia business in New York and began designing and producing identification badges, which were added to the front of Gratacap's fire helmets. When Gratacap retired in the 1850s the Cairns Brothers took over the fire helmet business, and helmets are still sold with this manufacturer's name today. Their iconic fire helmet design, the "New Yorker" is still used by many fire fighters. **FIGURE 3-7**. A video of the

FIGURE 3-6 This fire fighter from the Vigilant Fire Company in Baltimore is wearing a parade hat and holding a speaking trumpet.
Courtesy of the Library of Congress

FIGURE 3-7 Assortment of fire helmets, including leather, metal, fiberglass, and plastic.
© Jones & Bartlett Learning

FIGURE 3-8 Modern self-contained breathing apparatus (SCBA).
© Robert Klinoff/Jones & Bartlett Learning

making of a classic MSA Cairns leather helmet is available at http://www.core77.com/blog/object_culture /tradition_vs_progress_the_art_of_the_american _fire_helmet_19203.asp

Bunker gear, or turnout clothing, went through the same development process as the rest of the firefighting equipment. As the methods of attack became more aggressive, the fire fighters' personal protective equipment (PPE) had to keep up. Long canvas coats were worn with boots that came up to the thighs. These protected the fire fighter from falling embers. Water barriers were added to the coat in an attempt to keep the fire fighters dry; insulation was added to protect them from heat. The long boots were eventually replaced with knee-length boots and insulated pants. These better protected the fire fighter from fire underneath them. (See the Fire Department Resources chapter for more information.)

Today, self-contained breathing apparatus (SCBA) is used at almost all fires. Prior to breathing apparatus being invented, fire fighters wrapped wet rags around their faces or used their beards, dipped in water, to filter the air as best they could. One method used for getting fresh air in a smoky building was to place the fire fighter's face down by the nozzle because the nozzle draws air alongside the hose stream. The next improvement was the canister (gas mask), which protected the fire fighters from smoke particles but did nothing to protect them from toxic fire gases and low oxygen concentrations. When the modern SCBA was invented, around 1945, the ability to perform interior attack was greatly enhanced **FIGURE 3-8** (Hashagen 2013). The fire fighters no longer had to back off from the fire and "take a blow" to fill their lungs with fresh air. The use of breathing apparatus is also expected to reduce the number of fire fighters suffering cancer from fire-related exposures to toxic gases and soot.

SAFETY TIP

Today, a self-contained breathing apparatus is used for almost all fires, with the exception of wildland fires, to protect the fire fighter from toxic gases and superheated air.

History of Wildland Fire in America

To have a full appreciation of firefighting in America, one must have knowledge and understanding of the history of wildland fire. Today there is much political discussion and legal action among the various stakeholders as to how fire should be managed on public lands. With the anticipated effects of global warming and resultant climate change, it is important to understand the possibilities of what may happen should vegetation in normally wet areas become much drier and fire prone. Imagine if the northeastern and northern midwestern parts of the country began to burn more like the southwest with its major seasonal conflagration wildland fires. It is now common for Southern California to experience major wildland fires with resultant structure losses throughout the year, even in the winter months.

Strategic Burning and Land Clearing

The history of wildland fire in America is older than the nation. Prior to settlement and the removal or relocation of native tribes, fire was set by the natives to herd buffalo and to renew the vegetation and open up the forest floor by reducing undergrowth and downed materials. By burning this material, the larger trees had less competition for water and nutrients and the forests were populated with large trees and less undergrowth than they are now. This returned nutrients to the soil, which was tied up in smaller vegetation, leaves, and downed logs. It also reduced the tons per acre of fuel loading. The Native Americans also used wildland fire to herd bison (American buffalo) and other game so they were easier to hunt with bow and arrow. Wild game, such as deer and other browsing animals, eat the new shoots from younger vegetation. Therefore, a periodic wildland fire provides a healthy environment for the generation of new plant life within the reach of the browsers (Auburn University 2013).

As settlement moved into the formerly wild areas, trees were cut down for lumber and areas were cleared for farming. As the areas filled in with towns and homesteads, wildland fire became more of a threat to life and property. Fire was often used to remove underbrush and the leftovers of logging when clearing land. These fires, along with lightning-caused fires, were left untended and there was no organized fire protection. One research writer has postulated that more early settlers were killed by wildland fires than were killed by conflict with Native Americans (Holbrook 1943). The unrestricted alteration of the vegetation and land dramatically changed the environment. Areas along the banks of the major rivers, such as the Mississippi, were logged. This resulted in the river being wider and shallower, with more topsoil being washed downstream and deposited in other areas. Logging left the area along the river more prone to flooding as the banks of the river were less well defined.

The Impact of the Railroad

In 1869, the transcontinental railroad was completed and the East and West Coasts were connected by rail. The completion of the railroad came at a cost to the environment. Trees were logged for lumber to build trestles, snow sheds over the tracks in the Sierras, telegraph poles; to clear right of way for the tracks; as fuel for the steam boilers that propelled the trains; and for railroad ties. When the trees were cut, the limbs were left alongside the track bed (logging slash). There was no time for niceties, such as spreading, burning, or mulching the timber slash. The slash was left where it lay. As it dried out, it provided a receptive fuel bed for the ignition of fires. The trains were powered by wood fires in the boilers, and sparks were a constant danger to the dried fuels along the track bed and to the wooden cars of the trains themselves. The trains also ignited grass fires as they crossed the Great Plains (Ambrose 2000).

Loss of Life and Property

The greatest loss-of-life wildland fire in U.S. history occurred in Peshtigo, Wisconsin, in 1871. The Peshtigo Fire cost the lives of more than 1500 people and burned 2400 mi^2 in the vicinity of Green Bay, Wisconsin. The Peshtigo Fire occurred around the same time as the Great Chicago Fire but did not receive the same national attention (Wisconsin Historical Society 2013). Another large life loss wildfire occurred in Hinckley, Minnesota in August of 1894, killing over 400 people. (Brown 2006). Both Peshtigo and Hinckley were logging towns located in the woods; buildings and sidewalks were primarily made of wood and the streets were often covered with sawdust to reduce dust. It must be kept in mind that during these times there was no organized fire protection on a large scale and there was no easy or rapid way to let people know of the fire danger or the predicted weather. For many, the only way to escape was by rail. Escaping by train was highly risky as the trestle bridges and rail cars were all made of wood and prone to catching fire if the wildland fire was close enough.

Of the numerous wildland fires in U.S. history, two extreme wildland fires have had a great effect on national wildland fire suppression policy. The first, in 1910, was called the "Big Blowup." This fire occurred in the panhandle of Idaho and northwestern Montana. Fueled by dry forests and driven by near hurricane-force winds, the fire burned more than 3 million acres in two days. It leaped forward at a rate of 30–50 mi a day, and burned several towns, including Wallace, Idaho. The smoke from the fire reached Boston. At least 85 people were killed. Due to the extremely remote area of the fires, no definite death toll could be established. Of the 78 fire fighters confirmed killed, only 29 could be positively identified, with the rest burned beyond recognition. Snow eventually extinguished the fire, but dead trees were seen burning until at least the end of February of the next year (Pyne 2001). During this fire, Forest Ranger Ed Pulaski saved the lives of 45 men by holding them at gunpoint in a mine opening. He then went on to invent the commonly used wildland firefighting tool which still bears his name (Egan 2009).

After the Big Blowup, wildland fire was compared to war. The U.S. Forest Service was mandated by Congress to suppress wildland fires. Under the

Weeks Act (1911), funding was provided for the Forest Service to cooperate with states in fire protection. The policy was to suppress all fires. This resulted in the 10 AM policy, which required every fire to be put out by 10 AM the next day. Obviously, this was not possible in every case, but it was the stated goal.

Over time, a let-burn policy was established for fires occurring in national parks. The policy directed the fire management staff to let burn fires that were started by lightning, not human caused, and did not pose a threat to people. Areas would be closed if a fire occurred to prevent the exposure of campers to ongoing wildland fires. This policy allowed for the natural process of fuel reduction and provided for increased forest health. The public has pushed back on this, stating that no one goes to a national park to see scorched areas of dead timber. When people go to a park or forest, they expect to see green and growing vegetation. There is also the issue of cities having developed near federal forest lands. When a large wildland fire burns, it releases great amounts of smoke, negatively affecting air quality.

Other fires that dramatically changed federal wildland fire policy were the Yellowstone fires of 1988. This was a complex of 13 fires that eventually burned approximately 800,000 acres in Yellowstone National Park and 400,000 acres outside the park. The lightning-caused fires were managed under a let-burn policy, but greatly exceeded expectations as to their final size. The fires cost $112 million to suppress and were finally extinguished by winter snow.

The initial reaction to these fires was that the park was destroyed. Time has proven that the park has not only survived, but thrived due to allowing the natural process of fire to reduce fuels, return nutrients to the soil, and open up the forest to the survival of larger trees due to less competition from lesser trees for water, sunlight, and soil.

Fire Marks

Two American icons have evolved from wildland fire in the United States. The first is Bambi, from the classic 1942 Walt Disney movie by that name. In the movie, the careless hunter starts a wildland fire that causes great destruction. The other icon is Smokey Bear, first brought to national attention in 1944. Smokey's message of "Only *you* can prevent forest fires" still rings true today because 90 percent of wildland fires are human caused (Maclean 2003).*

Wildland Fire Management

The national discussion on wildland fire management now centers on fuel reduction and how to accomplish it. Fuel reduction is necessary to keep wildland fires from producing so much heat that they destroy all vegetation and sterilize the soil. Waxes from sap are released during the fire, making the soil shed water and thereby increasing runoff, which leads to flooding. It is a complicated problem that is still being researched.

From the viewpoint of the local fire fighter, where every fire is to be suppressed, the actions of the federal agencies can be somewhat confusing. When the federal government (U.S. Forest Service, National Park Service, Bureau of Land Management, and U.S. Fish and Wildlife Service) have a wildland fire on their lands, they "manage" it. To manage the fire, several things are taken into consideration, such as cost per acre to suppress the fire using various strategies, personnel safety due to terrain features, resource values to be protected, and historic site values (U.S. Geological Survey 2018). The fire is on their land, not private land. They will still protect private in-holdings such as church camps and cabins but may allow the fire to burn around them. There is also land declared by Congress to be "wilderness." These are remote and roadless areas. If there is a wildland fire on wilderness land, no motorized equipment, such as chainsaws, portable pumps, and bulldozers, are allowed to be used. Wilderness area fires that have been caused by lightning may just be monitored and not extinguished. If the fire was man-caused, it is extinguished.

In the national forests and other wildlands, lightning is the cause of many large-acreage fires each year. One of the reasons lightning accounts for so much of the acreage loss is that firefighting resources are soon overwhelmed when a large storm comes through and ignites numerous fires. In the summer, dry lightning storms occur, and **ground strikes** start fires. Federal and state wildland firefighting agencies reported an average of 9000 wildland fires started by lightning to the National Interagency Fire Center (NIFC) per year in 2008–2012. These fires tended to be larger than fires started by human causes. The average lightning-caused fire burned 402 acres, 9 times the average of 45 acres seen in human-caused wildland fires.

Another common cause of wildland fires is humans, sometimes referred to as "two-legged lightning." Careless campers who leave campfires unattended, burn trash carelessly, smoke in high-fire danger areas, operate faulty equipment, or drive in dry grass with catalytic convertors on their vehicles are the source of fire

---

* Modified from Maclean, John N. 2003. *Fire and Ashes.* New York: Henry Holt and Company LLC.; Maclean, John N. 2013. *The Esperanza Fire.* Berkeley, CA: Counterpoint

starts. As many as 90 percent of wildland fires in the United States are caused by humans.

Once these fires gain headway, vast areas of natural resources are lost, animals are killed, and valuable watersheds are destroyed. In time these areas will return to their previous condition, but this may take well over 100 years.

Equipment

Firefighting equipment evolved because of the need for more firefighting capability. As cities grew and the fire load increased, new ways were needed to apply water where and when it was needed. The first known fire pump, called the *siphona*, was invented in Egypt in the third century BCE **FIGURE 3-9**.

> ### Tip
>
> Firefighting equipment evolved because of the need for more firefighting capability. As buildings grew taller, for example, more powerful pumps were required to lift the water to the required height.

The siphona comprised two cylinders with pistons that sucked water in at the base as each piston rose and expelled it as the piston went down. The water was fed into an air-filled chamber, compressing the air in the chamber and forcing the water out a nozzle on the top. The purpose of the air space in the chamber was to give a steady stream. For centuries this information was lost (Kentucky Educational Television 2013). When rediscovered, it was used to design the engine

commonly referred to as the *hand pumper*. This concept carried into the hand-pumped engines that followed and continued to be used into the 20th century.

Fire hooks were another firefighting tool; they were designed to pull burning thatch from roofs and to pull down houses to create firebreaks **FIGURE 3-10**. In the 16th century, large syringes were developed to squirt water onto the fire. As you may well imagine, they lacked effectiveness.

Many different designs of hand-pumped engines were tried, and many were discarded due to inefficiency or poor operation. The most effective and longest lasting design used the principle of the siphona, with long handles that could be swung out from the sides and latched in place. The handles were hinged to allow the pumper to pass through narrow streets and alleys. These handles were called brakes. Up to 15 ft (4.572 m) long, the brakes were manned by volunteers, and the engine was pumped between 60 and 170 strokes a minute **FIGURE 3-11**. Pumping the engines was hard work, and at high rates of speed, the people pumping had to be relieved at regular intervals.

The basic design and principle of the engines were the same, but the execution varied greatly. Most were of the two-cylinder design, but at least one four-cylinder engine existed. Some were of the end-stroke design with the brakes perpendicular to the tongue; others were of the side-stroke design. To gain pumping efficiency in a short length, even double-deck pumpers that used men both on the ground and up on a platform to operate the brakes were made.

The first hand pumpers discharged water through a large nozzle, called a gooseneck, mounted on top. The limitation of this design was that the engine had to be positioned close to the fire, and this limited the

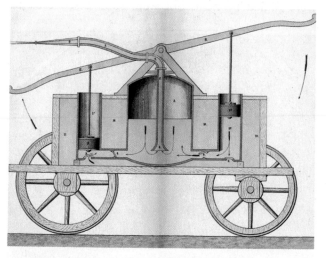

FIGURE 3-9 Siphona using double-action piston pump. This principle was used to create hand-operated fire pumps (hand pumpers).

FIGURE 3-10 Fire hook used for tearing down buildings to create a firebreak.

FIGURE 3-11 Hand pumper fire engine.
© Jones & Bartlett Learning

FIGURE 3-12 Hose carts were used for transporting hose to the fire scene. These are still used in some settings.
© Jones & Bartlett Learning

ability to aim the discharged water. Fire fighters, wanting to be able to position their water streams better, experimented with extending a hose from the gooseneck. This gave them the capability of attacking the fire from different angles without having to move the engine. It also allowed them to have the engine closer to the water source and farther from the heat, smoke, and embers from the fire. From this improvement, the concept of the interior attack was born. Fire fighters could apply the water directly to the seat of the fire, not just through a window or doorway.

Another limitation of the first hand pumpers was that they had to have the tub around the intake of the pump filled by a bucket brigade. When the design was changed so the engines could take suction (draft) from cisterns under the streets or be filled from hydrants, their efficiency was greatly increased as they did not have to be filled by buckets as they pumped. All hands were needed to man the brakes, whether the problem was a long hose lay to the fire, a need for a hose stream to be aimed into an upper floor window, or a large fire that required long hours of pumping. When this suction hose was permanently attached to the body of the pumper, it was called a squirrel tail.

The first hoses were made of sewn leather, which had a tendency to leak, reducing the hose stream at the nozzle. The next improvement was riveted leather hose, which was much tougher and leaked considerably less. These improvements continued through linen hose, rubber-lined cotton hose, and today's synthetic hose.

As water systems improved and hose became more popular, hose companies were formed. The hose was carried on a wheeled carriage called a hose cart **FIGURE 3-12**. If the water system was good enough, the hose companies could use the pressure from the

water mains to place nozzle streams used to attack the fire into service. If the water system was insufficient, the hose companies would respond with, and were often affiliated with, the hand pumper companies.

Until horse-drawn apparatus was introduced, almost every piece of wheeled firefighting equipment was pulled by hand—which could be very dangerous, especially in hilly areas. A fully rigged hand pumper is very heavy and the common practice was to run with it, to the fire, as fast as possible. When a pumper went out of control due to hills or sharp turns, the only thing to do was to get out of the way. This was very hard on the equipment and the fire fighters. Many a volunteer was run over or pinned to a building by the very apparatus he was pulling to the fire (Frady 1991).

The next major improvement in firefighting apparatus was the steamer **FIGURE 3-13**. The first steamers were pulled by hand but were soon discovered to

FIGURE 3-13 Steam-powered fire pumper.
© Jones & Bartlett Learning

be too heavy, so horses were used to pull them. The steamer took only a three-person crew to get it to the fire and place it into operation. The steamer could pump for as long as there was coal to keep it running, without having large numbers of volunteers to man the brakes. At first the volunteers saw the steamers as a threat to their existence and fought hard against their adoption. The steamer gave the cities the opportunity to have a regular paid fire department at low cost due to the small number of men required to operate it. This, along with the problems of fighting among the volunteer companies at fires, led to the extinction of the volunteer fire departments in many larger cities (McPhee 2008).

As buildings grew in height, a way had to be found to effect rescues from windows when the access through the inside of the building was untenable. This led to the introduction of the ladder company. The responsibility of the ladder company was to bring the ladder wagon to the fire. The ladder wagon was also equipped with hooks, ropes, and other equipment, giving us the term *hook and ladder*, which is still used today. The pike pole, looking much like a lancer's pike, was developed to pull down ceilings to get at concealed spaces **FIGURE 3-14**. The pike axe, a single blade axe with a point on the back side of the head, was designed for forcible entry. With these tools carried on the ladder wagons, the job of the ladder company evolved into one of forcible entry, rescue, and ventilation, which it still is today (ICMA 2012).

Extension ladders up to 75 ft (22.86 m) in length were brought to fires. These allowed access as high as the fifth floor but were extremely heavy and cumbersome. By 1870, the aerial ladder apparatus was developed. Its ladder was extended by hand cranks.

The next improvements were spring-assisted raising, compressed air, and, finally, hydraulic pressure.

The chemical wagon, invented in the late 1800s, carried two tanks—one of soda and water, and the other of acid—which, when mixed, created carbon dioxide gas. The gas created pressure in the vessel, forcing the water out through the hoses. Their effectiveness was limited to the amount of gas created to expel the water. These engines were smaller than steamers, weighed much less, and were easier to place into operation, but their effectiveness was limited to the water they carried and the amount of gas created to expel the water (Vintage Fire Museum 2013). Because of its speed and ease of operation, the chemical engine extinguished many fires in their first stages. However, a severe problem developed if the hose became kinked or plugged, because an explosion could occur.

The next major improvement in fire apparatus was the invention and use of the internal combustion engine **FIGURE 3-15**. Unlike horses, the gasoline engines did not get tired on long runs, nor were they subject to disease or being frightened by dogs or people. The first gasoline-powered engines were not very reliable, and steam pumpers continued to be used. Because the departments already had steam engines in use, gasoline-powered tractors were built to pull the steamers to the fire, replacing the horses. As improvements were made in design and reliability, the motorized apparatus finally dominated the scene, and other types of pumping apparatus and steam and chemical engines were eventually phased out. In the engine design in use today, the same motor that propels the apparatus drives the pump. This design further simplified the apparatus, reducing weight and cost.

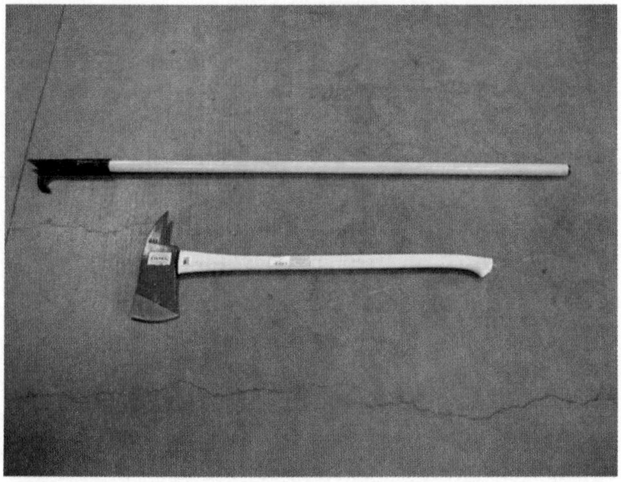

FIGURE 3-14 Pike pole and pike axe firefighting tools.
© Jones & Bartlett Learning

FIGURE 3-15 Restored 1928 Moreland gasoline-powered fire engine.
Courtesy of Kern County Fire Department

Fire Service Symbols

There are many symbols of the fire service recognizable to fire fighters and the public. One is the fire helmet with its distinctive shape, covered previously in this chapter. Another is the fire axe, or pike headed axe, commonly known as a "fire axe." The symbol of the red fire engine is common in artwork and toys, but is not universal. Fire engines come in many colors, including white, white with colored stripe(s), black over red, white over red, yellow, lime yellow, green, and purple. Each department makes its own determination as to the color of its engines. The primary factor should be visibility and recognition as fire apparatus. Smokey Bear (there is no "the" in Smokey Bear as this is his proper name) is the symbol of wildfire used by the Ad Council in television, billboard, and print media fire prevention advertising. Sparky the Fire Dog©, created by the National Fire Protection Association (NFPA), is the symbol commonly associated with urban and suburban fire agencies.

Maltese Cross

A common symbol in the fire service, the Maltese cross, is considered a symbol of protection and a badge of honor. The Maltese cross dates back to the days of the Crusades. When the Christian knights approached the cities held by the Saracens, they were assaulted by glass bombs containing naphtha, a flammable liquid. Once the naphtha had penetrated the knights' armor, the Saracens would throw flaming torches at them. Hundreds of the knights were burned alive, and many others perished or were injured coming to their aid. To recognize them for their bravery, the survivors were awarded a special cross as a badge of honor. Many of the knights were members of the Knights of St. John, which was headquartered on Malta. The badge became known as the Maltese cross. The Maltese cross continues to symbolize that a fire fighter will lay down his life for others. The Maltese cross remains the fire fighter's badge of honor, and in some departments the badges are this shape (Peabody Fire Department 1996).

Dalmatian Dogs

When horses were introduced into firehouses, Dalmatian dogs were also introduced as fire service mascots **FIGURE 3-16**. For centuries they were used as followers and guardians of horse-drawn vehicles. They were kept with the horses to keep them calm. When responding to fires, the dog ran ahead, behind, or under the fire apparatus and chased off other dogs and

FIGURE 3-16 Dalmatian dog fire service mascot.
© Susan Kuklin/Science Source

animals that would bother the horses. The Dalmatian is still the only recognized carriage dog in the world. The Dalmatian is also known for its friendliness and ability to work (Local 1259 IAFF 2013).

Fire Losses

Throughout history, fires have resulted in tremendous loss of life and property. In terms of property loss, one of the most disastrous fires ever to occur in the United States was the so-called Great Chicago Fire, which occurred on October 8–10, 1871. The fire burned for 3 days and destroyed $3\frac{1}{3}$ mi^2 of the city of Chicago, Illinois. The fire left 100,000 people homeless and killed approximately 300 (Chicago History Museum 1999). As a result of this and other fires, President Calvin Coolidge proclaimed the first National Fire Prevention Week to be October 4–10, 1925 (Fire Files Digital Library 2013). This tradition continues today with Fire Prevention Week occurring each October.

The city of Baltimore, Maryland, was stricken by a conflagration on February 7, 1904. A fire started in a storage area in the basement of the six-story Hurst Building. Soon the fire leapt up an unenclosed shaft and spread to the top floor. Nearby buildings with unprotected openings were soon afire. As the conflagration spread, the mayor called for help from other major cities in the area. Fire equipment and fire fighters were shipped in by train. Upon their arrival, they found that the threads on their hoses would not fit those on the Baltimore hydrants. The out-of-town fire fighters were forced to rip up the streets and create ponds of water around the hydrants to supply their engines. This problem eventually led to the development of standardized hose threads for firefighting equipment. By the time the fire was stopped, 155 acres of mercantile property, valued at $50 million, was destroyed, and 50,000 people lost their jobs (Ferris 1996).

On April 18, 1906, an earthquake rocked the city of San Francisco. The earthquake lasted only 1½ minutes, but the resulting fires lasted for 2 days. The fires destroyed 4.7 mi² of the city and burned over 25,000 buildings. Four hundred fifty people were killed in the disaster. The estimated loss was from $350 million to $1 billion (History.com 2013).

On April 12, 1908, Chelsea, Massachusetts, was devastated by a fire started in rags. High winds whipped the flames out of control and destroyed approximately one-half of the improved area of the city. In October 1973, 65 years later, the same thing happened: Chelsea had another conflagration with the same problems of fire spread and control (Resnek 2012).

In the early part of the 20th century, when combustible construction was prevalent, conflagrations occurred at an alarming rate. In 1913, a major fire hit Hot Springs, Arkansas, destroying over 500 buildings (ArkansasOnline 2013). Salem, Massachusetts, was ravaged by fire on June 25, 1914; 1376 buildings were destroyed (Sensagent Corporation 2012).

The list goes on and on. The common denominator in these conflagrations is **combustible construction**. The streets were narrow and there were ineffective building codes in place regarding **unprotected vertical shafts**, which led to rapid fire spread throughout the structures.

To put property from fire loss in perspective, consider the following: If something is stolen in a crime, it is lost, not destroyed. Usually only items considered to be of resale value are taken. If a house burns down, many of the family's belongings are destroyed and unrecoverable. Anyone who has ever had a home burn can tell you it is not the loss of the DVD player or the television that really matters, because these items can be replaced with money provided by insurance. It's the loss of the irreplaceable photos and family heirlooms that is devastating.

When a business burns or is severely damaged by fire, jobs and tax revenue are lost. Other businesses take over the burned-out business's customers. With the cost of rebuilding and regaining its previous customers, the burned-out business cannot compete and only a few reopen, which has a negative impact on the community as a whole.

Arson is another cause of high-loss fires. Many people have been killed as the result of arsonists setting fires to occupied buildings. Usually motivated by spite or revenge, the arsonist has no feeling for the consequence of his or her actions. Arson is especially common during riots. In periods of civil unrest, it is not uncommon for fire fighters and their equipment to be attacked. A Los Angeles city fire fighter, driving a fire engine down the street, was shot in the neck by a passing motorist during the 1992 riots (Lopez 2009). The 862 structure fires set during the Los Angeles riots caused three deaths, and damage was estimated at $1 billion.

More recently, a primary cause of large fires resulting in great property loss and numerous deaths has been terrorism. On September 11, 2001, the threat of large-scale terrorist attacks in the United States became reality. Two hijacked airliners were purposefully flown into the twin towers of the World Trade Center in New York City, a third airliner hit the Pentagon in Arlington County, Virginia, and a fourth plane crashed in Pennsylvania. As a result of these incidents, the life toll was over 3000 persons, including 343 fire fighters. The property loss was estimated to be in the tens of billions of dollars. There is a constant threat that these types of incidents will continue to occur.

Loss from Wildland Fires

An ever-increasing number of homes have been destroyed by wildland fire each year since the 1960s; the annual average has risen from 250 homes a year to 2700. There is an increasing number of homes in wildland fire-prone areas. Of the 17 million homes built from 1990 to 2008, 10 million were constructed in the wildland-urban interface area (Maclean 2013).

According to the Insurance Information Institute (2019) all of the top 10 costliest wildland fires in United States history occurred in California. The Camp Fire in 2018 resulted in approximately $10.5 billion in losses.

Part of the reason Southern California wildland fires reach conflagration proportions is the extreme weather experienced in the fall. The weather pattern builds up high pressure over the Great Basin, the area of eastern Washington, Idaho, and Utah, and is coupled with low pressure off the coast of Southern California. This difference in pressure causes hot, dry Santa Ana winds to blow from the Great Basin across Southern California to the Pacific Ocean. Powered by these winds and fueled by dry brush, wildland fires grow quickly to conflagration size. With combustible exterior construction in the form of wooden decks, wood shake roofs, and wood exteriors, the fire soon rages out of control. In the great fire siege of October 26 through November 7, 1993, 1152 structures and more than 200,000 acres were consumed by fire. Many of these structures were homes. Two of the fires alone were responsible for massive destruction: The Laguna Fire burned 366 structures and the Malibu Fire burned 323. Two factors that finally let the firefighting forces get the upper hand were the winds dying down and the fire burning to the Pacific Ocean.

In the fall of 2003, Southern California was stricken by a fire siege unequaled in previous experience. It lasted for 10 days, from October 21 to October 30, and consisted of 14 large fires that occurred in the following six counties: San Diego, Riverside, Los Angeles, San Bernardino, Ventura, and Santa Barbara. By the time the fires were controlled, they had burned 745,933 acres (1165 mi^2). Included in the losses were 3646 residences, 36 commercial structures, and 1169 outbuildings. The human toll included 22 fatalities—one fire fighter and 21 civilians. There were an additional 239 reportable injuries. The fires were fought by 15,631 personnel, 1898 engines, 296 hand crews, 181 bulldozers, 105 helicopters, and 43 air tankers (California Governor's Office of Emergency Services 1993).

A similar situation impacted Southern California in 2007, resulting in the deaths of five U.S. Forest Service fire fighters performing structure protection on the Esperanza Fire. Nationally, there were 85,705 wildland fires in 2007 with 9,328,045 acres burned. In 2008 there were 78,979 wildland fires. There were 5,292,468 acres burned, and suppression costs were over $1.3 billion. The number of people involved in fighting the fires, including fire fighters and support personnel, was more than 30,000. When major fire seasons occur, resources from across the United States, Canada, and Mexico have been mobilized.

Despite the availability of resources and improvements in firefighting equipment and technique, Los Angeles County experienced the largest fire in its recorded history (160,557 acres). The Station Fire in August, 2009, is the largest recorded fire in the history of the Angeles National Forest and the 10th largest fire in California history since records began being kept in 1933. The Station Fire resulted in the deaths of two fire fighters (InciWeb 2009).

The Mendocino Complex of fires, consisting of the River Fire and Ranch Fire in July, August, and September of 2018 is the largest fire in California's recorded history. These fires were managed as a Complex due to their close proximity and the prediction that they would burn into each other. As of August 23, 2018, the fires had burned 415,685 acres. They resulted in one fire fighter fatality, and burned 157 residences and 120 other structures. These fires were managed in unified command (Chapter 13) by CalFire (CALFIRE) and the U.S. Forest Service (INCIWEB, 2018).

Another large fire, burning at the same time in relatively close proximity, was the Carr Fire. This fire spawned a rarely seen phenomenon called a Firenado. This was a vertical vortex of burning material and fire gases that uprooted trees, ripped roofs off houses, and downed power lines. The National Weather Service estimates the winds were in excess of 143 mph. This wind speed rates as EF-3 on the Enhanced Fujita Scale. Some believe it was this same fire behavior that destroyed Peshtigo, Wisconsin in 1871.

The Camp Fire, in North Central California in November of 2018 burned 153,336 acres, destroyed 14,000 residences, 528 commercial buildings, and 4293 other buildings. The fire almost totally destroyed the town of Paradise. This fire resulted in 85 civilian casualties, the highest in recent history. Some of the fatalities occurred in homes while others perished in or out of their vehicles while trying to escape through the firestorm. The fire was fought by 1065 personnel, 73 fire engines, 11 hand crews, 2 helicopters, 3 bulldozers, and 3 water tenders (Inciweb 2018).

While the Camp Fire was burning, another fire in Southern California, the Woolsey Fire, ignited. This fire resulted in 1500 structures destroyed with an additional 341 damaged. Three civilians lost their lives in this incident. These types of fires are not limited to California. Similar fire experiences are occurring across the nation. In all, at least 39 of the 50 states have received federal disaster declarations due to wildland fires. The Wallow Fire, which burned from May 29 to July 4, 2011, burned approximately 540,000 acres, equaling 840 mi^2, and involved portions of Arizona and New Mexico (InciWeb 2011). In Texas in 2011, close to 4 million acres burned. Fire fighters from 25 states were brought to assist with the fires. The Caughlin Fire in Reno, Nevada, in November 2011 burned just under 2000 acres but destroyed or damaged 42 structures (City of Reno, Nevada 2011). In August 2012, at least 65 homes were destroyed in the rash of wildland fires occurring north and south of Oklahoma City and south of Tulsa, Oklahoma (UPI.com 2012).

Overall, the most active wildland fire year since 1960 was 2007, with 9,873,745 acres reported burned (85,705 fires). The federal fire suppression cost for these fires was nearly $1.8 billion. This illustrates that despite greatly improved firefighting techniques and resources, wildland fires are still a huge problem for the United States. The almost $2 billion number is not the true picture of what was lost in these fires. This number does not take into account the fire suppression funds expended by states and local government. Other costs that are not captured in this number are indirect costs, such as lost revenue due to the closing of highways, railways, and tourist attractions; burned residences, outbuildings, vehicles, power poles, fence lines, and grazing land; livestock killed by the fires; and timber and other natural resources damaged or destroyed. The cost of damage from wildland fires has not been lost on our enemies. The Japanese tried unsuccessfully to ignite forest fires on the West Coast during World War II. An al-Qaeda English-language magazine has supported the idea of starting forest fires, especially in the rapidly growing wildland-urban interface areas of western Montana (Gabbert 2012).

Wildland Fire Fighter Fatalities

From 1910 to 2017 there have been 1128 wildland fire fighter fatalities. Wildland fire fighter fatalities during the period of 2007–2016 were 24 percent heart attacks, 20 percent vehicle accidents, 18 percent aircraft accidents, 17 percent entrapments, 10 percent other medical, 7 percent falling trees/rolling rocks, and 4 percent miscellaneous (NIFC 2018).

Several notable fires resulted in major loss of lives by wildland fire fighters **TABLE 3-1**. These can easily be researched online and several have had books written about them.

This problem is not limited to the United States—Australia, Italy, Greece, and Israel are experiencing these problems as well. As more people leave the cities and

build their homes in the hills and mountains, urban interface and urban intermix fires will continue to grow in frequency and devastation. People are building homes at the heads of canyons and on ridge tops, offering homeowners a beautiful view but creating a firefighting nightmare. Often built at the end of narrow streets and culs-de-sac, these home sites are death traps for firefighting crews if the fire gains speed as it races uphill. As firefighting is done further down the hill and people water their roofs to protect their homes, the water supply at the higher elevations becomes depleted. Soon there is not enough pressure for engines to operate from the hydrant system, if there is one. This leaves the fire fighters with only the water in the tank on the apparatus to deal with an onrushing wall of 50-foot or higher flames.

Combustible roofs are a cause of conflagration fires. In the western United States, particularly in California, conflagration fires have started out as brush fires spreading to structures. This happened in the Berkeley hills in 1923 and again in 1991. On October 20, 1991, the East

TABLE 3-1 Notable Wildland Fires for Fire Fighter Fatalities

Fire	Year	Location	Fire Fighter Deaths
The Big Blowup	1910	Idaho, Montana, and Washington	78
Griffith Park Fire	1933	Los Angeles, California	25
Blackwater Fire	1937	Wyoming	15
Cleveland National Forest	1943	California	11
Mann Gulch Fire	1949	Gates of the Mountain Wild Area (now Gates of the Mountain Wilderness), Montana	13
Rattlesnake Fire	1953	Mendocino National Forest, California	15
Inaja Fire	1956	Cleveland National Forest, California	11
Loop Fire	1966	Angeles National Forest, California	12
South Canyon Fire (also known as Storm King Mountain Fire)	1994	Colorado	14
Esperanza Fire	2006	San Bernardino National Forest, California	5
Iron Complex	2008	Shasta Trinity National Forest, California	9
Yarnell Hill Fire	2013	Arizona	19

Data from National Interagency Fire Center. 2013. *Historical Wildland Firefighter Fatality Reports*

Bay Hills Fire took the lives of 23 civilians, one fire fighter, and one police officer. This fire destroyed an average of one home every 13 seconds for the first 9 hours of its existence, for a total of 2103 structures (Routley 1991).

Probable Future Problems Caused by Global Warming

The evidence that the earth is undergoing a dangerous warming trend becomes more apparent with each passing year. The warmest year measured to date was 2016. The 10 warmest Septembers, as measured by global land and ocean surface temperatures, have occurred since 2003 with the last five Septembers (2014, 2015, 2016, 2017, and 2018) ranking as the five warmest on record (NOAA.gov). The results of global warning are sure to be of concern to fire fighters and other emergency responders. North America is experiencing an increase in:

- *Increased warming, drought, and insect outbreaks*: These are all caused by or linked to climate change and have increased wildfires and impacts to people and ecosystems in the Southwest. Fire models project more wildfires and increased risks to communities across extensive areas (GlobalChange.gov).

- *Floods and flooding*: Flooding increases the need for emergency response to rescue and treat victims; it also results in property damage. During times of flood, buildings can still burn to the water level with fire fighter response hampered or prevented altogether. Houston, Texas has experienced three "500-year" floods in 3 years. In 2017, Hurricane Harvey left Houston flooded with torrential rains (WashingtonPost.com).

- *Heat-related deaths and illness*: Heat emergencies result in increased incidents requiring medical aid from emergency responders, as well as an increased heat exposure hazard to fire fighters as they try to fight fires in increasingly hotter air temperatures.

- *Disease and pests*: A wider spread of diseases, such as West Nile virus and dengue fever can arise due to a lack of freezing temperatures to kill off mosquitoes and other insect pests. Bark beetles will be more likely to attack and kill large areas of coniferous forest (pine and fir), resulting in an increased fuel load in areas preheated from higher air temperatures and drier from drought, making them more susceptible to large fires. Intermixed with these areas are increasing numbers of homes to be protected during a wildland fire.

- *Shorter winters*: Shorter winters and longer hot and dry seasons will cause naturally occurring vegetation fuels (grass, brush, and trees) to become drier and more likely to ignite and burn.

- *Droughts*: There is much discussion as to global warming's effect on drought. Several states have recently suffered extreme drought and an increase in acres burned in wildland fires. Large fires have occurred in Arizona, Texas, and Oklahoma, burning valuable grazing land, forested areas, outbuildings, and homes. Another negative effect of drought is the availability of water supply, both for domestic use and for firefighting. In rural areas, static sources of water dry up, making them unavailable for drafting with pumpers or to supply water for firefighting helicopters.

- *Rising sea levels and melting ice caps*: Sea levels are expected to rise 3 ft (0.9144 m) by the end of the century. Over half of the U.S. population lives within 50 mi of the coast. This makes beachfront and low-lying coastal properties more susceptible to storms, such as hurricanes, and storm surge flooding. Coastal flooding caused by rain and tidal surge were both evidenced in hurricanes such as Katrina (2005) in the Gulf Coast states and Sandy (2012) in the northeastern coastal states. These both led to loss of lives and property. During Superstorm Sandy, a neighborhood called Breezy Point in the New York City borough of Queens burned with the loss of more than 80 homes.

- *Tornadoes and severe thunderstorms*: A team from Purdue University (2007) produced a study that found by the end of this century the number of days that favor severe storms could more than double in locations such as Atlanta and New York. The study also found that the increase in storm conditions occurs during the typical storm seasons for these locations and not during dry seasons when such storms could be beneficial. As of this writing, 2019 is on course to be a record year for tornadoes.

Tip

The United States, Canada, and Hungary have the highest fire death rates per capita in the industrialized world.

The U.S. Fire Problem

Annually, the NFPA collects data from the National Fire Incident Reporting System (NFIRS) and develops a snapshot view of the national fire problem in the United States **TABLE 3-2**. Some of their findings include:

- On average, U.S. fire departments respond to:
  - A fire every 24 seconds.
  - A structure fire every 66 seconds.
  - A home fire every 90 seconds.
  - An outside or unclassified fire every 48 seconds.
  - A highway vehicle fire every 182 seconds.*

- The United States has one of the highest fire death rates per capita in the industrialized world. The fire death rate is 13.1 per million of population.

- Approximately 4000 people die every year in the United States as the result of fires, and another 22,000 are injured.

- About 100 U.S. fire fighters are killed annually in duty-related incidents, and another 87,000 are injured.

- Fire is the third-leading cause of accidental death in the home.

- An average of 1.9 million fires are reported each year. Keep in mind that not all fires are reported.

- Direct property loss caused by fires is estimated at more than $10 billion each year.

- The South has the highest fire incident and death rates in the nation. This statistic is related to income level, with poverty-stricken areas having the highest fire and death rates.

- Eighty-two percent of all civilian fire fatalities occur in the home.

- Cooking is the leading cause of home fires and injuries in the United States.

- Careless smoking is the leading cause of fire deaths.

- Heating equipment is the second leading cause of home fires and home fire deaths.

- In commercial properties, arson is the leading cause of deaths, injuries, and dollar loss.

- The fire death risk among senior citizens is more than double the average population.

- The fire death risk for children under age 5 is nearly double that of the average population.

- Children playing with fire start more than 50 percent of the fires that kill young children.

- Men die or are injured in fires twice as often as women.

- Approximately 90 percent of U.S. homes have at least one smoke alarm.

- More than 40 percent of residential fires and 60 percent of residential fatalities occur in homes with no smoke alarm.

TABLE 3-2 Fire Loss in the United States 2011–2017

Year	Number of Fires	Civilian Deaths	Fire Fighter Deaths	Dollar Loss (in billions)
2017	1,319,500	3,400	54	$10.7
2016	1,342,000	3,390	89	$10.6
2015	1,345,500	3,280	90	$14.3
2014	1,298,000	3,275	64	$11.6
2013	1,240,000	3,240	99	$11.5
2012	1,375,000	2,855	70	$12.4
2011	1,389,500	2,640	61	$9.7

Fire Loss in the United States (year). Quincy, MA: National Fire Protection Association

* NFPA Report: "An Overview of the U.S. Fire Problem"

- The U.S. fire service responds to a fire every 17 seconds.
- Twice as many fires occur in the kitchen than in any other area of the home.
- Nationwide, there is a fire death every 147 minutes and a fire injury every 24 minutes.
- The top five areas of fire origin in residences are: kitchen (26 percent); bedroom (13 percent); living room/den (8 percent); chimney (8 percent); and laundry area (5 percent).
- Residential sprinklers decrease the home fire death rate by 74 percent.

Purpose and Scope of Fire Agencies

The main purpose of the fire service is to manage community risk. By having trained personnel and specialized equipment on hand, emergency responders exemplify a community's response to risk. Most departments now focus on the control of fires in individual properties and the rescue of endangered occupants. Departments exist to limit the probable loss to the community when a fire occurs. Imagine the consequences if there were no fire departments.

A department with a mission statement such as "protecting life and property through effective public education, fire prevention, and emergency services," is focusing on risk reduction. Managing risk is done through reducing risk, the weighing of cost versus benefit. It would be impossible and too expensive to eliminate risk completely. This is expressed in terms of acceptable risk. The term *acceptable risk* describes the likelihood of an event whose probability of occurrence is small, whose consequences are so slight, or whose benefits (perceived or real) are so great, that individuals or groups in society are willing to take or be subjected to the risk that the event might occur. In his address to the U.S. Naval Postgraduate School in 1954, Admiral Hyman Rickover said, "It is a human inclination to hope things will work out, despite evidence or doubt to the contrary. A successful manager must resist this temptation"* (Rockwell 1992).

A community expresses its assessment (perceived threat) of its overall fire risk through the resources it is willing to commit to its fire department. If the fire department is unable to perform its fire control mission, the community's fire risk balance could be compromised.

The modern fire service manages more community risk than just that of fires. The fire service now provides services in the areas of emergency medical services (EMS) to reduce the consequences of medical emergencies. It also provides emergency management, providing services to reduce the effect and consequences of disasters such as earthquakes, floods, tornadoes, and ice storms. Rescue teams safely remove citizens from situations such as structural collapse, swift water, ice, trenches, and vehicle crashes. Hazardous materials response teams protect the population and the environment from the effects of uncontrolled releases of hazardous materials.

Emergency response is only one factor in the reduction of risk. Placing a fire station on every corner would still not prevent incidents from occurring. Often, once the incident has occurred, it is too late to save lives; therefore, departments engage in public education, prevention, and code enforcement (discussed further in the Fire Prevention chapter).

With this tremendous opportunity to be seen as heroes comes responsibility as well. Emergency response organizations are viewed as essential public services. As such, members need to ensure that they are always ready to respond and carry out their sometimes life-saving functions. When departments cannot respond, public outcry can be intense. Managers must always be aware of possible disruptions to service (equipment failure, unavailable personnel, and other reasons). It may not be your fault, but no one wants to hear excuses.

Another part of this responsibility is that safety organizations are also recipients of public funds. Resources must be managed carefully in a cost-effective manner to allow the utilization of funds where they are most needed. If fire fighters are careless and lose or destroy equipment, money must be diverted from other areas to replace it. If personnel are not safety conscious, injuries rise, and costs for workers' compensation, medical treatment, and replacement personnel rise as well.

Fire departments must manage financial, liability, and safety risks within three major categories:

1. Risk to the community—community risk [the National Fire Academy (NFA) offers a course in Community Risk Reduction]
2. Risk to the fire department organization—organizational risk
3. Risk during emergency operations—operational risk

The original purpose of fire fighters was to save lives and property from fire. Generally, that meant pulling people from burning buildings and spraying water until the fire was out. In doing so, much property was

---

* Quote by Admiral Hyman Rickover

lost from water damage and the demolition of buildings to stop the spread of fire through the creation of firebreaks. As time went on and the fire service became more sophisticated, salvage work became a fire department responsibility. It was no longer acceptable to extinguish fires by "washing it off the lot." The term *saving property from fire* took on a new meaning.

Fire fighters have always saved lives at fires when they could. As equipment has improved, this capability has been greatly enhanced. This mission of saving lives has expanded to the point that the modern fire department is responsible for almost all rescue activities, including dive teams, urban search and rescue, technical rescue, swiftwater rescue, and cliff rescue, as well as auto accident extrication and advanced life support **FIGURE 3-17**. The fire department is a readily available, trained professional force, making it the logical choice to assume increased rescue responsibility. As the population increases and people become bolder in their searches for excitement, the importance of the rescue capability of the fire service is sure to grow.

FIGURE 3-17 Fire fighters performing technical rescue training.
Courtesy of Kern County Fire Department

Fire Prevention

In the area of fire prevention, the original fire fighters were limited to inspecting chimneys and making sure fires were extinguished at curfew. Today fire prevention activities have increased to the point where the fire prevention bureau is involved in the design process of new buildings through codes, water flow requirements, and inspections. This process does not stop when the building is finished and occupied by a tenant. Periodic inspections are made to ensure compliance with codes and ordinances.

> **Tip**
>
> Public education for fire safety and prevention has become a fire department responsibility in many jurisdictions.

Importance of Smoke Detectors

Smoke detectors (also called smoke alarms) are inexpensive and easy to install. They have a tremendous impact on life safety in residential occupancies due to their ability to notify occupants of a fire. People who are poor, elderly, live in rural areas, smoke, or live in manufactured homes or substandard housing are especially at risk, according to the U.S. Centers for Disease Control and Prevention. Many fire departments have smoke detector programs where they will install a smoke detector for free in the homes of those who cannot afford them.

> **Tip**
>
> Senior citizens are at the highest risk of being killed in a fire. This is partially due to their inability to move quickly once they discover the fire.

Public Education

Public education for fire safety and prevention has become a fire department responsibility in many jurisdictions. Demonstrations are put on at shopping malls, service clubs, businesses, schools, and fairs. This education is not limited to fire safety; problems particular to an area are addressed as well. In the western United States, earthquake safety lectures and demonstrations are common. Phoenix, Arizona, has a fire department–sponsored home swimming pool safety program to prevent children from drowning. Fire departments in the Midwest present information on tornado safety, and on the East Coast, programs are presented on hurricane safety. Fire departments are now being requested to become involved with the proper installation of child passenger-safety seats.

Hazardous Materials

As hazardous materials have become more of a concern, fire departments have taken over responsibility for hazardous materials control functions, including inspecting businesses and keeping inventories. In some areas, when old underground fuel tanks are removed, the fire department issues the required permits and inspects the operation.

Working with Law Enforcement

Arson investigation has become an important function of the fire department. In conjunction with law enforcement personnel, fire investigators search for the cause of fires and gather evidence to prosecute criminals. As budgets have tightened, cost recovery has become more popular as a means of recouping money spent to combat fires caused by negligence or illegal activities. Some departments extend this concept to pursuing cost recovery on traffic incidents involving drunk drivers. Cost recovery requires the fire department, law enforcement, and prosecution services to work together in a close relationship.

With the increase in active shooter incidents, fire fighters in some jurisdictions have been provided body armor and are now being trained as part of entry teams for this type of incident. Fire fighter paramedics may also be assigned as part of SWAT teams to assist police department and civilian victims in SWAT situations.

Fire Defense Planning

To provide an effective, integrated fire defense system at the least cost, planning must take place. The fire chief must take into account all public and private resources at his or her disposal to achieve the desired goals. One of the first questions to consider is, "What is the acceptable level of loss due to fire?" Losses due to fire can be divided into three general categories: deaths, injuries,

and property. The fire chief and the fire department staff decide what they envision the community's fire defenses should be. In many jurisdictions, civilian advisory boards are included in this process.

The goals of fire protection are to prevent fire from starting, prevent loss of life in case fire does start, confine fire to its place of origin, and extinguish fire once it does start. The goals state what level of fire protection is desired. Once the goals are stated, the objectives can be set. **Objectives** are steps to be taken to achieve the goals. The objectives must be clearly stated and measurable. It must be possible to determine if what is being done is accomplishing the objectives.

To formulate any plan, it is important to know what the starting point is. Statistics are kept and analyzed to determine what is presently occurring. Reports are generated specifying type of loss, type of occupancy, time of day, ignition source, item first ignited, and direct cause of loss. Using these facts, the fire department can look for trends and major factors. Once these factors are identified, the department determines which areas to target, in the form of objectives in the overall plan.

Goals must be matched with jurisdiction and department philosophy. It must be determined whether the goals are economically and politically attainable. There are measures that, if taken, would not be acceptable to the public as a whole. An example of this is home fire prevention inspections. Although it seems like a good idea, under the U.S. Constitution people have the Fourth Amendment right not to have their homes invaded without reasonable cause. This program can be conducted only with the homeowner's approval. Another factor is the cost: How many fire fighters would it take to inspect a city of 40,000 homes on an annual basis?

The next step is to establish policies. **Policies** are general guidelines for how things will be done. An example of a policy would be that all fire fighters wear breathing apparatus during **overhaul** operations, or until declared unnecessary by the officer in charge. The policy clearly states the department's position on wearing breathing apparatus, while still leaving room for decision making on a case-by-case basis. The objective is to reduce fire fighter injuries; the policy provides a guideline to achieve the objective.

Personnel must be provided to attain the objectives. If reduced property loss is the objective, enough people need to be assigned to the fire in a timely fashion to control, extinguish, and perform salvage work. Personnel are the single greatest cost in a career fire department. Often up to 90 percent of the fire department budget is devoted to personnel costs. The decision on equipment staffing is one of the most often-contested in the nation. Some groups feel that the minimum effective engine company staffing is four people; others think three are sufficient. Costs must be balanced against benefits and available resources to come to a decision.

The NFPA has developed two standards to address the issue of adequate staffing: NFPA 1710: *Standard for the Organization and Deployment of Fire Suppression Operations, Emergency Medical Operations, and Special Operations to the Public by Career Fire Departments* and NFPA 1720: *Standard for the Organization and Deployment of Fire Suppression Operations, Emergency Medical Operations, and Special Operations to the Public by Volunteer Fire Departments, Organization and Deployment*. These may be used by fire departments to justify staffing patterns and deployment of resources.

Procedures are established to describe how employees will perform recurrent and important functions. Procedures differ from policies in their specific nature. A typical procedure is for fire fighters to inventory all of the equipment on their apparatus on a daily basis.

Facilities for training and housing employees must be provided. Facilities also include apparatus and equipment. Without tools, fire fighters cannot perform their jobs. Budgeting must take into account that tools become worn out, lost, and broken and must be replaced. A fire station may last 100 years or more, an engine for 20 years, PPE for 3–5 years, and flashlight batteries a matter of months.

Providing funding for the personnel, facilities, and equipment requires an excursion into the political arena. The chief and his or her staff must determine how much the system will cost. Cost analysis is performed from both within and outside the department. The department's business manager determines cost factors, which are reviewed by the jurisdiction's budget analyst. After the cost analysis is completed, the city council or other elected body is lobbied for the budget to be approved. This is done mostly behind the scenes. When the chief appears before the governing body to seek approval of the budget, it is mostly a formality. The appearance before the governing body gives the public a chance to review the requested budget and to comment on it. One must always keep in mind that there never seems to be enough money to satisfy the needs of all of the groups competing for funds. As a fire fighter,

it may be obvious to you that the department needs new equipment and more personnel, but this is not so obvious to the people who want longer library hours. In tight fiscal times, everyone has to make sacrifices.

Once the budget is approved, the chief is charged with applying the available resources to achieve the desired results. At every step of the way, results must be evaluated to see if the goals are being achieved. The process does not stop here. Planning is a never-ending cycle of setting goals, determining objectives, and evaluating results. The fire department exists in a dynamic political and social climate with ever-changing demands and expectations.

All-Hazard Planning

With the modern fire department responding to many incident types, it is important to look at all of them when planning. The fire chief and his or her staff must prepare the department to respond to incidents such as fires, floods, earthquakes, terrorism, medical incidents, swift water, and any other incident they may be charged with. Resources must be obtained, trained, and staged appropriately. Considerations must be made to comply with legal requirements for certain incident types. These are discussed further in the Codes and Ordinances chapter.

Much like fire defense planning, risk assessments and financial decisions must be made as to the likelihood and consequences of such incidents. It is difficult to maintain public focus on supporting the costs of all of these specialty resources when an incident of this type may not have occurred for a while. It is also difficult to maintain readiness on the part of the fire fighters when they may believe there is little likelihood of an occurrence.

Since multiple agencies may respond to these incidents, agreements must be reached as to which will be the lead among the agencies responding before one of these incident types occurs. Training on interagency cooperation and establishing communications interoperability must be completed beforehand.

Risk Management

The determining factor in evaluating risk is the probability that an undesired event might occur, with a harmful or undesirable consequence, and the severity of the harm that might result.

In describing risk, the probability of an occurrence happening can be in subjective terms, such as *rare* or *high*, or in numerical terms, such as *20 percent* or *one in three*. Harmful consequences are often expressed in descriptive terms such as *death*, *injury*, and *disaster*.

Probability and consequences can be combined and expressed mathematically as the product of loss and probability. An insurance company, for example, might describe a facility as a $10 million risk that has a 2 percent probability of loss. The probability of risk comprises two factors: The chance that something undesirable might happen and also the probable outcome as rated on a scale of negative consequences (FEMA 2006).

Consider how the two factors of probability and consequence aid in planning. Some risks are low probability and high consequence (e.g., a helicopter crash); others are high probability and low consequence (e.g., burning your fingers changing the oil in your car). What we have to really watch for are the situations that are high in both categories. One of these is firefighting.

The term risk management refers to any activity that involves the evaluation or comparison of risks and the development of approaches that change the probability or the consequences of a harmful action. Risk management comprises the entire process of identification and evaluation of risks as well as the identification, selection, and implementation of control measures that might alter risk (USFA 2018).

The three control measures can be categorized as administrative, engineering, and personnel protection. Administrative controls consist of guidelines, policies, and procedures established to limit losses. Their intention is to make the task safer for the worker. These controls include standard operating procedures (SOPs), training requirements, safe work practices, and regulations and standards.

Engineering measures are systems engineered to remove or limit hazards. Engineering measures include equipment design, traffic engineering (traffic signal preemptive devices), and mechanical ventilation. These measures are also designed to make tasks safer for the worker.

Personal protection consists of equipment, clothing, and devices designed to protect the worker. Examples of PPE include helmets, gloves, turnouts, hoods, SCBA, and tools. The purpose of these items is to protect the worker from the hazard(s).

Risk Management Plan

Revisions in NFPA 1500: *Standard on Fire Department Occupational Safety and Health Program* require that a written risk management plan be prepared by fire departments and be made part of their official policies and procedures. A risk management plan establishes policy. The plan serves as documentation that risks have been identified and evaluated and that a reasonable control plan has been implemented and followed.

In the federal government, this process is referred to as a job hazard analysis. The components of a risk management plan required by NFPA 1500 are:

- Risk identification
- Risk evaluation
- Risk control techniques
- Program evaluation and review

The elements of the plan are intended to apply to all aspects of a fire department's operations and activities, not just emergency activities.

Community Risk Reduction Planning

When planning to protect the community from risk, the process can be divided into four steps. Step one is preparation. Preparation includes assessing what types of risks are to be faced (e.g., fires, aircraft crashes, tornadoes, flooding, earthquakes, terrorism).

Step two is mitigation. Once the threats have been determined, mitigation steps are taken to reduce the threat. Mitigation may include training hospital and city personnel in emergency management concepts, applying engineering solutions to prevent minor disasters from becoming major disasters, implementing changes in building and fire codes, and locating and equipping fire fighters to deal with threats to the community.

The third step is response. This is when the fire department and other agencies apply their personnel, training, and equipment to lessen the damage from the incident.

In the fourth step, the recovery phase, things are returned to normal as quickly and as much as possible.

The Future of Fire Protection

The future holds many changes for those in the firefighting profession. The term *fire department* has become outmoded, though still used by many agencies for reasons of tradition. More properly, the term would be *emergency services agency*. With the sole exception of police functions, the fire department is responsible for most public safety emergency functions and many that are nonemergency. The organization, its responsibilities, and its equipment are much different than they were in the past. The only thing that has not changed is the public's need for emergency services and the fire fighter's willingness to do what it takes to serve and protect. The fire department of tomorrow will be vastly different from the one of today.

An area of increasing concern for the fire and rescue service is homeland security. Homeland security is defined as a "concerted national effort to prevent terrorist attacks within the United States, reduce America's vulnerability to terrorism, and minimize the damage and recover from attacks that do occur." As first responders the fire service most certainly has a role in this.

Another recent focus has been on protecting workers from job hazards. Vast bodies of law and regulation cover all aspects of the work environment. Many of these regulations are contained in the *Code of Federal Regulations* 29, Part 1910, the body of law that establishes the federal Occupational Safety and Health Administration (OSHA). As these regulations are more closely applied to the fire department, the risks taken by fire fighters in their jobs will be scrutinized for safety and necessity. This is just another area to which modern fire fighters are going to have to adapt.

Overall, the trend is expected to be toward fire prevention and away from manual fire suppression. Some people feel that the fire department of the future will be more of a regulatory agency for built-in protection systems. The case can be made for the fact that fires have become more difficult to extinguish because of building design changes and construction components. It is becoming harder and harder to apply water to a fire in a high-rise building. In a political atmosphere of user fees versus general taxation, it stands to reason that built-in suppression systems will take the place of a large firefighting force. This more directly places the cost of protection on the owner.

> **Tip**
>
> Overall the trend is expected to be toward fire prevention and away from manual fire suppression.

Wrap-Up

CHAPTER SUMMARY

- When people first learned to use fire, their culture and society changed dramatically, and fire became humanity's constant companion.
- The first known firefighting forces were in ancient Rome.
- The first settlement in the United States, Jamestown, was almost a failure due to a fire exacerbated by the building materials used in homes.
- After the Revolutionary War, the concept of volunteer fire companies spread across the nation. Men joined the local volunteer company for social reasons as well as to serve the public and to protect their towns.
- The modern fire department is proactive in the community in trying to minimize loss from fire and other disasters.
- As paid fire fighters became more common, the fire station needed to have sleeping quarters because the fire fighters were there 24 hours a day.
- In the early days of the volunteers, the fire fighters fought fire in whatever they happened to be wearing at the time. As pride in the companies grew, uniforms were used as a means of identifying the companies' members.
- To have a full appreciation of firefighting in America, one must have knowledge and understanding of the history of wildland fire.
- The early practice of burning trees and undergrowth returned nutrients to the soil, which was tied up in smaller vegetation, leaves, and downed logs.
- The completion of the transcontinental railroad came at a cost to the environment. Trees were logged for lumber to build trestles, for snow sheds over the tracks in the Sierras, for telegraph poles, to clear right of way for the tracks, for use as fuel for the steam boilers that propelled the trains, and for railroad ties. When the trees were cut, the limbs were left alongside the track bed, becoming a receptive fuel bed for the ignition of fires.
- The greatest loss-of-life wildland fire in U.S. history occurred in Peshtigo, Wisconsin, in 1871.
- Fuel reduction is necessary to keep wildland fires from producing so much heat that they destroy all vegetation and sterilize the soil.

- Firefighting equipment evolved because of the need for more firefighting capability. As cities grew and the fire load increased, new ways were needed to apply water where and when it was needed.

- A common symbol in the fire service, the Maltese cross, is considered a symbol of protection and a badge of honor.

- When horses were introduced into firehouses, Dalmatian dogs were also introduced as fire service mascots. They were kept with the horses to keep them calm.

- Throughout history, fires have resulted in tremendous loss of life and property.

- An ever-increasing number of homes have been destroyed by wildland fires each year since the 1960s; the annual average has risen from 250 homes a year to 2700.

- Due to global warming and climate change, the threat from wildland fires is going to increase greatly.

- Annually, the NFPA collects data from the NFIRS and develops a snapshot view of the national fire problem in the United States.

- The main purpose of the fire service is to manage community risk. By having trained personnel and specialized equipment on hand, emergency responders exemplify a community's response to risk.

- Modern fire departments approach community protection and risk reduction from an all-risk mindset. This includes many other incident types than just fires.

- In the area of fire prevention, the original fire fighters were limited to inspecting chimneys and making sure fires were extinguished at curfew.

- Public education for fire safety and prevention has become a fire department responsibility in many jurisdictions. This education is not limited to fire safety; problems particular to an area are addressed as well.

- As hazardous materials have become more of a concern, fire departments have taken over responsibility for hazardous materials control functions, including inspecting businesses and keeping inventories.

- Arson investigation has become an important function of the fire department. In conjunction with law enforcement personnel, fire investigators search for the cause of fires and gather evidence to prosecute criminals.

- The fire chief and the fire department staff decide what they envision the community's fire defenses should be. In many jurisdictions, civilian advisory boards are included in this process.

- With the modern fire department responding to many incident types, it is important to look at all of them when planning.

- The determining factor in evaluating risk is the probability that an undesired event might occur, with a harmful or undesirable consequence, and the severity of the harm that might result.

- When planning to protect the community from risk, the process can be divided into four steps:
  1. Preparation
  2. Mitigation
  3. Response
  4. Recovery

- With the sole exception of police functions, the fire department is responsible for most public safety emergency functions and many functions that are nonemergency. The organization, its responsibilities, and its equipment are much different than they were in the past.

KEY TERMS

Combustible construction The use of unprotected wood and wood by-products in building construction.

Conflagration Very large fire that defies control efforts and causes extensive damage over a large area.

Ground strikes When lightning bolts reach objects on the ground, often starting fires in trees when there is no rain received with the storm.

Interoperability The ability of different departments to communicate on common radio frequencies at incidents. This includes assisting fire departments and other departments, such as public works and law enforcement.

Inventory The act of accounting for all of the equipment and tools assigned to a piece of apparatus.

Objectives Steps to be taken to achieve goals.

Overhaul The operation performed to ensure that all embers are extinguished after a fire is controlled.

Policies General guidelines of how things will be done.

Risk management Any activity that involves the evaluation or comparison of risks and the development of approaches that change the probability or the consequences of a harmful action.

Unprotected vertical shafts Laundry chutes or elevator shafts. When not enclosed in fire-resistive construction, these openings act as chimneys, allowing rapid fire spread in multilevel buildings.

Urban interface The area where built-up areas of homes and businesses have little separation from the natural-growing wildland area.

Urban intermix Buildings interspersed in wildland areas; homes built in the woods.

Watersheds Complex geographic, geologic, and vegetative components that control runoff of rainwater and support varied ecosystems.

Wildland fire An unplanned fire burning in vegetation.

CASE STUDY

For as long as the American fire service has been around, the focus has been on serving the public. In the last decade the fire service has increased its focus on service to the point that the citizens it serves are now referred to as "customers." In an effort to determine the level of customer service provided and seek feedback, some jurisdictions are conducting either written or online customer surveys of people who receive their services.

1. Does the fire department need to provide excellent customer service?

    A. No, we are the only service available.
    B. No, it is more important to get ready for the next incident.
    C. Yes, that is why they pay fire fighters so much.
    D. Yes, it is the correct thing to do.

2. As a fire fighter, when there is an incident at an address you have the right to make entry when the people who live there are not home. It is appropriate to:

    A. go through their belonging.
    B. take items that you need at the fire station
    C. do as little damage as possible
    D. take a bottle of water out of their refrigerator if you are thirsty

3. Which is the best definition of customer service?

    A. We met their expectations, not just their needs.
    B. We knocked down that fire and did some damage to the property, but we needed to.
    C. They won't do that again, that's for sure. Taught those residents a lesson.
    D. That's going to cost some serious money to repair, but that's the way it is.

4. Everyone should be treated:

    A. as if they owe us for showing up and solving their problem
    B. with dignity and respect, no matter their situation
    C. I don't care how they're treated; they are often just in the way
    D. as if most fires are the customers' fault and they need to be reminded of that fact

5. If you see another member of the company destroying property unnecessarily, you should:

    A. talk to them about it
    B. let it go; maybe they have a reason for their behavior
    C. leave it alone; you are not the boss
    D. let them blow off steam; this is a stressful job

REVIEW QUESTIONS

1. Why was organized firefighting developed?

2. List several of the measures developed to prevent fires in the early United States.

3. What factors led to the demise of the volunteer fire companies in the large cities?

4. Trace the development of fire department pumping apparatus.

5. Which groups are at the highest risk of being killed in fires?

6. Where do most fatal fires occur?

7. What is the leading cause of residential fires?

8. What is the leading cause of wildland fires?

9. List the effects of global warming on the wildland fire environment.

10. What services are provided by the fire department in your area?

11. Describe the four goals of fire protection.

12. List the three control measures in risk management.

DISCUSSION QUESTIONS

1. What is the fire department's role in community risk reduction?

2. What is the one most effective thing that could be done to improve fire protection in your area?

3. What would the effects of global warming require the fire department to do to reduce risk in your community?

4. The level of fire protection provided to a municipality is a political decision. What can be done to improve the political climate to increase the fire protection level?

5. Why is preincident planning important in the fire service?

REFERENCES AND ADDITIONAL RESOURCES

Ambrose, Stephen E. 2000. *Nothing Like It in the World: The Men Who Built the Transcontinental Railroad 1863–1869.* New York, NY: Simon & Schuster.

ArkansasOnline. 2013. "Event to Commemorate Largest Urban Fire in Arkansas." *ArkansasOnline,* September 3. http://www.arkansasonline.com/news/2013/sep/03/event-commemorate-largest-urban-fire-arkansas/?f=news-arkansas

Auburn University. 2013. *Prescribed Fire History.* Auburn, AL: Auburn University. www.auburn.edu/academic/forestry_wildlife/fire/history.htm

Brown, Daniel. 2006. *Under a Flaming Sky.* New York, NY: Harper Perennial.

California Department of Forestry and Fire Protection. 2003. *California Fire Siege 2003: The Story.* Sacramento, CA: California Department of Forestry and Fire Protection. www.fire.ca.gov/downloads/2003FireStoryInternet.pdf

California Department of Forestry and Fire Protection. 2018. *Woolsey Fire.* Sacramento, CA: CALFIRE.

California Governor's Office of Emergency Services. 1993. *The Southern California Wildfire Siege October–November 1993.* Sacramento, CA: Office of Emergency Services.

Chicago History Museum. 1999. *The Great Chicago Fire.* Chicago, IL. Chicago History Museum.

City of Cincinnati Fire & EMS. 2013. *Cincinnati Fire Department: First in the Nation.* Cincinnati, OH: City of Cincinnati Fire & EMS.

City of Reno, Nevada. 2011. *Caughlin Fire Reno, November 2011.* Reno, NV: City of Reno.

Densmore, Frances. 1927. *United States National Museum.* Bulletin 136. Washington, DC: United States Government Printing Office.

DOI/USDA. 2018. *Wildland Fire Fatalities and Trend Analysis.* Washington, DC: DOI/USDA.

Egan, T. 2009. *The Big Burn.* Boston and New York: Mariner books, Houghton Mifflin Harcourt.

Evans, Bruce. 2003. "It's Time to Embrace 360° Customer Service." *Fire Chief Magazine,* July 1.

Federal Emergency Management Agency. 2014. *Hazards, Disasters and U.S. Emergency Management System: An Introduction.* Emmitsburg, MD: Federal Emergency Management Agency.

Federal Emergency Management Agency. 2018. *Are You Ready?* Jessup, MD: Federal Emergency Management Agency.

Ferris, Marc. 1996. *Baltimore's Fire History.* Fort Atkinson, WI: FIREHOUSE. http://www.firehouse.com/news/10545429/baltimores-fire-history

Fire Files Digital Library. 2013. *Presidential Proclamations—Fire Prevention Week.* http://fire.omeka.net/exhibits/show/presidential-proclamations---f/fpw

Frady, Steven R. 1991. *Red Shirts and Leather Helmets: Volunteer Fire Fighting on the Comstock Lode.* Reno: University of Nevada Press.

Gabbert, Bill. 2012. *Al Qaeda Magazine Encourages Forest Fire Arson in the US.* Hot Springs, SD: Wildfire Today. http://wildfiretoday.com/2012/05/02/al-qaeda-magazine-encourages-forest-fire-arson-in-the-us/

Global Change.gov. May, 2018. *Fourth National Climate Assessment Update: May 2018.* Global Change.Gov. www.globalchange.gov/news/fourth-national-climate-assessment-update-may-2018-0

Hasenmeier, Paul. 2008. *The History of Firefighter Personal Protective Equipment.* Tulsa, OK: Pennwell Publishing. http://www.fireengineering.com/articles/2008/06/the-history-of-firefighter-personal-protective-equipment.html

Hashagen, Paul. 2013. *SCBA History: The Development of Breathing Apparatus.* Freeport, NY. http://lishfd.org/History/scba_history.htm

History.com. 2013. *San Francisco Earthquake: April 18, 1906.* A&E Television Networks. History.com.

Holbrook, Stewart H. 1943. *Burning an Empire.* New York, NY: The Macmillan Company.

ICMA. 2012. *Managing Fire and Emergency Services*, 4th ed. Washington, DC: International City/County Management Association.

InciWeb. 2009. *Station Fire*. Boise, ID: National Interagency Fire Center.

InciWeb. 2011. *Wallow Fire*. Boise, ID: National Interagency Fire Center. http://inciweb.org/incident/2262

Inciweb. 2018. Camp Fire. Inciweb. Boise, ID: National Interagency Fire Center.

InciWeb. 2018. *Mendocino Complex Fire*. Boise, ID: National Interagency Fire Center. Insurance Information Institute. 2019. *Facts + Statistics: Wildfires*. https://www.iii.org/fact-statistic/facts-statistics-wildfires

Ingraham, Christopher. 2017. "Houston Is Experiencing Its Third '500-Year' Flood in 3 Years. How Is That Possible?" *The Washington Post,* August 29, 2017. www.washingtonpost.com/news/wonk/wp/2017/08/29/houston-is-experiencing-its-third-500-year-flood-in-3-years-how-is-that-possible/?utm_term=.5c6eec1b55c6

Insurance Information Institute. 2019. *Facts + Statistics: Wildfires*. https://www.iii.org/fact-statistic/facts-statistics-wildfires#Top%2010%20Costliest%20Wildland%20Fires%20In%20The%20United%20States%20(1)

Kentucky Educational Television. 2013. *The Origin of the Fire Engine*. Lexington, KY: Ket.org.

Local 1259 IAFF. 2013. *Dalmatians*. Texas City, TX: Local 1259 IAFF Texas City. http://www.local1259iaff.org/dalmatians.html

Lopez, Robert. 2009. "LA Firefighter Shot in 1992 Riots Makes Amazing Recovery." *Los Angeles Times,* May 4. https://www.latimes.com/archives/la-xpm-2009-may-04-me-firefighter4-story.html

Maclean, John N. 2003. *Fire and Ashes*. New York, NY: Henry Holt and Company LLC.

Maclean, John N. 2013. *The Esperanza Fire*. Berkeley, CA: COUNTERPOINT.

McPhee, Isaac. 2008. *The Steam-Powered Fire Engine*. Suite 101. https://suite101.com/a/the-steampowered-fire-engine-a43484

My Firefighter Nation. 2014. *History of the Fire Department Helmet*. https://my.firefighternation.com/profiles/blogs/history-of-the-fire-department

National Fire Protection Association. 2018. *The U.S. Fire Problem*. Quincy, MA: National Fire Protection Association.

NFPA. 2019. *Smoke Alarms*. Quincy, MA: National Fire Protection Association.

National Interagency Fire Center. 2018. *Historical Wildland Firefighter Fatality Reports*. Boise, ID: National Interagency Fire Center. http://www.nifc.gov/safety/safety_HistFatality_report.html

National Oceanic and Atmospheric Administration. September, 2018. *September 2018 and Year to Date Were 4th Hottest on Record for the Globe*. Washington, DC: U.S. Department of Commerce. www.noaa.gov/news/september-2018-and-year-to-date-were-4th-hottest-on-record-for-globe

National Oceanographic and Atmospheric Administration. 2013. *National Coastal Population Report*. Washington, DC: National Oceanographic and Atmospheric Administration. www.oceanservice.noaa.gov/facts/population.html

National Park Service, Fire and Aviation Management. 2013. *Wildfire Causes*. Washington, DC: National Park Service.

http://www.nps.gov/fire/wildland-fire/learning-center/fire-in-depth/wildfire-causes.cfm

Ocean City Fools. 2011. *History of the Leather Helmet*. https://bensenvillefpd.org/wp-content/uploads/2016/08/HISTORY-OF-THE-LEATHER-HELMET.pdf

Oracle Think Quest. 2003. *Chicago's History*. http://library.thinkquest.org/CR0215480/fire.htm

Peabody Fire Department. 1996. *Origin of the Maltese Cross*. Peabody, MA: Peabody Fire Department. https://www.datasync.com/nexus/fire/cross.htm

Purdue University. 2007. "Global Warming Likely to Increase Stormy Weather, Especially in Certain U.S. Locations." *Science Daily*, December 5. http://www.sciencedaily.com/releases/2007/12/071204121949.htm

Pyne, Stephen J. 2001. *The Big Blowup*. Paonia, CO: High Country News. April 23. http://www.hcn.org/issues/201/10428

Resnek, Josh. April 26, 2012. *The Great Chelsea Fire of 1908*. Chelsea, MA: Chelsea Record. http://www.chelsearecord.com/2012/04/26/the-great-chelsea-fire-of-1908/

Resnek, Josh. October 18, 2012. *Chelsea Fire of 1973: 39 Years Ago This Week*. Chelsea, MA: Chelsea Record. http://www.chelsearecord.com/2012/10/18/chelsea-fire-of-1973-39-years-ago-this-week/

Rivard, Ry. 2012. "Agency Says West Virginians Twice as Likely to Die in Fires." *Daily Mail*, March 26. http://www.dailymail.com/News/201203260224

Rockwell, Theodore. 1992. *The Rickover Effect*. Annapolis, MD: Naval Institute Press.

Routley, Gordon. 1991. *The East Bay Hills Fire*. Washington, DC: Federal Emergency Management Agency. http://www.usfa.fema.gov/downloads/pdf/publications/tr-060.pdf

Sensagent. 2012. *Great Salem Fire of 1914*. Sensagent Corporation. http://dictionary.sensagent.com/Great_Salem_Fire_of_1914/en-en/

Texas A&M Forest Service. 2011. *Texas Fires 2011*. College Station, TX: Texas A&M. www.txforestservice.tamu.edu/main/article.aspx?id=12888

Thiel, Adam, and Charles Jennings, eds. 2012. *Managing Fire and Emergency Services*. Washington, DC: International City/County Management Association.

United States Department of Agriculture Forest Service. 1977. *Fire Weather*. Washington, DC: U.S. Government Printing Office.

United States Fire Administration. 2018. *Risk Management Practices in the Fire and Rescue Service*. Washington, DC: Federal Emergency Management Agency.

United States Geological Survey. 2018. *Wildland Fire Decision Support System*. Washington, DC: US Geological Survey. http://wfdss.usgs.gov/wfdss/WFDSS_About.shtml

UPI.com. 2012. "52,000 Acres Burn in Oklahoma Wildfires." August 5. http://www.upi.com/Top_News/US/2012/08/05/52000-acres-burn-in-Oklahoma-wildfires/UPI-41191344170202/

Vintage Fire Museum. 2013. *Fire Engines*. New Albany, NY: Vintage Fire Museum. https://www.vintagefiremuseum.org/?page_id=10

Wisconsin Historical Society. 2013. *Peshtigo Fire*. Madison, WI: Wisconsin Historical Society. https://www.wisconsinhistory.org/Records/Article/CS1750

CHAPTER 4

Chemistry and Physics of Fire

OBJECTIVES

After studying this chapter, you should be able to:

- Discuss the difference between the fire triangle and the fire tetrahedron.
- Discuss the chemistry of fire in terms of oxidizers and fuels.
- Discuss the physics of fire in terms of the three states of matter.
- Differentiate heat and temperature.
- Identify and describe the four methods of heat transfer.
- Discuss the five classifications of fire.
- Identify and describe the four stages of fire.

Case Study

On the morning of November 21, 1980, 85 people died and more than 700 were injured as a result of a fire at the MGM Grand Hotel in Las Vegas, Nevada. This was the second largest life-loss hotel fire in the United States' history.

The fire spread rapidly throughout the rooms of the deli and surrounding areas, feeding on ornamental decorations on the walls and ceiling. The ceiling was covered with cellulose fiber acoustical tile affixed to gypsum wallboard with glue. The application of the glue left a small void space between the tile and the ceiling of the structure. The glue burned with thick black smoke. On the walls of the rooms were a variety of materials including plastics, wood veneer, and wallpaper. The floors and furniture were covered with volatile materials including plastics, polyurethane materials, and polyvinyl plastic. All of these materials released large volumes of thick, dense smoke. Also burned were areas of the casino with gaming tables, covered in felt and plastic, upholstered chairs, and floor coverings.

1. Which methods of heat transfer were evident in the spread of this fire?
2. Which manner of heat transfer was the most prominent in the spread of this fire?
3. What part did the furnishings and wall coverings play in terms of this fire?
4. Why was it a contributing factor to fire spread that there was a small air space between the ceiling tiles and the gypsum wallboard of the ceiling?

Modified from sources: Akasaka, 2010 and Clark County Fire Department, 2011

 *Access Navigate for more resources.*

Introduction

To combat hostile fires effectively, a fire fighter must have an understanding of the chemistry and physics of fire. Determining the correct extinguishing agent for a specific type of fuel involves many factors. The method of application for that agent, which may be in limited supply, is also an important consideration.

Selecting the wrong extinguishing agent could have disastrous results; for example, applying water to a burning magnesium fire will result in an explosion. Depending on the amount of burning magnesium and the amount of water, the explosion could be large enough to injure fire fighters and spread the fire. As another example, a burning gasoline spill can spread with the application of water because gasoline will float on water and continue burning. It is critical to choose the right extinguishing agent and application method for each fuel to avoid costly consequences.

Knowledge of the physics of fire is required to contain and control its spread. A lack of knowledge of how fire spreads through a structure can result in more damage, injuries, and deaths. In a wildland fire situation, the failure to take into account how fire spreads can result in wasted efforts and the entrapment of fire fighters, resulting in injuries and death. A comprehensive knowledge of both of these factors requires advanced study and experience.

This chapter provides an analysis of the basic components of fire as a chemical chain reaction, the major phases of fire, and the main factors that influence fire spread and fire behavior. These factors will be illustrated in both the structural and wildland fire environments.

Fire Defined

Fire is a rapid, self-sustaining oxidation process accompanied by the evolution of heat and light in varying intensities. Combustion is described as a chemical reaction that releases energy as heat and, usually, light. The process of combustion is considered an exothermic reaction because it releases heat. Conversely, an endothermic reaction is a reaction that absorbs heat. When a substance is undergoing combustion, we usually refer to it as being "on fire." Another very common oxidation reaction is the rusting of iron. This is not considered to be combustion because it involves insignificant amounts of heat, without light, and proceeds at a slow rate.

Fire Triangle

Fire was originally believed to be based on three elements being present: fuel, air, and heat. Fires could be controlled or prevented by removing one of these three elements. To be more accurate, in the discussion of fires and combustion from a chemical and physical standpoint, the air leg of the triangle will be replaced with oxidizer, and the heat leg will be replaced with energy; the fuel leg remains unchanged **FIGURE 4-1**.

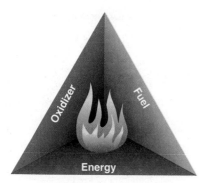

FIGURE 4-1 The fire triangle consists of fuel, an oxidizer, and energy.
© Jones & Bartlett Learning

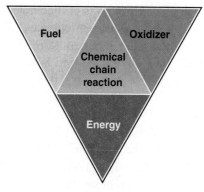

FIGURE 4-2 Fire tetrahedron.
© Jones & Bartlett Learning

Fire Tetrahedron

Scientists have determined that a fourth component, a chemical chain reaction, is present in the combustion process **FIGURE 4-2**. The chain reaction occurs when the fuel is broken down by heat. When paper burns, the molecules in the paper (fuel) are broken down by heat, producing chemically reactive species called free radicals, which then combine with the oxidizer. This recombination process releases more heat and chemical reactants, causing the fuel to break down further and the process to continue. As long as there are fuel, oxidizer, and energy, in the appropriate amounts, and the chemical chain reaction is not interrupted, the combustion process will continue.

Chemistry of Fire

There are two basic components to the chemistry of fire: oxidizer and fuel. Without either of them, the fire cannot occur. (The third component of the triangle, energy, falls into the area of physics.)

Oxidizer

Oxidizers are substances that evolve or generate oxygen, either at ambient temperatures or when exposed to heat. Oxygen is the most commonly occurring oxidizer. It is a naturally occurring element in air. Oxygen in its natural state is a gas. Air is a mixture of approximately 21 percent oxygen and 78 percent nitrogen, with the remaining 1 percent other elements. Air is a mixture, not a compound, which leaves the oxygen readily available to act as the oxidizer in the combustion process. Most fires we see are burning in air at 21 percent oxygen. In an oxygen-enriched atmosphere (over 21 percent), the fire intensifies. These atmospheres can occur when certain chemical compounds are mixed with, or are proximal to, the fuel.

An example of a substance that can release oxygen when heated is a common swimming pool chlorinating compound, calcium hypochlorite. When energy is added to this compound, usually from the heat of a fire, the oxygen it contains is released and can enrich the atmosphere. Another oxidizer is ammonium nitrate, which is used in fertilizers and as an oxidizer in blasting agents. Conversely, in an oxygen-deficient atmosphere, the rate of combustion will decrease and the fire will possibly go out on its own.

Other elements can take the place of oxygen in the combustion reaction. Two of these, fluorine and chlorine, are found on the periodic table of the elements, listed under the halogen family. Both naturally occur as gases, as does oxygen. Fluorine is a much stronger oxidizer than oxygen, and a fire in a fluorine atmosphere would burn more rapidly than one occurring in air.

Fuel

A fuel is described as anything that will burn. Carbon and hydrogen are the two most common elements in fuels. These fuels are referred to as hydrocarbons. Carbon in its natural state is a solid. It is the main element in organic fuels; that is, fuels that at one time were living things, such as petroleum products (gasoline, natural gas, plastics), wood, paper, cotton, and other natural fibers. Hydrogen is a flammable gas. Other elements and compounds can act as fuels as well, including metals that will combust, such as sodium, aluminum, and magnesium.

The most common fire involves a fuel composed of mostly carbon and hydrogen in an atmosphere with oxygen present. The oxygen combines with hydrogen to produce water vapor and with the carbon to form carbon dioxide (CO_2). These are the two by-products of complete combustion. Complete combustion rarely occurs because of outside factors. These factors may include, but are not limited to, fuel size, arrangement, contaminants, and the availability of the oxidizer.

This gives us the by-products of incomplete combustion, such as smoke, carbon monoxide (CO), carbon dioxide, and other fire gases. A term that has only recently been officially recognized is black fire. This describes a situation where heavy, dense, black smoke is being emitted by a fire. This smoke will be of high velocity, high volume, turbulent, and extremely dense. It will also be extremely hot. For all practical purposes this is a dense, superheated cloud of fuel that is too rich to ignite. This smoke may be doing as much damage as fire. It can also be a sign of imminent flashover.

> **Note**
>
> Carbon and hydrogen are the two most common elements in fuels.

Physics of Fire

Fuel

Fuel may exist in any one of the three states of matter—solid, liquid, and gas **FIGURE 4-3**.

All molecules vibrate at temperatures above absolute zero. The molecules in liquids vibrate faster than those in solids, and gas molecules vibrate the fastest of the three. The state of the fuel is temperature dependent. As we increase the amount of energy in the molecule by applying heat, and raising the temperature, the molecules vibrate faster and faster. As this vibration increases, solids become liquids and liquids become gases.

Combustion usually occurs when the fuel has been converted to the vapor or gaseous state because the oxidizer occurs as a gas, and it takes both oxidizer and fuel in the gaseous state for the recombination to occur. Solid and liquid fuels are converted to the gaseous state by the application of energy. This can be observed by carefully watching a candle burn. The flame seems to float a small distance from the wick. This process is called pyrolysis **FIGURE 4-4**. Pyrolysis is the chemical decomposition of matter through

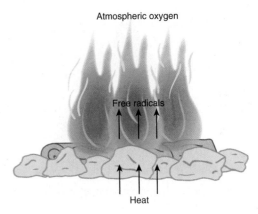

FIGURE 4-4 Pyrolysis is the release of free radicals as a result of input heat.
© Jones & Bartlett Learning

the action of heat. Some fuels, such as pure carbon or the combustible metals, do not have to convert to the vapor state to burn. They do not have to decompose or vaporize; the union with oxygen takes place directly. This may take place slowly, as in rusting with iron or rapidly, such as the glowing combustion of magnesium. When the fuel is brought to the temperature at which it continues combustion without any external input of heat and becomes self-sustaining, it has reached its ignition temperature.

As heat is added to a substance, the molecules are broken down into shorter and shorter chains. The longer molecules break into shorter molecules, and the shorter ones break into what are called free radicals. The free radicals are the by-products of the fuel that combine directly with the oxidizer, leading to combustion.

For example, let us look at the process in a piece of wood. Wood is composed of long chains of molecules that contain hydrogen and carbon. As heat is applied, the molecules receiving the heat start to break down into shorter and shorter molecules. The end result is the shortest hydrocarbon molecule: methane, a flammable gas. Several other short-chain hydrocarbon molecules resulting from this process will burn as well, including ethane and propane. When these molecules are further broken down by heat into free radicals, they combine with the oxidizer, and combustion takes place. Sufficient amounts of oxidizer (usually oxygen) and an ignition source of sufficient temperature must be present. In the absence of either of these, pyrolysis can produce flammable gases that are only waiting for the missing component—sufficient oxygen or another oxidizer. When the oxidizer is added, combustion begins.

Solid Fuels

Several factors affect the rate at which solid fuels are pyrolyzed—mass, arrangement, continuity, and moisture content.

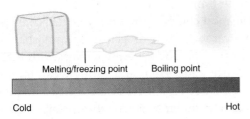

FIGURE 4-3 States of matter, in many cases, are temperature dependent.
© Jones & Bartlett Learning

FIGURE 4-5 Relative fuel sizes, showing fire spreading through grass, brush, and trees.
Courtesy of Kern County Fire Department

Mass affects the fuel, in that the smaller or more finely divided the fuel, such as dust or chips, the less heat is required to get it to pyrolyze because the fuel has a greater surface area in relation to its mass **FIGURE 4-5**. Ignition capability is based on the mass and surface area being ignited. A wooden 2 inch × 4 inch has greater mass vs. surface area than sawdust, which has less mass vs. surface area per particle. So, when lighting a campfire, paper and small twigs are easier to ignite than larger logs.

Arrangement is based on the spacing of the fuel particles, regardless of their size. A house in the framing stage of construction will ignite more easily and burn faster than a stack of lumber, even though both are basically the same fuel **FIGURE 4-6**. When gasoline is ignited in a container, it burns rapidly on the surface; when atomized into a fine spray, it burns explosively.

Continuity is the grouping of fuel over a prescribed area. This may be vertical, horizontal, or both. An example of vertical arrangement is dry grass spreading fire upward into brush and then into trees. An example of breaking the horizontal continuity is the same fire's spread being stopped by a road or fire break **FIGURE 4-7**.

Moisture content is the amount of moisture contained in a fuel. The moisture content of a fuel affects its ease of ignition. On a humid or rainy day, forest fires are not much of a problem. The fuels involved

Tip

Fuel moisture in brush and other organic fuels is expressed as a percentage of weight. The fuel is weighed and then dried to remove the moisture and weighed again. The difference in weight is recorded and expressed as a percentage. These percentages apply to both live and dead fuels.

FIGURE 4-6 A. Structure in framing stage of construction, illustrating loose fuel arrangement, which is easy to ignite and burns rapidly. B. Stacked lumber, illustrating tight fuel arrangement, which is difficult to ignite.
© Jones & Bartlett Learning

FIGURE 4-7 Fire fighters widening a fire break to break fuel continuity.
Courtesy of Kern County Fire Department

can absorb moisture directly from the air and will have too high moisture content to ignite easily or burn readily. For pyrolysis of the fuel to take place and burning to begin, the moisture must be turned to vapor and removed. This process absorbs large amounts of energy. If a large enough fire with sufficient energy is started, it can overcome the high moisture content of the fuel and burn rapidly.

Flame spread. The rate of surface-burning characteristic of a material is calculated based on test results, which indicates the relative rate at which flame will spread over the surface of the material as compared with flame spread on asbestos-cement board, which is rated 0, and on red oak, which is rated 100. Note that this rating is not the rate at which the flame actually spreads along the surface and is not at all an indication of the fire resistance of the material.

The Steiner Tunnel (ASTM E-84) is frequently referenced as a method to assess flame spread and smoke density and is a mandated test for many commercial building materials. The three attributes measured are flame spread, temperature, and smoke density. The test consists of a 25-foot vented tunnel, lined with firebrick, with the test material mounted to the top of the chamber. At one end of the chamber, the sample is subjected to a high-energy flame for 10 minutes. A fan draws the flame across the surface of the material being tested. Flame spread is determined visually through windows built into the tunnel. An optical cell mounted at the tunnel exhaust measures smoke density. Provision may also be made for the measurement and analysis of combustion gases.

Many fires have spread very rapidly across interior finishes, leading to loss of life and property. Although not the only factor, flame spread was a contributing factor in these fires. Smoke is a factor in life safety because of its toxic characteristics and its ability to obscure vision.

Liquid Fuels

Molecules of liquids move freely and flow like water but do not readily separate. Liquids will assume the shape of their container. Liquid fuels have physical properties that make them difficult to extinguish, and they increase the hazards to persons and property.

When a liquid fuel is spilled, it flows across the ground, increasing the size of the spill. It also flows downhill, pooling in low-lying areas. Several other properties of liquids are important to fire fighters. They include the following:

- *Miscibility.* Miscibility is the ability of a substance to mix with water. Most flammable liquids do not mix with water (nonmixable), making them more difficult to extinguish. Because they cannot be diluted, they float on the surface of the water and continue burning. Adding water to a floating fuel fire just spreads the fire. Fuels that mix with water can be diluted to the point that they will no longer burn. An example of a mixable fuel is alcohol.

- *Specific gravity (liquid).* Specific gravity is the weight of a liquid as compared to the weight of an equal volume of water. Water is assigned a value of 1.0. Nonmixable liquids with a specific gravity less than 1.0 will float on water. Nonmixable liquids with a specific gravity higher than 1.0 will sink in water **FIGURE 4-8**. Most flammable liquids, such as gasoline, have a specific gravity less than 1.0 and will float, and burn, on water.

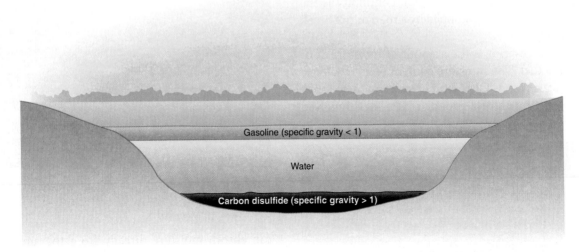

Gasoline (specific gravity < 1)

Water

Carbon disulfide (specific gravity > 1)

FIGURE 4-8 Specific gravity. Liquids with a specific gravity greater than 1.0 will sink in water, and those with a specific gravity less than 1.0 will float on water.

- *Volatility*. Volatility is the ease with which a fuel gives off vapors at ambient temperature. This property has a definite effect on the flammability, ease of extinguishment, and fire prevention considerations of the fuel.

- *Vapor pressure*. All liquids have a vapor pressure, which is the pressure exerted by vapor molecules on the sides of a container, at equilibrium. Vapor pressure is caused by the molecules of the liquid escaping the surface through evaporation. When the number of molecules leaving the surface of the liquid is the same as the number returning, the liquid is at equilibrium. Vapor pressure is temperature (energy) dependent. As the temperature of the liquid rises, the molecules have more energy to overcome atmospheric pressure, and the vapor pressure of the liquid increases **FIGURE 4-9**. This can easily be observed when steam occurs over heated water. The input heat excites the water molecules and increases the vapor pressure, which enables the water molecules to overcome the atmospheric pressure and escape the surface.

- *Boiling point*. When the vapor pressure equals atmospheric pressure at the surface of the liquid, the boiling point is reached. At this point, the molecules of the liquid have enough energy to actively leave the surface of the liquid. Liquids at their boiling point evolve vapors at a much more rapid pace than through evaporation.

This can easily be observed with a covered pan of water on the stove. As energy is added in the form of heat, the water begins to boil and the vapor pressure increases (steam). If the pan has a lid, the steam will soon develop enough vapor pressure to lift the lid off the pan. If the pan of water were left unheated sitting on the counter, without a lid, the water would eventually evaporate, but at a much slower rate.

- *Vapor density*. The relative density of a vapor or gas as compared to air is the vapor density. Air is assigned a value of 1.0. This is important in that a vapor with a higher density than air will sink and lie along the ground, seeking low spots in which to pool. A vapor with a lower density than air will tend to rise and dissipate **FIGURE 4-10**. Gasoline vapors, which are very heavy, will flow downhill and pool, like a liquid. Liquefied petroleum gas vapors act in much the same way. Natural gas vapors will rise and dissipate in the slightest breeze. The exceptions to this occur when the ambient temperature is very high or very low. If it is cold enough, a vapor that is lighter than air at normal temperatures may act as if it is heavier than air. On a hot day, especially when spilled on hot asphalt, heavy vapors can be heated to the point that they will rise. Relative humidity can also have an effect on how a vapor reacts. This effect occurs because humid air contains more moisture than dry air and is therefore heavier.

- *Flash point*. The flash point is the minimum temperature of a liquid at which it gives off vapors sufficient to form an ignitable mixture

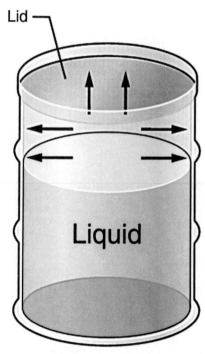

FIGURE 4-9 Vapor pressure.
© Jones & Bartlett Learning

Low vapor density

High vapor density

FIGURE 4-10 Vapor density. Vapors heavier than air will sink, and those lighter than air will rise.
© Jones & Bartlett Learning

with air, but ignition is not sustained (in other words, it will flash and then quit burning). Above the surface of a liquid, because of evaporation or boiling, there is a vapor space. If an ignition source is introduced into this vapor space and ignition occurs, the liquid is above its flash point. When liquids have a flash point below 100°F, they are called *flammable* liquids **TABLE 4-1**. When the flash point is at or above 100°F, they are called *combustible* liquids.

- *Fire point.* The lowest temperature at which a volatile combustible substance continues to burn in air after its vapors have been ignited. This temperature is higher than the flash point.

TABLE 4-1 Selected Physical Properties of Some Flammable and Combustible Substances

Name	Boiling Point (°F)	Flash Point (°F)	Flammability Range (percent) in Air	Specific Gravity	Vapor Density (RGasD)*
Acetone	133	0	2.5–12.8	0.79	
Acetylene[a]			2.5–100		0.91
Ammonia	−28		15–28		0.60
Benzene	176	12	1.2–7.8	0.88	
n-Butane	31		1.6–8.4	0.6 (@ 31°)	2.11
Carbon monoxide	−313		12.5–74		0.97
Ethyl alcohol	173	55	3.3–19	0.79	
Ethyl ether	94	−49	1.9–36	0.71	
Ethylene oxide[a]	51	−20	3.0–100	0.82 (liquid @ 50°)	1.49 gas
Formaldehyde	−6		7.0–73		1.04
Gasoline (average)	102	−45	1.4–7.6	0.72–0.76 (@ 60°)	
Hydrogen sulfide	−77		4.0–44		1.19
Isopropyl alcohol	181	53	2.0–12.7	0.79	
Isopropyl ether	154	−18	1.4–7.9	0.73	
Kerosene	347–617	100–162	0.7–5.0	0.81	
Methyl alcohol	147	52	6.0–36	0.79	
Methyl bromide	38		10–16	1.73 (liquid @ 32°)	3.36
Toluene	232	40	1.1–7.1	0.87	
Turpentine	309–338	95	0.8–?	0.86	

* RGasD is the relative density of gas referenced to air.

[a] Notice that some of the substances have flammable ranges that go as high as a 100 percent concentration. This makes them extremely dangerous, even when in their containers. All of them can be classified as fuels.

Data from National Institutes of Safety and Health. 2007. *NIOSH Pocket Guide to Chemical Hazards.* Washington, DC: Department of Health and Human Services, pp. 3–325

Gas/Vapor Fuels

Gas is the state of matter defined as a fluid that has neither independent shape nor volume but tends to expand indefinitely. Gases always fill the container in which they are stored. Materials that are normally gases, when contained as liquids, are at or above their boiling points at ambient temperature and convert to gas upon their release. For example, when butane and propane are released from their containers, they revert to the gaseous state.

Classification of Gases. Gases can be categorized as flammable and nonflammable. There are gases that are not flammable that support combustion—the best example of which is oxygen. Oxygen itself does not burn, but if the oxygen concentration is increased, most fires will burn more rapidly.

There are several definitions important to understand when referring to flammable gases. These definitions describe the gases' characteristics.

- *Upper flammable limit.* The maximum concentration of gas or vapor in air above which it is not possible to ignite the vapors is its upper flammable limit. A vapor that is above its upper flammable limit is said to be "too rich" to burn.

- *Lower flammable limit.* The minimum concentration of gas or vapor in air below which it is not possible to ignite the vapors is its lower flammable limit. A vapor that is below its lower flammable limit is said to be "too lean" to burn.

- *Flammable range.* The proportion of gas or vapor in air between the upper and lower flammable limits is its flammable range **FIGURE 4-11**. This proportion of vapor in air is measured as a percent. At the liquid surface, the vapor will be at 100 percent concentration. The further the limits are from the surface, the more the concentration will decrease until it reaches 0 percent.

A word of caution about gaseous fuels: When a compressed gas, such as butane, is released, the vapor cloud you see indicates that the gas is colder than the air temperature, condensing the moisture in the air. It appears much like fog. This is not the extent of the

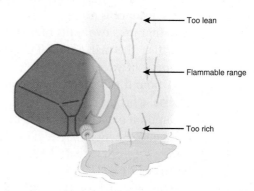

FIGURE 4-11 Flammable range. Too rich to burn at liquid's surface, within the flammable range, and too lean to burn at increased distance from liquid's surface.
© Jones & Bartlett Learning

vapor cloud. Once the vapor warms up, it disappears from view but may still be lingering in the area. It is possible to stand in a vapor cloud with a concentration that is within its flammable range. If the cloud were to ignite, the person would be burned severely, if not killed. If a vehicle with a diesel engine is driven through a vapor cloud, it can act just like stepping on the accelerator. Letting off the throttle would have no effect. The governor would not limit engine speed, because it limits fuel flow, not air flow. The engine would rev to the point that it would probably come apart. There would more than likely be ignition of the vapor cloud from the heat of the turbocharger or pieces of hot carbon being blown out of the exhaust.

Heat and Temperature

Heat is a form of energy. There are several possible sources of heat: chemical (the breaking down and recombination of molecules), mechanical (friction, friction sparks, compression), electrical (arc or spark, static electricity, lightning), and nuclear energy (fission and fusion). When heat energy is absorbed by a substance, its temperature rises.

In the fire service, heat energy's ability to change the temperature of a substance is expressed in several ways. The English system uses two units of measurement. One is the British thermal unit (BTU). This is defined as the amount of energy required to raise the temperature of 1 pound of water 1°F. The other expression is the calorie, which is from the metric system. A calorie is the amount of heat required to raise 1 gram of water 1°C. The International System of Units system uses the joule. A calorie equals 4.187 joules, and there are 1054 joules in 1 BTU.

Temperature is the measure of the hotness or coldness of an object. A more scientific description is the measurement of the average speed of vibration of the molecules in a substance. Temperature is measured on four scales: Fahrenheit, Celsius, Kelvin, and Rankine **FIGURE 4-12**. On the Fahrenheit scale, water freezes at 32°F and boils at 212°F. On the Celsius scale, water freezes at 0°C and boils at 100°C. On the Kelvin scale, water freezes at 273K and boils at 373K. (Note: The Kelvin scale does not use the symbol ° to represent degrees.) On the Rankine scale, water freezes at 492°R and boils at 672°R. On the Kelvin and Rankine scales, there are no negative numbers because 0K or 0°R represent the lowest attainable temperature.

Scientific theory and experiments show that there is a limit below which it is impossible to cool matter. For the purpose of this book, temperatures will be given in degrees Fahrenheit (°F) and degrees Celsius (°C).

Heat and temperature are not to be confused. Heat is a measurement of energy, and temperature is a measurement of how much energy a material retains. To experience this point, go outside on a sunny day. The roof of a dark-colored car will be much warmer than the roof of a light-colored car. Both are subjected to the same amount of heat energy radiated by the sun. The dark roof absorbs and retains more of that energy and feels warmer and, consequently, higher in temperature.

Heat Transfer

For pyrolysis to take place there must be a way for the heat (energy) to be transferred from the heat source to the fuel. There are three methods of heat transfer **FIGURE 4-13**:

1. *Conduction*. This is the transfer of heat through a solid medium without visible motion. This can be observed by placing a pan on a hot stove. The pan soon becomes heated (a rise in temperature), but there is no visible movement of the pan. The denser a material is, the better conductor it makes. Metals are good conductors of heat or electrical energy. This is why gold is used for electrical contacts in computers and spacecraft. Wood is a poor conductor of energy. Air is a very poor conductor of energy.

2. *Convection*. This is the transfer of heat through a circulating medium, such as liquids and gases. When a fire occurs, the smoke and heated gases rise because of convection **FIGURE 4-14**. As a gas is heated, it expands and becomes less dense, making it lighter than the surrounding gases, and it rises. The hotter the gases, the faster they rise. This is evident in structure fires. As they burn with more intensity, the smoke rises at a more rapid rate. As the gas cools, it returns to its former elevation, completing the circulation.

3. *Radiation*. This is the transfer of heat by wavelengths of energy. Heat in the form of rays is the infrared portion of the electromagnetic spectrum. As these wavelengths of energy

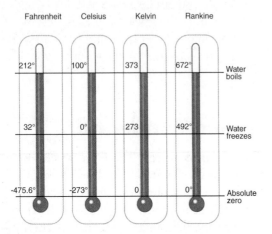

FIGURE 4-12 Relationship of temperature measurement scales.
© Jones & Bartlett Learning

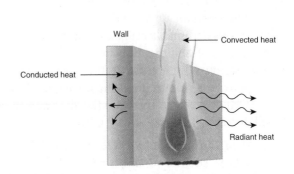

FIGURE 4-13 Transfer of heat through conduction, convection, and radiation.
© Jones & Bartlett Learning

FIGURE 4-14 Convection column developing over wildland fire as a result of rising heated air produced by the fire.
© Jones & Bartlett Learning

FIGURE 4-15 Transfer of heat through radiation, causing fire to spread.
Courtesy of Kern County Fire Department

FIGURE 4-16 Transfer of heat through direct flame impingement. Fire is lapping onto porch structure from openings.
Courtesy of Kern County Fire Department

travel through space, they eventually come into contact with an object. As the object absorbs the energy, it becomes heated, resulting in a rise in temperature. When you walk outside on a sunny day, the air temperature is warm, but objects in the sun are much warmer because they have absorbed the wavelengths of energy generated by the sun. The darker the object, the more of the wavelengths it can absorb, and it will become warmer. One thing that must be kept in mind when dealing with radiated heat is that it is not affected by wind. Radiated heat can set fire to objects upwind of the fire, whereas convection-caused fires are much more likely to occur on the downwind side **FIGURE 4-15**.

Some fire service professionals describe a fourth method of heat transfer, known as direct contact, direct flame impingement, or autoexposure, which is the transfer of heat through direct flame contact. This method of heat transfer can be observed in high-rise building fires. As the windows on the lower floors are broken out, the flames lap onto the windows of the upper floors, breaking them also. The flames then come into direct contact with materials on the upper floor, spreading the fire. This fourth method of heat transfer combines elements of convection and radiation.

The convection current makes the flames rise, and when materials are bathed in flame, heat is radiated onto them. This method of heat transfer is not accepted by everyone in the scientific community but illustrates fire spread in many instances **FIGURE 4-16**.

The example of a structure fire can illustrate the four methods of heat transfer. A fire starts in a stack of boxes in a warehouse, and the heat and smoke rise because of convection. When enough convected heat builds up, the underside of the roof or ceiling catches fire. The boxes at the bottom of the stack were the first to ignite and the fire spreads upward through the stack as a result of direct flame impingement. After a while, radiated heat spreads the fire to piled stock several feet away. As the fire gains headway, a pipe running through the wall is heated to the point that stock piled around it on the other side of the wall ignites. Rarely

Ordinary combustibles

Flammable liquids

Electrical equipment

Combustible metals

Combustible cooking

FIGURE 4-17 Classification of fire symbols.
© Jones & Bartlett Learning

SAFETY TIP

Due to extreme temperatures, fire fighters cannot survive in a room with a flashover.

does a fire spread because of only one method of heat transfer; usually several methods are involved.

Classification of Fires

Fires have been divided into five basic classifications, based on the type of fuel involved **FIGURE 4-17**. In some instances, more than one type of fuel may be involved in a single incident. The classifications are as follows:

1. *Class A*. Ordinary combustibles. This class includes materials composed of primarily natural fibers, such as wood, paper, and cotton.
2. *Class B*. Flammable liquids and combustible liquids. These include materials that can flow while burning, such as gasoline, kerosene, alcohol, and cooking oil. This definition is often shortened to "flammable liquids" and should not be confused with the rating of flammable versus combustible, based on flash point. Most plastics fall into this category as they are made from petroleum products and can melt and flow as they burn
3. *Class C*. Energized electrical. The key word here is *energized*. If the power is turned off, the fire is reclassified into what other materials are

burning. One must be cautious because some appliances, such as older television sets and other electrical equipment, may retain stored energy, even years after their last use.
4. *Class D*. Combustible metals. These include magnesium, titanium, zirconium, sodium, and potassium. Most metals burn with a brilliant white flame that can damage the eyes.
5. *Class K*. Cooking materials (cooking oils/fats/grease). This is a newer classification of fire. Class K extinguishers are specifically designed to supplement fire suppression systems in commercial kitchens. These extinguishers are for cooking oil, fat, and grease fires.

It is not uncommon for more than one class of fire to be involved in an incident. In a typical automobile fire, there are Class A fuels (vehicle contents), Class B fuels (gasoline and plastics), and some vehicles have Class D components (usually magnesium and/or aluminum). If the fire has resulted from the vehicle crashing into a structure or power pole, or in the case of a hybrid or electric vehicle, there may be Class C components also involved in the fire.

SAFETY TIP

When oxygen is introduced to the room, the fire gases present and all of the fuels above their ignition temperature can burn with explosive force.

Stages of Fire

There is some disagreement in fire-related literature as to exactly how many stages of fire there are; some say four and others say five. For the purposes of this text, fire is descried as occurring in four stages. In some older texts these may be descried as "phases of fire."

Incipient Stage

In the incipient stage, the fire has ignited and reaches a point where it no longer needs input heat from outside sources to continue burning. In this stage, oxygen in the surrounding air is approximately 21 percent, heat in larger quantities is starting to be produced, and the combustion reaction begins to accelerate.

Growth Stage

In the growth stage the fire increases in intensity dependent on the elements of the fire tetrahedron being available in sufficient quantities. Fire is spread from the materials first ignited to surrounding materials.

Free-Burning Stage

In the free-burning stage, the fire releases increasing heat, bringing more fuel to its ignition temperature. As the additional fuels start to burn, the fire spreads and gains in intensity. Heat is transferred onto nearby surfaces though radiation. Convected heat rises and preheats fuels above the fire, bringing them to their ignition temperature. In this manner, the fire begins to spread more and more rapidly.

If a fire is burning in a confined area, the ceiling temperature increases rapidly as heated air rises. Ceiling temperatures can easily reach 1000°F. As this process continues, insufficient amounts of oxygen are available, causing incomplete combustion. Residual fire gases start to accumulate at the ceiling level. This accumulation of fire gases from time to time will ignite if sufficient oxygen is present, causing a **flameover** or **rollover** situation to occur (Figure 1-4). As soon as these gases are consumed, the flame is no longer seen and the process will repeat itself as temperatures in the enclosed areas continue to increase. The presence of this condition lets fire fighters know that they are approaching the seat of the fire and need to start applying water at the

FIGURE 4-18 Fire in free-burning stage.
Courtesy of Kern County Fire Department

ceiling level to reduce temperature and prevent a flashover. If they fail to cool the environment, conduction, radiation, and direct flame contact will cause other combustibles in the room to pyrolyze **FIGURE 4-18**.

Flashover

Flashover is not a stage of fire, but is very important to interior firefighting, so is mentioned here. When the contents in the room [which may include Personal Protective Equipment (PPE)] are brought to their ignition temperature and sufficient oxygen is present, flashover can occur. If a flashover were to take place, temperatures, even at floor level, would rise dramatically. Fire fighters cannot survive in a room with a flashover, even wearing full PPE and SCBA. The only chance they have for survival is to be very close to an exit that they can access quickly. Escape up and over a windowsill is unlikely, as this will place fire fighters into an even higher temperature zone. Should they escape, their SCBA will most likely have failed and they will sustain serious burn injuries. Without functional SCBA, fire fighters are fully exposed to fire gases, smoke particles, elevated temperatures, and low oxygen concentrations and can only survive in a room with temperatures of 300°F of dry heat for only a short period of time.

Smoldering/Decay Stage

The decay stage occurs when the fire has run out of available fuel or oxygen. In an automobile or structure fire, for example, the fuel can be consumed to

FIGURE 4-19 Fire in decay stage.
Courtesy of Kern County Fire Department

the point where the fire runs out of things to burn (fuel), causing it to go out on its own. The fire can also go into the decay stage when suppression action has reduced the fire to smoldering embers **FIGURE 4-19**.

In a sealed environment, the fire may run out of sufficient oxygen to sustain combustion. If the room involved is well sealed from outside air, the oxygen in the room is consumed. As the oxygen content drops below 15 percent, combustion is slowed and the fire enters the fourth phase. In this example of the fourth phase, flame may die out, and glowing combustion takes place. Pyrolysis continues to occur with amounts of combustible gases produced. The room is superheated and charged with smoke and combustible fire gases. The gases and room contents are above their ignition temperatures, and the only item lacking is sufficient oxygen. The fire gases and smoke are alternately forced out and sucked back into the structure. The windows become blackened by smoke stain baked on by the heat of the fire. When oxygen is introduced to the room, the fire gases present and all of the fuels above their ignition temperature can burn with explosive force. This event is referred to as a **backdraft** or smoke explosion.

Wrap-Up

CHAPTER SUMMARY

- To combat hostile fires effectively, a fire fighter must have an understanding of the chemistry and physics of fire.
- Combustion is a chemical reaction that releases energy as heat and usually as light. When a substance is undergoing combustion, we usually refer to it as being "on fire."
- The fire triangle consists of fuel, oxidizer, and energy.
- The fire tetrahedron adds a chemical chain reaction to the fire triangle.
- Oxidizers are substances that evolve or generate oxygen, either at ambient temperatures or when exposed to heat.
- A fuel is anything that will burn. Carbon and hydrogen are the two most common elements in fuels.
- Fuel may exist in any one of the three states of matter: solid, liquid, and gas.
- There are several possible sources of heat—chemical (the breaking down and recombination of molecules), mechanical (friction, friction sparks, compression), electrical (arc or spark, static electricity, lightning), and nuclear (fission and fusion).
- For pyrolysis to take place, there must be a way for the heat (energy) to be transferred from the heat source to the fuel. There are three methods of heat transfer:
    1. Conduction
    2. Convection
    3. Radiation
- Some fire service professionals describe a fourth method of heat transfer, known as direct contact, direct flame impingement, or autoexposure, which is the transfer of heat through direct flame contact.

- Fires have been divided into five basic classifications, based on the type of fuel involved:
    1. *Class A*: Ordinary combustibles
    2. *Class B*: Flammable liquids and combustible liquids, including plastics
    3. *Class C*: Energized electrical
    4. *Class D*: Combustible metals
    5. *Class K*: Cooking materials (oils, fats, grease)
- Fire development is divided into four stages:
    1. *Incipient*: The fire has ignited and reaches a point where it no longer needs input heat from outside sources to continue burning.
    2. *Growth*: The fire is extending beyond the materials that first ignited it.
    3. *Freeburning*: The fire releases heat, bringing more fuel to its ignition temperature.
    4. *Decay*: The fire has run out of available fuel or oxygen.

KEY TERMS

Absolute zero The lowest temperature that is theoretically possible. The temperature at which all molecular motion ceases. Absolute zero is expressed as −459.67°F, −273.15°C, 0K, and 0°R.

Ambient temperature The temperature surrounding an object; air temperature.

Atmospheric pressure The pressure of the atmosphere exerted on any point, which is 14.7 psi at sea level.

Backdraft A type of explosion caused by the sudden influx of air into a mixture of gases, which have been heated above the ignition temperature of at least one of them.

Black fire A situation where heavy, dense, black smoke is being forcefully emitted by a fire.

Continuity The manner in which a fuel is spread across an area. Horizontal continuity is expressed as either uniform or patchy.

Endothermic reaction Reaction that absorbs heat.

Energy The capacity for doing work.

Evaporation The changing of liquid to a vapor.

Exothermic reaction Reaction that results in the release of energy in the form of heat.

Flameover A condition occurring in a structure fire where the fire extends across the ceiling consuming heated gases. Same as rollover.

Flashover A condition during a fire in a room when the contents are heated to their ignition temperature and flames break out over the entire area almost simultaneously.

Free radicals An atom or group of atoms (molecule) that is unstable and must combine with other atoms to achieve stability.

Fuel Anything that will burn.

Ignition temperature The minimum temperature to which a substance must be raised before it will ignite. The piloted ignition temperature is usually much lower than the autoignition temperature. Piloted ignition may be provided by a spark or flame or by raising the general temperature.

Mixable Capable of mixing without separation.

Mixture A substance made up of two or more substances physically mixed together.

Moisture content A description of the amount of moisture contained in a natural fuel, such as brush, grass, or other natural fiber. It is usually expressed as a percentage by weight.

Molecules Combined groups of atoms. Molecules composed of two or more different kinds of atoms are called compounds.

Nonmixable Not capable of mixing; will separate.

Oxidation The chemical combination of any substance with an oxidizer.

Oxidizer A substance that gains electrons in a chemical reaction.

Pyrolysis The chemical decomposition of matter through the action of heat.

Rollover A condition occurring in a structure fire where the fire extends across the ceiling consuming heated gases. Same as flameover.

Vertical arrangement The manner in which a fuel is arranged vertically above ground, divided into ground fuels, surface fuels, and aerial fuels.

CASE STUDY

On a hot August morning, a wildland fire began at the base of a slope in the western mountains. The heat of the day had commenced, and the wind was blowing upslope at 10 mph. The fire spread from grass to brush to trees and began to travel rapidly up the slope. The fire developed a large smoke column that was lifting hot embers in the rising air and dropping them ¼–½ mi in advance of the spreading fire front. The embers were dropping into a continuous receptive fuel bed of dry grass. This caused additional ignitions to occur, and soon the fire was jumping every natural barrier in its path. Roads could not stop it.

At the top of the slope, a subdivision of 500 homes had been developed 20 years previously. Many of the homes had wooden exteriors and decks overlooking the slope. There was inadequate clearance between the homes and the wildland fuels. The embers from the fire rained down on the homes. A few of the roofs had not been properly cleared of pine needles and leaves. Ignitions of these items began to occur on the roofs. Flames from below swept upwards under the decks and came into contact with their wooden structures. The decks began to burn as well. When one home would begin to burn, the heat produced would set the homes on either side on fire. The fire soon reached conflagration proportions and burned every home in the subdivision.

1. In which fuel is the fire most likely to start?

A. Grass
B. Brush
C. Trees
D. Downed logs

2. Which method of heat transfer caused the fire to lift embers into the air?

A. Conduction
B. Convection
C. Radiation
D. Direct flame contact

3. Why were the homes so susceptible to the fire?

A. Exterior covering
B. Lack of fire hydrants
C. Poor fire response
D. Location

4. Why did the roofs catch fire so easily?

A. Construction type
B. Slope of the roofs
C. Roof covering
D. Fuel on the roofs

5. Why did the decks ignite?

A. Conduction
B. Radiation
C. Convection
D. Direct flame contact

REVIEW QUESTIONS

1. List the three legs of the fire triangle.

2. When making the triangle a tetrahedron, what is the fourth side?

3. What is the most commonly occurring oxidizer?

4. Fuel may occur in any of the three states of matter. What are these states?

5. List the four factors affecting the burning rate of solid fuels.

6. What is meant by the term *black fire*?

7. Would an insoluble liquid fuel with a specific gravity of 0.8 sink or float when water is added? Explain.

8. Will a gas with a vapor density of 1.2 float or sink in air? Explain.

9. What effect does the ambient temperature have on the ignitability of a liquid?

10. List the four methods of heat transfer. How do they affect firefighting operations?

11. An out of control forest fire is burning in which phase of fire?

12. Taking into account the chemistry and physics of fire, what can be done to reduce the risk of fire spread in a building?

13. Which gas in (Table 4-1) is the most dangerous under normal conditions? Why?

DISCUSSION QUESTIONS

1. Is direct flame impingement truly a fourth method of heat transfer?

2. Why is the burning rate of fuel dependent on its physical state?

3. Which method of heat transfer is the most important in a structure fire? Justify your answer.

4. Which of the four phases of fire poses the greatest threat to a fire fighter's safety?

REFERENCES AND ADDITIONAL RESOURCES

Akasaka, Rio, ed. 2010. "30 Years Ago Today: MGM Grand Hotel Fire, November 21, 1980." *Firefighter_EMT*. http://www .firefighter-emt.com/archives/30-years-ago-today-mgm -grand-fire.php.

ASTM. 2013. *ASTM E84–13a Standard Test Method for Surface Burning Characteristics of Building Materials*. West Consho-hocken, PA: ASTM International. http://www.astm.org /Standards/E84.htm.

Clark County Fire Department. 2011. *MGM Hotel Fire November 21, 1980*. Las Vegas, NV: Clark County Fire Department. http://www.clarkcountynv.gov/depts/fire/Pages /MGMHotelFire.aspx.

Drysdale, D. D. 2003. "Chemistry and Physics of Fire". In *Fire Protection Handbook (19th ed., Vol. 1)*, 2–51 through 2–71. Quincy, MA: National Fire Protection Association.

Fire, Frank L. 2004. *The Common Sense Approach to Hazardous Materials*. New York, NY: Fire Engineering.

Friedman, Raymond. 1998. *Principles of Fire Protection Chemistry and Physics*. Quincy, MA: National Fire Protection Association.

Grimwood, Paul. 2013. "Black Fire." London, UK, Firetactics. com. http://www.firetactics.com/BLACK_FIRE.htm

Meyer, Eugene. 2013. *Chemistry of Hazardous Materials*. Upper Saddle River, NJ: Brady.

National Fire Protection Association. 2008. *Fire Protection Handbook* (20th ed.). Quincy, MA: National Fire Protection Association.

National Fire Protection Association. *NFPA 10: Standard for Portable Fire Extinguishers*. Quincy, MA: National Fire Pro-tection Association.

National Institute for Occupational Safety and Health. 2016. *NIOSH Pocket Guide to Chemical Hazards*. Washington, DC: Department of Health and Human Services. http://www.cdc .gov/niosh/docs/2005-149/pdfs/2005-149.pdf.

National Wildfire Coordinating Group. 2012. *Glossary of Wild-land Fire Terminology*. Boise, ID: National Wildfire Coordi-nating Group. http://www.nwcg.gov/pms/pubs/glossary /index.htm.

CHAPTER 5

Public and Private Support Organizations

OBJECTIVES

After studying this chapter, you should be able to:

- Identify and describe national and international support organizations.
- Identify and describe federal support organizations.
- Identify and describe state support organizations.
- Identify and describe local support organizations.
- Discuss the value of periodical publications.

Case Study

In 1992, the City of Lawrence, Massachusetts, was experiencing a large number of arson-caused fires. The fires were so numerous that they were reported on by the news media on a national level. To address the issue of the fires, a multiagency and multijurisdictional arson task force consisting of investigators from the Lawrence fire and police departments; Massachusetts State Fire Marshal's Office; and the Federal Bureau of Alcohol, Tobacco and Firearms (now called the Bureau of Alcohol, Tobacco, Firearms and Explosives) was formed.

By pooling their resources and taking a team approach, the task force was successful in reducing the number of arson-related fires to nearly zero in 6 months. The task force approach was so successful that it has been adopted to address every serious fire or explosion investigation.

This type of task force is active in many other cities and jurisdictions across the country.

1. Why would the agencies listed all be invited to participate in the task force?
2. List nongovernmental agencies or associations that may be invited to participate in an arson task force and discuss why they would be invited.
3. List organizations and resources that could be utilized to provide arson awareness information to the public.

 JONES & BARTLETT LEARNING NAVIGATE 2   *Access Navigate for more resources.*

Introduction

Numerous and varied organizations and associations related to the fire service exist on the national, state, and local levels. Their purposes vary widely, as do their memberships. Some of them are fire organizations and others are in direct or indirect support of the fire service. Presented in this chapter are a handful of these organizations and their effect on the fire service. As many of them have activities that apply in several areas, they have been divided into their main areas of effort. More information about each of these organizations and associations is available online, including mission statements and contact information.

Almost every one of these organizations provides training and materials to further the knowledge and capabilities of personnel seeking to understand and apply their subject matter. They are a readily available resource for research into fire- and safety-related fields of study. Not all of the associations and agencies that affect the fire service and the delivery of fire protection are mentioned here, as they are too numerous to list.

Additionally, at the end of the chapter is a listing of fire service–related publications. These are resources for the student and fire service professional. They provide information on the latest in fire and emergency service–related strategies and tactics, tools and equipment, training, incident reviews, etc. You can learn a lot from the experiences of others. Firefighting is a dangerous occupation, and studying the results of others' actions could save you from harm in the future. The sources of information may serve as the starting point for research or a new idea you wish to pursue.

National and International Organizations

There are numerous associations and organizations, made up of interested parties, such as manufacturers, industry, professionals in the field or area of interest, and representatives from government agencies on all levels from local to federal. Their goals are to further the standardization, regulation, and provision of fire safety and fire loss reduction in the United States and other countries. The membership of the associations and organizations bring a global approach to problem solving in that their members come from a wide background and have differing approaches. All of these differing perspectives can be considered when addressing a particular issue or problem.

National Fire Protection Association

The one association that has arguably the greatest impact on the fire service is the National Fire Protection Association (NFPA). Due to the breadth of its scope, the NFPA could appear in just about all of the categories in this chapter. Organized in 1896, the NFPA has members from the fire service and private organizations. The NFPA is recognized for its efforts in developing standards on fire fighter safety and equipment, as well as professional standards.

The NFPA has developed several fire and injury prevention programs aimed at the most vulnerable citizens: the very young and senior citizens. The first program developed was *Learn Not to Burn*. The program was developed to reduce fire injuries and deaths through education for children aged 4 and 5. The three messages developed for this program are:

1. When you hear a smoke alarm, get out and stay out.
2. Practice your escape plan (for child care providers).
3. Stay away from hot things that hurt you.*

An expansion of *Learn Not to Burn* is *Risk Watch*. This program is directed at children aged 14 and under. *Risk Watch* focuses on the areas of injuries most common in children, including motor vehicle accidents; fire and burns; choking, strangling, and suffocation; poisoning; falls; firearms; bicycle and pedestrian injury; and drowning.

The program developed for adults aged 65 and older is *Remembering When*. The program focuses on safety messages regarding smoking; space heater use; kitchen dangers; stop, drop, and roll; use of smoke alarms (detectors); escape routes; emergency numbers; and planning your escape around your abilities.

One of the NFPA's nationwide projects is the *Fire Sprinkler Initiative–Faces of Fire Campaign*. This campaign features stories of people impacted by fire and demonstrates the need for home fire sprinklers.

Transportation and Storage of Hazardous Materials

There are numerous incidents fire fighters respond to that involve hazardous materials and other dangerous substances. These occur in either transportation or storage. There are several government agencies and associations related to transportation by rail, highways, aircraft, and pipeline, which are the primary means of transporting hazardous materials and transportation-related emergency incidents.

U.S. Department of Transportation

The federal government agency that regulates transportation in the United States is the Department of Transportation (DOT). This agency regulates shipping of hazardous materials in trucking and railroads and on aircraft and waterways in the United States.

The DOT specifies requirements for placards and labels for hazardous materials being transported. It publishes the *Emergency Response Guidebook (ERG)*, which is carried on fire apparatus for reference in case of shipping accidents. Using placards and shipping

papers, the *Guidebook* provides information on initial actions to be taken in case of a spill or suspected spill of hazardous materials in a transportation accident. The *Guidebook* is provided free of charge to fire departments and is also available on the Internet as a downloadable .pdf and as a smartphone app.

Association of American Railroads

An association of the railroad companies in the United States, the Association of American Railroads (AAR) provides training and reference materials for the fire service. The AAR publishes the *Pocket Guide to Tank Cars*, a handy guide for training and reference for tank car types and their fittings.

The AAR/Bureau of Explosives publishes *Emergency Handling of Hazardous Materials in Surface Transportation*, which is provided to aid emergency responders in handling hazardous materials incidents.

American Petroleum Institute

The American Petroleum Institute (API) represents the oil and gas industry. API has an e-learning training catalog that offers approximately 60 safety-related courses. Some of them may be of interest to fire fighters working in or around the oil and gas industry. The API can also provide wildlife cleanup after an oil-related emergency, such as a pipeline or tanker spill.

Chemical Transportation Emergency Center

The Chemical Transportation Emergency Center (CHEMTREC), a service provider by the chemical industry, provides a 24-hour emergency number to call in case of a chemical emergency incident. Upon receipt of the initial call, with the name of the product, CHEMTREC provides immediate advice on the nature of the product and the steps to be taken in handling the early stages of a problem. CHEMTREC then promptly contacts the shipper of the material involved for more detailed information and on-scene assistance when available. Calls to CHEMTREC are to be limited to emergencies only. The emergency phone number is 800-424-9300.

Chlorine is widely used in water purification and other processes. It is transported in containers (sizes ranging from rail cars to small tanks used by swimming pool service personnel). Due to the highly toxic and corrosive nature of chlorine and the amount in transportation and storage, CHEMTREC has a special subunit called the Chlorine Emergency Plan (CHLOREP). When an incident involving the

---

* Modified from *NFPA - Learn Not to Burn*

release or threatened release of chlorine is reported to CHEMTREC, CHLOREP is activated. The nearest manufacturer is notified and a representative makes contact with the agency in charge of the incident to assist in hazard mitigation operations.

The Chlorine Institute

To ensure public safety in the handling of chlorine, the Chlorine Institute maintains a scientific and technical organization to aid the chlorine industry. This includes the development of regulations, both governmental and voluntary. The Institute also provides technical and safety-related information and emergency response procedures to prevent and mitigate the release of chlorine and other related chemicals.

Federal Aviation Administration

The Federal Aviation Administration (FAA) regulates fire protection at general aviation airports where commercial carriers operate and also aboard aircraft. Part 139 of FAA regulations establishes the requirements for aircraft rescue firefighting (ARFF) services at airports. These include requirements for training, vehicles, and other fire and rescue services. The FAA also controls the movement of hazardous materials by air. When an aviation accident occurs, the FAA investigates.

National Transportation Safety Board

Should there be a plane crash or other transportation-for-hire–related incident, the National Transportation Safety Board (NTSB), a federal agency, may perform the investigation. The NTSB also maintains vehicle accident information, including fire apparatus, and initiates vehicle recalls if deemed warranted.

Propane Education and Research Council

The Propane Education and Research Council (PERC) was established by the U. S. Congress in 1996. The PERC comprises a membership board that includes representatives from the propane industry and the public. PERC's focus is on safety, training, and research and development for all those who work with, sell, buy, and depend on propane.

Fire Protection Systems and Equipment

There are numerous trade groups, associations, and organizations involved in the development and promotion of fire detection and suppression devices used in industry and residential properties. These devices are further examined in the Fire Protection Systems and Equipment chapter.

The most common built-in fire suppression device used is automatic fire sprinklers. They have been used for over 100 years and are proven to be highly effective in suppressing fires. There are numerous organizations that promote their use in manufacturing, storage, assembly, hotels, motels, hospitals, correctional facilities, restaurants, general business, and residential settings. The organizations and associations listed here are some of those that advocate for or act as trade groups for the automatic fire sprinkler industry.

American Fire Sprinkler Association

Membership in the American Fire Sprinkler Association (AFSA) consists of contractors, manufacturers, dealers, and distributors of automatic fire sprinklers. AFSA is the publisher of *Sprinkler Age* magazine, a trade publication.

ASTM International

ASTM International (formerly known as the American Society for Testing and Materials) is an organization that develops and delivers international voluntary consensus standards. The purpose of the standards is to improve product quality, enhance safety, and increase consumer confidence in the products that meet ASTM standards.

Automatic Fire Alarm Association

The Automatic Fire Alarm Association (AFAA) is a trade group with membership including manufacturers, distributors, state/regional associations, engineers, users, and fire and building officials. The goal of the AFAA is to improve life safety through use of properly designed, installed, and maintained automatic fire detectors and early warning systems. The AFAA advocates in the model fire and building code development process (discussed in the Codes and Ordinances chapter). It also provides training programs and seminars to serve its industry.

FM Global

Resources available through FM Global (formerly Factory Mutual) include consulting services, property inspection, water supply and sprinkler system evaluation, safe operation of industrial processes, research, and numerous other property-related factors relating to insurance. The FM Global laboratories were developed for evaluating fire protection devices and equipment, primarily in industrial fire protection. The laboratories are involved with the evaluation of fire protection equipment used by fire service personnel, including the systems installed in buildings, such as sprinkler systems, detection and alarm systems, and portable equipment, including extinguishers. FM Global publishes *A Pocket Guide to Automatic Sprinklers*, a reference source for firefighting personnel for training, inspections, and prefire planning.

Fire Suppression Systems Association

The Fire Suppression Systems Association (FSSA) consists of suppliers, manufacturers, and installers of automatic fire systems. The FSSA is an advocacy group that promotes the use of fire suppression systems in the overall fire protection industry.

National Fire Sprinkler Association

National Fire Sprinkler Association members include fire sprinkler contractors, manufacturers and distributors of fire sprinkler equipment, fire and building officials, and insurance authorities. Their mission is to create a market for the widespread acceptance of fire sprinkler systems in both new and existing construction.

Codes and Standards

There are governmental and nongovernmental groups that are involved in the code and standards development process. They also advocate for the adoption and application of codes and standards to promote fire and life safety.

American National Standards Institute

The American National Standards Institute (ANSI) is a private, nonprofit organization that identifies public requirements for national safety, engineering, and industrial standards and coordinates voluntary standardization activities of concerned organizations. It also coordinates the standardization of items on the international scale, so a piece of equipment that works in the United States will work in other countries as well.

Building Officials and Code Administrators

Building Officials and Code Administrators (BOCA) is an organization that produces model codes for adoption by jurisdictions throughout the United States. The organization also sponsors training, testing, and certification for building officials and code administrators.

International Conference of Building Officials

The International Conference of Building Officials (ICBO) is an organization that develops model building codes for adoption by state and local agencies. ICBO created the *Uniform Building Code* and is a member of the International Fire Code Institute.

The Center for Campus Fire Safety

The Center for Campus Fire Safety represents over 4000 colleges and universities. Their purpose is to provide resources and support to individuals responsible for fire and life safety on college campuses and to fire departments that have college and/or university campuses in their communities.

The International Fire Code Institute

The International Fire Code Institute (IFCI) is a consensus code organization. Its membership includes fire department personnel, fire service professionals, building officials, design engineers, architects, product manufacturers, and others who are interested in increasing fire and life safety to establish safer communities. IFCI is the first organization in the United States to focus on the development and publication of a model fire code, namely the *Uniform Fire Code*. IFCI now publishes the *International Urban–Wildland Interface Code* as well, and is a participant in the international fire process. The organization provides training seminars and certification targeted toward fire department and building department personnel, architects, and engineers involved in building design, plan review, and fire/life safety inspections. The IFCI is now part of the International Code Council.

Southern Building Code Congress International

The Southern Building Code Congress International (SBCCI) created the *Southern Standard Building Code* and sponsors training, testing, and certification for building officials. SBCCI is now part of the International Code Council.

National Fire Protection Association

In addition to their activities listed previously, the NFPA has a series of codes and standards used nationally, titled the *National Fire Codes*. Standards that directly affect fire fighters from the first day on the job are:

- NFPA 1001: *Standard for Fire Fighter Professional Qualifications*
- NFPA 1404: *Standard for Fire Service Respiratory Protection Training*
- NFPA 1500: *Standard on Fire Department Occupational Safety, Health, and Wellness Program*
- NFPA 1581: *Standard on Fire Department Infection Control Program*
- NFPA 1901: *Standard for Automotive Fire Apparatus*
- NFPA 1961: *Standard on Fire Hose*
- NFPA 1971: *Standard on Protective Ensembles for Structural Fire Fighting and Proximity Fire Fighting*
- NFPA 1975: *Standard on Emergency Services Work Clothing Elements*

As you can see from this list, the NFPA has developed standards on all manner of firefighting equipment and programs. These standards are available through the National Fire Codes Subscription Service. The NFPA is also involved in research, technical advisory services, education, and other fire-related services. The NFPA has published books on a wide range of fire-related subjects, including the *Fire Protection Handbook*, considered by many to be the ultimate source of fire protection information.

National Safety Council

The National Safety Council is a private organization whose mission is to educate and influence society to adopt safety and health policies, practices, and procedures that mitigate human and economic losses arising from accidental causes and adverse occupational and environmental health exposures. Members include industry, labor unions, associations, hospitals, community service organizations, traffic safety associations, fire service, local safety councils, and commissions. The council provides training, awards, seminars, speakers, public education, and conferences; it also compiles statistics.

Disaster Assistance and Recovery
American Red Cross

The American Red Cross (ARC) is a quasi-governmental agency listed in the National Response Plan for disasters.

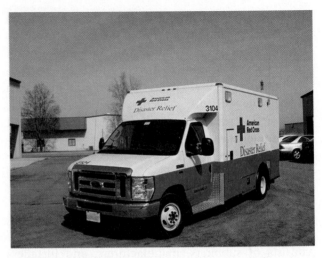

FIGURE 5-1 Red Cross emergency response vehicle.
© Jones & Bartlett Learning

The ARC provides assistance to the victims of disasters on a large scale, setting up shelters for evacuated persons, or on a small scale, assisting one family that has lost its home due to fire or other disaster **FIGURE 5-1**. The ARC is also involved in providing training and certification to emergency responders in the area of infectious disease control through its course, AIDS Education for Emergency Workers.

Salvation Army

The Salvation Army is an international, church-based, nonprofit organization that assists persons in need. They may be called upon to provide canteen services on an as-needed basis in some communities. This is usually limited to coffee and donuts or sandwiches on an extended incident. They can be accessed through the local office.

American Rescue Dog Association

The American Rescue Dog Association provides volunteer search and rescue dogs and handlers free of charge to requesting public agencies to search for lost or missing persons. This can make a life-or-death difference to trapped victims at incidents, such as earthquakes, train crashes, aircraft crashes, hurricanes, and tornadoes. Any incident where occupants are trapped by debris and not easily visible could benefit from this association.

Professional Accreditation and Certification
Board of Certified Safety Professionals

The Board of Certified Safety Professionals (BCSP) is an association for safety engineers, industrial hygienists, safety managers, and fire protection engineers. A person may apply for certification both nationally and internationally through the BCSP. It also provides online and college-based training for safety professionals.

Fire Department Safety Officers Association

The mission of the Fire Department Safety Officers Association (FDSOA) is to promote safety standards and practices in the fire, rescue, and emergency services community. Professional certification as a safety officer is considered to be desirable and may soon be required. The FDSOA offers an online course and certification exam to become certified as an Incident Safety Officer.

International Fire Service Accreditation Congress

The International Fire Service Accreditation Congress (IFSAC) is a peer-driven, self-governing system that accredits both public fire service certification programs and higher education fire-related degree programs. The mission of the IFSAC is to increase the level of professionalism of the fire service through the accreditation of certifying entities and degree-granting institutions.

International Municipal Signal Association

The International Municipal Signal Association (IMSA) is organized to assist its members with technical knowledge and information regarding fire and police alarms and traffic control signals. IMSA members include persons employed by governmental organizations and private corporations who are interested in promoting public safety.

National Board of Fire Service Professional Qualifications

The mission of the National Board of Fire Service Professional Qualifications is to establish a recognized means of acknowledging professional achievement in the fire service and related fields by accrediting organizations that certify uniform members of public fire departments, both career and volunteer. This is the parent organization of the Pro Board, described next.

Pro Board Fire Service Professional Qualifications System

The purpose of the Pro Board is to establish an internationally recognized means of acknowledging professional achievement in the fire service and related fields. The primary goal is the accreditation of organizations that certify uniform members of public fire departments, both career and volunteer. The Pro Board accredits organizations that use the National Fire Protection Association's (NFPA's) professional qualification standards. Accreditation is generally provided at the state or provincial level to the empowered certifying authority of that jurisdiction.

Professional Research

Building and Fire Research Laboratory

The Building and Fire Research Laboratory provides the scientific and technical basis for reducing fire losses and the costs of fire protection.

International Association for Fire Safety Science

The International Association for Fire Safety Science promotes research into the science of preventing and mitigating the adverse effects of fires. Membership includes fire research scientists.

Society of Fire Protection Engineers

The Society of Fire Protection Engineers is a professional society for fire protection engineers. It advances the science of fire protection engineering and its allied fields; promotes education; and provides awards, training programs, seminars, and conferences.

Underwriters Laboratories

The goal of Underwriters Laboratories, Inc. (UL), a nonprofit organization, is to promote public safety through scientific investigation of various materials in regard to the hazard they present by their use. Once the items are tested and certified as safe, the organization lists and marks the material as having passed its rigorous tests. The UL reserves the right to test a representative sample of a manufacturer's listed product at any time to assure compliance. A problem encountered in the marketplace is fake UL labels appearing on untested devices. A UL label can be verified using the UL online certification directory.

UL has two subunits that deal directly with fire fighter safety; these are the Fire Safety Council and the UL Firefighter Safety Research Institute (UL FSRI). UL FSRI is dedicated to increasing fire fighter knowledge to reduce injuries and deaths in the fire service and in the communities they serve. Another service of UL is the Modern Fire Behavior training and information and modern building construction research site—modernfirebehavior.com.

Fire Apparatus and Equipment

Fire Apparatus Manufacturer's Association

The Fire Apparatus Manufacturer's Association (FAMA) comprises manufacturers of fire apparatus and equipment. FAMA is a trade group that works for the betterment of the fire service industry through the manufacture of safe and efficient fire apparatus and equipment. It also works with such organizations as the NFPA in standards development. An example is *NFPA 1901: Automotive Fire Apparatus*.

Personnel and Training Support

Center for Radiological/Nuclear Training

The Center for Radiological/Nuclear Training (CTOS) provides training at no cost to eligible participants for first responders to prevent, mitigate, or respond to terrorist use of radiological or nuclear weapons of mass destruction. Training is conducted at both the Nevada National Security Site and through mobile training.

Firefighter Behavioral Health Alliance

The Firefighter Behavioral Health Alliance's (FBHA's) goal is to provide behavioral health workshops to fire departments and emergency medical services (EMS) organizations across the globe, focusing on behavioral health awareness with a strong emphasis towards suicide prevention and promoting resources available to fire fighters/EMS personnel and their families. This organization gathers information on fire fighter/EMS suicides and has a self-assessment screening tool for suicide ideations for fire fighters/EMTs. It is financially supported by numerous fire-related organizations and manufacturers.

Firefighter Cancer Support Network

The purpose of this organization is to provide support to fire and EMS personnel suffering from cancer and their families. The organization works through a network of mentors who have already experienced the type of cancer the person is experiencing. They also conduct awareness and educational campaigns using local, state, and regional partners. This organization is based on the premise that fire fighters are at a high risk of suffering from cancer.

International Association of Fire Chiefs

The International Association of Fire Chiefs (IAFC) represents the interests of officers of the chief rank from throughout the United States and other countries. This organization furthers the professional advancement of the fire service. The IAFC is one of the founding members of the IFCI.

International Association of Fire Fighters

The International Association of Fire Fighters (IAFF) represents fire fighters and their interests. It is the largest union organization in the Northern Hemisphere, with approximately 200,000 members from the United States and Canada. IAFF has local and state offices as well as a national organization and is politically active on all levels of government. The IAFF promotes the safety and welfare of fire fighters through fire prevention, protection, research, and legislation. The IAFF is also a major contributor to the Muscular Dystrophy Association and has its own foundation, the Burn Foundation, to promote burn care research. Membership is limited to active and retired paid fire fighters. The IAFF is affiliated with the American Federation of Labor and Congress of Industrial Organizations.

International City/County Management Association

The International City/County Management Association (ICMA) is an association of local-government management professionals whose purposes are to strengthen urban government through quality professional management and to develop and disseminate new approaches through training programs, information services, and publications. ICMA publishes the book *Managing Fire and Rescue Services*.

International Society of Fire Service Instructors

The International Society of Fire Service Instructors (ISFSI) serves to further the exchange of educational materials and techniques. ISFSI sponsors the Fire Department Instructors Conference and provides training, programs, seminars, and conferences to further fire service education.

Leary Firefighter Foundation

The Leary Firefighters Foundation was established in 2000 by actor Denis Leary in response to a tragic fire in Worcester, Massachusetts, that claimed the lives of

Denis's cousin, a childhood friend, and four other fire fighters. The Leary Firefighters Foundation's mission is to provide funding and resources for Fire Departments to obtain the best available equipment, technology, and training. It is dedicated to helping maintain the highest level of public safety in communities.

Mutual Aid Box Alarm System

The Mutual Aid Box Alarm System (MABAS) is a mutual aid system that has been in existence since the late 1960s. MABAS operates throughout the states of Illinois and Wisconsin as well as parts of Indiana, Iowa, Michigan, and Missouri. MABAS preplans and organizes mutual aid response to its member departments. MABAS includes approximately 1000 of the involved states' 1200 fire departments and is organized within 67 divisions. MABAS divisions geographically span an area from Lake Michigan to Iowa's border and south almost into Kentucky. The cities of Chicago, St. Louis, and Milwaukee are also MABAS member agencies (MABAS 2019).

National Fallen Firefighters Foundation

The United States Congress created the National Fallen Firefighters Foundation (NFFF) in 1992 to lead a nationwide effort to remember America's fallen fire fighters. This tax-exempt, nonprofit foundation has developed and expanded programs to honor our fallen fire heroes and assist their families and coworkers. The National Fallen Firefighters memorial is located on the grounds of the National Fire Academy in Emmitsburg, Maryland. The NFFF is also the administrator of the *16 Life Safety Initiatives*, the *Courage to be Safe*, and the *Everyone Goes Home* programs.

National Fire Heritage Center

The National Fire Heritage Center (NFHC) is a nonprofit organization working to preserve the written history of "fire in America." They support individuals, fire departments, libraries, museums, and others who wish to collect, catalog, and preserve fire-related writings.

The NFHC is both a fire history archive and a documentation preservation project. Their collection of over 14,000 cataloged items consists of written documents and three-dimensional items donated by the public and private sectors. Materials are cataloged, preserved, and made available to NFHC visitors and online researchers.

The NFHC is co-located with the Frederick County Fire-Rescue Museum at 300 South Seton Avenue in Emmitsburg, Maryland, within walking distance of the National Emergency Training Center.

National Volunteer Fire Council

The National Volunteer Fire Council (NVFC) is an organization of various state fire fighter associations that represent and pursue the interests of volunteer fire fighters and volunteer fire departments. The NVFC advocates for the interests of volunteer fire fighters in the U.S. Congress and with various federal agencies involved in the preservation of life and property. The NVFC provides training programs, public education, and conferences, and it compiles statistics useful to the furtherance of its advocacy.

National Wildfire Coordinating Group

The National Wildfire Coordinating Group (NWCG) primarily develops training and reference materials for wildland firefighting. It has now joined with the National Fire Academy in rewriting its Incident Management System (Chapter 13) positional courses to include "all risk" incident types. One of its publications is the *Wildland Fire Incident Management Field Guide* (which replaced the *Fireline Handbook* in 2013), a reference source that contains material on safety, resource use and capabilities, organization, command, and wildland fire behavior. Another widely used publication of the NWCG is the PMS-461 *Incident Response Pocket Guide* (IRPG), a safety and operational information resource that is required to be carried by all federal wildland fire fighters, which is also applicable to other incident types. Both of these documents are available on the Internet as free downloads, and the IRPG is also available as a smartphone app.

Incident Management
FIRESCOPE

The **Fi**refighting **Re**sources of **S**outhern **C**alifornia **O**rganized for **P**otential **E**mergencies (FIRESCOPE) was formed because of the need for a standardized system of fire management on large, multiagency incidents. As a result of its efforts, the Incident Command System (ICS) was developed. The ICS is the model that the National Incident Management System (NIMS), adopted by the federal government, is based on. Members include fire fighters and fire marshals from a wide range of agencies and jurisdictions. Part of their task is to produce standard job descriptions for positions and command structure at an incident. The incident types addressed so far are wildland, hazardous materials, multi-casualty, and urban search and rescue.

National Response Center

The National Response Center (NRC) provides a one-call notification service for hazardous materials emergencies. The NRC is notified by the state in which the incident is occurring if the property loss is over $50,000, if serious injury or death has occurred, or if there is a continuing danger to the public from the hazardous materials. When notified of an emergency, the NRC notifies the DOT, the Environmental Protection Agency, and the U.S. Coast Guard.

Insurance Providers

Insurance Committee for Arson Control

The Insurance Committee for Arson Control (ICAC) represents property/casualty insurance companies and their trade associations. It serves as the focus group for the insurance industry's overall anti-arson efforts and a liaison with other groups devoted to arson control. The ICAC provides public education and conferences and compiles statistics.

Insurance Services Office

The Insurance Services Office (ISO), a part of Verisk Corporation, is a voluntary, nonprofit, unincorporated association of insurers formed to gather information to assist in setting fire insurance rates. The ISO rates the fire protection capabilities of the jurisdictions its insurance company members insure. Utilizing the *Fire Suppression Rating Schedule (FSRS)*, the ISO provides the insurance industry with a means of identifying, for insurance purposes only, a relative analysis of what may be expected from public fire services. The results of an ISO audit of fire protection resources are expressed as a public protection classification (PPC). The classification spread of 1–10 is a scale of relative values whereby a municipal fire protection system may be compared with others. It is also indicative of a system's ability to defend against the major fires that may be expected in any given community. Where Class 10 is assigned, there is usually no protection. However, very minimal protection in a community of extensive development may also result in Class 10. Protection Class 1 represents a fire protection system of extreme capability. The rating process involves looking at the whole fire protection system of a given area, including the fire department, water supply, fire service communications, fire safety control, community risk reduction, and system of guidance.

Most large corporations automatically qualify for the highly protected risk insurance designation, which is shaped by the following six guiding principles:

1. A concerned management interested in implementing an aggressive program of loss prevention and control
2. Acceptable construction in good repair with adequate exposure protection
3. Interior protection with appropriate automatic extinguishing or suppression systems installed wherever there is combustible occupancy
4. Special hazard protection based on the identification and evaluation of all hazards peculiar to the class of occupancy and with the provision of proper supplementary protection
5. Exterior protection with the provision for an adequate water supply for automatic extinguishing systems and ancillary equipment and the availability of a private emergency organization
6. Adequate surveillance, interior and exterior, using guard patrols, surveillance systems, continuous occupancy, or a combination of these

Emergency Medical Services

International Rescue and Emergency Care Association

International Rescue and Emergency Care Association membership consists of prehospital care providers including fire, ambulance, and rescue squad members. It is organized to provide educational programs to share and evaluate the knowledge, skills, and abilities of emergency responders.

National Association of Emergency Medical Technicians

The National Association of Emergency Medical Technicians (NAEMT) promotes professionalism among emergency medical technicians (EMTs) and paramedics and provides training programs. The NAEMT offers several courses directed at emergency medical responders.

National Registry of Emergency Medical Technicians

The National Registry of Emergency Medical Technicians (NREMT) serves as the national EMS certification organization by providing a valid, uniform process to assess the knowledge and skills required for competent practice by EMS professionals throughout their careers and by maintaining a registry of certification status. All new EMTs must now take the NREMT exam to achieve their initial EMT certification.

Federal Organizations

There are numerous publicly funded agencies of the federal government concerned with protecting the public from fire. These organizations are involved in all aspects of fire fighter training, research, and development. Many of the organizations have resources that can assist in emergency situations.

All of the federal agencies, including those previously listed, are organized under the National Response Plan, established under the guidance of the Department of Homeland Security. The plan is all discipline and all risk, organized to provide a coordinated response of federal agencies for domestic incidents. The plan groups capabilities and resources into functions (called emergency service functions or ESFs) and describes agency responsibilities within each ESF.

Department of Defense

The Department of Defense assists fire departments with antiterrorism training and response. The training is called domestic preparedness and deals with weapons of mass destruction. It covers nuclear, biological, and chemical threats. Many military bases make resources available to local departments on a call-as-needed basis. They may also have training props that can be utilized.

It is becoming more common to see large numbers of armed forces personnel who are trained to fight fires activated on large wildland fire incidents. Some units of the army, Marine Corps, and reserves are now being given this training before they are needed. With their numerous personnel, command structure, and self-contained logistical support, they are a tremendous asset in times of need. Military bases usually have their own fire departments that can respond when requested by the local fire department. They may even have specialized apparatus, such as water tenders, heavy equipment, foam units, and aircraft rescue firefighting apparatus. Helicopters from military bases may also be available to perform emergency transport of victims in rescue situations.

Department of Labor

The Department of Labor (DOL) administers and enforces the Occupational Safety and Health Act, through the Occupational Safety and Health Administration (OSHA), to ensure safety in the workplace. The DOL compiles the national occupational injury and illness data for the Bureau of Labor Statistics.

Emergency Management Institute

The Emergency Management Institute (EMI) is authorized under the Civil Defense Act of 1950 to provide training to public sector managers to prepare for, mitigate, respond to, and recover from all types of emergencies. The EMI provides programs in emergency management, technical development, and professional development. The parent organization for the EMI is the Federal Emergency Management Agency (FEMA).

National Aeronautics and Space Administration

The National Aeronautics and Space Administration (NASA) is developing an artificial intelligence (AI) system named Audrey. The system takes data gathered in real time from fire fighters through sensors. The AI can detect changes in temperature, gases present, and other threats. The system is cloud-based and will automatically send warnings to personnel deployed at the incident to aid in hazard detection and risk management decision making.

National Emergency Training Center

The National Emergency Training Center (NETC) is sponsored by the EMI. At the NETC, representatives of the fire service, law enforcement, emergency medical services, civil defense, public works, state and local government, and public interest groups can meet and seek training to perform coordinated emergency response. The NETC is on the same campus as the National Fire Academy (NFA).

National Institute for Occupational Safety and Health

The National Institute for Occupational Safety and Health (NIOSH) is a part of the Centers for Disease Control and Prevention in the Department of Health and Human Services. NIOSH conducts research and provides educational functions to support OSHA. It also recommends occupational safety and health standards. NIOSH publishes the *Pocket Guide to Hazardous Materials*.

In 1998, NIOSH undertook the National Fire Fighter Fatality Investigation and Prevention Program, the overall goal of which is to define better the magnitude and characteristics of work-related deaths and severe injuries among fire fighters, to develop recommendations for the prevention of these injuries and deaths, and to implement and disseminate prevention efforts. The program consists of the following five parts and is centered on the field investigation of fire fighter fatalities:

1. Fire fighter fatality investigations
2. Cardiovascular disease fatality investigations
3. Fire fighter fatality database project
4. Intervention research project
5. Information dissemination project

National Incident Management System

The National Incident Management System (NIMS) is designed to standardize the participation of departments and agencies at all governmental levels, non-governmental organizations, and organizations in the private sector to work together in preventing and reducing the loss of life and property and damage to the environment from all incident types. The use of NIMS can be applied to all steps in incident-related activities, including prevention, response, mitigation, and recovery. NIMS is applicable to all types of incidents—all sizes, complexity, causes, locations, or jurisdictions.

NIMS works hand in hand with the National Response Framework (NRF). NIMS provides the template for the establishment of the incident management organization, while the NRF provides the structural guidelines and mechanisms for national-level policy for incident management.

National Integration Center

The National Integration Center (NIC) was established under the secretary of Homeland Security. The NIC publishes the standards, guidelines, and compliance protocols for determining whether a federal, state, tribal, or local government has properly implemented NIMS. The NIC manages publication of and, working with other departments and agencies, develops standards, guidelines, compliance procedures, and protocols for all aspects of NIMS.

National Institute of Standards and Technology

Founded in 1901, the National Institute of Standards and Technology (NIST) supports standardization of items and methods both nationally and internationally. NIST is a part of the Commerce Department and is an advisory agency with no regulatory authority. In April of 2010, NIST's Building and Fire Research Laboratory released the *Report on Residential Fireground Field Experiments* (NIST TN—1661), a landmark fire study on how crew sizes and arrival times influence the saving of lives and property.

NIST's work on Wildland Urban Interface (WUI) fires is part of its statutory responsibilities for enhancing disaster resilience by reducing the risks of fires, earthquakes, windstorms, and coastal inundation on buildings, infrastructure, and communities, including facility occupants/users and emergency responders.

National Weather Service

The National Weather Service (NWS) provides weather information to the public on a daily basis. This information is disseminated through media outlets, such as newspapers, television, and the Internet. They can also be requested to provide a much more localized "spot weather forecast" for a particular incident location. For major incidents, a meteorologist from the NWS is assigned to the Incident Management Team and works closely with the Fire Behavior Analyst to predict fire behavior. This is a great aid in planning resource needs, deployment, and control line location. This information is also important when planning hazardous materials incidents where evacuation downwind is a key element in the objective of providing for responder and public safety.

U.S. Department of Agriculture Forest Service

The U.S. Department of Agriculture Forest Service (USFS) is a nationwide organization charged with the management of the national forests **FIGURE 5-2**. This agency provides fire protection for national forest lands and assistance and resources to agencies protecting areas bordering on national forests. In the summer, this is one of the largest firefighting organizations in the country.

The federal wildfire agencies, USFS, and the Department of the Interior agencies listed below all work together seamlessly in suppressing fire on federal

FIGURE 5-2 USFS fire engine and fire fighters at a wildland fire.
© Jones & Bartlett Learning

FIGURE 5-3 BLM engine at wildland fire.
© Jones & Bartlett Learning

lands. This includes National Forests, National Parks, Wildlife Refuges, Indian Reservations, etc.

U.S. Department of Homeland Security

The U.S. Department of Homeland Security (DHS) was established to integrate resources from federal, state, and local governments to establish an agency focused on protecting the American people and their homeland. The national strategy seeks to develop a complementary system connecting all levels of government without duplicating effort.

The DHS has five major divisions or directorates, which include:

1. Border and Transportation Security
2. Emergency Preparedness and Response
3. Science and Technology
4. Information Analysis and Infrastructure Protection
5. Management

The directorate that interacts the most with fire departments is Emergency Preparedness and Response. It assists departments by providing grant funding for equipment and training.

U.S. Department of the Interior

There are four agencies with firefighting duties that fall under the U.S. Department of the Interior (DOI). Much like the USFS, the Bureau of Land Management (BLM) provides fire suppression services on lands under the control of the Department of the Interior, outside of the national parks, and adjoining private

lands **FIGURE 5-3**. Local government fire departments often come into contact with these agencies when protecting structures within federal jurisdiction or when training, and when rendering or receiving mutual aid on incidents that occur on the border between federal and local lands.

- The BLM provides fire suppression services on lands designated as BLM protection under the control of the DOI, outside of the national parks, and adjoining private lands (**FIGURE 5-3**).
- The U.S. National Park Service (NPS), much like the BLM, provides fire protection in the national parks and bordering lands where a fire may be a threat to the park.
- The U.S. Fish and Wildlife Service (FWS) provides fire protection on national wildlife refuges and assists the other federal firefighting agencies (USFS and BLM) in suppressing fires on federal lands. It may also assist local fire agencies with mutual aid.
- The Bureau of Indian Affairs (BIA) has responsibility for wildfires on Native American lands. This primarily involves tribal lands (reservations).

National Firefighting Equipment System

The National Firefighting Equipment System (NFES) publishes reference and training manuals on all aspects of wildland firefighting. The NFES also develops the courses used in the qualification rating system for wildland fire fighters. All of the publications available through this system are referenced by NFES number.

These publications are available through the NWCG NFES Catalog–Part 2: Publications. An example of this is the book *Basic Aviation Safety*, NFES #2097. These publications are available through the National Interagency Fire Center.

National Interagency Fire Center

The National Interagency Fire Center (NIFC) is a cooperative effort of the Department of the Interior and the Department of Agriculture. The NIFC is the location of the National Interagency Coordination Center and the National Multi-Agency Coordination Center (NMAC). The NIFC is the central nationwide supply point for resources required on large wildland firefighting incidents. By contacting NIFC, incident commanders can order everything from shower units to fire hoses to hand crews and other personnel.

Federal Emergency Management Agency

FEMA was created in 1978 by then-President Jimmy Carter, with the intent of placing federal disaster response coordination and services under one agency. In the 1980s, FEMA was heavily focused on civil defense planning. FEMA's performance was highly regarded during the Loma Prieta Earthquake in the San Francisco Bay area in 1989. After the Hurricane Andrew incident in 1992, FEMA was reorganized and focused more on natural disasters. FEMA maintains several Type 1 incident management teams across the United States. These teams are set up with personnel from various agencies and are called into action when the need arises. They are used when fires, floods, hurricanes, earthquakes, and other disasters occur. FEMA also maintains a list of urban search and rescue teams (USARs) activated for disasters with major structural collapse, such as earthquakes and explosions. These teams must meet specified minimum requirements for personnel and materials. When the Type 1 or USAR teams are on the top of the rotation list, the personnel assigned to the particular team must be able to respond in 2 hours.

United States Fire Administration

The United States Fire Administration (USFA) was established by the Federal Fire Prevention and Control Act of 1974 (Public Law 93-498). Its purposes are to reduce the nation's losses from fire through better fire prevention and control; to supplement existing programs of research, training, and education; and to encourage new, improved programs and activities by state and local governments.

National Fire Academy

The NFA, established under the USFA, is an organization dedicated to the professional development of fire fighters and related professionals across the United States. The NFA has a campus in Emmitsburg, Maryland; courses offered are in the areas of executive development, fire prevention, leadership, incident management, public education, fire service education, arson, infection control, and hazardous materials. These courses are designed to be presented on or off campus. Some of them are available online. The NFA also maintains a large library called the Learning Resources Center.

Fire and Emergency Services Higher Education Conference

The Fire and Emergency Services Higher Education (FESHE) conference is a gathering of college fire service program directors, state training directors, and representatives from organizations that have an interest in fire service higher education. Out of this conference came the FESHE guidelines. These guidelines include a professional development model as well as a model curriculum for fire science degree programs at the associate's and bachelor's degree levels (see the Fire Science Education and the Fire Fighter Selection Process chapter).

Department of Justice Bureau of Alcohol, Tobacco, Firearms, and Explosives

The Department of Justice Bureau of Alcohol, Tobacco, Firearms, and Explosives (ATF) assists in the investigation of arson and bomb incidents by gathering and processing evidence.

Occupational Safety and Health Administration

OSHA was established under federal law to ensure safe working conditions. It is a part of the Department of Labor. Some states have their own state-level agencies that enforce federal regulations. OSHA investigates workplace accidents and has the authority to levy fines against employers for unsafe working conditions. This includes fire departments that are found to have violated safe work practices.

Environmental Protection Agency

The Environmental Protection Agency (EPA) responds to, and acts as the coordinating agency on, large hazardous materials incidents. The person responding is called the on-scene coordinator and has the authority to ensure the spill is contained and cleaned up properly. The EPA has funding available under the Superfund to provide cleanup services at hazardous materials sites that are deemed a threat to the public.

Nuclear Regulatory Commission

The Nuclear Regulatory Commission (NRC) regulates all aspects of the nuclear power generating facilities in the nation, including fire protection.

U.S. Coast Guard

The U.S. Coast Guard (USCG) provides on-scene coordination on hazardous materials incidents that involve or threaten navigable or coastal waterways. The USCG publishes the *Chemical Hazards Response Information System (CHRIS)* manual and other hazardous materials information sources.

State Organizations

Much like the federal government, each state has numerous agencies and departments that interact with the fire service either directly or indirectly.

Office of the State Fire Marshal

The office of the state fire marshal receives its authority to enforce the state's fire laws from the legislature. In most states, the state fire marshal is appointed by the governor. The state fire marshal's responsibilities vary by state. Responsibilities usually include:

- The review and approval of construction plans for fire safety
- The investigation and determination of fire cause
- The investigation of fire deaths
- The aggressive attack on arson through investigation and prosecution
- Regulation and control of the storage and use of explosives and other hazardous materials

- Preparation and promotion of fire-related legislation
- Promotion and enforcement of the state's fire codes*

State Fire Training

Many states have a training organization. The purpose of this organization is to develop, coordinate, and deliver training throughout the state. Full-time management staff and full- and part-time instructors are state certified to present the curriculum. Often the training programs are presented under the sponsorship of a local college, which provides a way to present certifications and college credits and provides a means to pay the instructors. For more information, contact the state fire marshal's office in your state.

State Emergency Management Agency

The state emergency management agency provides training and assistance to local governments in the preparation, response, recovery, and mitigation of disasters. The office can provide funding, personnel, vehicles, and equipment to assist local agencies in combating the problems due to the emergency. This agency may also coordinate the state Master Mutual Aid agreement.

Fire Commissions

In some states, fire commissions are set up to act as advisory panels to the state fire training organization. They oversee the expenditure of public funds for training purposes. They also adopt the standards used in certification.

State Fire Chiefs Associations

Fire chiefs organize on the state level to gather and find solutions to common problems. They also promote professional development and legislation.

State Fire Fighter Associations

Members from various firefighting organizations in the state organize into fire fighter associations to lobby for legislation to further their interests. Some legislation that has been passed in certain states concerns cancer presumption and heart presumption, meaning

---

* Modified from U.S.Legistative

that if a fire fighter has cancer or heart disease, it is presumed to be job connected, which allows for a job-connected disability pension. These organizations also provide training to their members in the areas of contract negotiation and wage and hour issues.

State Police

The police and fire department join forces in their efforts to investigate arson and apprehend and convict arsonists. The police have systems in place to check license plates on burned vehicles, have access to crime labs, and are experts on gathering and preserving evidence. If the state has a state-level highway patrol, the state and federal highway system is under its jurisdiction.

State Environmental Protection Agency

State environmental protection agencies become involved when hazardous material incidents threaten the environment. This responsibility is placed under the wildlife management agency in some states. When there is a state-level superfund, money may be available to clean up spills in some jurisdictions.

State Occupational Safety and Health Administrations

In the states that have OSHA-like agencies, the agencies enforce state and federal regulations pertaining to worker safety in the workplace.

State Forestry Department

Lands in the state that are not incorporated into cities or under federal jurisdiction, although private property, are often the responsibility of the state for fire protection. By using state department of forestry resources and contracting with local fire protection agencies, fires are suppressed.

Special Task Forces

When there are specific fire-related problems that need to be addressed, it is common for the governor to set up a special task force to deal with them. A good example of this would be an arson task force comprising fire fighters, law enforcement, insurance companies, and state agency representatives to address the arson issue affecting a community. The task force would conduct a study and make recommendations based on its findings.

National Guard

The National Guard can be activated by the governor of a state. In some areas, National Guard units are trained in wildland firefighting and activated when local resources are overwhelmed. In times of major wildland fires, National Guard planes may be set up as air tankers, called mobile aerial firefighting system (MAFFS) units, to combat fires. The National Guard also has numerous types of heavy equipment and trucks that can be used to assist in mitigating fires, floods, earthquakes, and other incidents when activated.

Local Organizations

Burn Foundations

Burn foundations are set up to raise and disseminate funds to the victims of burn injuries and their families. The money is also used for burn injury research and to purchase burn treatment equipment. Many fire department personnel are involved, on their off-duty time, in these activities. Burn foundations are not only good public relations for fire fighters but may also make the difference in having the technology available for treating fire fighters if they are injured.

Local Government

Most career fire departments are entities of some level of local government. The locally elected officials establish the fire department and delegate the authority to fight fires and enforce fire protection laws through ordinances. The local government raises the funds necessary to purchase equipment and pay the fire fighters. This may be done through general taxation, special districts, or fire taxes. Local government is responsible for adopting fire codes through local ordinances.

Health Department

After fires or incidents that may affect the public health, the health department may become involved in determining what must be done to correct the problem. It may condemn (as unsafe for consumption) a load of fruit spilled on the road in a vehicle accident or condemn a freezer of food at a restaurant after a fire.

Health departments become involved in hazardous materials incidents by determining whether a hazardous materials spill has been cleaned up sufficiently. If there is an outbreak of infectious disease, such as measles or chicken pox, the health department tracks the spread of the disease and notifies the public, including emergency medical responders, about the threat.

Law Enforcement

Fire and law enforcement departments find themselves working together on many incidents. When evacuation is necessary due to hazardous materials incidents or wildland fires, law enforcement usually evacuates people and denies entry to the affected area. In some states, the county sheriff serves as the officer in charge of wildland fire incidents.

Law enforcement provides traffic control at incident scenes and may be trained in emergency first aid. The goal of law enforcement and fire departments are much the same—to save lives and protect property. A positive working relationship with local law enforcement is a definite plus in accomplishing the fire department's mission.

Building Department

The fire department and the building department work together to ensure fire and life safety in buildings. The building department is more involved when the building is being constructed and when major remodeling that requires a building permit takes place. The building department should also notify the fire department when new buildings, other than homes, are certified ready to occupy. By working closely with the building department, the fire department can ensure that fire-related codes are enforced during the construction phase.

Water Department

The water department controls the jurisdiction's water system to a great degree. If there is to be adequate water available for firefighting purposes, there must be a working relationship with the water department. If damage or repairs have affected part of the water system, such as turning off the water to fire hydrants, it is to the fire department's advantage to be notified of the location, duration, and extent of the area affected. In times of high water demand, due to required fire flow to combat fires, a representative of the water company may have to be contacted to make adjustments to the water system.

Zoning/Planning Commission

The local planning commission decides what types of occupancies will be built where. It decides where the residential areas will be and what the density will be.

These areas can be divided into single- and multiple-family structures. The planning commission also decides where commercial and industrial centers will be located. In the planning of future fire station sites, this information becomes extremely important. A fire station may be in one place for 50 years or more. The surrounding area and its fire protection needs will dictate the size of the station in terms of room needed for a large or small crew and the type of apparatus it must house. The fire department should be included in the planning process in the interest of providing optimum fire protection.

Street Department

The local street department (may also be called the highways and bridges department or public works) designs and maintains streets and bridges in the jurisdiction, which concerns the fire department in terms of being able to get the apparatus into all areas and across all of the bridges **FIGURE 5-4**. If culs-de-sac are too tight or bridges cannot handle the weight of apparatus, it severely limits response routes and times and is dangerous to fire fighters. In the case of a large brush fire, a negotiable escape route for fire apparatus must be available. Many fire departments have had their apparatus and the traffic lights equipped with a system that allows them to change the lights to green in their direction of travel; this is called a traffic preemption system. Another reason for maintaining a good relationship with the road department is that it has heavy equipment and operators that may be required at an emergency scene. A skip loader and a dump truck are very handy to have when cleaning up after a semitrailer wreck or when sand needs to be laid to contain a hazardous materials spill. The road department should also

FIGURE 5-4 Public works/roads department water tenders.
© Jones & Bartlett Learning

notify the fire department if roads or bridges are closed due to maintenance or repair.

Judicial System

In the U.S. legal system, people are innocent until proven guilty and are entitled to due process. A working relationship must be maintained with the judicial system, both the courts and the prosecutor's office, to ensure that these rights are protected. If they are not, every citation for violation of the fire code or arson case will be dismissed.

It is not uncommon for fire fighters to be subpoenaed into court to testify in lawsuits and criminal prosecutions arising from emergencies that they responded to. They may just be witnesses to what the scene looked like when they got there, or they may be defendants due to the actions they took. A positive working relationship with the local prosecutor's office needs to be maintained to assist in preparing fire fighters when they are called on to present testimony.

Office of Emergency Management

The local office of emergency management can come to the aid of the fire department in times of major emergency by providing resources in the form of personnel and equipment. It also serves as the liaison between local and state emergency management offices.

Emergency Medical Service Agency

The local EMS agency ensures that ambulance and fire department EMTs and paramedics are provided training and meet licensing requirements. It also regulates private ambulance companies in its jurisdiction. In the case of a mass casualty incident or other specific circumstances, it may be responsible for polling the local hospitals to determine the availability of beds in their emergency rooms for victims.

Emergency Operations Center

Many jurisdictions, on local and state levels, have an emergency operations center (EOC) to which different functional representatives respond during an emergency activation. Once activated, the center is the focal point of decision making on a regional basis. Resource needs are addressed and prioritized in cases of multiple incidents. Cooperators in an EOC may be government and private agencies, including law

FIGURE 5-5 Union local headquarters.
© Donna McNicol, donnamcnicol.com

enforcement, fire, human services, waste management, mental health care providers, the Red Cross, and others as needed. On the departmental level, the center is called a department operations center, and it performs such functions as staffing fire stations that are vacant due to commitment to an incident, accessing resources from departments outside the jurisdiction, and other similar tasks.

Fire Fighter Union Locals

Fire fighters who have joined state or national union organizations often have a fire fighter union local office **FIGURE 5-5**. This allows them to address grievances, negotiate contracts, and deal with other matters on a local level. These organizations are the focal point for political activity in local elections. Many fire fighter union locals are heavily involved in fund-raising activities for burn foundations and other charities.

Community Service Organizations

Local community service organizations are sources of revenue and equipment for fire organizations **FIGURE 5-6**. They also provide opportunities for community outreach by allowing fire fighters to attend their meetings and make fire safety presentations.

As an example of their community support and involvement, many of these groups have purchased smoke detectors to be given out to the public. By working together, the service groups and the fire department can improve public relations as well as provide valuable public safety services. Some examples of these service groups are Elks, Kiwanis, Lions, Optimists, and Rotary clubs.

FIGURE 5-6 Community service organizations' logos posted on a city sign.
© Jones & Bartlett Learning

Periodical Publications

Another source of current information is emergency service–related magazines. These magazines come in both printed and electronic formats. They aid fire fighters in staying up to date on the latest developments in fire and emergency medical service equipment, methods, and activities. They are an aid to the

fire science student in that they are a source for finding out about the latest in firefighting and medical technology, safety practices, strategy and tactics, equipment design, and incident review. They can also be an aid in developing critical thinking skills. Combined with the agencies discussed earlier in the chapter, they can be used to conduct research for term papers and other projects. By studying the results of the actions of others, you can develop yourself to be a better educated, safer, and more efficient fire fighter.

The following is a listing of some of the leading fire service–related magazines:

- *Firehouse*
- *Fire Apparatus Journal*
- *Fire Chief*
- *Fire Engineering*
- *Firefighter Nation*
- *Firefighting News*
- *FireRescue1*
- *Industrial Fire World*
- *Journal of Emergency Medical Services (JEMS)*
- *NFPA Journal*
- *Wildland Fire fighter*

Wrap-Up

CHAPTER REVIEW

- Numerous and varied organizations and associations related to the fire service exist on the national, state, and local levels.
- National and international organizations include the National Fire Protection Association (NFPA) and organizations in the following fields:
  - Transportation and storage of hazardous materials
  - Fire protection systems and equipment
  - Codes and standards
  - Disaster assistance and recovery
  - Professional accreditation and certification
  - Professional research
  - Fire apparatus and equipment
  - Personnel and training support
  - Incident management
  - Insurance providers
  - Emergency medical services
- There are numerous publicly funded agencies of the federal government concerned with protecting the public from fire.

- All of the federal agencies are organized under the National Response Plan, established under the guidance of the Department of Homeland Security (DHS).
- Much like the federal government, each state has numerous agencies and departments that interact with the fire service either directly or indirectly.
- Local organizations that may work with or benefit the fire department include the following:
  - Burn foundations
  - Local government
  - Health department
  - Law enforcement
  - Building department
  - Water department
  - Zoning/planning commission
  - Street department
  - Judicial system
  - Office of emergency management
  - Emergency medical services agency
  - Emergency operations center
  - Fire fighter union locals
  - Community service organizations
- Another source of current information is emergency service–related magazines, which come in both printed and electronic formats and aid fire fighters in staying up to date on the latest developments in fire and emergency medical service equipment, methods, and activities.

KEY TERMS

Accreditation A process in which certification of competency, authority, or credibility is presented.

Certify A formal, written document issued and undersigned by an official authority to recognize an individual or a group possessing certain qualifications or meeting certain standards

Consensus standards Standards that are developed through the consensus process. Usually representatives from government and industry meet to determine the language of the standard. Input is sought and then meetings are held to determine the final language used in the standard. This process is often used in the creation of codes, such as the NFPA's National Fire Codes.

Corrosive Able to destroy and damage other substances with which it comes into contact.

Fire flow The total volume of water required to control the fire incident expressed in gallons per minute.

National Response Plan The plan that delineates the all-discipline, all-hazards response and responsibilities of all federal agencies for the management of domestic incidents in the United States.

Placards Signs or notices for display in a public place.

Toxic A substance that is poisonous and possibly lethal.

CASE STUDY

A fire department decides to bring down the fire death loss in the community through an outreach program. One of the goals is to engage the public in the conversation. The objective of "each one teach one" regarding home fire safety will be accomplished through going to elementary schools and presenting information to children, who will then take the information home and share it with their parents and siblings. Another objective is to ensure there is a working smoke detector in every dwelling.

The public education specialist from the fire department goes online and researches programs that address home fire safety. Several are available and some of them are aimed at elementary school–aged children.

1. Which of the listed agencies gathers and disseminates civilian fire death data?

    A. IAFF
    B. NFPA
    C. DOT
    D. NIOSH

2. A good option to approach local civilian groups to assist in funding a smoke detector giveaway program would be to contact:

    A. the National Integration Center.
    B. the National Safety Council.
    C. the International Fire Code Institute.
    D. service clubs.

3. What are the names of two NFPA safety education programs for children?

    A. *Learn Not to Burn*
    B. *Hot Spot Spotters*
    C. *Fire Out!*
    D. *Risk Watch*

4. A contact on the state level to seek assistance with putting the program together would be the:

    A. office of state fire marshal.
    B. state fire training department.
    C. office of emergency services.
    D. state forester.

REVIEW QUESTIONS

1. What organization would be a source of information about fire equipment?

2. What organization would be a source of information about fire sprinklers?

3. What organization would be a source of information about labor relations?

4. What organization would be a source of information about model codes and their adoption?

5. When you arrive at a hazardous materials incident, you see a placard on a truck involved and need information as a guide as to how to proceed. What is your information source?

6. Suppose you are at the scene of a hazardous materials spill and need information on the product. Whom do you contact?

7. At the same incident, the spill is about to enter a navigable waterway. Whom do you contact to alert the proper federal agencies?

8. The responsible federal agencies have been alerted. Which ones are likely to respond?

9. Suppose your department needs training manuals on a variety of firefighting subjects, and you are asked to find a source for them. Whom would you contact?

10. Suppose there has been a large-scale natural disaster, other than a fire, in your area. Which federal agency takes responsibility for providing assistance?

11. Which state-level agency is activated to assist?

12. There is a wildland fire of extreme proportions in the national forest in the area. Which federal agencies can be called on to assist?

13. For the preceding fire, state resources are needed; which agencies are called upon to assist?

DISCUSSION QUESTIONS

1. Suppose you are trying to raise money to purchase a rescue tool for your department. What are some of the local service clubs you can contact?

2. Suppose arson has become a growing problem in your jurisdiction. What public and private groups can be activated to form an arson task force?

3. There are on average 100 fire fighter line-of-duty deaths a year. Which of the organizations discussed in this chapter are heavily involved in addressing this issue?

4. List sources to be used in preparing a plan of action to search for job opportunities in the fire service.

REFERENCES AND ADDITIONAL RESOURCES

Association of American Railroads. 2017. *Field Guide to Tank Cars*. Washington, DC: Association of American Railroads.

Association of American Railroads/Bureau of Explosives. 2011. *Emergency Handling of Hazardous Materials in Surface Transportation*. Sewickley, PA: Association of American Railroads/Bureau of Explosives.

Averill, J., Lori Moore-Merrell, Adam Barowy, Robert Santos, Richard Peacock, Kathy A. Notarianni, and Doug Wissoker. 2010. *Report on Residential Fireground Field Experiments*. Washington, DC: National Institute of Standards and Technology.

Insurance Services Office. 2013. *Fire Suppression Rating Schedule*. Jersey City, NJ: Verisk Analytics.

International Code Council. 2015. *International Urban–Wildland Interface Code*. Whittier, CA: International Fire Code Institute.

Mutual Aid Box Alarm System. 2019. Mutual Aid Box Alarm System – IL. http://www.mabas-il.org/Pages/default.aspx

National Fire Protection Association. *NFPA 1001: Standard for Fire Fighter Professional Qualifications*. Quincy, MA: National Fire Protection Association.

National Fire Protection Association. *NFPA 1404: Standard for Fire Service Respiratory Protection Training*. Quincy, MA: National Fire Protection Association.

National Fire Protection Association. *NFPA 1500: Standard on Fire Department Occupational Safety, Health and Wellness Program*. Quincy, MA: National Fire Protection Association.

National Fire Protection Association. *NFPA 1581: Standard on Fire Department Infection Control Program*. Quincy, MA: National Fire Protection Association.

National Fire Protection Association. *NFPA 1901: Standard for Automotive Fire Apparatus*. Quincy, MA: National Fire Protection Association.

National Fire Protection Association. *NFPA 1961: Standard on Fire Hose*. Quincy, MA: National Fire Protection Association.

National Fire Protection Association. *NFPA 1971: Standard on Protective Ensembles for Structural Fire Fighting and Proximity Fire Fighting*. Quincy, MA: National Fire Protection Association.

National Fire Protection Association. *NFPA 1975: Standard on Emergency Services Work Clothing Elements*. Quincy, MA: National Fire Protection Association.

National Fire Protection Association. 2017. *Fire Protection Handbook*. 21st ed. Quincy, MA: National Fire Protection Association.

National Wildfire Coordinating Group. 2014. *PMS 210 Wildland Fire Incident Management Field Guide*. Boise, ID: National Wildfire Coordinating Group.

National Wildfire Coordinating Group. 2018. *PMS 461 Incident Response Pocket Guide*. Boise, ID: National Wildfire Coordinating Group.

U.S. Coast Guard. 1999. *Chemical Hazards Response Information System (CHRIS)*. Washington, DC: United States Coast Guard.

U.S. Department of Transportation. 2016. *Emergency Response Guidebook*. Washington, DC: U.S. Department of Transportation.

CHAPTER **6**

Fire Department Resources

OBJECTIVES

After studying this chapter, you should be able to:

- List and describe facilities in modern fire departments.
- List and describe common fire apparatus.
- List and describe fire tools and appliances.
- List and describe heavy equipment used in the fire service.
- List and describe personal protective equipment used in the fire service.
- Describe the types and uses of aircraft in firefighting.

Case Study

On Father's Day, June 17, 2001, a fire began in the Long Island General Supply store in the Long Island City section of Queens in New York, NY. The store contained flammable liquids and propane tanks. At 2:48 PM (1448 hours), a small explosion was heard from the building, followed by a huge blast. The explosion knocked down every fire fighter working the fire from the street and knocked down a wall of the building.

A call came over the radio from one fire fighter stating that he was trapped in the basement: "I'm trapped in the basement by the stairs. Come get me." The chief at the scene immediately ordered a fourth alarm, which brought additional apparatus and personnel to the scene. In addition, numerous special calls for apparatus and personnel were made. Eventually there were 144 pieces of apparatus at the scene: 46 engines, 33 ladders, 16 battalion chiefs, 2 deputy chiefs, all 5 rescues, 7 squads, and many more pieces of specialized equipment. When the fire was finally extinguished, 3 fire fighters had been killed.

1. What resources would be used to remove tons of rubble from a building collapse and extinguish a fire with fire fighters trapped?
2. How long would your self-contained breathing apparatus (SCBA) provide you with breathing air if you were trapped in a building?
3. What safety devices are incorporated into your full structural personal protective equipment (PPE) to protect you from falling debris and heat exposure? How effective would they be in this situation?
4. What resources are available at your local fire department to deal with this type of incident?
5. What resources are available from local construction companies to deal with this type of incident?

 *Access Navigate for more resources.*

Introduction

The modern fire department relies on many types of resources—such as facilities, apparatus, and equipment—to do its job. The facilities described in this chapter are not available at every fire department due to need and budget constraints; they represent a sample of the facilities at fire departments across the country.

The modern fire service requires a variety of specialized equipment and facilities to address the myriad of tasks it is expected to perform. The modern fire engine and the equipment it carries can perform many functions, but not all. For example, it can be effective at carrying an extinguishing agent to the fire, allowing the placement of that agent where it is needed, and applying the agent in the correct manner.

In addition to the standard pumper (also called an engine; in this text, the terms are used interchangeably), specialized equipment is utilized by the fire service to deal with many specific circumstances often faced by fire fighters. These include a vehicle crash with extrication of victims, refinery fires, trench rescues, and hazardous materials or medical aid incidents.

In this chapter, the common types of fire and emergency service facilities, equipment, and apparatus will be described (some additional facilities will be introduced in Chapter 8, *Support Functions*). The descriptions are broad in scope because each department has its own thoughts on what is required to meet its needs. A visit to a fire apparatus factory will bear this out. Engines and trucks are specified by the purchasing fire department in many configurations, such as body design, color, pump, tank size, and emergency lights.

The apparatus and equipment described in this chapter have evolved over a period of many years to fulfill specific functions for particular firefighting situations or methods. Not all of the apparatus and equipment is operated or carried by every fire department because situations and types of incidents vary. It is important to be aware of the differing types of apparatus and equipment when operating in conjunction with other departments and agencies on large or complex incidents. As you study this chapter, try to think of some apparatus and equipment available in your area that could be adapted to firefighting use. As you look at the chapter, you should realize that this is exactly what has been done with many of the tools in use today.

The most important resource, however, is the fire fighter. Without intelligent, trained, aggressive fire fighters, the fanciest and most expensive facilities, apparatus, and equipment in the world are worth nothing. Always remember that skill and knowledge are what makes the whole system work. The best fire fighters are capable not only of using all of the tools available to them, but also of adapting those tools to new uses.

Fire Apparatus

The modern fire service requires many types of apparatus to perform its duties in protecting the community. These types of apparatus vary widely in their design and application and what they are called in various parts of the country (i.e., wagon, pump, engine). In many instances, apparatus has been modified or specially designed to perform the required work better. There are several basic designs for the specialized apparatus used today.

Cab and Chassis

Fire apparatus manufacturers start out with a cab and chassis **FIGURE 6-1**. Depending on the needs and specifications of the buyer, these can vary greatly. The cab and chassis can be either two- or four-wheel drive. Fire apparatus are designed to meet the specifications of *NFPA 1901: Standard for Automotive Fire Apparatus*. To meet these specifications, apparatus must provide inside seating for all personnel. Wearing seat belts is law in most states and is department policy, as well as good safety practice. Today's new engines are likely to be of the four-door cab variety. This development in safety has mostly done away with the practice of fire fighters riding on the tail step or running boards of the apparatus. Even in pumpers of the semi-closed cab type, fire fighters should never release their seat belts and stand up until the apparatus is stopped and the brake is set. It is very easy to fall from a moving apparatus if you are standing and the driver hits a bump or swerves to miss an obstacle.

Let us now take a tour of the cab area of a standard new piece of fire apparatus. The vehicle is equipped

with large mirrors to aid in safe operation. There should also be a fish-eye mirror for pulling up close alongside objects like curbs. Inside the cab is the driver's seat with a large steering wheel **FIGURE 6-2**. Mounted on the dash, in front of the driver, are gauges for fuel, temperature, and air pressure in the air brake system. There is a speedometer and tachometer as well. Also on the dash is a push/pull switch that operates the air brakes and a headlight switch.

The vehicle is equipped with either a manual or automatic transmission. When equipped with a manual transmission, there is a switch or lever that disengages the drive train from the rear wheels and transfers the power output of the motor to the pump. The power is redirected through a pump transfer transmission. After the power is transferred to the pump gearing, the road transmission is returned to top gear. When equipped with an automatic transmission, there is a pump switch that engages the pump transfer

FIGURE 6-1 Fire engine chassis prior to buildup. Pump is located behind cab. Water tank is mounted to frame rails behind the pump.
© Jones & Bartlett Learning

FIGURE 6-2 Interior of fire engine cab.
© Jones & Bartlett Learning

transmission. Once engaged, the road transmission is returned to top gear. With either type of transmission, if—either through oversight or mechanical failure of the switch—the power is not redirected through the pump transmission, the vehicle may lurch forward when the clutch is engaged or the throttle is opened.

Vehicles with a manual transmission have clutch, brake, and throttle pedals. An automatic transmission vehicle has brake and throttle pedals. If the power plant of the pumper is a diesel, it may also have an engine brake of the Jake brake type. The use of these devices greatly reduces brake fade and extends the life of brake components on a 25,000–40,000-pound vehicle, the weight of a typical pumper. Ladder trucks are even heavier. Fire engines and other fire vehicles lead a tough life, accelerating and stopping repeatedly on the way to emergencies in metropolitan areas and when operated in hilly terrain.

In the center of the dash are the switches that control the lights, electronic siren, and/or electric siren. There are switches for the light bar on the roof, as well as the warning lights on the rear and sides. In most pumpers there is a so-called master switch that permits all of the warning lights to be controlled by one switch, allowing the individual switches to be left in the on position. If there are alley lights mounted in the light bar, there should be two switches: one for the right and one for the left. Another light bar–mounted device is for preemption of traffic signals. This allows the changing of traffic lights to green in the direction of travel of the apparatus. There is another switch for the lights on the rear, called pickup lights or hose lights, which are used at night to illuminate the area around the rear of the vehicle for reloading hose or backing up. The electronic siren is equipped with an on/off switch and settings for public address, radio outside speaker, manual, yelp, high/low, and wail. In the manual position, pushing the horn button in the middle of the steering wheel activates the siren. This manual use is handy when making long rural responses in which the siren is needed only intermittently. Some sirens have an electronic air horn capability, which is usually specified on vehicles that do not have an onboard air compressor and therefore cannot support real air horns. The air horns are mounted in the front bumper to increase their effectiveness. They are placed close to the height of most automobile windows **FIGURE 6-3**.

The radio system for the apparatus to maintain contact with the dispatch center and the other apparatus is located in the cab. Modern radios have multichannel capability. The use of multiple channels allows the fire department to operate at several incidents at one time without developing overcrowding on one frequency.

FIGURE 6-3 Front view of fire vehicle showing location of air horns in front bumper.
© Jones & Bartlett Learning

There may also be an intercom system, with headsets and microphones for the crew. These allow the members of the crew to talk to each other easily over the sound of the motor, siren, and air horn when responding. As an added benefit, they protect the fire fighters' hearing. On a piece of apparatus with the fire fighters separated from the officer by the back wall of the cab, the headsets allow the officer to give instructions to the fire fighters and allow the fire fighters to hear the radio traffic and the at-scene description given by the officer. There may be a connection for a headset on the pump panel so the pump operator can hear the radio over the roar of the motor when operating at scene **FIGURE 6-4**.

Many fire apparatus are designed with a breathing apparatus mounted between the driver and the officer.

FIGURE 6-4 Operator at pump panel maintaining communications through use of headphones.
© Jones & Bartlett Learning

This allows the officer to don the breathing apparatus quickly before leaving the cab.

In some models of enclosed cab pumpers, the fire fighters ride facing rearward; in others they face forward. The seats in this area can be designed with the back cut out, allowing a breathing apparatus to be mounted where it can be quickly donned when needed.

In areas where summer heat is a factor, the apparatus may be equipped with a built-in air-conditioning system, which allows the fire fighters in full turnout gear to stay cool when making long responses. It also makes riding in the apparatus much more comfortable on routine assignments.

The cab portions vary widely depending on the specifications and financial resources of the purchaser. Some fire departments have custom pumpers with an **ambulance gurney** mounted crosswise in the cab, giving the engine patient transport capability. Other vehicles have a cab with a walk-through design and enough room to contain a mobile command post.

Many fire department vehicle cabs are now equipped with **mobile data computers (MDCs) FIGURE 6-5**. These computers allow personnel to connect to the computer-aided dispatch (CAD) system. The units provide a connection to gather and update incident information. They can provide information about addresses, such as known hazards, owner contact information, and so forth. An onboard computer can contain a basic street map with overlays of hydrant locations, preincident plans, sewers and storm drains, and other information. To make the information more user friendly, the layers can be turned on and off as needed. The units are usually equipped with touch screens so the operators can provide availability information

FIGURE 6-5 Mobile data computer (MDC) terminal for retrieving information while en route and at scene.
© Jones & Bartlett Learning

at scene, without using the radio—thereby minimizing voice traffic on sometimes crowded channels and reducing the possibility of a message being missed or having to be repeated.

In addition, **automatic vehicle location (AVL)** systems are often used in conjunction with MDCs. The unit in the apparatus receives GPS signals, and once its location is computed, it is provided as a graphic display to the dispatch center CAD system and in the apparatus. This allows the closest resource response to be generated. AVLs also come with routing capability that displays the route to the scene. The route information can include and be based on one-way streets, school zones, construction zones, highway divider walls, speed limits, and other features that may restrict response to a location. When the incident is entered into the CAD system, the computer utilizes the AVL and routing information to determine the closest appropriate resource.

Both MDC and AVL systems assist units in getting to the scene more quickly. Fires grow very rapidly, and in the case of medical emergencies and rescues, time can be the difference between life and death.

With the high number of accidents that happen when backing up, video cameras aimed over the rear, like those mounted on motor homes, are coming into use. Some ladder trucks with tiller steering are equipped with side view cameras for the tiller operator as an aid in steering. At the fire scene, one of the greatest resources the officer has is pertinent and up-to-date information. As more sophisticated devices become available for the storage and retrieval of information and communications, they will find their way into the cabs of fire apparatus.

The standard for fire engines includes two individual battery sets of the truck type, allowing for additional starting amperage as well as a backup in case one set goes dead. It also allows for increased storage capacity for the tremendous draw placed on the electrical system due to the warning lights. A switch in the cab allows the batteries to be turned off and one or both sets of batteries to be used at a time.

Motor

Fire apparatus today are mostly powered by diesel motors, noted for their long life and durability under tough conditions **FIGURE 6-6**. Diesel motors are selected for their abundance of torque. It takes a lot of power to operate a large gallon-per-minute pump and to supply effective hose streams. The motor must also be able to propel a heavy vehicle to operating speed in a short period of time. It is not uncommon for vehicles

FIGURE 6-6 Diesel motor—with turbocharger to boost horsepower output—installed in fire engine.
© Jones & Bartlett Learning

used in hilly terrain to have a diesel motor with both a turbocharger and a supercharger.

The motor is often equipped with an oversized alternator to supply power for all of the extra lights used as warning devices. Leaving the motor at idle with all of the emergency lights operating for any length of time can drain the batteries. The oversize alternator must be turning at around 1000 rpm to develop enough amperage to operate all of the additional lights and other electrical equipment on the vehicle. Apparatus are equipped with a high-idle switch that automatically increases the idle speed when engaged. Most apparatus are also supplied with an **inverter** that allows 110-volt lighting to be used without starting the onboard generator.

Modular Apparatus

Some departments use modular apparatus. By having different modules that are mountable on the chassis, the department gains flexibility. An example would be a cab and chassis with a dismountable body stocked with hazardous-materials or heavy-rescue equipment, giving the department the capability of having less of the expensive parts of a truck—the cab and chassis—and several choices as to which bodies to mount as the need arises. The U.S. Forest Service and Bureau of Land Management have employed this concept for many years by utilizing flatbed trucks with self-contained pumper bodies mounted on them. This concept does not work with a regular pumper, however, because of the plumbing and mounting of the pump to the chassis. In areas where a large water tank is carried, it also tends to raise the center of gravity to the point that the apparatus is limited to on-road use only.

Pumper/Engine

The basic piece of motorized apparatus in the fire service is the pumper. These apparatus are designed to meet *NFPA 1901: Standard for Automotive Fire Apparatus* specifications. Most pumpers in service are of the triple-combination type. A triple-combination pumper is so named because it carries hose and a pump and has a water tank. Other apparatus is carried as the situation dictates. All of these attributes, as well as the cab and chassis, can be combined in various forms according to need. Some jurisdictions have purchased specially designed pumpers with an extendable mounted ladder or boom that gives them the capability of applying elevated streams.

> **Tip**
>
> Most pumpers in service are of the triple-combination type.

Water Tank

The water tank on apparatus is located under the hose bed, behind the cab and pump panel. Water tanks on fire pumpers vary in size. The NFPA requires a minimum size of 300 gallons to be certified although it is quite common to have a 500 gallon capacity. On small apparatus and metropolitan engines, where water is readily available from the hydrant system, tanks tend to be around 200 to 500 gallons. In rural areas, tanks range from 750 to 1500 gallons **FIGURE 6-7**. If it carries any more water than this, the apparatus is considered to be a water tender/tanker. If the tank exceeds 1000 gallons, the vehicle usually rides on three axles due to vehicle weight. The tanks are equipped with **baffles** to prevent the water from shifting around and causing the vehicle to become unstable.

As water tanks increase in size, and vehicle length and width stay the same, the only way to go is up.

> **SAFETY TIP**
>
> The large tank adds to the overall weight of the apparatus, and going up in height raises the center of gravity, making the engines more top heavy and likely to tip over when operating in sidehill situations or when performing evasive maneuvers.

FIGURE 6-7 Combination water tender and fire engine with preconnected attack lines and large-volume pump.
Courtesy of Jeff Riechmann

FIGURE 6-8 Contrast of hose bed heights due to tank size.
© Jones & Bartlett Learning

Standard-size engines with large tanks tend to have hose beds high in the air, making the hose harder to access **FIGURE 6-8**. A roll of wet 2½-in. hose weighs approximately 60 pounds and is hard to lift into the hose bed to carry it back to the station. The higher the hose bed, the harder it is to load the wet hose. The large tank adds to the overall weight of the apparatus, and going up in height raises the center of gravity, making the engines more top heavy and likely to tip over when operating in sidehill situations or when performing evasive maneuvers.

Plastic, galvanized steel, and stainless steel are popular materials for water tank construction. Untreated metal tanks tend to corrode over time and develop leaks. Plastic, galvanized steel, and stainless steel tanks are corrosion resistant and stand up better to wetting agents and foaming agents added to water to improve firefighting characteristics.

Foam Systems

More and more apparatus are being equipped with built-in foam systems. These can be either class A, class B, or both. A built-in class A system provides superior knockdown on ordinary combustible materials. Class B systems are for use on hydrocarbon fuels (e.g., gasoline, diesel fuel, and jet fuel). The systems consist of a built-in foam concentrate tank, an injector, and a means of adjusting the concentrate level delivered into the water stream. Many foam systems also come with a flow meter to keep track of how much water has flowed and how much foam concentrate has been used.

Another variation of the class A foam system is the compressed air foam system. This system utilizes an air compressor to inject air into the hose stream as it leaves the pump. Combined with class A foam, this provides a light and airy foam that can stick to vertical surfaces. It is often used to pretreat structures and trees in the path of oncoming wildland fires, giving personnel the opportunity to protect structures without actually remaining at the site, directly in the path of an advancing fire. A word of caution here is that the nozzle reaction from a hose connected to a compressed air foam system pumper is much more than that of a regular pumper at the same pump pressure due to the release of air pressure in the hose line when the nozzle is opened.

Another type of compressed air foam system used is one that utilizes compressed air stored in tanks on the apparatus to expel the extinguishing agent. This system is lightweight and simple in design. It is used for small fires and may be installed on a ladder truck to save the weight and complexity of having a centrifugal fire pump on board.

Pumps

The main purpose of any pump is to lift water or to add pressure to the water so it can flow through hose and nozzles and be applied to the fire away from the pump. To deliver the contents of a reservoir or water tank to the third floor of a building requires some device to force the water through the hose and appliances.

Centrifugal Pump

The most commonly used main pump on fire apparatus is the centrifugal pump. The pump is usually located directly behind the cab of the engine (see Figure 6-1). It has many desirable features from the firefighting standpoint. The centrifugal pump consists of one or more vaned wheels, called impellers,

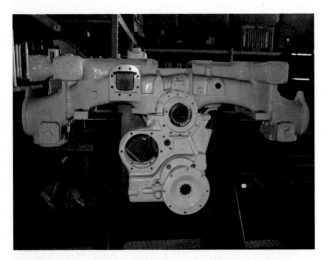

FIGURE 6-9 Fire pump cutaway mounted for display.
© Jones & Bartlett Learning

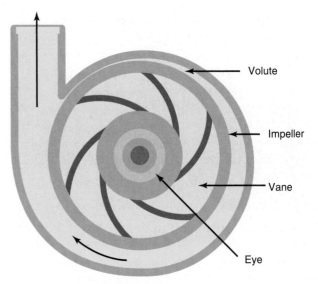

FIGURE 6-10 Interior design of centrifugal pump.
© Jones & Bartlett Learning

mounted on a shaft **FIGURE 6-9**. Power is supplied to the pump from the motor of the pumper through a pump transmission. The transmission can consist of a transfer case or a power takeoff (PTO) unit. If the transmission is equipped with a transfer case, the power is redirected from the rear wheels to the pump, and the pumper stays stationary when the pump is operated. In the case of a PTO unit, the pumper can pump and roll at the same time. This type of construction is often found on wildland fire pumpers so mobile attacks can be made. The limitation with this setup is that the pump operates at engine speed, so driving at low speeds reduces pump pressure as rpm is kept low. A third option, often found on engines used for wildland firefighting, is a separate motor carried for operating the pump. This option also allows for the use of the pump while moving. The pump speed is determined by a separate throttle allowing for high pressure at low vehicle speed.

> **Tip**
>
> The main purpose of any pump is to lift water, and the most commonly used main pump on fire apparatus is the centrifugal pump.

The pump casing has one or more suction inlets where water can enter the pump. The water is then directed into the center of the impeller, called the eye. As the pump impeller spins on its axle, it directs the water to the outside of the casing, imparting centrifugal energy to the water, hence the term *centrifugal pump*. The principle is the same as a merry-go-round. As you stand in the middle of the disk and it spins

faster and faster, you are pushed toward the outside. The centrifugal pump does the same thing. The impeller consists of a two-sided disk with vanes between the disks. This design better aids the impeller in transferring the centrifugal energy to the water. When the water leaves the impeller, it collides with the pump casing, resulting in increased pressure in the pump. The casing, called the volute, is designed as a modified circle with a wider clearance in one area. As the water builds up pressure, it is forced out of the volute and into the discharge plumbing. The outflow of the water creates a partial vacuum at the eye of the impeller, drawing more water into the pump through the suction plumbing. The discharge plumbing extends out through the pump panel on the side of the apparatus and has valves and threaded ends for the connection of fire hoses **FIGURE 6-10**.

The advantages of the centrifugal pump are numerous. The clearances between the impeller and the pump casing allow the pump to do several things. It can spin at high rates of speed and build up large amounts of pressure without discharging any water. This situation occurs when discharge valves and nozzles are turned off and on at the fire scene. There is not always someone available to stand by the pumper and operate the throttle as the volume and pressure

> **SAFETY TIP**
>
> Water can become quite hot if it is not circulated through the pump for a while. High water temperature may damage the pump or cause a scald burn to personnel when a nozzle is first opened.

demand increase and decrease. One must be careful because the water can become quite hot if no water is circulated through the pump for a while, depending on the speed, possibly damaging the pump or causing a scald burn to personnel when a nozzle is first opened. Another feature is that a centrifugal pump can take advantage of any pressure coming in on the suction side, either from another pumper or a hydrant, effectively allowing the motor driving the pump to work less hard. Centrifugal pumps can also tolerate the pumping of trash and dirty water to a certain extent. They are equipped with suction screens to keep out debris and rocks that can damage the pump **FIGURE 6-11**.

The driver/operator controls pump pressure, water flow, and other functions from the pump panel position (see Figure 6-4). Attached to the pump is a device provided for the fire fighter's safety known as the *relief valve* or *pressure governor*. The pressure governor reads the pressure provided to the hose lines. The maximum pressure is preset at the pump panel. If a nozzle is shut down or flow is otherwise restricted, the pressure governor adjusts the throttle setting to the pump motor to keep the remaining lines from exceeding the preset pressure.

The relief valve is another type of semiautomatic pressure-regulating device. This spring-operated valve is set at the desired pressure on the pump operator's panel. When actuated, some of the water is redirected from the discharge side of the pump back to the suction side. When the pump is operated with two or more lines coming from it, and if one of the lines were to be shut down, all of the water would try to exit through the line remaining open, which would cause a sudden pressure surge in the open line. A pressure surge could cause a fire fighter to lose footing or fall from a ladder. The relief valve effectively reduces this pressure surge.

> **Tip**
>
> The centrifugal pump can act only on the water that enters it, but it can take advantage of any pressure coming in on the suction side.

The disadvantage of the centrifugal pump is that it can act only on the water that enters it. It cannot draw water into itself from a static water source. The water must be introduced under slight pressure, from a hydrant or other pumper, or another type of pump must be used to create a vacuum in the pump casing, causing the water to enter. Once the water does enter, the discharge of water from the pump can keep the vacuum going, and the suction is self-sustaining. If the pump is driven too hard and the suction capability is exceeded, the pump will start to cavitate (to form small vapor bubbles in the interior), causing damage to the impeller.

Main fire pumps of the centrifugal type come in sizes ranging from 250 to 2500 gpm, in increments of 250 gpm. The minimum recognized fire pump is 250 gpm. In addition, the pump must deliver 100 percent rated capacity at 150 psi, 70 percent rated capacity at 200 psi, and 50 percent rated capacity at 250 psi. For example, should a fire department specify a 1000 gpm pump in its new apparatus, the pump must be able to deliver 1000 gpm at 150 psi, 700 gpm at 200 psi, and 500 gpm at 200 psi.

Positive Displacement Pumps

The other type of pump mounted on fire apparatus is the positive displacement pump **FIGURE 6-12**. This type of pump can come in several forms: gear pumps, piston pumps (like the hand pumpers), and diaphragm pumps. The principle is that every time the pump cycles, a specified amount of fluid is taken in and discharged. If 1 gallon of water enters on the suction side at the start of a cycle, 1 gallon will be discharged from the pressure side at the end of the cycle. As the rate of speed increases, the volume will increase in direct proportion.

These pumps have the advantage of being able to pump air, thus making them self-priming. When piggybacked onto a centrifugal pump, the positive displacement pump can evacuate the air in the centrifugal pump. When used for this purpose, it is called

FIGURE 6-11 Pump suction inlet screen to prevent ingestion of rocks and other debris into pump.

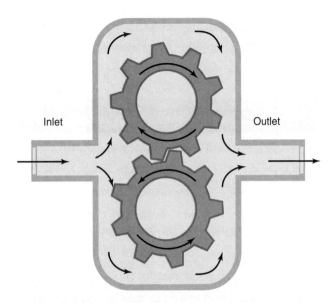

Inlet

Outlet

FIGURE 6-13 Pumper operating at draft, taking suction from a static water source.
© Jones & Bartlett Learning

FIGURE 6-12 Gear-type positive displacement pump. A. Interior components. B. Priming pump mounted on pump body.
© Jones & Bartlett Learning

a priming pump because it primes the centrifugal pump.

This action creates a reduction of pressure in the centrifugal pump, which allows water to be drawn up the suction hose and into the pump. Once the water enters the pump, pressure is added by the pump and a fire stream can be developed **FIGURE 6-13**. This occurs because the weight of the atmosphere over the earth exerts approximately 14.7 psi on the surface at sea level. When the atmospheric pressure inside the pump is reduced below 14.7 psi, the water is forced up the suction hose and into the pump.

Because the centrifugal pump cannot pump air due to its loose tolerances, the positive displacement pump is needed to create the vacuum. The positive displacement pump is very small in relation to the centrifugal pump and is typically driven by an electric motor the

size of an automobile starter motor. On some wildland firefighting engines, the priming pump is hand operated. Another advantage of positive displacement pumps is that they can create tremendous pressure when pumping fluids. For applications where high pressure is necessary, such as pressure washers, they are the pump of choice.

The disadvantages of positive displacement pumps are enough that they are not used as main fire pumps. If a positive displacement pump were being used and the nozzle were turned off, the pressure in the hose would increase until something blew out. The tolerances are so close that very small debris can jam the pump. They also do not gain any benefit from water forced into them. Because of its design, a positive displacement pump of the same gallons per minute would be much heavier and more expensive than a centrifugal pump.

Aerial Ladder and Elevating Platform Apparatus

Aerial ladder apparatus come in two configurations. One is the tractor trailer type with tiller steering. The tiller operator sits in a small cab at the rear of the apparatus and can steer the rear wheels **FIGURE 6-14** (see also Figure 6-56). The advantages of the tractor trailer and tiller types are that they can maneuver in tight spaces and make sharp corners. The disadvantages are that if the driver goes too fast, the tiller operator can lose control and collide with parked cars or, even worse, people standing on the sidewalk. The other configuration is the straight chassis, which carries the ladder on the chassis, not in a tractor trailer design (see Figure 6-59). This configuration may be

FIGURE 6-14 Aerial ladder truck with ladder extended.
Courtesy of Kern County Fire Department

limited to ladder length due to front axle weight restrictions. Another variation is the all-steer, where the rear wheels of a straight chassis are steerable.

SAFETY TIP

If the driver goes too fast, the tiller operator can lose control and collide with parked cars or, even worse, people standing on the sidewalk.

The aerial portion of the apparatus can come in various configurations as well. There are the extendable ladder types, in which the ladder raises from the bed and a set of fly sections is extended. These ladders are advantageous in that personnel can ascend and descend the ladder when it is raised. On this type of apparatus, the ladder placement is controlled from the operator's platform at the base of the ladder. Some of these apparatus are equipped with an enclosed platform at the ladder tip from which personnel operate. This type is known as an aerial ladder platform apparatus. In another design, the articulated boom type, the boom is raised hydraulically and extended through adjusting the angle of a knuckle joint **FIGURE 6-15**. This type of aerial apparatus, while very strong and easy to place from elevated platform controls, is limited in that personnel have to move the basket to the ground to enter or exit. Elevated platforms are good for using heavy tools from the basket, such as a modified jackhammer for breaching walls. In addition to the aerial ladder carried on the truck, the NFPA requires that 115 ft of ground ladders be carried for the apparatus to be classified as a ladder truck.

The ladder truck is often equipped with an intercom system connecting the person at the tip or in the basket with the operator. Some apparatus come with breathing air cylinders mounted to the basket so the

FIGURE 6-15 Articulated boom ladder truck operating at structure fire.
Courtesy of Rick Steines, Dubuque (IA) Fire Department

personnel can use them as a supply instead of SCBA bottles, which gives them a much longer supply of air. Some aerials have plumbing supplied so water can be pumped to the tip and applied through elevated streams. This design is called a water tower.

Aerials may be equipped with a fire pump on the apparatus; on others, an engine is used to pump the fire stream.

Quint

An apparatus equipped with pump, water tank, ground ladders, hose bed, and an aerial device is called a quint. Such apparatus are used as a pumper and ladder truck combination. The quint has advantages in that it can be used as a pumper. The addition of the pump, tank, and hose does add quite a bit of weight to the chassis. An aerial ladder truck is already a very complicated system with many moving parts. The addition of a pump adds to the complexity of the vehicle, requiring more maintenance. It is a good practice for departments operating quints to have a standard operating procedure/standard operating guideline (SOP)/(SOG) in place dictating when it is to be used as an engine or ladder.

Squads

Squad vehicles are the specialty vehicles of the fire service. Just about any time some special configuration is needed for a specific purpose, the vehicle is called a squad. Squads are usually strategically located and respond upon request. In departments that provide advanced life support medical functions without the transportation capability of an ambulance, the vehicle may very well be called a medical squad.

Another form of squad is the special lighting vehicle equipped with a high-wattage generator and numerous removable lights and extension cords. Hazardous materials vans and vehicles, designed for a specific purpose and outfitted with equipment for a specific function, fit the description of squads **FIGURE 6-16**.

Special air units are also squads. They are equipped with extra SCBA bottles and equipment. Some are equipped with the special compressors for breathing air or **cascade systems**. A squad with a compressor or cascade system allows the filling of SCBA bottles at the scene. These vehicles are now often combined with a large wattage electrical generator, and the combination is often called an air and light squad or unit.

A mobile command post, activated on large assignments, can be called a squad. **Tactical support** and **rehab** vehicles fit the same criteria **FIGURE 6-17**.

A squad-type vehicle used in wildland firefighting is a terra torch **FIGURE 6-18**. This unit has a tank full of jellied gasoline, similar to napalm, that is squirted from a special nozzle equipped with an igniter. The terra torch is used to light **backfires** and perform **burning out** operations.

FIGURE 6-16 Hazardous materials response team vehicle.
© Jones & Bartlett Learning

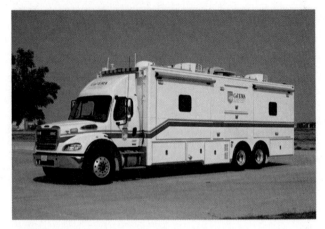

FIGURE 6-17 Tactical support (mobile command post) vehicle.
Courtesy of Charles Madderom

FIGURE 6-18 Terra torch igniting ground fuels at a wildland fire.
Courtesy of Kern County Fire Department

Ambulances

In many departments, ambulances are staffed by the fire department. Ambulances come in various shapes and sizes.

Aircraft Rescue Firefighting Apparatus

A type of apparatus specially designed for airport firefighting is the aircraft rescue firefighting (ARFF) unit, also called crash fire rescue (CFR) apparatus **FIGURE 6-19**. These apparatus are designed with large water tanks and built-in foam tanks and systems, and are all-wheel drive for going off runways. They are equipped with **turret nozzles** on the roof, forward-facing nozzles on the front bumper, and **ground sweep nozzles** to keep fire from beneath them when driving through burning fuel. All of these nozzles can be remotely controlled from inside the cab, making them a very effective firefighting

FIGURE 6-19 Aircraft rescue firefighting (ARFF) specialty vehicle.
Courtesy of Jeff Reichman

FIGURE 6-20 Hose-drying tower.
© Jones & Bartlett Learning

combination with only one person on board. They are also equipped with attack lines that can be pulled for firefighting away from the vehicle, such as interior attack in a large aircraft. The apparatus has the capability to pump and roll at the same time. There are suction inlets on the side of the vehicle that allow them to connect to fire hydrants or fire engines as the need arises. They are equipped with a minimum of ladders that are used to gain access to aircraft wing surfaces and interiors.

Another type of aircraft firefighting apparatus is mounted on a standard truck chassis and is equipped with a twinned system that allows foam and dry chemical extinguishing agent to be applied at the same time. The system has large tanks of dry chemical agent and expellant gas mounted on the apparatus. There is a water tank, foam concentrate tank, and pump for the foam system. These two extinguishing agents are discharged through hose mounted on a reel with two nozzles connected.

Fire Tools and Appliances

The fire service utilizes many types of tools and appliances to combat fires and perform rescues and other tasks. Those illustrated here are the more common items, but are not a complete listing.

Hose

Fire hose is used for getting the required water from the source of supply to where it is needed to control the fire. Fire hose is constructed in several different ways. The older type is rubber lined with one or more cotton jackets. The rubber liner prevents

leaks, and the cotton jacket(s) give the hose resistance to rupture under high pressure, provide abrasion resistance, and allow it to maintain flexibility. The problem is that hose of this type is heavy and requires thorough drying inside and out after use **FIGURE 6-20**. When used, the hose must be rolled up, loaded on the engine, and returned to the station to be dried to prevent acids forming on the inside and mildew on the outside. This condition makes the hose labor intensive and requires a complete hose change for the engine to be held in reserve to replace the hose that is drying.

Now in common use is synthetic hose that has no rubber liner and will not mildew. This hose is lighter in weight, more flexible, and can be reloaded on the engine at the fire scene. Using this type of hose reduces hose inventories required and saves time; the hose weighs less and takes up less space in the hose bed of the engine (allowing more to be carried). When fire fighters return from a fire with cotton hose, the hose is in rolls in the hose bed; synthetic hose is reloaded and ready for the next assignment. Many times, an engine has been diverted to another incident before it was able to return to the station for a fresh hose load.

To connect the hose together, each end has a coupling. One end has exposed threads and is called the male end. The other has a swivel with interior threads and is called the female end. Couplings have traditionally been made of brass but are being replaced by pyrolite. Pyrolite is a lighter-weight metal than brass and is more resistant to bending. Couplings are most commonly of either the threaded type with national standard thread or of the quick-connect (Storz) type **FIGURE 6-21**. The exception to this is the 1-in. hose used in wildland firefighting. It has either national standard or, most commonly, iron pipe thread. Hose comes in either 50- or 100-ft lengths.

The hose load carried on a pumper is determined by the type and size of fires expected to be encountered. The hose is carried in the area of the pumper known as the hose bed. It is designed for easy access

FIGURE 6-22 Fire fighters advancing hose lines to attack a structure fire.
Courtesy of Kern County Fire Department

FIGURE 6-21 Storz quick-connect fire hose coupling on large-diameter hose (LDH).
© Jones & Bartlett Learning

and laid out according to purchaser specification. The hose bed is often equipped with a cover to keep water, burning embers, and other debris off the hose.

Attack lines must be able to supply sufficient amounts of water (gallons per minute (gpm)), while not being so heavy or rigid that they cannot be maneuvered **FIGURE 6-22**. Standard attack lines for structural firefighting are 1½- or 1¾-in. with 1½-in. couplings. Hose of this size offers good flow, in the range of 100–200 gpm, while retaining ease of mobility. Attack lines are laid on the pumper so that they can be pulled by one person, advanced, and placed into operation as rapidly as possible. Supply hose is laid in the hose bed so that it can be fed onto the ground as the engine drives forward **FIGURE 6-23**. This makes it easier to establish a hose lay in a short period of time.

Supply line hose is designed to be laid out and not moved around often, especially when full of water. The past standard was the 2½-in. hose. Its attributes were that it could flow respectable amounts of water as supply line as well as having the capability of being used as an attack line when necessary. A jurisdiction with mostly rural areas would more likely carry 2½-in. supply lines. This allows for more linear feet of hose to be carried to facilitate the longer hose lays needed in an area with long distances between water

FIGURE 6-23 Fire fighters deploying supply line from hose bed.
© Jones & Bartlett Learning

FIGURE 6-24 Hose laid in hose bed (left to right): 1¾-, 2½-, and 4-in. hose. The hose is laid in the hose bed in a flat lay.
© Jones & Bartlett Learning

supplies. A standard pumper should be able to carry around 2000 ft or more of 2½-in. hose in its hose bed. With a three-pumper relay, one pumper at the water source, one in the middle of the hose lay, and the third pumper at the fire, it would be possible to supply water for over a mile. As more pumpers are added, it is theoretically possible to extend the hose lay indefinitely. The relay pumpers are needed to boost the pressure as a 1000-ft hose lay at 250 gpm would require approximately 125 psi to overcome the friction of the water going through the hose. A modern compromise in this area is the 3-in. hose with 2½-in. couplings. It can flow more water with less friction loss, without being too large to manage when charged. The 2½-in. couplings allow it to be used with standard 2½-in. hose **FIGURE 6-24**.

In a metropolitan or industrial setting, the emphasis is on large-diameter hose (LDH) (4–5 in.) for supply lines to supply large volumes of water to large fires better. This hose takes up more space per linear foot than 2½-in. hose, therefore reducing the total length of hose that can be carried on a standard pumper. With 4-in. diameter hose, 1000 gallons of water per minute can be pumped a distance of 1000 ft with only 20 psi lost due to friction of the water against the inside of the hose. A hose 5 in. or more in diameter would have even less pressure lost to friction. LDH comes in 100-ft lengths, and a length of it wet and rolled up may weigh over 100 pounds, making it very hard to hand up into the hose bed. You can imagine how much work it would be to pick up a 2000-ft hose lay. These size lines are where synthetic hose is the most appreciated. It is rolled to remove the air and then unrolled as it is loaded back into the hose bed. There is no need to lift the full rolls into the hose bed for transport.

On engines that are used primarily for wildland firefighting, 1- and 1½-in. hoses are carried. The smaller diameter is used because the hose lays must often be put in by hand over rough terrain. The hose is carried in rolls or packs, allowing for easier carrying **FIGURE 6-25**. Wildland fire fighters must master the

FIGURE 6-25 Hose packs and rolled hose set up for quick deployment at wildland fires.
© Jones & Bartlett Learning

FIGURE 6-26 Fire fighter deploying hard rubber line from hose reel.
© Jones & Bartlett Learning

skill of extending a hose lay by adding lengths of hose while actively fighting the fire as they advance.

Some pumpers are equipped with a hose reel that contains either ¾- or 1-in. hard rubber line **FIGURE 6-26**. This is used for controlling small fires and saves time because it can just be rerolled on the reel and is ready to go.

The hard suction hose is rubber or plastic with wire wrapping on the inside that makes the walls stiff **FIGURE 6-27**. When a vacuum is created inside the hose, it will not collapse. When drafting water from a static source, an engine requires a hose that will stay rigid and let the water through. Hard suction hose is carried on apparatus for taking water from swimming pools, reservoirs, and other static sources. It comes in 10-ft lengths and is typically 2½ or 4½ in. in diameter. For 1500-gpm pumpers, it is required to be 6 in. in diameter. This size is rarely carried because it

is extremely heavy and hard to use. A certain amount of volume is sacrificed for ease of operation, and 4½-in. hose is often carried instead.

Another method of taking water from static sources is the floating pump. It is a small gasoline-powered pump mounted on a floating housing that will pump a 1½-in. line at around 90 gpm (see Figure 6-56). Some apparatus also carry portable pumps that are set up at the site and draft water, pumping it through a hose line into waiting vehicles or portable tanks.

Nozzles

After the water leaves the pump and travels through the hose, it is applied to the fire through the use of nozzles. There are nozzles for most sizes of hose, ranging from garden hose to master streams. The nozzles used to fight wildland firefighting are for 1-in. diameter hoses. They turn on and off and adjust the stream by rotating the nozzle head in relation to the base. They flow approximately 23 gpm and can be used in a straight stream or fog pattern **FIGURE 6-28**. A quick shutoff can be added for conserving water. They are made of aluminum to save weight. Another hose used in wildland firefighting is ⅝ in. in diameter (garden hose sized) and is commonly referred to as pencil hose. It allows the distribution of small amounts of water while being light in weight (synthetic) and easily carried and moved. It is often used for mop-up operations as it does not flow enough water for attacking flames of any great intensity. An added advantage is that it requires standard garden hose fittings and nozzles, which are lightweight and inexpensive.

The nozzles used on 1½-, 1¾-, and 2½-in. attack lines are made of chrome-plated brass or forged aluminum. They are called combination nozzles if they have the capability of both straight stream and fog patterns.

FIGURE 6-27 Hard suction hoses, 2½- and 4-in. diameters.
© Jones & Bartlett Learning

FIGURE 6-28 Fire fighters attacking fire from a distance with a straight stream nozzle pattern.
Courtesy of Kern County Fire Department

FIGURE 6-29 Assortment of hand-held nozzles.
© Jones & Bartlett Learning

Combination nozzles may have an adjustment ring on them for controlling the amount of water they will flow per minute. This adjustment would typically be from 60 to 125 gpm on a 1½-in. nozzle and 125 to 250 gpm on a 2½-in. nozzle. Combination nozzles have rubber around the head so they will not be damaged if bumped into things **FIGURE 6-29.** Nozzles are not to be used to break out windows. This can embed glass in the rubber and cut your hands when you adjust the stream. A better way is to spray water on a hot window pane, causing the pane to break, or use the proper tool for the job, such as an axe or pike pole.

SAFETY TIP

The nozzle is shut off when the bail is moved toward the front of the nozzle. This safety feature causes the nozzle to shut itself off if dropped.

Other nozzles have straight-bore tips that project a solid stream. To generate a broken stream of water from a straight tip nozzle, emulating a fog pattern, the fire fighter would have to bounce the stream off a wall or ceiling or place his or her finger over the opening to break up the stream. Fire departments differ in whether they carry combination or straight stream nozzles on their attack lines. There is a bail handle on the top for opening and closing the nozzle. This is designed so that it is shut off when the bail is moved toward the front of the nozzle. This safety feature causes the nozzle to shut itself off if dropped. The nozzle is equipped with a swivel female coupling for connecting it to the fire hose.

Nozzles can vary in appearance and performance. Some hand-held nozzles are equipped with a pistol grip handle. Some combination nozzles have a plastic ring on the tip that spins when the nozzle is set for the fog pattern. This spinning action breaks up the fingers of water into a finer fog. There is another class of nozzles called automatic nozzles. They are designed to give the same reach at different pressures. They depend on a set of springs inside that adjust the flow rate to the pressure.

A recent innovation is the low back pressure nozzle. As the water leaves the nozzle, the laws of physics dictate that there will be an equal and opposite reaction, felt as back pressure. This makes it hard to control even average amounts of flow at normal pressures. If you are flowing 125 gpm at 100 pounds nozzle pressure, you can definitely feel it. These nozzles reduce the amount of back pressure to the point that a normal- to smaller-sized person can effectively control and maneuver a fire stream.

As the demand for fire flow to suppress large fires increases (measured in gallons per minute) so do the sizes of the nozzles. The larger nozzles are not designed for hand-held use and come in two configurations. There is the adjustable-flow, adjustable stream type, and a set of straight tips of assorted sizes. The adjustable type commonly flows from 350 to 1000 gpm. These are combination nozzles that can be used for straight or fog streams. The straight tips typically range in size from 1 to 2 in. in diameter. They are arranged in such a manner that if the smallest one on the end is unscrewed the next size is available and so on. The advantage to straight tips is their reach. At the same pressure and flow, a straight tip will far outreach a combination nozzle. Straight-tip nozzles also give better penetration when directed into interior fires or used to knock out windows or ceiling panels. The general rule of thumb is that fire streams directed from the street are effective only to the third floor; for this reason, these large-flow nozzles are also mounted on aerial apparatus where they can be directed through windows, onto roofs, or over walls; or they can be used to cool convection columns.

Tip

As the demand for fire flow increases, so do the sizes of the nozzles.

Pumper apparatus carry a **master stream appliance** that is built in and/or removable. If built in, the appliance is mounted on the top of the apparatus **FIGURE 6-30**. Some of these are detachable and come with a mounting base that is attached for operation remote from the

FIGURE 6-30 Pumper-mounted master stream device in operation at a structure fire.
Courtesy of Kern County Fire Department

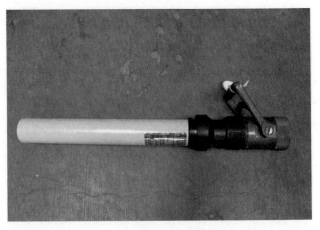

FIGURE 6-31 Foam nozzle attachment.
© Jones & Bartlett Learning

pumper. The advantage to having the nozzle built in is that it can be used for a quick, massive attack on a fire. The problem lies in that it may be a one-shot deal. If supply lines are not laid and the pumper is working solely from its tank, a nozzle set at 500 gpm will empty the tank, depending on size, in 1–2 minutes. This can leave you with no water and a fire that is still out of control. However, it may be able to stop a fire before it has a chance to spread.

Another creative use of this type of setup is to equip the top-mounted nozzle with a straight tip for attacking roadside grass and brush fires, or for situations with steep terrain or wind where a ground attack will not catch the fire. The master stream is used to try to snuff the head of the fire before it gets out of reach. The reasoning is that if you do not catch the fire now, it will grow beyond control in a few minutes. You might as well take your best shot.

A detachable, remotely operated master stream, called a monitor nozzle, is used when the pumper cannot gain access to the location where the nozzle is needed or it would be unsafe to locate the pumper and personnel close to the fire. The nozzle base is equipped with several 2½-in. inlets that are equipped with clapper valves. Clapper valves are one-way check valves that allow water into the base from the hose lines, but will close any hose inlet that is not in use, so not all of the inlets need to be used. This allows the monitor nozzle to be placed into operation while only one hose line is attached and others are being laid. To operate in a remotely located situation, the monitor nozzle is placed and supply lines are laid between the pumper and the nozzle. This operation is commonly used in oil refinery fires. In this situation the nozzle is placed and secured, and the personnel withdraw to a safe location while fire control operations are performed.

A type of nozzle that comes in all of the previously described sizes is the foam nozzle. This nozzle is designed to aerate the foam solution coming through the hose, giving the foam a light, fluffy appearance that makes it easier to see and, in some situations, making the foam more effective **FIGURE 6-31**.

SAFETY TIP

When using in-line foam eductors, it is extremely important to follow the manufacturer's requirements as to the pump pressure, flow rate set on the nozzle, length of line after the eductor, and elevation of the nozzle over the eductor. A caution here is that an in-line eductor may not work with an automatic nozzle as nozzle flow rate is variable due to hose line pressure and not manually selected by the operator.

All of the types of nozzles available to today's fire fighter are too numerous to list. There are wall-piercing nozzles, cellar nozzles, distributor nozzles, and many others. Basically, if someone thought of a better way of applying water, a nozzle was developed to fill the need. As building construction methods and materials change, nozzles will, too.

Another appliance that has found its way onto most fire pumpers is the foam eductor. This device is equipped with a female and male coupling and is inserted into the hose line. It has a suction tube that is inserted into a 5-gallon container or 55-gallon drum of foam concentrate. Through the use of the Venturi principle, the foam concentrate is drawn up through the suction hose and enters the hose line to be discharged as foam solution at the nozzle. When using in-line foam eductors, it is extremely important to follow the manufacturer's requirements as to the pump pressure, flow rate set on the nozzle, length of line after the eductor, and elevation of the nozzle over the eductor. All of these factors influence the quality of the foam produced. A caution here is that an in-line eductor may not work with an automatic nozzle because nozzle flow rate is variable due to hose line pressure and not manually selected by the operator.

Fittings

To give fire fighters versatility in constructing hose lays and accessing water supplies, pumpers carry a wide variety of fittings **FIGURE 6-32**. There are double male and double female fittings in all of the hose sizes carried on the pumper. These allow the fire fighter to connect two hose lays that were laid in different directions. Otherwise one of the hose lays would have to be reversed. There are reducers and increasers. Fittings are described by the female fitting first. If a fitting were adapted from 2½-in. female to 1½-in. male, it would be a reducer; and if it were a 1½-in. female to 2½-in. male, it would be an increaser. Some adapters are for changing the thread. If we wanted to extend a 1-in. line of the end of a 1½-in. line, we may have to change from national standard thread to iron pipe thread as well as changing the coupling diameter. This fitting could be called a reducer/adapter. A very common adapter is used for attaching the 4- or 4½-in. front mount suction hose on a pumper to a hydrant with a 2½-in. outlet. In areas where water is available from plumbing systems on wells and tanks, such as rural areas, a well-equipped pumper carries a set of adapters to attach its hose fittings to iron pipe thread fittings in the 1½-, 2-, and 3-in. pipe sizes. In areas where vacuum trucks are prevalent, many pumpers carry fittings that allow the adaptation from cam lock

FIGURE 6-33 Wyes and siamese for splitting and joining hose lines. Wyes with valves are considered gated wyes.
© Robert Klinoff/Jones & Bartlett Learning

fittings to the thread type carried by the department. This gives fire fighters the capability of using the vacuum truck as a water tender.

Other fittings are used to divide and combine hose layouts **FIGURE 6-33**. A wye is used to divide a hose line into two hose lines. The wye is equipped with a female fitting on the incoming side and male fittings on the discharge side. These can be the same size, as in one 2½-in. to two 2½-in. or, more commonly, one 2½-in. to two 1½-in.. If the wye is equipped with shut-offs, it is called a gated wye; without shut-offs, it is a straight wye.

The fitting used for combining hose lines is called a siamese. The siamese is equipped with two female fittings on the incoming side and a male fitting on the discharge side. It can be equipped with clapper valves that close automatically under pressure. These clappers allow water to enter from one female end without leaking out of the other female–male end if only one line is attached or charged.

SAFETY TIP

The ladders carried on all apparatus should meet the required specifications for firefighting use.

FIGURE 6-32 Reducers and adapters for connecting various sizes of hose lines.
© Jones & Bartlett Learning

Ladders

NFPA 1901: Standard for Automotive Fire Apparatus requires that pumpers carry a minimum of one straight ladder at least 14 ft in length with roof hooks. The standard also requires an extension ladder of at

FIGURE 6-34 Portable ladders (left to right): extension ladder, roof ladder, and folding attic ladder. In this photo, the attic ladder is in the unfolded (ready for use) position. Ladders of this type that are made of wood may be used in some departments.

© Jones & Bartlett Learning

least 24 ft in length and a folding ladder—commonly called an attic ladder—that is 10 ft in length. The attic ladder is used primarily for gaining access to the access hole to the attic in structures. The ladders carried on all apparatus should meet the required specifications for firefighting use **FIGURE 6-34**.

Self-Contained Breathing Apparatus

SCBA are required equipment on pumpers and are designed for firefighting (see Figure 3-8). SCBA allow the fire fighters to work safely in environments with inhalation hazards, such as toxic smoke, elevated air temperature, and oxygen deficiency. The SCBA consists of a backpack, air bottle, face mask, and regulator. It is designed to operate in the positive pressure mode so if there is a face mask seal leak, the fire fighter does not breathe any harmful products due to positive pressure from the air bottle flowing into the face mask.

SAFETY TIP

SCBA regulators are designed to operate in the positive pressure mode so if there is a face mask seal leak, the fire fighter does not breathe any harmful products.

The backpack is designed to be donned quickly. It has adjustable shoulder and waist straps to fit different-sized users. The air bottle carries the compressed air

that the fire fighter will be breathing. It may be steel, carbon fiber, or fiberglass-wrapped aluminum. It is equipped with an air gauge that shows how much air it contains and a handwheel for turning it on and off. The mask covers the face with a clear plastic face piece for visibility. Straps on the mask hold it securely to the head.

A new trend is a built-in communication system that allows the person wearing the mask to communicate on the radio and/or through a speaking diaphragm. Another innovation is the in-mask heads up display that gives the wearer up-to-date information on the amount of air remaining in the air bottle. The regulator adjusts the pressure of the air from that in the tank to a pressure that can safely be breathed. It is equipped with a low-pressure warning bell or whistle to let fire fighters know when they are close to running out of air and should leave the work area for a safe area.

Hand Tools

Fire fighters are always looking for ways to do their jobs with greater speed, safety, and efficiency with fewer than the necessary people at scene to get the job done. Often lives and much property are at stake. Fire fighters follow the Boy Scout motto, "Be prepared." A question one must keep in mind is, "If you are at a scene and you need certain equipment immediately, and you did not bring the equipment with you, how exactly are you going to get it?"

Fire fighters carry just about every type of hand tool imaginable. In addition to regular hand tools such as wrenches and screwdrivers, the fire engine is equipped with tools specially designed for firefighting needs. Hose wrenches that aid in the tightening and loosening of fittings are called spanners. Hydrant wrenches are carried for opening and closing fire hydrants **FIGURE 6-35**.

For vehicle extrication and rescue, fire apparatus carry specially designed rescue tools. Designed to be placed in the gap between the car door and the body, these tools exert up to 60,000 pounds of force to pop the door open. They can also be equipped with cutters to cut the pillars that attach the roof **FIGURE 6-36**. Equipped with special high-strength chains, they can be used to pull the steering column up and away from the victim. Many departments are now carrying hydraulically powered rams, specially designed to be used with their rescue tools, to push car bodies apart to gain access to victims. Gas-powered circular saws with metal cutting blades are commonly used for vehicle rescue operations. After several accidents involving flammable vapors and sparks produced

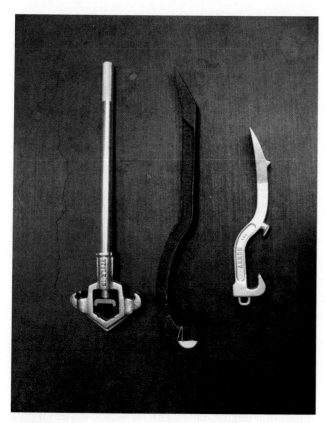

FIGURE 6-35 Spanners and hydrant wrench.
© Jones & Bartlett Learning

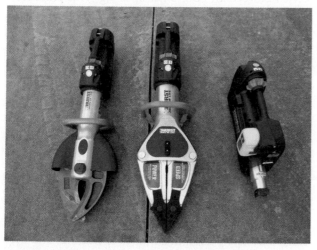

FIGURE 6-36 Rescue tools with spreader, ram, and shears. These tools are battery operated for ease of transport and use. Previously these tools were gasoline engine or electrically powered.
© Jones & Bartlett Learning

FIGURE 6-37 Air bags used for lifting heavy objects.
Courtesy of Kern County Fire Department

by cutting operations, which injured and killed fire fighters and victims, their use has been curtailed. For taking out the side windows on a car for victim access, the spring-loaded hand-held punch works very well. It shatters the glass without spreading it all over the victims as striking the window with an axe would.

Another rescue tool is the air bag system **FIGURE 6-37**. Bags are inflated through their own regulator from an SCBA bottle and, depending on size, are capable of lifting from 12 to 70 tons. They are used for lifting vehicles and heavy objects off of victims and for other lifting jobs. Their advantage is in their thinness—less than an inch when deflated—which allows them to be easily slipped into narrow spaces. They are effective in soft dirt where a rescue tool would dig into the ground, and they are not a source of ignition when flammable vapors are a danger to the operation. They can also be used to lift an automobile dashboard trapping a victim's legs. A rescue-rated chain is attached to a stable point at the front of the vehicle on one end, stretched over the air bag, and attached to the steering column at the other end. When the air bag is inflated it tightens the chain and lifts the steering column, pulling the dashboard upward. This procedure is sometimes referred to as *rolling the dash*.

For ventilating roofs, several tools are available. For working on composition roofing, the axe was a traditional choice. The next development in this area was the gas-powered circular saw with a wood-cutting blade. The gas-powered circular saw with a metal-cutting blade is effective on metal sheeting roofs. The tool of choice for today's fire fighter is the carbide chain–equipped chain saw. Quick, lightweight, and easily used on different thicknesses of roofing, it is gaining in popularity **FIGURE 6-38**. When clay or concrete tile roofing needs to be removed in order to get at the wooden roof sheeting with a saw, a sledge-hammer or pike axe is used to break the tiles. Once the cuts are made, an axe, rubbish hook, or pike pole can be used to remove the desired material **FIGURE 6-39**. The ventilation crew should always take a pike pole with them to poke out the ceiling below the roof to complete ventilation to the outside (see Figure 3-14).

A tool that has made great inroads into the firefighting field is the power fan. Traditional smoke ejectors

FIGURE 6-38 Gasoline powered circular and chain saws.
© Robert Klinoff/Jones & Bartlett Learning

FIGURE 6-39 New versions of the standard pick axe for forcible entry and ventilation work. On the right is a Halligan tool set up in "set of irons" configuration with pike axe.

SAFETY TIP

In today's lightweight construction, roofs are collapsing sooner and fire fighters are in great danger of falling through and being injured or killed while performing roof ventilation. There are several ways to avoid this, such as sounding the roof with a tool as you proceed, staying over main structural components (such as bearing walls and ridges), and using a ladder on the roof to distribute your weight.

FIGURE 6-40 Gasoline-powered ventilation fan.
Courtesy of Jeff Riechmann

were electric fans that drew the smoke from the building. They had to be hung in windows and doorways and supplied with electricity to operate. They tended to be inefficient and often got in the way. The new tool for this purpose is the power fan that is powered by a gasoline engine, electricity, or water, depending on type **FIGURE 6-40**. When used, it is placed back from the doorway on the outside of the building so that its cone of forced air covers the opening. This placement leaves the doorway open for access by the fire fighters. These fans move so much air that, when used properly, they can evacuate the smoke from a three-bedroom home in a few minutes. They can also be used to pressurize stairwells in high-rise buildings to keep them free of smoke when firefighting operations require doors onto fire floors to be opened. Their greatest advantage is that they help reduce the number of times that roof ventilation is required at structure fires. In today's lightweight construction, roofs are collapsing sooner, and fire fighters are in great danger of falling through and being injured or killed while performing roof ventilation. A power fan can be placed and started by one person in a matter of minutes and then left to operate. In contrast, a roof ventilation crew usually consisted of at least three people going above the fire. It took them a while to place their ladders, evaluate the roof, and effect ventilation.

Salvage covers are tarps used by fire fighters to protect a building's contents from damage due to water and falling debris **FIGURE 6-41**. A salvage cover spread on the floor makes quick work of cleaning up a room after pulling the ceiling and spilling the insulation all over the floor. It can also be used to channel water down stairs and to make temporary catch basins. A salvage cover laid over ladders can be used to make a sump for drafting.

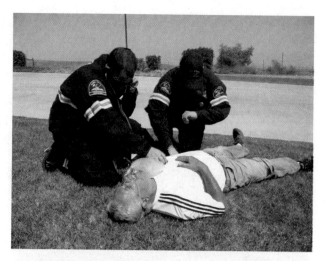

FIGURE 6-42 Oxygen and first aid equipment being used by fire fighters to treat a victim at a medical aid incident.
© Jones & Bartlett Learning

FIGURE 6-41 A. Results of proper salvage operations. B. Results of improper salvage operations.
© Jones & Bartlett Learning. Photographed by Glen E. Ellman
© Joseph Sohm/Shutterstock

FIGURE 6-43 Automatic External Defibrillator.
© Robert Klinoff/Jones & Bartlett Learning

Almost all fire apparatus carries a fire extinguisher of one type or another. Many fire departments have a person detailed to carry a 2½-gallon water extinguisher into structure fires. This amount of water is often sufficient to extinguish a small fire without pulling hoses inside.

Most fire engines are equipped with a gasoline-powered generator, power cords, and detachable lights. These lights are used to illuminate the scene when the electricity has been turned off or there is none available. If the engine carries electrically operated tools, the generator should be capable of powering them also.

Medical Aid Equipment

The amount and complexity of the medical aid equipment carried on fire apparatus depends in large part on the level of life support provided. The two basic components of a life-support system are the oxygen administration equipment and the medical aid kit **FIGURE 6-42**. The oxygen administration equipment allows the fire fighters to administer oxygen to

patients. It can also be used to ventilate victims during cardiopulmonary resuscitation and has a suction device for removing vomit and other material from the patient's mouth. The medical aid trauma kit or bag is a first aid kit containing an assortment of bandages and tools necessary to stop bleeding along with splinting equipment for broken bones. The next step above basic life support (BLS) is the automated external defibrillator (AED) **FIGURE 6-43.** It is used to convert the heart rhythm of a patient in fibrillation to an organized rhythm. In a department where advanced cardiac life support (ACLS) is provided, the equipment and the treatment available are more complicated.

Thermal Imaging Cameras

Thermal imaging cameras are used by the fire service for rescue and finding heat sources **FIGURE 6-44**. They

FIGURE 6-45 Wildland firefighting hand tools.
© Jones & Bartlett Learning

FIGURE 6-44 A. Thermal imaging camera in use. B. View of screen on a thermal imaging camera.
© Jones & Bartlett Learning. Courtesy of MIEMSS
© Jones & Bartlett Learning

are battery operated and either hand held or helmet mounted. Looking through the viewfinder, the fire fighter sees sources of higher heat as a brighter color against a darker background. The use of the camera allows the operator to determine heat sources in limited-visibility situations. They are very effective at locating victims in smoky atmospheres or people ejected from vehicles in tall vegetation. They can also be used to find fire hidden in walls, hot electrical items, and other heat sources. Their use speeds up rescue and fire source location and minimizes damage in tearing open walls, floors, and ceilings to look for hidden fire. Fire fighters must be properly trained and have experience in thermal imaging operation and use to ensure accuracy in identifying heat source images.

Wildland Firefighting Hand Tools

In areas where wildland firefighting is a function of the department, specially designed tools will be carried **FIGURE 6-45**. One of these tools is the McLeod,

a tool that has a scraping blade on one side of the head and a rake on the other. The scraper is used to remove fuel, such as grass, down to mineral soil to form a firebreak. The rake side of the McLeod is used to remove heavy **duff** and **forest litter**. Another tool, the Pulaski, consists of a grubbing blade on one side of the head and an axe on the other. The grubbing blade is used for digging out roots and removing sage and other types of brush and for loosening the ground in preparation for the scraping tools that follow. The axe side of the head is used for removing the lower limbs of trees ("lollipopping") and chopping heavier fuels. The Pulaski, due to its awkward balance, is not a very good chopping tool compared to an axe, and great care should be taken in its use. The shovel carried for wildland firefighting is shorter than a standard shovel and has a more pointed head. This makes it a better scraping tool. It can also be used for throwing dirt to cool down hot spots. When properly sharpened, it can also be used to lollipop tree limbs. Lollipopping prevents fire from climbing the dead limbs at the lower levels of trees to prevent **crown fires**.

A new tool is the combi, which has a swivel head much like a trenching shovel. It can be used as a shovel, grub hoe, and chopping tool. Chain saws are also used for wildland firefighting. They are used to fell trees, to cut up limbs and logs, and to create firebreaks in brush. Wildland fire fighters prefer to keep their tools very sharp, and great care must be exercised when operating around or with these tools as they can easily cause injuries.

Portable tanks and portable pumps are often set up at forest fires **FIGURE 6-46**. Through the use of the pumps, tanks, and hose lays, water supplies can be established in areas that are inaccessible to vehicles.

FIGURE 6-46 Portable pumps (left to right): backpack pump and floatable pump.
© Jones & Bartlett Learning

The water is sometimes used to fill backpack pumps. A backpack pump consists of a 5-gallon water bag or can with shoulder straps and a hand-operated pump wand that forces the water out of a nozzle. Backpack pumps are primarily used for mop-up due to their limited carrying capacity.

> **SAFETY TIP**
>
> Wildland fire fighters prefer to keep their tools very sharp, and great care must be exercised when operating around or with these tools as they can cause very bad cuts.

Heavy Equipment

Fire departments that have wildland areas or special needs have heavy equipment in the form of bulldozers, commonly referred to as dozers, to establish fire control lines in brush, timber, and grass. In the western United States, dozers are equipped with blades and discs. In the southeastern United States, they are often equipped with a plow for creating fire lines in heavy tangle undergrowth and boggy areas. Some of these dozers are equipped with built-in water tanks and pumps with hose reels (pumper cat). They are used to provide water in areas inaccessible to wheeled apparatus. To get the heavy equipment to the fire, the department has to have heavy transports, consisting of truck tractors and low-bed trailers.

In most areas, dozers of the D-5 or D-6 (Caterpillar Inc. size reference; other manufacturers use other designations) size are used. Bigger dozers are so large and have such a wide blade that it is hard to transport them on narrow mountain roads. They are also so heavy that they tend to destroy old roads that were originally designed to carry stagecoaches. The exception to this is areas with heavy concentrations of mature brush. The Los Angeles County Fire Department utilizes bulldozers in the D-9 category to push through dense brush, and these dozers can also cut fire line uphill at a faster rate of speed than the smaller dozers.

In areas where oil production and refining operations are prevalent, a special foam unit is an asset. The foam unit consists of a large tank for the foam concentrate and a pump with a motor to add the concentrate to the fire stream. The foam unit may also carry special foam nozzles and appliances not carried on other apparatus. When met at the scene by several engines or a water tender, the foam unit has firefighting capability away from established water supplies.

> **SAFETY TIP**
>
> Any material made from nylon or related synthetic fibers has the possibility of melting to your skin.

Personal Protective Equipment

To protect fire fighters in various types of dangerous environments, many types of specialized protective clothing are required. The layering of protection provides a greater margin of safety.

Station/Work Uniforms

Modern fire fighter work uniforms are made primarily of fire-resistant fabrics, which provide another layer of protection under the PPE worn for firefighting. The station/work uniform is not a substitute for wearing the proper PPE. For any kind of firefighting, cotton underwear should be worn. Any material made from nylon or nylon-related synthetic fibers has the possibility of melting to your skin. You should never get in a situation with this kind of heat, but fires can be unpredictable and accidents do happen. The more layers you have on, the better you are protected from external heat. Some departments are experimenting with long sleeve T-shirts under wildland fire shirts for added protection. There is even discussion about whether plastic watchbands should be worn as they may melt under extreme conditions.

Structure Fire PPE

The fire fighter's first line of defense against the harmful effects of heat and flame is PPE. This equipment is designed as a system for your safety, and all components should be properly worn when the situation dictates. For structural firefighting, the turnout uniform and SCBA are used **FIGURE 6-47**. The turnout uniform consists of a helmet with a hood or other ear protection, turnout coat, turnout pants, gloves, and boots. Designed and tested to resist impacts, the helmet is equipped with an internal suspension system to help absorb impact forces. The helmet can be constructed of one of several materials, with high-impact plastic becoming the norm. It should have reflective material on the exterior. It has a long brim on the back to keep hot water and melting roofing tar from going down the back of your neck. Eye protection in the form of a face shield or goggles should be included for use when an SCBA is not used. The NFPA has multiple standards that apply to PPE from station uniforms to structural and special-use firefighting gear.

The turnout coat is constructed of an outer shell of a material that will not support combustion and has reflective stripes. There are two layers of inner liner: one a vapor barrier to keep water and other liquids out and the other an insulating material to protect from heat. The coat should never be worn without the inner liner in place. The coat is equipped with either snaps and hooks or Velcro hook and loop fastener and a zipper to secure the front flap, and a high collar that is raised and snapped in place. The pants are constructed in the same manner. Gloves are necessary to protect your hands from heat and debris. For a time, plastic-coated gloves were used; they kept the hands dry but tended to melt to the skin under high temperatures. Firefighting gloves are now made of leather with long knit cuffs, unless the cuffs are built into the jacket, to protect the wrists.

FIGURE 6-47 Fire fighter in full structural firefighting PPE pulling 1¾-in. hose to attack a structure fire.

Courtesy of Kern County Fire Department

Remember, interior attack structural firefighting is performed in a hot, dark, smoky environment, often crawling around on your hands and knees, and all of your body needs protection. Firefighting boots are made of rubber or leather with tread soles and have steel toe and sole protection. Several items that often come in handy and should be carried in your pockets are a knife, a piece of rope, and a flashlight.

Personal Alarm/Personal Alert Safety System

According to *NFPA 1982: Standard on Personal Alert Safety Systems (PASS)*, all fire fighters are required to carry, either on their turnout coat or SCBA, a personal alarm locator device (PAL or PASS),. These devices emit a loud alarm signal when the person wearing them does not move for approximately 30 seconds. Their purpose is to help others find you if you become trapped or are rendered unconscious. It is your responsibility to make sure to turn it on before engaging in firefighting operations. All new SCBA models have PASS that activates automatically when the air is turned on. With accounting for personnel being such an important part of firefighting safety, some departments are equipping members with removable helmet tags or other identifiers. These are gathered at a point at the incident to assist in keeping track of who is assigned where. In the future, each person may be equipped with a bar code, and a portable scanner would be used to record their passing a certain point at the scene.

Proximity Suits

The PPE worn primarily by aircraft fire fighters is called a proximity suit. It consists of a system that includes boots, pants, a coat, gloves, and a hood with an internal helmet. The suit is made of an aluminized material to reflect heat, and the face piece is specially coated to reflect heat as well. The key point is that it is a proximity suit; it is not designed for walking through fire (see Figure 6-32).

SAFETY TIP

The fire fighter's first line of defense against the harmful effects of heat and flame is PPE. Interior attack structural firefighting is performed in a hot, dark, smoky environment, often crawling around on your hands and knees, and all of your body needs protection.

Wildland PPE

The wildland firefighting uniform consists of a hard hat–type helmet with goggles, and ear and neck protection in the form of a shroud attached inside the helmet **FIGURE 6-48**. The helmet is plastic, so as not to attract lightning. A fire shirt of fire-resistant material is worn to protect from heat and hot embers. The pants worn should be made of fire-resistive material. Boots, at least 8 in. in height, made from heavy leather (no Kevlar side panels) with laces (not zippers) and lug soles should be worn. Wildland fire fighter boots should not be equipped with steel toes. When walking on hot materials the steel toe can become heated and steel toed boots can cause toe impact injuries when walking downhill for long periods. Dry grass is slippery, and ankle protection from hot material on the ground is a must. When stumps and roots burn under the ground, they are not always

FIGURE 6-48 Fire fighters in full wildland firefighting PPE cutting a handline on a wildland fire.
Courtesy of Kern County Fire Department

FIGURE 6-49 Fire shelter deployed on a road. The tools are a modified shovel and a fire line pack.
© Jones & Bartlett Learning

evident. It is relatively easy to have the ground collapse and drop your foot into a bed of hot coals. You should always carry a compass and canteens of water with you.

Another critical part of the wildland PPE is the fire shelter **FIGURE 6-49**. It is a small tent folded up in a container that you carry on your belt or on the outside of your line pack (backpack). It is made of an aluminized fabric and can be quickly unfolded. You should never leave the limited safety of your apparatus in a wildland fire situation without your fire shelter with you. Some fire fighters have gotten into the habit of carrying the fire shelter in their backpack instead of having it readily available. The South Canyon incident of Storm King Mountain in Colorado in the summer of 1994 showed the importance of having the fire shelter ready and being able to deploy it quickly, as a fire can change directions and/or increase in intensity very rapidly. At this fire, the situation changed in a matter of seconds and 14 fire fighters were overrun and killed by the fire (Bureau of Land Management 1994).

SAFETY TIP

In a wildland fire situation, you should never leave the limited safety of your apparatus without your fire shelter. Your fire shelter should always be mounted on the outside of your pack for easy access and quick deployment.

Emergency Medical PPE

For responding to EMS incidents, fire fighters need to be protected as well. A disposable, long-sleeved shirt of moisture-resistant material should be worn. Latex, nitrile, or vinyl gloves are worn on the hands. Eye protection and face mask protection are also a good idea. Use of these items prevents exposure to blood and other body fluids that may be splashed or dripped on you (see Figure 6-42). The basic rule is "if it's wet and it's not yours don't get it on you."

Aircraft

Aircraft are assuming a larger role in firefighting operations. There are basically two types of aircraft: fixed-wing (airplanes) and rotary wing (helicopters). Fixed-wing aircraft are used to transport crews to fires, sometimes across the country. Another method of crew delivery is the smoke jumper plane. These fire fighters board the plane with their tools, supplies, and equipment and parachute into the area near the fire. When they are done with the fire, they either hike out to a pickup spot with their equipment or are picked up by helicopter.

Fixed-Wing Aircraft

Fixed-wing aircraft are also used as air tankers **FIGURE 6-50**. Tankers range in size from single-engine airplanes to four-engine DC-7s and DC-10s. Depending on the aircraft size they can deliver from 350 to 20,000 gallons of fire retardant per load to be dropped on the fire. The largest being used is a converted 747 (Global Supertanker). This plane is capable of carrying up to 19,600 gallons of retardant or water. It is designed with an internal pressurized liquid drop system. Flying at a height of 400–800 ft the system can make drops at varying **coverage levels**. In a **trail drop** configuration the plane can drop for 3 mi 50 yards wide. This type of airtanker is referred to as a VLAT, which is the acronym for very large airtanker.

The limitation of airtankers is that they cannot operate at night, in high winds, or in heavy smoke. In areas of steep, narrow canyons they may not be able to fly low enough to drop on the fire effectively. Their turnaround time to return to their base and refill slows them down. Their main advantage is their ability to cover long distances relatively quickly. If the only air tanker available is 300 mi away, it can still be over your fire in a fairly short period of time. Another concept in aircraft is the Canadair CL-215 air tanker. This aircraft can drop down over the surface of a large lake or the ocean and fill itself as it skims the surface. It looks much like a sea plane with high-mounted engines, pontoons on the wings, and a boat hull–shaped fuselage.

Rotary Wing Aircraft

Rotary wing aircraft (helicopters) are rapidly gaining in use by firefighting forces. They can transport crews to remote areas and drop them off at **helispots** that are nothing more than openings in the forest canopy, either meadows or ridge tops. Helitack crews are trained to **rappel** from the helicopter to the ground if no landing area is available. Helicopters can **sling load** supplies and equipment to and from the fire line. When dropping water or fire retardant, helicopters are capable of pinpoint accuracy on burning snags, buildings, and hot spots. They do not need an airport and runways to operate. Some are equipped with fixed tanks, and others are equipped with buckets **FIGURE 6-51**. The fixed tank craft either land and are refilled from a pumper or have a built-in pump with a tube hanging down that is dipped into a water source. The pump fills the tank while the helicopter hovers. These pumps are hydraulically driven and can fill the 2000-gallon tank in a few minutes **FIGURE 6-52**. They are capable of operating out of a water source only 18 in. deep. The bucket-equipped helicopter hovers

FIGURE 6-50 Air tanker making retardant drop on wildland fire.
© TFoxFoto/Shutterstock

FIGURE 6-51 Helicopter making water drop on wildland fire, supporting ground forces.
Courtesy of Andrea Booher/FEMA

FIGURE 6-52 Helicopter filling belly tank with snorkel.
Courtesy of Erickson Air-Crane, Inc

over the water source and dangles the bucket into the water. Helicopter operations are hampered by high winds and heavy smoke, just like the air tankers. Several departments are now looking into the feasibility of using helicopters to attack fires at night.

Overall, the advantages of aircraft in direct firefighting operations are the delivery of fire retardant or water in areas that are inaccessible by pumpers or that are untenable on the ground. They can slow down a fast-moving fire enough to allow ground forces to gain the upper hand. They have also saved lives in their ability to place a drop when and where it is needed to keep ground forces from being overrun. Aircraft of both types are also used to fly reconnaissance missions. By flying over the fire area with infrared sensing devices and GPS, the personnel can determine where hot spots exist and map the fire edge through smoke and poor visibility. They can also determine where fire control lines have been completed and where they need to be reinforced or extended.

Tip

The advantages of aircraft in direct firefighting operations are the delivery of fire retardant or water in areas that are inaccessible by pumpers or that are untenable on the ground.

Fire Department Facilities

The modern fire department requires numerous types of facilities for response, support, and administrative functions. They are illustrated here. Not every department will need all of them due to size of response area and the size of the department.

Fire Marks

Honeywell Corporation is now developing a GPS system called the GLANSER fire fighter locating and tracking system. GLANSER is an acronym for Geo-spatial Location, Accountability, and Navigation System for Emergency Responders. Every fire fighter wears a device that communicates with every other fire fighter's device, as well as the incident commander, and they all communicate back to a computer, monitored by the incident commander or the designated accountability officer.

Another item currently under development is the 3-D Immersive Video Imaging Network (IVIN), software that shows the inside of critical buildings in a jurisdiction. The software creates 360-degree models of the buildings in its database that public safety professionals can then use to simulate responses to those buildings, thus facilitating response planning.

Headquarters

The fire department headquarters is where the managerial staff of the fire department is located. The fire chief, administrative officer, and their staffs work out of the headquarters. The heads of the various bureaus—fire prevention, training, investigation, and others—have their offices at this facility. By having all of the top staff in one location, it is much easier to perform unified planning. The whole staff, or selected personnel, can be gathered on short notice to confer on matters requiring immediate attention.

The headquarters may be located at the main fire station or at a separate location. There are advantages and disadvantages to either location. Having the headquarters located at the main fire station helps the staff keep a finger on the pulse of the organization **FIGURE 6-53**. It brings them closer to the troops in the field. It also provides the staff with personnel who can be used to perform tasks, such as running errands, when necessary. The disadvantages are that the fire fighters at the main fire station will be assigned many of the minor jobs that the staff needs done, which is often disruptive to the routine work that the company officer has planned. Fire fighters tend to be quite social, and a certain amount of productive time will be lost with the station crew talking to the staff personnel and fire fighters from other stations as they come and go during the day. Additionally, having apparatus responding from the headquarters is disruptive to

FIGURE 6-53 Fire headquarters colocated with fire station.
© Jones & Bartlett Learning

FIGURE 6-54 Fire headquarters building separate from fire station.
© Jones & Bartlett Learning

the headquarters staff. When apparatus leaves, there is usually a certain amount of noise in the revving of engines, sirens, and air horns. The apparatus pulling out leaves behind a cloud of diesel smoke that can find its way into the office spaces unless the station is equipped with exhaust recovery systems.

There are advantages to working at the main fire station. The fire fighters working at the main fire station are usually the best informed as to what is going on in the department, which can be very beneficial at promotional testing time. Because they are the most visible to the top staff, however, they are also usually held to a higher standard. In a large department with widespread stations, the old adage, "out of sight, out of mind" may well apply. The headquarters fire fighters may also get choice assignments. Even if this is not true, it is often the perception of the fire fighters at the other stations. Headquarters fire fighters are right there when things become available and they can be the first to get their names on the list for classes and other events.

Having the headquarters remotely located also has advantages and disadvantages **FIGURE 6-54**. A site can be chosen that allows for future expansion as the department increases in size. The role of the fire department has grown immensely during the last few years, and the number of personnel needed to administrate and perform the new functions has grown along with it. Having the headquarters located separately reduces some of the problems noted in the previous paragraphs. The staff personnel are left alone to perform their duties without the interference and disruption of station activities. The remote location also cuts down on the production of rumors from eavesdropping and from personnel accidentally (or purposely) seeing memos and other confidential

communications or suggestions. When someone reports to headquarters for disciplinary reasons, he or she does not need a whole station crew watching and then spreading the news. Having the office away from the unofficial communication system allows the staff the luxury of brainstorming and other creative thinking without the fear that anything placed on a dry erase board or paper will get to the field as a done deal, not just an option under discussion that may be accepted or rejected.

Some of the disadvantages are that someone has to be found to carry out errands (like heavy lifting or moving office equipment) that office personnel cannot perform themselves. This may even require having to detail a crew from a station over to headquarters to move things on an occasional basis. When new equipment—such as nozzles or turnouts—is to be tested, it needs to be taken to a fire station, not just sent downstairs for evaluation. This tends to formalize the contact of the staff officers with the field.

Automotive Repair Facility

Mechanics are needed to maintain a fleet of apparatus and all of the other motorized equipment used by the fire department. They are hired for their expertise in working with the types of apparatus operated by the fire department. A complete facility has hoists that can handle the weight of a fire engine for service from underneath **FIGURE 6-55**. The facility should be heated and cooled for the comfort of the mechanics in winter and summer. Each mechanic requires a complete set of hand tools. The shop should be equipped with a set of specialty tools for work on certain parts of the apparatus. Heavy tools, such as lathes and presses, are needed to perform certain jobs and to fabricate

FIGURE 6-55 Fire department vehicle repair facility.
© Jones & Bartlett Learning

FIGURE 6-56 Drill tower being used for aerial ladder training.
© Jones & Bartlett Learning

FIGURE 6-57 Burn building used for live fire training.
© Jones & Bartlett Learning

parts when vehicles have a new motor installed, or when some part that can no longer be purchased needs to be replaced. A lube and oil change bay should be included, as should tire-servicing equipment. A separate area of the shop may be set up for welding and fabrication. Many manufacturers of fire apparatus have not remained in business for the service life of the apparatus, making parts, such as compartment doors, unavailable.

Training Center

One of the most important facilities a fire department can have is a training facility. It need not be overly fancy or expensive. Many training props can be created from donated items.

A drill tower is effective for training personnel in the use of ground ladders and aerial apparatus **FIGURE 6-56**. It can also be used for training in rappelling and high-angle rescue. If the training tower is equipped with interior stairwells and a standpipe system, it can be used for training in high-rise firefighting. Most drill towers are constructed of concrete, brick, or steel. Training towers have been made from wood, with the uprights made out of telephone poles. Burning is not usually done in drill towers because of the damaging effects of heat and smoke.

An effective way to fill the tower with smoke is with a smoke machine. These machines leave no harmful or unsightly residue.

A burn building or prop is effective for training fire fighters under hot and/or smoky conditions **FIGURE 6-57**. Demonstrations showing the first two phases of a fire and development of smoke and heat in an interior fire environment can be safely performed. These types of buildings are especially good for training

fire fighters in interior attack because a backdraft or flashover is not likely to occur. The building should be constructed of noncombustible materials, allowing it to last through many training fires without damage. If the fires are kept to a few palettes or small amounts of ordinary combustibles, the effect of the heat can be obtained without damage to the building. Several handfuls of damp straw with a road flare stuck in the center will make all the smoke required. The fires should be kept as small as possible, and flammable liquids, tires, or other highly flammable substances should not be used. Replaceable ceiling panels with drywall nailed to a wooden frame and roof panels using plywood are effective for ventilation training.

Some districts have purchased mobile burn trailers that can be transported to the various departments in the district, thus allowing for the local fire fighters to remain in the community while receiving the required training. This has help reduce training costs for travel and back fill in many areas.

Whenever live fire is part of the drill, a safety officer, full turnouts, and self-contained breathing apparatus are requirements. The best course of action is to adhere to *NFPA 1403: Standard on Live Fire Training Evolutions.*

Burn buildings are also useful for demonstrating the dangers of various household materials in a fire situation. A very effective demonstration is to place a dried-out Christmas tree in the living room of the burn building. The area surrounding the tree is set up with furniture and wrapped boxes. The tree is then set on fire, demonstrating the dangers of allowing the tree to dry out.

A motivational drill to build speed in firefighting salvage operations in a burn building with a sprinkler system is to have one crew hooking up to the sprinkler system while another is inside spreading salvage covers. If the inside crew is fast enough, it will complete its work and get out before getting wet.

The burn building can also be used for hazardous materials training by setting up a simulated clandestine drug lab and having the team enter. The room may contain examples of the common booby traps to promote awareness of the dangers present. Law enforcement may be interested in using these facilities to practice hostage rescue and other skills.

No training center would be complete without classrooms. These can be plain or fancy. If the money is available, they can include multimedia DVD players, televisions, satellite reception, and numerous other audiovisual training aids. Chalkboards or dry erase boards are required for drawing out hose lays or doing computations. The advantage of designated classrooms is that they are designed to be used in that way. An apparatus room at a station tends to smell like diesel smoke and is not very well heated or cooled. Tables and chairs are required, and students need to be able to take notes and sit through some classes that may last all day. Adequate lighting is necessary to lessen eye strain. If the local department cannot afford to build its own classrooms, it may be able to borrow space

from recreation centers, veterans' halls, or schools on weekends. When fire department classrooms are available, other agencies are often amenable to giving fire department personnel free tuition to classes in return for use of the facilities. These kinds of arrangements are often made with such agencies as the U.S. Forest Service and the Bureau of Land Management.

The training facility requires a storage area or engine house to store apparatus and equipment used for training. The training center needs to be furnished with a variety of equipment, such as ladders, hose, and other items. To borrow this equipment from frontline engines every time a drill or academy is held is inconvenient, disruptive, and hard to manage. The engine house is also a good place for training with salvage covers and other pieces of equipment that take up large areas. If fire fighters are wet from drilling, the engine house is a good place for them to gather during or after drills to get out of the weather. If they were to gather in the classroom, they would get dirty water and mud on the floors.

As an integral part of the training center, there need to be several hydrants and a **drafting pit**. The fire hydrants can be used for operator training **FIGURE 6-58**. Having the hydrants within the training center allows the persons being trained to practice and be tested on **hydrant hookups** without worrying about traffic or adversely affecting an area's water supply. The hydrants can be used in performing drills at the drill tower and the burn building. When training with heavy stream appliances, there needs to be somewhere to discharge upward of 1000 gpm without causing a traffic accident or other damage.

The drafting pit allows for operator training in drafting operations **FIGURE 6-59**. The pit is designed so that the water discharged from the engine is directed back into the pit. This feature allows for long periods

FIGURE 6-58 Hydrant hookup practice.
© Jones & Bartlett Learning

FIGURE 6-59 Drafting pit being used for operator testing.
© Jones & Bartlett Learning

of pumping without wasting water or creating runoff problems. This pit is used for testing fire engines at draft both annually and after pump repairs.

If the training center is on enough acreage, it can contain a driver training/testing course, often referred to as an emergency vehicle operations course (EVOC). It is always better to train people on tasks such as emergency stopping and high-speed lane changes, somewhere away from other traffic **FIGURE 6-60**.

As a function of the extra space, an area can be set aside for drills in trench rescue and structural collapse. Having these props inside a fenced and locked facility allows them to be left set up without fear of children getting injured playing around them.

The fire department can usually come up with enough donated material to set up props for hazardous materials training **FIGURE 6-61**. This would include plumbing props, railroad tank cars, and large

tanks. The large tanks can be used for confined space rescue training as an added benefit. Many businesses will donate materials in return for access to the props to perform their own training.

In departments spread over large geographic areas, training is often accomplished by sending out training in an online format or a monthly CD or DVD. Many fine training programs are commercially available in a video format as well. When the training center is equipped with a studio and an audiovisual specialist, the department can make its own videos **FIGURE 6-62**. This allows department-specific training programs to be created and duplicated for distribution. Some departments even have their own closed-circuit television channels for presenting training and other information.

The training facility may have offices specifically for the training staff. These offices require copying and word-processing equipment for developing and disseminating training programs and information. Having the staff present at the facility also allows for improved instructor support. Some training facilities have complete

FIGURE 6-60 Driver training on obstacle course, alley dock exercise.
Courtesy of Kern County Fire Department

FIGURE 6-61 Hazardous materials operations training.
Courtesy of Kern County Fire Department

FIGURE 6-62 Electronic media studio for production of training materials.
© Jones & Bartlett Learning

FIGURE 6-63 Specialized air compressor used for filling SCBA bottles.
© Jones & Bartlett Learning

firefighting-related libraries, which allow fire fighters, instructors, and students to come to one central location to check out books, videos, and other materials.

Warehouse/Central Stores

The fire department warehouse/central stores center is designed to stock most of the day-to-day needs of the fire department administration, fire stations, and fire fighters. All of the materials required, from toilet paper to turnouts, are stored while waiting to be issued to the personnel. By having materials at hand, the department's supply orders can be filled in a timely fashion. The warehouse is also a place to store extra turnouts and other items not currently in use. A stock of these items must be maintained because it may take several months to obtain them from the manufacturer.

The central warehouse facility is also a good place to locate the repair facility for SCBA. In larger departments with many SCBAs, a technician is employed with this as his or her main function. This can reduce the cost of going outside for service and can provide for shorter downtime of the SCBA because the fire department will be the number one priority. The SCBA for other governmental agencies that use them, such as the corrections department and environmental health department, can also be serviced at this facility. The repair facility may also contain a specialized breathing air compressor for the on-site filling of SCBA bottles **FIGURE 6-63**. A regular air compressor of the home or industrial type is not acceptable because it uses oil to lubricate the compressor pistons and makes the compressed air unfit to be used as a source of breathing air. The bottles for the SCUBA of the search and rescue dive team can be filled here as well.

Communications Center

The fire department receives calls for emergency assistance at the communications center **FIGURE 6-64**. Most of the United States now has a 911 system in place. In a typical situation, the 911 calls are received by either the local law enforcement agency or the fire department. The dispatcher asks the type of emergency. If the incident requires fire department response, it is routed to the fire department dispatcher. The dispatcher, through the use of a keyboard at the console, enters the incident's location and type into a CAD computer system. The CAD system then places the information necessary to dispatch the required units on the dispatcher's console screen. Depending on the nature of the incident, the situation may require one or more pieces of apparatus. A vehicle accident may require an engine and an ambulance. An accident with pinned-in victims may be dispatched as the nearest engine, aerial ladder truck, or specialty vehicle with rescue equipment, and an ambulance. If an air ambulance (medical helicopter) or other apparatus is available, it can be dispatched as well.

An enhanced 911 system is a great improvement over the older systems in several ways. It used to be that the dispatcher received the call and looked up the location on the map. In large or complicated jurisdictions, such as large cities or departments with vast geographic areas, dispatchers were required to have an extensive knowledge of the jurisdiction. Often children and people in distress are not sure where they are and provide either the wrong address or none at all. An enhanced 911 system is programmed with the address of the land-line phone being used to make the call. This system does not work as well when a cellular

FIGURE 6-64 Emergency services dispatch center.
© Jones & Bartlett Learning

phone is used to make the call. The dispatch center usually has the capability of tracking the call to the nearest cell phone transmission tower to the phone being used to make the call, which can provide them with at least a close approximation of the location of the origin of the call.

The older system also required the dispatcher to determine which apparatus/station(s) to dispatch from run cards. These cards had to be looked through to find the proper one for the location and type of incident. If the responsible station was out on another assignment, the dispatcher had to determine who was second in. This old system could be very time consuming, especially in a department with a large geographical area, high incident volume, and numerous stations. With the development of modern CAD systems, dispatchers already have the location and the computer tells them whose station area the incident is in, and, depending on the type of incident, what apparatus to dispatch. Some systems are also equipped with GPS locators on all of the apparatus so the closest apparatus to the incident can be dispatched, regardless of station location. These GPS systems are called AVL systems and were previously described in this chapter.

The next evolutionary step in the development of the 911 emergency communications system is NG911. NG911 will eventually replace the current system, known as E911, which was implemented in the 1970s. NG911 is a system comprising managed Internet protocol (IP)-based networks and elements that augment present-day E911 features and functions and add new capabilities. NG911 is designed to provide access to emergency services from all sources and to provide multimedia data capabilities for public safety access points (PSAPs) and other emergency service

organizations. Transition to NG911 will take several years to accomplish. Work has already begun in some regions that are putting in place IP infrastructures.

Fire Stations

All of the facilities described so far are for the support of the fire fighters in the fire station. Fire stations started out as nothing more than a shed to house the firefighting apparatus and equipment. They then evolved into a place to house the apparatus and equipment *and* a social hall for the volunteers to gather. With the advent of the paid fire department and permanently assigned personnel, the stations were equipped with living quarters for the fire fighters. Today's fire station serves many functions **FIGURE 6-65**. There is the apparatus room for the apparatus and equipment, a kitchen for cooking meals, sleeping quarters for the crew, an office for paperwork and maintaining files, a workroom for maintaining equipment, an area with physical fitness equipment, and restrooms and showers.

Many of the changes being incorporated into the design of the modern fire station are due to the inclusion of women in the fire service and laws relating to handicapped access. Other factors are diesel exhaust exclusion and reduction of contamination from firefighting PPE from the living area. A newly constructed fire station will probably include separate bedrooms for the crew members instead of the old barracks-style format. At least one of the restrooms will be equipped for handicapped access. The station may be located in an industrial area or in a residential neighborhood, but it is often designed to fit in as well as possible with the surrounding structures. A modern, professional-looking office area is included to make the public feel more welcome when it visits the station for permits or other information. One of the main requirements

FIGURE 6-65 Publicity photo for the grand opening of a modern fire station.
Courtesy of Kern County Fire Department

is that the station be well maintained and clean. The public expects fire fighters to be professionals, and a run-down–looking fire station does nothing to enhance a professional image.

A well-designed fire station is situated on a large lot with enough room for maneuvering fire apparatus and performing training evolutions. The lot may have to be secured to keep people from entering when the fire crews are absent. The apparatus room should be equipped with automatic doors that can be closed by a remote control in the apparatus when the fire fighters leave. The apparatus room may also have electric reels for the **motor block heaters** and air hose reels for inflating tires. Some departments are installing exhaust smoke removal systems in apparatus rooms. Ventilated storage cabinets need to be provided for the storage of turnouts of the off-duty personnel. Many departments are now supplying their fire stations with special washing machines so fire fighter can maintain the cleanliness of the PPE to assist in preventing the spread of carcinogens contaminating their gear from operating at incidents. There is also a hose rack for storing extra hose to replace that on the apparatus. There may also be a special air compressor for filling SCBA bottles. Out back is a hose tower for drying hose before storing or reloading on the apparatus.

One of the most iconic elements of the multi-story fire station is the fire pole. The fire pole was invented and popularized by fire fighters in the city of Chicago. The fire fighters of Station 21 could respond more quickly than others through the use of the sliding pole to get to the ground floor (Walcott 2018).

Tip

The public expects fire fighters to be professionals, and a run-down–looking fire station or apparatus does nothing to enhance a department's professional image.

Wrap-Up

CHAPTER SUMMARY

- The modern fire department relies on many types of resources—such as facilities, apparatus, and equipment—to do its job.
- The modern fire service requires many types of apparatus to perform its duties in protecting the community. These types of apparatus vary widely in their design and application.
- Fire apparatus manufacturers start out with a cab and chassis. Depending on the needs and specifications of the buyer, these can vary greatly.
- Fire apparatus today are mostly powered by diesel motors, noted for their long life and durability under tough conditions.
- Some departments use modular apparatus. By having different modules that are mountable on the chassis, the department gains flexibility.
- The basic piece of motorized apparatus in the fire service is the pumper.
- Water tanks on fire pumpers vary in size. On small apparatus and metropolitan engines, where water is readily available from the hydrant system, tanks tend to be around 200 to 500 gallons. In rural areas, tanks range from 750 to 1500 gallons.
- More and more apparatus are being equipped with built-in foam systems. These can be either class A, class B, or both.
- The main purpose of any pump is to lift water or to add pressure to the water so it can flow through hose and nozzles and be applied to the fire away from the pump.
- The most commonly used main pump on fire apparatus is the centrifugal pump, which consists of one or more vaned wheels, called impellers, mounted on a shaft.
- The positive displacement pump can come in several forms: gear pumps, piston pumps (like the hand pumpers), and diaphragm pumps.
- Aerial ladder apparatus comes in two configurations: the tractor trailer type with tiller steering and the straight chassis.

- A quint is an apparatus equipped with pump, water tank, ground ladders, hose bed, and an aerial device, and it is used as a pumper and ladder truck combination.

- Squad vehicles are the specialty vehicles of the fire service. Just about any time some special configuration is needed for a specific purpose, the vehicle is called a squad.

- Aircraft rescue firefighting (ARFF) apparatus are designed with large water tanks and built-in foam tanks and systems, and are all-wheel drive for going off runways.

- The fire service utilizes many types of tools and appliances to combat fires and perform rescues and other tasks.

- Fire hose is used for getting the required water from the source of supply to where it is needed to control the fire and can be constructed in several different ways.

- After the water leaves the pump and travels through the hose it is applied to the fire through the use of nozzles. There are nozzles for most sizes of hose, ranging from garden hose to master streams.

- To give fire fighters versatility in constructing hose lays and accessing water supplies, pumpers carry a wide variety of fittings.

- The ladders carried on all apparatus should meet the required specifications for firefighting use.

- SCBA allows the fire fighters to work safely in environments with inhalation hazards, such as toxic smoke, elevated air temperature, and oxygen deficiency.

- In addition to regular hand tools, such as wrenches and screwdrivers, the fire engine is equipped with tools specially designed for firefighting needs.

- Thermal imaging cameras, which are battery powered and either hand-held or helmet mounted, are used by the fire service for rescue and finding heat sources.

- In areas where wildland firefighting is a function of the department, specially designed tools will be carried.

- Fire departments that have wildland areas or special needs have heavy equipment in the form of bulldozers, commonly referred to as dozers, to establish fire control lines in brush, timber, and grass.

- To protect fire fighters in various types of dangerous environments many types of specialized protective clothing are required.

- Modern fire fighter work uniforms are made primarily of fire-resistant fabrics, which provide another layer of protection under the PPE worn for firefighting.

- For structural firefighting, the turnout uniform consists of a helmet with a hood or other ear protection, turnout coat, turnout pants, gloves, and boots.

- According to *NFPA 1982: Standard on Personal Alert Safety Systems (PASS)*, all fire fighters are required to carry, either on their turnout coat or SCBA, a personal alarm device (PAL or PASS).

- Aircraft fire fighters wear a proximity suit consisting of boots, pants, a coat, gloves, and a hood with an internal helmet. The suit is made of an aluminized material to reflect heat, and the face piece is specially coated to reflect heat as well.

- The wildland firefighting uniform consists of a hard hat–type helmet with goggles, and ear and neck protection in the form of a shroud attached inside the helmet.

- For responding to medical aid incidents, fire fighters need to wear a disposable, long-sleeved shirt of moisture-resistant material; latex, nitrile, or vinyl gloves; eye protection; and face mask protection.

- Aircraft are assuming a larger role in firefighting operations. There are basically two types of aircraft: fixed-wing (airplanes) and rotary wing (helicopters).

- Fixed-wing aircraft are also used as air tankers. Tankers range in size from single engine airplanes to four-engine DC-7s and DC-10s.

- Rotary wing aircraft are rapidly gaining in use by firefighting forces. They can transport crews to remote areas and drop them off at helispots that are nothing more than openings in the forest canopy, either meadows or ridge tops.

- The modern fire department requires numerous types of facilities for response, support, and administrative functions, but not every department will need all of them due to size of response area and the size of the department.

- The fire department headquarters is where the managerial staff of the fire department is located. By having all of the top staff in one location, it is much easier to perform unified planning.

- Mechanics are needed to maintain a fleet of apparatus and all of the other motorized equipment used by the fire department.
- One of the most important facilities a fire department can have is a training facility.
- The fire department warehouse/central stores center is designed to stock most of the day-to-day needs of the fire department administration, fire stations, and fire fighters. By having materials at hand, the department's supply orders can be filled in a timely fashion.
- The fire department receives calls for emergency assistance at the communications center.
- Today's fire station serves many functions and therefore requires several different areas, including an apparatus room, a kitchen, sleeping quarters, an office, a workroom for equipment maintenance, a physical fitness area, storage for PPE and other items, and restrooms and showers.

KEY TERMS

Accountability officer The person at an incident scene responsible for tracking the location of personnel operating at the incident. This may be the incident commander or a designee.

Alley lights Lights mounted in a light bar that shine to the side of the vehicle; commonly used for spotting addresses on structures at night.

Ambulance gurney The wheeled cot that patients are placed on prior to transport in an ambulance.

Articulated boom Elevating device consisting of a boom that is hinged in the middle.

Automatic vehicle location (AVL) A means for automatically determining and transmitting the geographic location of a vehicle.

Backfires Fires lit in front of an advancing fire to remove fuel and widen control lines.

Baffles Partitions placed in tanks that prevent the water from sloshing and making the vehicle unstable when turning corners.

Burning out Lighting a fire to remove fuel along the flanks of a fire; also used to remove unburned islands that remain as the fire advances.

Cascade systems Systems of large compressed gas cylinders connected to a manifold.

Cavitate To form small vapor bubbles in the interior of a pump.

Composition roofing Tar paper and shingles or tar paper covered with roofing asphalt.

Confined space A space that is not designed to be occupied on a regular basis that is lacking in natural ventilation.

Coverage levels The number of gallons applied per 100 square feet. Light fuels, such as grass would require a coverage level of 1 and heavier fuels (i.e., timber and brush) would require higher coverage levels.

Crown fires Fires in the tops of trees. These fires move very rapidly and defy control efforts.

Drafting pit An open-topped underground tank that is used for drafting operations and pump testing.

Duff Leaves, pine needles, and other dead forest material.

Fire stream The flow of water projected from a fire nozzle (may also be called a hose stream as it is coming from the nozzle attached to the end of a fire hose).

Foam eductor An in-line device that draws foam concentrate from a container into the hose stream.

Forest litter The components of duff, including tree limbs.

Ground sweep nozzles Nozzles mounted underneath apparatus to sweep fire from under the vehicle.

Helispots Natural or improved take-off and landing areas intended for temporary or occasional helicopter use.

High-angle rescue Rescue utilizing ropes and other equipment. Examples are removing persons from smokestacks, wind turbines, or water towers.

Hydrant hookups Attaching the suction hose from the pumper to the hydrant.

Infrared sensing devices Devices that can detect heat energy through smoke and clouds. Used for aerial mapping of fire edges and locating hot spots.

Inverter An electrical device that converts 12-volt current to 110 volt. Used to operate lights and tools from a vehicle's charging system.

Jake brake Common name for the Jacobs Engine Brake and other brakes of that type. Used on diesel motors.

Light bar Roof-mounted unit containing emergency warning lights.

Master stream appliance Large-bore nozzle equipped with a base. Not designed for hand-held use.

Mobile data computer (MDC) A computer mounted in the apparatus and connected to an antenna to provide and receive CAD information.

Motor block heaters Electrical devices that keep oil in the motor warm, make for easier starting, and help prevent damage when the motor is started in cold weather.

Positive pressure mode SCBA regulator function that keeps positive pressure in the mask face piece at all times.

Rappel Descend by means of a rope.

Rehab Short for rehabilitation. A time in which fire fighters rest, cool off, and rehydrate.

Salvage covers Tarps carried on fire apparatus used to cover building contents to prevent damage from water and falling debris. They may also be used to create water chutes to remove water from a building or to build a temporary sump when combined with ladders.

Sling load Material transported by being placed in a net suspended underneath a helicopter.

Standpipe system Plumbing system installed in multistory buildings for fire department use with outlets on each floor for attaching fire hose.

Static water source Pond, lake, or tank used to supply fire engines.

Tactical support A vehicle equipped to provide the needs of fire fighters at the emergency scene. See *rehab*.

Trail drop This is when the aircraft drops a specified coverage level for a long distance. Versus a salvo drop when a large amount of retardant or water is dropped on a small area, such as a patch of dead trees. Another type of drop is the segmented drop when the tanker drops in one area, without releasing its full load, then flies on and drops in other areas.

Turret nozzles Roof- or bumper-mounted nozzles remotely controlled from inside the cab.

Vacuum trucks Tank trucks equipped with a pump that evacuates the air from inside the tank, causing it to draw a vacuum. Used for picking up liquids from spills or tanks, commonly used at crude oil production facilities.

CASE STUDY

The fire department receives a report of a fire in the local oil refinery. Fire fighters don their PPE and board their apparatus to respond to the fire. There is a large column of black smoke rising as they approach. After determining a safe route of entry, and what is burning, the fire fighters begin their attack on the fire. It is burning in a 20,000-barrel (840,000-gallon) crude oil tank that is three-fourths full. The tank is surrounded by other tanks inside the refinery grounds.

The decision is made to attack the fire. The tank is 75 ft in height. Water and foam must be projected over the top of the sides of the tank to land on the burning surface. Additional lines are needed to cool the tank shell and that of the surrounding tanks to keep radiant heat from causing one or more of the exposed tanks from catching fire as well. The water supply is on the perimeter of the refinery. The nearest fire hydrant is approximately 600 ft away.

1. What size supply hose is going to be needed to get sufficient water from the hydrant to the pumper?

A. 1½ in.
B. 1¾ in.
C. 2½ in.
D. LDH (4 or 5 in.)

2. The attack lines from the pumper to the nozzles, used to project water onto the exposures, are probably going to be which diameter?

A. 1½ in.
B. 1¾ in.
C. 2½ in.
D. LDH (4 or 5 in.)

3. Which type of PPE should be worn by the fire fighters fighting the fire?

A. Wildland
B. Structural
C. EMS
D. Hazardous materials

4. Foam is going to be used to fight the fire. Which device is needed to create the foam solution in the hose stream?

A. Power fan
B. Wye
C. Reducer
D. Eductor

5. To place the foam solution where it needs to be, on the burning surface, the apparatus of choice would be a/an:

A. squad.
B. triple-combination pumper.
C. dozer.
D. aerial ladder truck.

REVIEW QUESTIONS

1. List several advantages of a fire department having its headquarters separate from a fire station.

2. List three structures that may be located at the training facility and their uses.

3. What is meant by an enhanced 911 system?

4. Why is the diesel motor chosen for most pumper apparatus?

5. Explain the difference between centrifugal and positive displacement pumps.

6. List three types of squad vehicles and their functions.

7. What is the difference between attack and supply hose lines? Give examples of each.

8. List the four main components of a SCBA.

9. List three wildland firefighting tools and their uses.

10. What are the components of structural PPE?

11. What are the components of wildland PPE?

12. What are the components of EMS PPE?

DISCUSSION QUESTIONS

1. Why is the centrifugal pump chosen as the main fire pump for most fire apparatus?

2. Why does the fire service need a variety of vehicle types to perform its function efficiently?

3. Why is it important to wear all of your PPE for a specific fire (structure, wildland, ARFF, etc.) every time?

REFERENCES AND ADDITIONAL RESOURCES

Bureau of Land Management. 1994. *South Canyon Fire Investigation*. Washington, DC: U.S. Government Printing Office.

Hutchinson, Bill. 2001. "3 Firefighters Killed in Qns. Store Blast." *New York Daily News*, June 18. http://www.nydailynews.com/archives/news/3-firefighters-killed-qns-store-blast-article-1.912112

National Emergency Number Association. 2011. *NG9-1-1 Transition Planning Committee*. Alexandria, VA: National Emergency Number Association. Posted December 5, 2011. Accessed August 19, 2013. http://www.nena.org/?page=NG911_TransPlanning.

National Fire Protection Association. *NFPA 414: Standard for Aircraft Rescue and Fire-Fighting Vehicles*. Quincy, MA: National Fire Protection Association.

National Fire Protection Association. *NFPA 1221: Standard for the Installation, Maintenance, and Use of Emergency Services Communications Systems*. Quincy, MA: National Fire Protection Association.

National Fire Protection Association. *NFPA 1402: Guide to Building Fire Service Training Centers*. Quincy, MA: National Fire Protection Association.

National Fire Protection Association. *NFPA 1403: Standard on Live Fire Training Evolutions*. Quincy, MA: National Fire Protection Association.

National Fire Protection Association. *NFPA 1901: Standard for Automotive Fire Apparatus*. Quincy, MA: National Fire Protection Association.

National Fire Protection Association. *NFPA 1911: Standard for the Inspection, Maintenance, Testing, and Retirement of In-Service Automotive Fire Apparatus*. Quincy, MA: National Fire Protection Association.

National Fire Protection Association. *NFPA 1932: Standard on Use, Maintenance, and Service Testing of In-Service Fire Department Ground Ladders*. Quincy, MA: National Fire Protection Association.

National Fire Protection Association. *NFPA 1961: Standard on Fire Hose*. Quincy, MA: National Fire Protection Association.

National Fire Protection Association. *NFPA 1963: Standard for Fire Hose Connections*. Quincy, MA: National Fire Protection Association.

National Fire Protection Association. *NFPA 1971: Standard on Protective Ensembles for Structural Fire Fighting and Proximity Fire Fighting*. Quincy, MA: National Fire Protection Association.

National Fire Protection Association. *NFPA 1975: Standard on Emergency Services Work Clothing Elements*. Quincy, MA: National Fire Protection Association.

National Fire Protection Association. *NFPA 1977: Standard on Protective Clothing and Equipment for Wildland Fire Fighting*. Quincy, MA: National Fire Protection Association.

National Fire Protection Association. *NFPA 1981: Standard on Open-Circuit Self-Contained Breathing Apparatus (SCBA) for Emergency Services*. Quincy, MA: National Fire Protection Association.

National Fire Protection Association. *NFPA 1982: Standard on Personal Alert Safety Systems (PASS)*. Quincy, MA: National Fire Protection Association.

National Fire Protection Association. *NFPA 1999: Standard on Protective Clothing and Ensembles for Medical Emergency Operations.* Quincy, MA: National Fire Protection Association.

National Interagency Fire Center. 2002. *National Fire Equipment System Course Manual, S-270 Basic Air Operations.* Boise, ID: National Interagency Fire Center.

Naum, Christopher. 2010. "Learning the Lessons from the Past." *TheCompanyOfficer.com.* http://thecompanyofficer.com/tag/fathers-day-fire/

Rafols, Tess. 2013. "Potentially Life-Saving Technology to Track, Find Firefighters in Testing Phase." *AZFamily.com.* Phoenix, AZ: Meredith Corporation.

Schumm, Bill. 2010. "Explosion at Texas Oil Refinery Triggers Large Fire." *Firegeezer.com,* May 5. http://firegeezer.com/2010/05/05/explosion-at-texas-oil-refinery-triggers-large-fire/

Sullivan, Lucas. 2012. "3-D Maps Designed to Help First Responders." *The Columbus Dispatch,* April 30. http://www.dispatch.com/content/stories/local/2012/04/30/ 3-d-maps-designed-to-help-first-responders.html

Walcott, Dekalb. 2018. *Black Heroes of Fire, the History of the First African-American Fire Company in Chicago.* Chicago, IL: Black Heroes of Fire Publishing Company.

CHAPTER **7**

Fire Department Administration

OBJECTIVES

After studying this chapter, you should be able to:

- Identify and describe the six principles of command.
- Identify and list management concepts as they apply to the fire service.
- List items used to evaluate a fire department's effectiveness.
- Identify different fire department types.
- Identify the four methods of communication.

Case Study

In 2013, the Miami-Dade Fire Rescue Department exceeded its overtime budget. This led to what are termed as *rolling brownouts*, which are the temporary closing of fire stations or individual units, such as engines and ambulances. Instead of closing an engine company or station permanently, the stations or engines are closed for a day at a time on a rotating basis among the stations. "The proposed budget for the fiscal year beginning Oct. 1 calls for a $15 million shortfall for the fire rescue department's budget. The department will also need to lay off 149 fire fighters" (Candea 2013). This is a story that is happening all across the country due to budget shortfalls.

Most fire departments are funded through property taxes. When the housing bubble burst and the economy declined, it caused a severe drop in revenue to cities, counties, and states. The result was forced belt tightening for all departments. The fire and rescue services are not exempt from these cuts. Provision of fire service is part of a system of service delivery to the public, along with health services, police, parks and recreation, libraries, etc. All of these affect the quality of life for residents.

"Moving forward, most local governments either will not have the funds to sustain traditional fire and EMS delivery models or will not be willing to allocate such a large portion of available community dollars to the fire department" (Snook 2013).* Fire chiefs must be prepared to innovate and embrace the changes that are occurring. In the future, the core mission of the fire department must take the priority within the department and fit in with community priorities as well.

1. What happens to the community's level of risk with the reduction of staffing on engine/truck companies (e.g., from four to three or three to two persons) or the closing of stations or engine/truck companies, even on a temporary basis?
2. How should it be decided where to reduce staffing, shut down companies, or close stations?
3. Who ultimately makes the decision as to the fire department's budget?
4. What role does the fire chief play in the decisions concerning the fire department's budget?

Modified from Candea, Benjamin. 2013. "Miami-Dade Fire Rescue Chief Laments Budget Cut." Local 10.com. http://www.local10.com/news/miamidade-fire-rescue-chief-laments-budget-cuts/-/1717324/21151732/-/1joyvp/-/index.html

* Snook, Jack. 2013. "Budget, Fire Leadership Advice to Manage Departments in the New Reality." Fire Chief. January 20. http://firechief.com/budgets/budget-fire-leadership-advice-manage-departments-new-reality

NAVIGATE 2 Access Navigate for more resources.

Introduction

Without an administration, the fire department cannot function properly. Like any other entity, public or private, the fire department must comply with laws and provide services to its personnel. The fire department includes a functional division, called staff, to provide various support services. The staff arranges to pay department personnel; manages worker health and wellness programs; complies with local, state, and national reporting requirements; budgets; and plans for resources. Public entities are bound to many layers of governmental regulations, policies, and procedures. Failure to follow these can result in sanctions and lawsuits.

Staffs vary as to size and structure depending on the needs of the entity for which they perform. A small organization may require only a chief who interacts with the city or district to manage the fire protection organization. In large municipalities and counties, the department may require numerous staff officer positions to manage the organization. The same is true of incident management. The size of the management team must be determined by the number of resources to be managed and the complexity of the situation.

> **Tip**
>
> The fire chief must balance the needs of the community and the department with the resources available in line with what the community determines is **acceptable risk**.

To understand how the fire department functions, it is necessary to understand the management principles behind the organization. It is also necessary to understand the command structure and how it operates on a day-to-day basis and under emergency incident conditions.

This chapter is about administration, but some of the descriptions used are based on firefighting operations to illustrate the points more clearly.

Principles of Command

The guiding principles for the organization of a fire department can be referred to as the principles of command. These principles are not used in the same manner in every fire department, because a department should be organized to best serve the needs of the jurisdiction it protects. These principles are general guidelines. As you review these principles, you will see that they are used in both emergency and nonemergency organizations. The principles of command are divided into six areas in this text, with accountability and responsibility grouped under delegation of authority. Some texts refer to them as eight principles because they single out accountability and responsibility and address them separately.

Unity of Command

The first command principle is **unity of command**. This means that each person in the organization has only one boss. This concept also requires that everyone in the organization has a clear understanding of who supervises them and whom they supervise. It is a proven fact that personnel work best when they are supervised by only one person **FIGURE 7-1**. When this principle is violated, the person receives orders, often conflicting, from several people at the same time. Obviously, you cannot satisfy all of those making requests of you. This leads to confusion, inefficiency, and sometimes nothing gets done or the wrong thing gets done. To address this issue in incident command situations, the National Incident Management System (NIMS), described in Chapter 13, *Emergency Incident Management*, is based on unity of command as its organizational structure.

This command principle is probably the most violated, especially at incident scenes, when proper procedure is not followed. For example, when operating at the scene of an incident, the fire fighter may be involved in performing a task for the company officer. The chief approaches and sends the fire fighter to perform another assignment. The fire fighter knows that the chief outranks the company officer and is placed in the position of not knowing exactly what to do. This question is commonly asked on oral interview

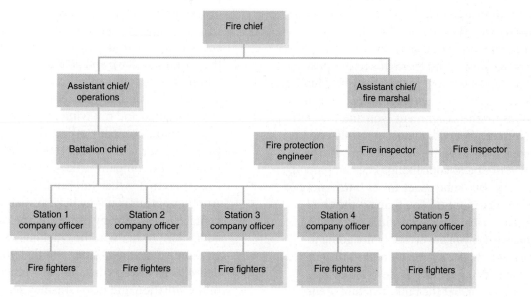

FIGURE 7-1 Fire department organization.

examinations and requires judgment on the part of the prospective fire fighter:

You are operating at a structure fire and your company officer sends you to the engine for a couple of salvage covers. While you are at the engine, the battalion chief approaches you and wants you to go and turn off the utilities to the fire building. What will you do? Justify your answer.

You already know that the chief outranks the company officer and has a very good reason for asking you to turn off the utilities. Possible answers are:

1. Do what the chief requests without question or comment.
2. Tell the chief you will get to it when you have time.
3. Ignore the chief's request and continue with your original assignment.
4. Return to your company officer's location and ask him or her which task you should perform first.
5. Tell the chief that you already have an assignment from your company officer and ask the chief to let the company officer know that you will be delayed.

Which one did you choose? The correct answer is the last one. The reasons why the answers are correct or incorrect are listed in the same order as the possible answers.

1. If you do as the chief requests without question, he or she may feel that you are performing as ordered; however, your company officer is going to be understandably upset when you do not return. The situation may be such that the company officer becomes concerned about your whereabouts and leaves his or her position on the assignment to look for you. This leads to two persons being removed from their positions on the assignment (you and the company officer). You will almost certainly end up in a consultation with your company officer that will not be to your liking. The company officer is going to be upset and is much more likely to take it out on you than on the chief.
2. If you tell the chief that you will get to it when you have time, you are certainly not going to impress the chief. As a matter of fact, the chief is probably going to have a one-sided discussion with you about the rank structure of the fire department and where you fit into that

structure. The chief is then most likely going to tell you to go and do as you were told. After the incident is completed, the chief is probably going to have a discussion with your company officer about your attitude and training needs. The company officer is already going to be unhappy with you because you did not show up with the salvage covers as requested. The fact that the company officer had to come and check on your safety will compound the situation.

3. If your choice is to ignore the chief and return to your company officer with the salvage covers, the chief is going to be very dissatisfied with your performance. The chief expected you to turn off the utilities. When this is not done, someone may come into contact with an energized wire and receive an injury. When things like this happen, chiefs tend not to want to hear your excuses for not doing as you were told. The chief is going to come down on you and your company officer, making problems for both of you.

4. If you choose to return to your company officer's location and ask for clarification, you are going to be considered to be of somewhat less than average intelligence. You have not brought the salvage covers, nor have you done as the chief requested. No one is going to be particularly happy with you at this time. You have now become so ineffective that neither task is accomplished.

5. When you tell the chief that you are already doing a job for your company officer and request that he notify the company officer that you will be delayed or not return at all, you are placing the responsibility for changing your orders on the chief. The chief can then decide which assignment is more important. He or she will have to find someone else to turn off the utilities, or have you go ahead and do it, advising your company officer of the decision. This lets the company officer know that you are reassigned and not lying in a hole somewhere, so the company officer does not have to come and check on you. If the chief forgets to notify the company officer, you have at least tried to do what you can in the situation. Another option you may have available is to contact your company officer on the radio and advise him or her that the chief has requested you to turn off the utilities and that you will be delayed. This keeps the company officer on top of the situation and he or she may send someone else after the salvage covers.

The purpose of this was not to portray company officers and chiefs as unreasonable people with the poor fire fighter caught in the middle. It was to illustrate how things can go wrong, and often do, when the principle of unity of command is violated. This most often happens in the heat of battle at incident scenes, and you must be prepared to deal with the situation should it arise. There are right and wrong ways to act in situations such as the one presented, and it is up to you to perform appropriately.

Chain of Command

The second command principle is **chain of command**. The term *chain* is used because there is a link between the different levels of the organization. The levels do not operate independently of one another. The decisions made at the top affect the levels below to a greater or lesser degree. When the chief makes a decision, it is presented to the assistant chiefs, then to the battalion chiefs, and then to the company officers and fire fighters.

From the viewpoint of the new fire fighter, the chain of command pretty much ends with the company officer. The company officer is the company commander and the first-level supervisor. If the fire fighter has a question or problem, the company officer is usually the one to deal with it. As the fire fighter rises in rank, the chain of command dealt with extends further upward through the organization. The exception to this is when the fire fighter is assigned to a special project, such as public education. He or she may then answer directly to a higher-level officer, such as the chief of prevention.

The chain of command also represents the formal path of communication through the organization. Communication flows down from the top and up from the bottom through the chain. An example would be that if the fire fighter has a **policy** question, he or she should first ask the company officer. The company officer may not be able to answer the question. The fire fighter is then asked to write the question on a memo and forward it to the battalion chief. If it is a question that the battalion chief does not have the answer for, it can be passed further up the chain of command.

This method of operation may seem rather cumbersome and inefficient at times, but it serves a very good purpose: It allows the command structure of the organization to maintain control. If fire fighters went around their company officer when they wanted something and the battalion chief allowed it to happen, the company officer would have less authority over the fire fighters. Every time the company officer wanted something to happen that the fire fighters did not like, they would question the orders to the battalion chief and work would not get done in a timely fashion, if at all.

In every organization there is also an informal communication system called the grapevine or rumor mill. The problem with the informal communication system is that it often disseminates information that is either wholly or partially untrue or is based on speculation. Many times, it has turned out that the information has hurt someone's career or feelings and was totally unfounded. Everyone should be cautious about spreading rumors about others or what is going to happen in the department. Gossiping is not a trait that others are going to respect.

In an organization that has it, a clear chain of command is to be strictly adhered to. No one in the organization appreciates someone going around them or over their head. When this happens, the person bypassed usually feels slighted and may react accordingly. The only time the chain of command may be violated is in extreme circumstances or in emergency situations that require immediate action, such as a safety matter that requires immediate attention.

The chain of command also applies to the concept of unity of command. When viewing an organization chart, it is obvious who answers to whom in the organization. It also specifies the relationships of personnel to those above them, below them, and across from them on the chart. Do not think that the organization chart specifies all of the relationships between personnel in the organization. Not all personnel on any particular level are perceived as equals. The probationary fire fighter may be on the same level of the organization chart as the 10-year fire fighter, but they are not viewed the same by the command structure or their peers. When the senior fire fighter tells you to do something, it is not the time to tell him that you are his equal and take orders only from the company officer. Keep this in mind, as every organization has an informal organization as well as an informal communications structure. If it becomes a problem for you, discuss it with your company officer.

The organization chart of a small department will be much less complicated than that of a large department. Large departments tend to have many different bureaus and activities not performed by the smaller departments, in which more activities are grouped under one person because they are less complex in nature. Fire department organizations tend to be pyramid shaped. The organization is narrow at the top, usually beginning with one person, called the chief, chief engineer, commissioner, or other similar title.

Regardless of the title, the person at the top has the ultimate responsibility for what goes on in the organization. When a fire fighter is injured or killed or a major fire occurs, this person has to answer to the public and to the families affected.

The chain of command provides for the transfer of authority at incidents. When a higher-ranking officer arrives at scene, it is the policy in most departments that the higher-ranking officer has the option of taking over command of the incident. He or she will do this if he or she feels it is necessary. On a major fire, the chief officer may take over command from the company officer. This is not done to infer that the company officer is incapable. It is done because the higher-ranking officer has more experience and training in managing large incidents and diverse operations.

> **Tip**
>
> The chain of command provides for the transfer of authority at incidents.

It also frees the company officer to return to managing his or her own company or a portion of the incident. The chief officer may very well use the company officer as an information source if the incident is in the company officer's first-in area. On a small incident, the chief officer may arrive at the scene and just observe the operations being performed. There is no reason to take command from the company officer. The chief will be observing the actions of the at-scene crew to see that they are performing efficiently and safely. If either is seen as lacking, the company officer is going to receive some guidance as to the training needs of the crew.

This system works in reverse as well. When an incident has escalated to the point where the chief officer has taken charge, it will eventually be controlled. As the incident is brought under control and resources are released, the chief officer will return command of the incident to the company officer. The company officer retains the responsibility of seeing the incident to completion. The company officer also has the final responsibility of making sure that the fire is completely out or that the incident is properly terminated before all resources are released.

Under the chain of command concept, organizations are divided into line and staff functions. Line functions are those directly related to the goals and objectives of the fire department. Department goals and objectives include the preservation of life and property. The methods used to achieve these goals are prevention, suppression, and rescue.

Staff functions are all the other functions necessary to support the primary goals and objectives. Functions include training, vehicle repair, station maintenance, supply, and personnel. These functions indirectly support the goals and objectives of the line personnel. Without a personnel section there would be no one to make sure that new fire fighters were properly selected or that the currently employed personnel received their pay and benefits. Once hired, the new fire fighters must be trained and outfitted. Without vehicle repair, the equipment would not be able to leave the station to respond to incidents.

In many departments, the division between line and staff becomes blurred in the area of fire prevention. The primary goal of the fire department is the preservation of lives and property. The best way to do this is not to have fires—thus prevention. Personnel working in the fire prevention section of the department may not respond to fires, so they are considered staff. They do contribute directly to the primary objective, so they could be considered line.

The fire department organization cannot survive without both line and staff functions. In times of tight budgets, there is often a demand on the part of the line personnel to cut the staff. Their reasoning is that without fire fighters on the equipment, you do not have a fire department. In real life this is an oversimplification. Unless the fire fighters in the stations are capable of performing all of the functions of the staff in addition to their other duties, the staff is necessary.

Span of Control

The third management principle is that of **span of control**. Regardless of rank or education, a supervisor can effectively supervise only a certain number of personnel. Some supervise more than others, due to the nature or complexity of the work being performed. In firefighting, the effective span of control is considered to be three to seven personnel per supervisor, with five being optimal in most situations. This illustrates the need for intermediate levels of management between the chief and the fire fighters. The chief of a large fire department may indirectly supervise more than 3000 personnel. If you were to look at the organization chart, you would see that the chief has from three to seven personnel that confer directly with him or her. The chief's decisions are dispersed through the chain of command through these key personnel, and the chief receives input from them in return. If the same department were to have a chief and fire fighters with no intermediate levels, the chief would be concerned with and receive input directly from the other 2999 personnel in the department. As you can see from this extreme example, the chief could quickly suffer information overload. Using vacation requests as an example, the chief would have to coordinate the requests of all of the personnel at one time. This would be impossible.

> **Tip**
>
> Regardless of rank or education, a supervisor can effectively supervise only a certain number of personnel.

The same concept applies to the management of incidents. At incidents, the situation is not only dynamic, but dangerous as well. Limiting the span of control prevents the incident from being so subdivided that a coordinated attack cannot be made. It also prevents information overload on the part of the management personnel. An incident manager needs information as to what is happening overall but does not need to know that fire fighter Johnson was sent to get more chain saw fuel. The object is for the subordinates to pass on the important and required information without supplying too much that is unnecessary.

Division of Labor

The fourth management principle is **division of labor**. The work to be performed needs to be divided into specific areas to prevent duplication of effort, to apply the most appropriate resources to the job, and to determine responsibility for completion of the assigned work. Division of labor is also assigned based on area, skill, and complexity.

The division of labor is clearly seen in the classification of functions into line and staff. Some positions have responsibility for both, especially in small- to medium-sized departments. Assigning certain functions to the staff, such as record keeping and research, allows the direct application of effort without the distractions incurred by line personnel responding to incidents. It would be extremely difficult for line personnel to schedule and attend meetings if they were sent out on incidents on a regular basis.

Assigning areas of responsibility avoids duplication of effort and promotes efficient use of resources. The fire prevention bureau performs plan checking and building approval, whereas the arson unit investigates suspicious fires and assists in prosecution of arsonists.

The appropriate resources are applied because the fire prevention personnel are trained in plan checking and inspection and the arson personnel are trained in investigation, collection of evidence, and arrest procedures.

> **Tip**
>
> Assigning areas of responsibility avoids duplication of effort and promotes efficient use of resources.

Fire stations are located so that they each have an area of direct responsibility. They are required to perform fire prevention and preincident planning in their area along with other tasks. When needed, they can respond out of their area and assist adjoining companies in their areas of responsibility. This division of labor by area provides for a more efficient use of resources. The first-in company knows its area and its complexities better than any other company. The increased knowledge is a great advantage in structure fires or wildland fires. The adjoining companies coming in to assist can utilize the knowledge of the first-in company when attacking the fire. If all of the companies worked out of one central fire station in a large jurisdiction, little time would be spent in the outlying areas.

In other functions of the fire department, specialized skills become important. The hazardous materials team responds to hazardous materials incidents anywhere in the jurisdiction. This allows the department to train a few select personnel to a high level and maintain their skill level, saving the expense of training all of the department personnel to that level. Any one station may be involved in only one hazardous materials incident a year, whereas the hazardous materials team personnel respond on these incidents on a regular basis.

The complexity of certain fire department jobs requires the efforts of personnel assigned to that function alone. Dispatching in a busy department requires specialized skills in using the dispatch center equipment and possibly **emerge ncy medical dispatch** training. Examples of other specialized skills, used on major wildland fires, are those of weather experts and fire behavior specialists.

Delegation of Authority

The fifth management principle is that of **delegation of authority**. For work to get done, the manager must delegate authority to the subordinates. Managers who cannot do this spend so much of their time managing and reviewing what their subordinates are doing that they have no time for planning or other important functions. The fire department hiring process is selective enough that managers should be able to assume that their subordinates are capable of handling the jobs that they are trained to perform. It is the supervisor's responsibility to ensure that the personnel have the required training to perform their jobs. It is considered normal and good practice to check on the job being done once in a while, but it is not necessary or desirable to monitor personnel closely when they are performing work for which they are qualified.

Delegation of authority requires that the manager give the subordinate the authority and responsibility, while holding them accountable, for taking action on a specific mission. It also requires that the mission be broken down into segments that are assignable. When the fire chief places the fire marshal in charge of prevention, the fire marshal is given the mission of reducing the loss of life and property due to fires by preventing ignition of hostile fires. The fire marshal is not capable of personally inspecting every business in the jurisdiction. The fire marshal then empowers the inspectors to make inspections and implement the necessary changes to make the businesses more fire safe. The fire chief has delegated authority—along with responsibility and accountability—to the fire marshal, and the fire marshal has delegated it to the inspectors.

Tip

Delegation of authority requires that the manager give the subordinate the authority, along with responsibility and accountability, to take action on a specific mission.

Authority for decision making can be delegated, but overall responsibility cannot.

The mission is broken into assignable segments by the fire chief, specifying that the fire marshal prevents fires and does not have to suppress them as well. The fire marshal divides the workload among the various inspectors.

This concept is often used at an incident scene. In the case of a train wreck with mass casualties, the incident commander (IC) has responsibility for the whole incident. The IC determines the objectives to be achieved during the incident. These would include making sure that people are rescued, treated, and transported; that fires are suppressed; that any spilled diesel fuel is contained; and that the safety of fire fighters and civilians is provided for during the operations. The IC then assigns an operations section chief (OSC), who implements the tactical operations to accomplish the objectives. The OSC assigns units to rescue and remove people from the wreckage. The OSC makes sure those rescued are treated and transported, provides personnel to suppress any fires that may have occurred as a result of the incident, assigns others to control any diesel fuel leaks or spills, and will probably assign a qualified person to act as the safety officer to ensure that the operations are conducted safely.

All of the operations just listed are occurring at once, and it is easy to see that the IC cannot control all of them at the same time. There are too many different activities competing for the IC's time and attention. Through delegation, he or she is releasing the authority of making the needed decisions to the lowest possible level. With competent and properly trained personnel this works very well. If the personnel at the scene were untrained, it would not work well and would be unsafe.

The alternative to delegation is that the IC requires that every decision be cleared through him or her. This would bring the operation to a standstill. Every individual rescue, medical treatment decision, suppression action, and spill control operation would require consultation before action was taken.

Keep in mind, however, that authority for decision making can be delegated, but overall responsibility cannot. The fire chief is ultimately responsible for the actions of the subordinates. The fire chief must answer to the public when things go wrong. As personnel rise in rank, they assume more responsibility for achieving the organization's objectives. The company officer is responsible for seeing that subordinates are properly trained and prepared to perform emergency functions. He or she is also responsible for seeing that the program work assigned to his or her shift at the station is completed in a timely and satisfactory fashion. As a fire fighter, you are on the bottom of the organizational ladder and may think that you do not have much in the way of responsibility, but this is incorrect. You have the responsibility of coming to work prepared to perform your duties. This includes showing up in the proper uniform, remaining physically fit, and training yourself to perform your job. You are the one visible to the public on a day-to-day basis. To the people you assist in your job and the public, you are the fire department.

Program work, such as hydrant maintenance, is tedious and not nearly as exciting and rewarding as firefighting, but it is important to the overall mission of the fire department. When the company officer sends you out to service hydrants, you have been delegated the authority and responsibility to service the hydrants as you deem necessary. If what you deem necessary is sitting somewhere drinking coffee and filling out paperwork, without doing the required work, you are asking for trouble. It may not be a problem this year or even next year, but sooner or later there is going to be a fire in which an un-serviced hydrant is needed for water supply. The engine crew tries to make a hookup and cannot remove the caps that you indicated you removed and lubricated just last month. As it turns out, they are rusted on. The fire fighters return to the station, look in the hydrant maintenance book to see who serviced the hydrant last, and there is your name or initials. At this point accountability for the failure is going to be placed. At the least, you are going to get a severe verbal reprimand. If the fire cost major property damage, or even worse, loss of life, you are partially to blame. How will you feel about that? The company officer will be in trouble as well because he or she was responsible for ensuring that you were doing your work as assigned. If the press acquires the information that the fire escaped initial control due to a hydrant being unavailable because of poor maintenance, the whole department takes a public beating.

As previously stated, authority can be delegated, but overall responsibility cannot—that lies with the fire chief. It is the obligation of the person assuming the authority to take responsibility for seeing that the function is performed correctly within department guidelines. You must hold yourself and the others around you accountable to do as they are directed.

Exception Principle

The sixth management principle is the **exception principle**. This principle states that the person delegating authority wants to be informed in situations of major importance. To put it another way, the supervisor could say, "I only like surprises on my birthday." The principle refers to the fact that certain situations are going to arise that the supervisor needs to be informed about, even if the subordinate handles them. These types of situations could include personnel matters, major incidents, and incidents involving major expense to the department.

There may be certain situations that the supervisor wants to be consulted on before the subordinate commits the department to a course of action. This is certainly the case when personnel matters are involved. If the company officer is forced to remove a subordinate from duty, the supervisor should be notified. If this process is done incorrectly, the company officer has opened the department up to a tremendous amount of liability for damages from the person relieved from duty.

Another situation that could easily be encountered is a citizen volunteering a piece of equipment at a fire, for example, a water tender. The good citizen sees the need for the fire department to use his water tender and volunteers it to the company officer working at the fire. The citizen then finds out that the department is paying the local construction company $2000 a day for the use of its water tender. The citizen then decides that he should also be paid the same amount for his water tender that he initially volunteered for free. After all, it is only fair to pay for both of the water tenders, isn't it? This then places the department in the position of having to decide what to do about the volunteered–water tender owner's claim. In most cases, the fire department is not going to be able to pay the claim because there was no preexisting contractual relationship between the water tender owner and the department. The whole process could have been avoided if the principle of exception was applied and the company officer had checked with the supervisor before approving the use of the water tender in the first place.

A more drastic example of the exception principle is a scenario that happens at incidents when someone is trapped under a heavy object, such as a car that has rolled over. The tow truck arrives and volunteers to hook up to the car and remove it from on top of the trapped person. In the haste to remove the car, the company officer approves the operation. In so doing, the tow truck operator has an equipment failure and the car falls back on top of the injured victim and kills her. When this goes to court, as it most likely will, the company officer will be asked why he did not wait for the proper rescue equipment to arrive at the scene to perform the operation. He will also be asked, "What is standard department procedure and was it followed?" His answer will most likely be that, "No, that is not standard procedure, but the situation required removing the car as soon as possible." The company officer is so close to the scene that his judgment may be affected by the situation in front of him. It is easy to get caught up in the heat of the moment and lose proper perspective. When in doubt as to the proper course of action to be taken in a situation, follow department policy. Had he contacted his supervisor and asked him or her what to do, the answer probably would have been to wait the extra time for the proper rescue equipment to arrive. In some instances, the principle of exception can be used as a system of checks and balances to either approve or disapprove a course of action.

Management Concepts

Department Goals

A department must have clearly defined goals. These are often described overall in a mission statement, such as the following: "The goal of the fire department, through its members, is to provide effective life safety and emergency services in the most efficient and cost-effective manner possible to ensure public safety and minimize economic loss" (KCFD 2013a).*

Goals tend to be nonspecific and immeasurable in that they are too broad. The goal stated previously is to provide "effective life safety." Just what exactly is effective life safety? The goals must be supported with statements that specify just what exactly the department can expect to accomplish. Not all fire deaths can be avoided. Accidents are going to occur, no matter how effective the firefighting force and no matter how much the jurisdiction spends on fire protection. There could be a fire station on every corner in town and fires would still occur. People will die from medical and other emergencies. The department must determine where its resources can best be applied to provide the most good for the most people.

One method of doing this is for the chief to concentrate resources in the high-value districts of the jurisdiction. If there is a downtown area, that is where most of the resources will be stationed. In the outlying, even rural areas, the staffing can be reduced and the response areas for individual fire stations increased.

Objectives

Objectives are statements of measurable results to be achieved with the resources available. All objectives must be specific, measurable, attainable, realistic, and timely (SMART). The objectives must be oriented toward the stated goal. The fire department must have the resources available or arrange for their acquisition. The resources can be in the form of money, personnel, or equipment. An objective would be to reduce deaths from fires to so many per 1000 of population. This may sound cold and calculating, but it is the only method against which progress can be measured.

Goal and objective achievement require the use of policies. These may be referred to as **Standard Operating Procedures (SOP)** or **Standard Operating Guidelines (SOG)**. Policies are described as a definite course or method of action. As not all possibilities can be addressed individually, policies tend to be broad, encompassing different situations. The department's policies are to be written and clearly understood by all members of the fire department. In many organizations, policies were determined on an assignment and live on in the culture of the organization. An example of a policy that could have been initially developed this way—but that should be written down—would be that all personnel are to wear breathing apparatus on incidents when smoke is showing. By writing and maintaining clear and applicable policies, the management of the department will aid in the smooth running of the organization and avoid conflict.

The following is an example of a policy:

To guard against incurring injuries during drills and the handling of basic firefighting tools, the following rule will be adhered to:

1. Engine companies participating in drills shall wear the structure helmet, turnout coat, and gloves, when protective clothing would normally be worn, while performing the same evolutions in emergency situations.

An example of a policy/procedure would be that the oncoming shift is to make a face-to-face shift change with the off-going shift to discuss any problems with the apparatus/equipment or other matters that need to be passed on. Procedures are often more specific than policies. A specific description of how a form is to be filled out, such as a fire report or prevention inspection form, would be a procedure. The everyday operations of the department are mostly guided by procedures. Without the procedures, there would not be a standardized way of doing things, and each shift and station would be like its own little fire department. A large part of studying, as a new fire fighter and for promotion, is learning the policies and procedures applicable to the position in particular and the rules of the organization as a whole.

---

* Kern County Fire Department (KCFD). 2013b. *Manual of Operations: Administration*. Bakersfield, CA: Kern County Fire Department

The following is an example of a procedure:

The procedure for purchasing an off-duty badge shall be as follows:

1. Any member wishing to purchase an off-duty badge should have the vendor submit a request for verification of employment to the administrative deputy chief.
2. Upon verification of employment, the administrative deputy chief will send the verification back to the vendor. A copy of the verification will be retained by the department.
3. The department is to be notified immediately should the badge be lost or stolen or in the event another badge is to be ordered.*

Using objectives, policies, and procedures, the management sets the course of the organization to reach the desired goals. All of the items must remain within the confines of the mission statement.

Rules set the limits for behavior on and sometimes off the job. A standard rule would be, "No person shall absent themselves from duty without the notification and approval of their supervisor." This rule makes very clear what is considered acceptable behavior. Rules give guidance for what is required as well as a set of guidelines for determining when discipline is required. Within the framework of the rules, there can be circumstances that provide for leeway. The supervisor may be empowered to allow the subordinate to be absent for up to an hour in certain circumstances. This would be a policy modifying the rule. The trouble is it is often not written down and could cause problems if discipline were applied against one person and not another.

Job descriptions are an important part of any organization and are used from the very beginning of the selection process for new fire fighters (see the Fire Protection Career Opportunities chapter for job descriptions of various fire department positions). Through job descriptions, the department decides what is to be included in the testing and training processes for positions. They are also used to validate the testing process. The fire department usually does not ask management-level questions of a fire fighter applicant because management of other personnel is not part of a fire fighter's job description. Job descriptions let people know what is expected of them in the position they are assuming and define the limits of authority and responsibility of the assigned position. A job description for battalion chief is therefore quite different from one for fire fighter.

One phrase included in many job descriptions is "and other duties as assigned." Job descriptions are often used as a minimum standard of what is expected, not the maximum. A fire fighter performing a line function would have much different duties than one assigned to staff.

> **Tip**
>
> Job descriptions are often used as a minimum standard of what is expected, not the maximum.

Staffing

Staffing is the assignment of resources to the needs that have been identified in achieving the objectives. Staffing applies to both the line and staff functions.

One of the biggest debates in the fire service is the one over what constitutes adequate staffing on an engine company. In 1995, a staffing study was done using three-, four-, and five-person companies to perform fireground tasks. Some of the results of the study were as follows:

1. Tasks can be completed more rapidly with larger numbers of personnel.
2. A more complex task was completed more quickly when performed by more fire fighters in a carefully coordinated fashion.
3. The location of the tasks to be performed made a difference. Tasks performed above or below grade complicated the task. Requiring the fire fighters to carry items up or down ladders or stairs fatigued them more rapidly.
4. The ambient (air) temperature is a factor. If it was too hot or too cold, it more rapidly fatigued the fire fighters.
5. The quality of direction made a difference. A carefully coordinated and directed attack went smoother and was more efficient than one not well directed.
6. The level of training, fitness, and competence of the fire fighters is a determining factor. The most fit, best trained, and most competent companies performed the tasks the most efficiently (Lawrence 1995).**

When one applies the management cycle to this study, several conclusions can be drawn. A planned, organized, staffed, directed, and controlled operation is quicker and more efficient than one left to chance. Well-trained, physically fit fire fighters are more competent and can perform

* Kern County Fire Department (KCFD). 2013. *Administrative Procedure 100.15, Off Duty Badge*. Bakersfield, CA: Kern County Fire Department, p. 1
** LawrenceCortez. 1995. *Company Staffing: The Proof Is in Your Numbers*. Fire Engineering. Tulsa, OK: PennWell Publishing

Tip

Conditioned fire fighters who know their jobs and equipment can perform the task with efficiency and safety with little time lost in deciding what to do and how to do it.

better under such adverse conditions as complex, above and below grade tasks in wide temperature variations.

A study completed in 2010 by the National Institute of Standards and Technology drew the following conclusions when comparing two-person companies to three-, four-, and five-person companies (Averill et al. 2010):

- *Overall scene time*. The four-person crews operating on a low-hazard (one-, two-, or three-family dwellings and some small businesses) structure fire completed all the tasks on the fireground (on average) 7 minutes faster—nearly 30 percent—than the two-person crews. The four-person crews completed the same number of fireground tasks (on average) 5.1 minutes faster—nearly 25 percent—than the three-person crews. On the low-hazard residential structure fires, adding a fifth person to the crews did not decrease overall fireground task times.

- *Time to water on fire*. There was a 10 percent difference in the water-on-fire time between the two- and three-person crews. There was an additional 6 percent difference in the water-on-fire time between the three- and four-person crews (i.e., four-person crews put water on the fire 16 percent faster than two-person crews). There was an additional 6 percent difference in the water-on-fire time between the four- and five-person crews (i.e., five-person crews put water on the fire 22 percent faster than two-person crews).

- *Ground ladders and ventilation*. The four-person crews operating on a low-hazard structure fire completed laddering and ventilation (for life safety and rescue) 30 percent faster than the two-person crews and 25 percent faster than the three-person crews.

- *Primary search*. The three-person crews started and completed a primary search and rescue 25 percent faster than the two-person crews. The four- and five-person crews started and completed a primary search 6 percent faster than the three-person crews and 30 percent faster than the two-person crews. A 10 percent difference was equivalent to just over 1 minute.

- *Hose stretch time*. In comparing four- and five-person crews to two- and three-person crews collectively, the time difference to stretch a line was 76 seconds. In conducting more specific analysis comparing all crew sizes to the two-person crews, the differences are more distinct. Two-person crews took 57 seconds longer than three-person crews to stretch a line. Two-person crews took 87 seconds longer than four-person crews to complete the same tasks. Finally, the most notable comparison was between two-person crews and five-person crews—with more than 2 minutes (122 seconds) difference in task completion time (Averill et al. 2010, 10–11).*

Large fire departments with numerous companies available have implemented the practice of staging a tactical reserve, called rapid intervention crews (RICs) (discussed further in Chapter 14, *Emergency Operations*). The tactical reserve is used to stand by at the fire scene. They are in full personal protective equipment (PPE), equipped with the necessary tools and ready to go in, but uncommitted unless there is a fire fighter emergency. What qualifies as a fire fighter emergency is a fire fighter(s) trapped, lost, low on air, or injured either due to structural collapse or other situations that endanger fire fighters' lives. To many in smaller departments, such reserves seem like a luxury. Some fire fighters may even question the need, feeling that everyone at the scene should be involved in controlling the fire. The added margin of safety in having the RIC available at a moment's notice is not to be discounted.

The staff positions must be adequately filled as well. If effective fire prevention is to take place, enough personnel must be assigned to that function. They are better trained in fire prevention and committed to it as a means of achieving the organizational goals. In addition, personnel must be assigned to training on a full-time basis to keep a department trained on the new developments and techniques.

Evaluation of Effectiveness

The evaluation of the fire department is carried out both internally and externally. Internally, progress reports, program records, personnel evaluations, and

---

* Averill, Jason D., Lori Moore-Merrell, Adam Barowy, Robert Santos, Richard D. Peacock, Kathy Notarianni, and Doug Wissoker. 2010. *Report on Residential Fireground Field Experiments*. April 27. Washington, DC: National Institute of Standards and Technology. http://www.nist.gov/manuscript-publication-search.cfm?pub_id=904607

apparatus and station inspections are used. Externally, the ultimate evaluator is the public. Is it satisfied with the performance of the fire department? Its opinions are expressed through direct contact and through elected officials. Proactive fire departments may go to the extent of sending questionnaires out to the public to determine whether it is satisfied with the department's effectiveness and to identify areas where the department can improve.

The budget is also a valuable tool in evaluating the fire department. If the fire prevention budget was increased 50 percent and there is no reduction in the number of accidental fire starts, further study needs to take place. The study is performed to determine whether the program is effective, and if not, why not.

Evaluation must be carried out objectively. One fire in which a fire fighter made a spectacular rescue does not mean that the whole department is doing its job well.

Evaluation is to be carried out in comparison with accepted standards. If fire department A has five persons per engine and fire department B has only three, the losses in comparable fires should be less for department A.

From the first time you fill out a job application for fire fighter, you are being evaluated. All during your career, no matter what rank you attain, you will be evaluated. Those below you will evaluate you as a leader and those above you will evaluate you as a subordinate. Everyone answers to someone else. The fire fighter is evaluated by the company officer, the company officer is evaluated by the battalion chief, and so on. Even the chief of the department must answer to the fire commission, the city manager, or the directly elected officials of the jurisdiction.

Evaluation of employees is an ongoing process. It does not occur only when your annual evaluation form arrives at the station. The company officer and the remainder of the crew are also evaluating you for your strengths and weaknesses every time you are on duty. The ability to accept evaluation for what it is—a tool to aid in attaining the goals and objectives—will assist you in becoming a better fire fighter and an asset to the organization.

Do not overlook the first phrase in the mission statement, "The goal of the fire department, through its members..." (KCFD 2013b).* Never forget that what makes a great fire department is not money; it is the people who work there. When people are trained and motivated, they can do great things. The lack of the latest in equipment and facilities should not severely limit the department's abilities. Many a life has been saved and untold numbers of fires have been controlled without the latest equipment. What really gets the job done are the desire, resourcefulness, skills, knowledge, and abilities of the personnel on the fire engine.

> **Tip**
>
> What makes a great fire department is not money; it is the people who work there.

Fire Department Types

The U.S. Fire Administration estimates that there are more than 30,000 fire departments across the United States involving 1.1 million fire fighters, of which approximately 300,000 are career and 700,000 are volunteer. There are numerous types of fire departments throughout the nation. Depending on the needs and resources of the jurisdiction, the departments vary in size. As departments increase in size and number of persons served, they increase in complexity.

Volunteer Fire Department

The first fire departments in the United States were volunteer. Their spirit lives on in the personnel who give their time and effort to suppress fires and perform other necessary functions without pay. Of all fire departments in the United States, 87 percent are composed entirely or mostly of volunteers (USFA 2012). Most of the paid fire departments are in communities that protect 25,000 or more people. Most of the volunteers (94 percent) are in departments that protect fewer than 25,000 people, and more than half are located in small, rural departments that protect fewer than 2500 people (Karter and Stein 2012).

The volunteers must often participate in fundraising activities to buy the necessary equipment. Fire engines and equipment are not cheap. Often older engines can be purchased from paid departments at a nominal fee. Other times equipment is donated outright. Departments are becoming more reluctant to donate equipment because they will be named in the lawsuit if injury results from use of the equipment.

The number of volunteers and volunteer fire departments is dwindling. Unfortunately, many volunteer fire departments in the United States are facing a crisis due to reduced enrollment and retention of membership, causing some fire companies to close their doors or cities and towns have had to consolidate fire companies. This crisis is especially apparent during daytime hours when members are at work and unable to respond. Some companies have resorted to hiring limited staff during weekdays to cover calls until volunteers are available after they get out of work.

---

* Kern County Fire Department (KCFD). 2013b. *Manual of Operations: Administration.* Bakersfield, CA: Kern County Fire Department

Just because a department is made up of volunteers is no reason for it to lack in effectiveness. There are numerous training programs available from the National Fire Academy and state fire-training organizations. One advantage volunteer departments have is their ability to spend most of their money on equipment. A career (fully paid) fire department might spend upwards of 90 percent of its funding on personnel costs alone (ICMA 2002).

Within the realm of volunteer fire departments, there are differences. Some volunteer departments are totally volunteer; these have no paid members. Others, called combination departments, have several paid members supported by volunteers. A common paid position in a primarily volunteer department is that of driver **FIGURE 7-2**. By having a paid driver on each shift, the department has someone who is responsible for maintaining and ensuring the readiness of the equipment. The paid person also maintains the fire station. In areas with long responses, the paid person can respond in the engine and the volunteers can meet him or her at the scene. This prevents the problem of all of the volunteers responding to the station, finding that the equipment has already left with the volunteers who got there first. The other problem scenario is that all of the volunteers respond directly to the incident and no one brings the engine.

In the past, the person who received the alarm would transmit it through a system of bells or sirens. The sequence of the siren sounding, a series of long and short blasts, notified the volunteers as to which part of town to respond. Now most volunteers have voice or alphanumeric pagers or texting capability and can be notified at home or work as to the exact address of the incident.

The obvious limitation is that it is not predetermined how many personnel are going to arrive at the incident.

When a fire officer has a regular crew that he or she works with, the officer knows the crew members' strengths and weaknesses. The officer also has a good idea of the next-in company's response time and capabilities.

There are also variables among volunteers. Some volunteer fire fighters are in good physical shape and spend lots of their free time becoming trained; others are members, but lack training and commitment.

The term *volunteer* is being replaced with *on call* or *paid call* in many states. Labor law court rulings have determined that volunteer fire fighters are employees of the jurisdiction they serve and must be compensated for their work as well as be covered by worker's compensation insurance should they be injured in the line of duty.

Combination Fire Department

In combination departments, a large part of the staff is paid and the volunteers are used as supplemental personnel **FIGURE 7-3**. They are either used to cover the station when the regular crew is at an incident or they respond directly to the incident. This method of staffing saves the cost of paying a large staff of fire fighters when they are often not needed. One advantage is that the community receives protection at a prescribed minimum level all of the time. Only when circumstances dictate are the volunteers paged to respond. This is much less disruptive to the volunteers' jobs and businesses. They do not need to respond to every small incident that can be handled by the paid personnel. When the inevitable large incident occurs, there is a reserve force of personnel to assist in mitigating it.

Another advantage is that the paid fire fighters are available to receive training and pass it on to the volunteers.

FIGURE 7-2 Having a paid driver in a primarily volunteer department helps clarify who will respond to an incident with the equipment and who will meet the engine at the scene.

Courtesy of Mike Legeros

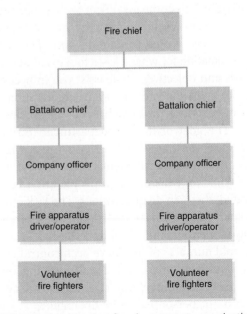

FIGURE 7-3 Combination fire department organization.

© Jones & Bartlett Learning

They are also available to receive specialized training and perform fire prevention activities. In this type of organization, the position of chief is usually paid because the paid staff is increased to the point where a professional manager is required for planning and budgeting.

Many departments use this concept in the form of a force of reserves. The reserves usually comprise young people with a career in the fire service as their goal. This program serves both parties well because the reserves seek training and experience, and the department gains a group of trained personnel at little or no cost.

Another variation on volunteers is paid-call fire fighters, reserve personnel who are paid by the incident. If they are paged to respond to cover a station, an incident, or a drill, they are paid a certain amount. This program is often used by departments to supplement forces in small towns and outlying areas. The ability to pay the reserve force is an incentive for the reserve personnel to attend the necessary drills and covers.

A problem that needs to be addressed in a proactive manner is the training mandated by state and federal law. From a liability standpoint, the jurisdiction is required to train the personnel to at least minimum levels. Some of this training is very time consuming and is hard for volunteers to complete due to their regular job commitments. As time goes on, jurisdictions are going to have to find ways to address this issue or suffer the consequences when someone is hurt or killed.

Public Safety Department

In an effort to save money and better utilize personnel, some jurisdictions have set up public safety departments **FIGURE 7-4**. As of April 2013, the city of Ypsilanti, Michigan, is projecting it will save $2.1 million over the next 5 years with the creation of a hybrid police and fire department (Stafford 2013). A more recent example of this department type is Cedar Falls, Iowa (established July 1, 2017). Under this concept, the police and fire departments are under the same department administrator. They may go as far as to have the personnel cross trained so that they can go out on patrol as police officers and respond to fires as needed. The advantages are that the jurisdiction gets productive time out of their personnel most of the time that they are on duty. The disadvantage is that if there is a crime or major fire, most of the resources are going to be tied up on one or the other. Another disadvantage is that both police work and firefighting require a lot of time spent on specialized training, and a person who is cross trained is not likely to have the training required to be very proficient at either job, or he or she will concentrate on one over the other.

Career Fire Department

The basic premise of a paid fire department is that all of the personnel are paid a salary, or in the case of the U.S. Forest Service, Bureau of Land Management, and Park Service, paid by the hour. In the large cities, such as Boston, Los Angeles, Chicago, Nashville, and New York, the departments are of such a size that a fully paid department is a necessity. Their operations are too large and complex to be performed by volunteers.

One advantage of a paid fire department is that the jurisdiction has much more control over the fire personnel. The jurisdiction has the ability to hire and fire as necessary. When people are hired, they are offered a rewarding career with pay and benefits. The person is expected, in return, to abide by the rules, procedures, and policies of the department.

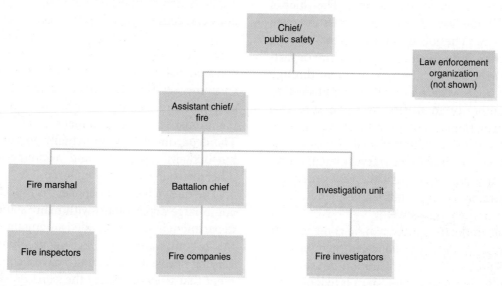

FIGURE 7-4 Public safety department organization.
© Jones & Bartlett Learning

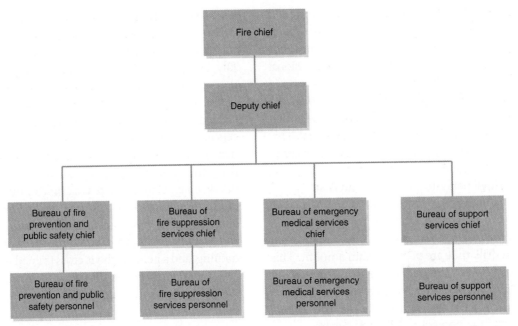

FIGURE 7-5 Large paid fire department organization. Operations personnel are below the bureau of fire suppression and rescue chief position.

© Jones & Bartlett Learning

Paid fire departments, especially the large ones, require expert management **FIGURE 7-5**. It is becoming less common for the top officer to be from a purely firefighting background. The best fire fighter in the department, from an operations standpoint, is not necessarily a good administrator. The chief of department, in the large departments, does not even respond to fires. The job of administrator requires his full attention. Administrators of large fire departments set the tone for the organization through policies and goals. Their work is accomplished by delegating authority to others to make sure that rules and regulations are followed and the mission of the department is accomplished. By referring back to the national professional development model (Figure 1-1) it is easy to see that the new generation fire administrator requires both training and education to perform the job effectively.

The people in the top jobs may have come to work as fire fighters, but they have educated themselves in the areas of public or fire administration. An easy way to see what is required for the top spots is to look at the job offers for fire chief in the back of magazines, such as *Fire Engineering, Fire Chief,* or *Firehouse.* These job offerings require administrative backgrounds with an advanced education, usually a minimum of a bachelor's degree, with a master's degree preferred. Often successful completion of the Executive Fire Officer Program at the National Fire Academy is required.

One of the most valuable attributes of a chief is the ability to communicate. When you are in a position where you must make it clear to others what you want done, without performing the function yourself, it is very important to be able to communicate what exactly it is you want—face-to-face, on paper, and over the phone. The chief also needs to be somewhat of a diplomat to present the needs of the fire department to the governing body and to interpret the wishes of the governing body and convert them into fire department action.

Some of the management concepts that are popular with chiefs are customer service, the one-department concept, team building, and incident effectiveness. Customer service refers to the fact that the public is the customer and, like any other business, the fire department must serve the customer's needs or be replaced. The department will not actually go out of business, but budgets can be cut and personnel replaced, or other alternatives can be explored.

The one-department concept is especially important in large departments with many stations. The fire department is already divided among the shifts. When this is compounded in large geographical areas with varying needs, it is difficult to have a cohesive department that works basically the same on all shifts at all stations. Standardization is sought as much as possible. In a department that covers municipal and rural areas, there will be some differences in operations and activities. Standardization is often established as far as practical so that a fire fighter working overtime at another station does not have to learn a whole new set of procedures used at incidents. The one-department concept also applies to attitude. The personnel who work in a metropolitan area closer to headquarters or at a battalion headquarters station are liable to be held to a higher standard of dress and performance than those working in stations in outlying areas. You must remember that you are paid to act and look professional no matter what station you are assigned to.

Team building occurs when people are willing to work together within department guidelines. For team building to succeed, each person must know his or her duties and responsibilities and where they fit in the overall scheme of things. They must also know the desired results of their actions. Some people will fill certain positions better than others. All of the personnel involved must be able to conform to the plan to work for the common good. All of the personnel on duty at the station must be willing to help each other, even in simple tasks, such as equipment maintenance and meal preparation. Each shift must be able to support the others in seeing that things get done at the station regardless of who is on duty. If the other shift had a fire just before they went off duty and had to leave a bunch of dirty tools, the oncoming shift should take care of them without complaint. Sooner or later you will be leaving it for them when you have a fire. This is not just at the single-station level. All of the stations that respond to incidents together need to be able to work together as an integrated unit. This concept carries right on up to the top of the organization.

Incident effectiveness can be defined as the fire department's ability to function quickly and efficiently when called to act. Incident effectiveness is achieved through focus on the four guiding principles of drill, equipment maintenance, programs, and station maintenance. If everyone in the department performs his or her job well, the department will be effective. Performing training drills raises the proficiency of individuals and the team as a whole. Equipment maintenance ensures that the apparatus will make it to the scene and that all of the tools are in working order. Programs make the personnel more aware of the problems in their response area and contribute to the accomplishment of the department's mission. Station maintenance leads to a more pleasant work environment. If one shift leaves the station a mess every time it goes off duty, it will create tension with the other shifts. Something as simple as emptying the trash cans and wiping out the sinks after use makes the station a more pleasant place for everyone.

In the modern fire department, more and more of the upper management positions are being filled by civilians. These people are chosen for their particular knowledge, skills, and abilities. They are not directly responsible for firefighting activity, so do not need firefighting skills. The jurisdiction derives a cost savings by not having to pay these people a safety section retirement. The common complaint is that they are "bean counters" and do not understand fire fighter needs and priorities. The downside of this is outweighed by the fact that these personnel are trained in administration and do their jobs very well. They are experts at budgeting and resource management. Their ability to squeeze the maximum benefit from every budget dollar positively affects the fire fighter in an indirect way. Most departments have recognized the need to keep the line functions under the control of managers with firefighting experience and limit the hiring of management personnel with no firefighting experience to staff positions.

In adhering to the principle of span of control, depending on the size and complexity of the department, there are layers of management between the chief and the fire fighter. Not all fire departments have the same name for the positions, but the duties and responsibilities of the positions are much the same.

Under the chief are the various assistant and deputy chiefs. These personnel are not required to perform the same functions as the chief but are required to have a high degree of ability as administrators. A deputy chief in charge of operations in a major department may have 2500 employees under his or her direct and indirect control. The deputy chief's communication powers must be of the highest ability as well. This chief will not know all of the fire fighters in the stations, but he or she is responsible for their safety and efficiency.

The person in charge of several stations is the battalion or district chief. This person usually works the same duty schedule as the fire fighters and is responsible for the activities of one shift, whereas staff personnel usually work a 40-hour week (typically, an 8 AM–5 PM workday). The battalion chief responds to emergencies when necessary. The battalion chief probably does not respond to medical aids, car fires, and other minor incidents but does respond to any major incidents, such as a structure fire, and other incidents that could become major or complex. It is the battalion chief's responsibility to ensure that the company officers and their crews are trained and performing to the expected level of proficiency. This is done through company officers' meetings and evaluating engine company proficiency by observing training drills and performance at incidents. The battalion chief evaluates the company officer on the performance evaluation and assists the company officer in determining goals for the next rating period.

When the battalion chief performs a station inspection, he or she inspects all of the station's assigned programs, records and reports, maintenance of the station, and condition of the apparatus. The station personnel are inspected for the proper work uniform, PPE, conformity to the grooming standard, and possession of a valid driver's license. The station inspection is a perfect opportunity for the battalion chief to ask the crew questions about the apparatus and department policy to make sure that they have been studying and keeping up on policy and procedure changes.

At the station level, the company officer is the first-level supervisor. He or she is responsible for ensuring that the crew is trained and equipped to respond to emergencies. The company officer evaluates the driver/operator and the fire fighter on their performance evaluations and assists them with determining goals for the next rating period. In the case of a probationary fire fighter there may be an evaluation every 3 or 6 months until probation is completed.

As a training tool, the company officer may delegate authority to complete some of the station-level programs to the driver/operator and fire fighter. This requires the crew members to be more familiar with the program and teaches them management skills. It also helps to prepare the crew members to take promotional examinations. The driver/operator is responsible for the equipment and knowing how to operate it. Most driver/operators will take the time to train the fire fighters in the use and operation of all of the equipment. (The fire fighter position's responsibilities are discussed in Chapter 2, *Fire Protection Career Opportunities*.)

Industrial Fire Brigade

Many businesses have organized private fire brigades. These may be organized to protect a manufacturing plant, oil refinery, or other location. The brigades are made up of personnel hired by the company. Often the personnel have other jobs, and the fire brigade is an additional duty that is performed when necessary. They receive equipment from their employer. The training they receive may be provided by the employer, a contracted service, the local fire department, or a combination of these. They are often organized into teams. They may respond to alarms in all parts of the property, or each geographical or functional area may have a separate fire brigade organization according to the needs of the property. The organization is such that a fire brigade is on duty on each working shift and at periods when the plant is shut down or idle.

Contract Fire Protection Service

There are private sector companies that provide fire prevention and suppression services. These companies arose in sprawling suburban and rural communities and are involved in every aspect of the fire prevention and suppression business. They provide services in national parks and forests, airports, nuclear plants, commercial businesses and industrial firms, and rural and residential neighborhoods.

The private fire protection companies provide fire protection service either by contract or subscription. Contract service is offered to local governments or special fire districts; subscription service is offered to residents or property owners through the payment of a subscription fee.

In the case of contract fire protection service, the officially designated representatives of a local jurisdiction (usually a town, city, or special tax district) award a private fire protection company the right to service that jurisdiction for a specified time period. The company is paid a fixed and contractually agreed-upon sum, either through the jurisdiction's general tax revenues or through a special fire tax levied by the jurisdiction.

In the case of subscription fire protection service, individual property owners or residential associations contract directly with a private company for fire protection service. When a fire occurs, the people who have a subscription on their property are not charged. If they do not have the subscription coverage, they are charged suppression costs, or, in some cases, the subscription fire protection company's personnel may protect other properties and let the non-subscribing property burn.

The private fire protection companies are able to generate cost savings in a number of ways:

- *By reducing personnel costs.* These companies rely heavily on trained volunteers and reservists. They often pay salaries and benefits commensurate with market rates, which are typically less than those earned by public sector fire department employees.

- *By making productive use of otherwise idle time.* When not actually fighting fires, private fire fighters are engaged in a number of other endeavors, including building and refurbishing fire apparatus, providing combination fire and security patrols, training industrial fire brigades, operating alarm monitoring and installation services, and sponsoring fire prevention campaigns and other educational activities.

- *By using innovative strategies and technologies to prevent and combat fires.* As an example, some of the private sector companies are now actively encouraging the use of sprinkler systems in residential buildings.

Communications

There are four basic methods of communication in common use in the fire service. These are face-to-face, radio/telephone, written, and electronic. The best method is face-to-face. Messages passed in this way are the most likely to be understood. Through this method the sender and receiver can see how each other reacts as clues to how and if the message was properly understood. You can often tell by the look on someone's face and body language that they did not really understand what you told them even if they are saying they did. It allows the asking of questions and clarification of misunderstanding. This is the method of choice for transferring command at incidents.

Radio communications are essential because personnel and equipment are spread out at an emergency scene and the personnel are too busy or the message is too time-sensitive for them to meet. A problem is that personnel may be embarrassed to ask for clarification of an order because they will feel stupid. It is better to ask for clarification than to misunderstand the order and do something that endangers other personnel. Discipline must be maintained when using radios to communicate. Fire department radios are not CBs, and there are specified procedures as to how to talk on the air. Another factor is that many people, especially the news media, have scanners and listen in on fire department frequencies. Radios should not be used to converse on unimportant matters. Many fire fighters have been in dangerous situations and were unable to call for help because someone was tying up the air with nonessential messages. A disadvantage of radios is that they are hard to use when wearing a self-contained breathing apparatus (SCBA) mask, unless it is equipped with a built-in radio interface, and may have poor transmission and reception when working in remote or hilly areas. They also do not work in some buildings. In high-rise buildings, the method of choice is the internal telephone system.

Telephones are a method of communication that, much like radios, depersonalizes the transmission and receipt of messages. Their advantage is that they provide a good connection without static or other interruption. A tool that is in common use is the cellular phone. It is good for transmitting and receiving information in that you can have a more or less private conversation. Persons with the proper equipment can listen in on cellular phone conversations, however.

Written communications are used when time is not a critical factor. They are used to record the policies and procedures of the department. They are also used to maintain a hard-copy record of communications for future reference and to reduce the possibility of error in interpretation. The problem is that hard copies of written messages either need to be handled by courier or mailed, which slows their delivery.

With the invention of fax machines and electronic mail, the transfer of written messages has been sped up. Through modems and networks, computers can be linked and messages can be quickly transmitted, with a hard copy produced when desired.

With use of the Internet, communications have greatly sped up. Many fire departments have their own web pages. Now departments are using social media platforms to get out their messages to the public. This can be used to announce evacuations, threats to a neighborhood or area, etc. There may also be internal, employees only, pages with limited access. One of the problems with this is that rumors can spread faster than ever on social media. To prevent issues with lawsuits, many fire departments are now prohibiting their personnel from taking or posting photos taken at incidents. This may involve medical privacy laws or other issues.

Wrap-Up

CHAPTER SUMMARY

- Without an administration, the fire department cannot function properly. Like any other entity, public or private, the fire department must comply with laws and provide services to its personnel.
- The guiding principles for the organization of a fire department can be referred to as the principles of command: unity of command, chain of command, span of control, division of labor, delegation of authority, and exception principle.

- The concept of unity of command means that each person in the organization has only one boss.
- The chain of command indicates the hierarchy of the organization.
- In firefighting, the effective span of control is considered to be three to seven personnel per supervisor, with five being optimal in most situations.
- Work to be performed needs to be divided into specific areas to prevent duplication of effort, to apply the most appropriate resources to the job, and to determine responsibility for completion of the assigned work.
- For work to get done, the manager must delegate authority to the subordinates.
- The exception principle states that the person delegating authority wants to be informed in situations of major importance.
- Goals must be supported with statements that specify what exactly the department can expect to accomplish.
- Objectives are statements of measurable results to be achieved with the resources available. All objectives must be specific, measurable, attainable, realistic, and timely (SMART).
- Policies, procedures, and guidelines are created and published to let members know expectations in various situations and standardize operations.
- Rules are established to describe acceptable and unacceptable behavior of fire department personnel.
- There are numerous types of fire departments throughout the nation. As departments increase in size and number of persons served, they increase in complexity.
- Volunteer fire protection service serves as the preliminary first-step fire service, to be followed, as the jurisdiction grows in size and population, by some form of paid service.
- In combination departments, a large part of the staff is paid and the volunteers are used as supplemental personnel.
- The basic premise of a paid fire department is that all of the personnel are paid a salary. One advantage of a paid fire department is that the jurisdiction has much more control over the fire personnel.
- Many businesses have organized private fire brigades. These may be organized to protect a manufacturing plant, oil refinery, or other location. The brigades are made up of personnel hired by the company.
- There are private sector companies that provide fire prevention and suppression services. They provide services in national parks and forests, airports, nuclear plants, commercial businesses and industrial firms, and rural and residential neighborhoods.
- The best method of communication is face-to-face. Through this method, the sender and receiver can see how each other reacts as clues to how and if the message was properly understood.

KEY TERMS

Acceptable risk A risk that is considered to be of low enough severity or frequency that it is considered acceptable.

Chain of command The formal path of communication through an organization. Individual members take orders from only one superior and give orders to a defined group of people immediately below them.

Cover To move resources into a fire station when the regular crew is assigned to an incident. In some departments this is called a *move up*.

Delegation of authority The action by which a commander assigns part of his or her authority commensurate with the assigned task to a subordinate commander. While ultimate responsibility cannot be relinquished, delegation of authority carries with it the imposition of a measure of responsibility.

Division of labor The assignment of work to those most qualified to carry it out or the division of a complex task into several less complex tasks.

Emergency medical dispatch A system in which dispatchers are trained to give medical advice to the persons at the incident, such as CPR instructions, until emergency help arrives.

Exception principle A method or plan of supervision under which only significant deviations from normally expected results or conditions are brought to the attention of a supervisor for consideration and decision.

Policy General guideline of how things will be done.

Rules One of a set of explicit or understood regulations or principles governing conduct within a particular activity or sphere.

Span of control The number of subordinates a manager can directly control. The number varies with the complexity of the operation and the skill of the subordinates.

Standard Operating Guidelines (SOG) Another term for Standard Operating Procedure (SOP).

Standard Operating Procedures (SOP) A particular way of accomplishing something or acting in a specified situation.

Unity of command The organizational principle in which every individual is accountable to only one designated supervisor to whom he or she reports at the scene of an incident.

Validate To make sure that the items included in the testing process are actual requirements of the job.

CASE STUDY

To close a budget shortfall, the fire department has decided to close fire stations and shut down ambulances. In addition to these cuts, supervisory positions are also being cut. One of the problems immediately associated with these cuts is an increase in response times to emergency incidents.

As response times increase, medical incident deaths increase, due to a delay in applying life-saving efforts. In addition, structure and other fires cause greater loss due to a delay in applying fire suppression efforts. Fires are given more of a head start. Delay in the arrival of firefighting resources creates a situation that is more difficult and less safe for fire fighters. A structure degrades as it burns, and a delay in applying water to the seat of the fire means that the structure is further weakened before a primary search for victims can be conducted. It is more than likely that fire deaths and fire losses will increase with longer response times and less resources arriving promptly to incidents.

1. The most significant impact to the public when reducing staffing resources is fewer:

   A. personnel at the station to perform maintenance.
   B. personnel at scene to mitigate the incident.
   C. fire fighters with government jobs.
   D. people on the city payroll, reducing money for taxes.

2. Initial units arriving later to a structure fire increases all *except*:

   A. fire loss.
   B. need for supervision.
   C. ability to rescue live occupants.
   D. ability to provide adequate overhaul.

3. The lack of adequate supervisory positions reduces:

   A. oversight of personnel.
   B. levels of management.
   C. paperwork processing delays.
   D. increased workload.

4. With the core mission of the fire department being saving lives and property, how does reducing staffing or closing companies/stations affect the mission?

   A. It has little effect.
   B. It depends on the amount of the cuts.
   C. It has a great effect.
   D. It cannot be determined.

REVIEW QUESTIONS

1. Summarize the six principles of command.
2. List the six components of the management cycle.
3. Why is it necessary for the management cycle to be a continuing process?
4. List the four forms of communication and their strengths and weaknesses.
5. Which form of communication is the most effective? Why?
6. In the chain of command, whom does the fire fighter answer to?
7. Draw and fill in the blanks on a simple organization chart with at least five levels of rank represented.

8. What are the general responsibilities of the fire fighter?

9. What are the general administrative responsibilities of the company officer?

10. Who is the first-level supervisor in the fire department?

11. If you had a question about policy or procedure, who should you ask?

12. What is the difference between line and staff functions?

13. Give an example of division of labor in the structure of the fire department.

DISCUSSION QUESTIONS

1. What is meant by the term *incident effectiveness*?

2. Why is customer service such an important concept in the modern fire service?

3. When issued conflicting orders by supervisors, what should you do?

REFERENCES AND ADDITIONAL RESOURCES

Averill, Jason D., Lori Moore-Merrell, Adam Barowy, Robert Santos, Richard D. Peacock, Kathy Notarianni, and Doug Wissoker. 2010. *Report on Residential Fireground Field Experiments*, April 27. Washington, DC: National Institute of Standards and Technology. http://www.nist.gov /manuscript-publication-search.cfm?pub_id=904607

Candea, Benjamin. 2013. "Miami-Dade Fire Rescue Chief Laments Budget Cut." *Local 10.com*. http://www.local10 .com/news/miamidade-fire-rescue-chief-laments-budget -cuts/-/1717324/21151732/-/1joyvp/-/index.html

City of Cedar Falls, Iowa. 2019. *Public Safety Services*. http: //www.cedarfalls.com/1145/Public-Safety-Services

Fire Protection Publications. 1983. *Incident Command System*. Stillwater, OK: Oklahoma State University.

Grad, Shelby. 2009. "L.A. Fire Department Cutbacks Begin Today; Response Times to Increase." *Los Angeles Times*, August 6. http://latimesblogs.latimes.com/lanow/2009/08 /la-fire-department-cutbacks-begin-today-response-times -to-increase.html

Guardino, John R., David Haarmeyer, and Robert W. Poole, Jr. 1993. *Fire Protection Privatization: A Cost Effective Approach to Public Safety*. Los Angeles, CA: Reason Foundation.

International City Management Association. 2002. *Managing Fire and Rescue Services*. Washington, DC: International City Management Association.

Karter, Michael, and Gary Stein. 2012. "U.S. Fire Department Profile through 2011." National Fire Protection Association.

http://www.nfpa.org/research/statistical-reports /fire-service-statistics/us-fire-department-profile

Kern County Fire Department (KCFD). 2013a. *Administrative Procedure 100.15, Off Duty Badge*. Bakersfield, CA: Kern County Fire Department.

Kern County Fire Department (KCFD). 2013b. *Manual of Operations: Administration*. Bakersfield, CA: Kern County Fire Department.

Lawrence, Cortez. 1995. *Company Staffing: The Proof Is in Your Numbers*. Fire Engineering. Tulsa, OK: PennWell Publishing.

National Fire Education System Course. 2006. *S-430 Operations Chief*. Boise, ID: National Interagency Fire Center.

National Fire Protection Association. 2008. *Fire Protection Handbook*. 20th ed. Quincy, MA: National Fire Protection Association.

Snook, Jack. 2013. "Budget, Fire Leadership Advice to Manage Departments in the New Reality." *Fire Chief*, January 20. http://firechief.com/budgets/budget-fire-leadership- advice -manage-departments-new-reality

Stafford, Katrease. 2013. "Hybrid Police and Fire Department Projected to Save Ypsilanti $2.1M, Add 8 New Officers." *Ypsilanti Reporter*, March 8. www.annarbor.com/news /ypsilanti/.ypsilanti

United States Fire Administration. 2012. *National Fire Department Census Quick Facts*. Emmitsburg, MD: National Fire Administration. http://apps.usfa.fema.gov/census /summary.cfm#e

CHAPTER 8

Support Functions

OBJECTIVES

After studying this chapter, you should be able to:

- Discuss the role of dispatch.
- Discuss the transmission of alarms.
- Identify the resources in the fire investigation unit.
- Identify the resources in the hazardous materials control unit.
- Discuss the role of an adjutant, or aide.
- Identify technical support groups and discuss their roles.
- Discuss resources available for information systems.
- Discuss the role of personnel/human resources.
- Discuss the role of a business manager.
- Discuss the role of incident business management.
- Discuss the role of warehouse/central stores.
- Discuss the role of a repair garage.
- Discuss the role of a radio shop.

Case Study

Whenever a large-scale incident occurs, be it wildland fire, flood, tornado, hurricane, or other incident, one of the positions necessary is that of the geographic information systems (GIS) technician. These technicians map the perimeter of the incident. They also develop information displayed on maps based on the requests of the incident. These maps include location of structures within wildland fire perimeters, aerial hazard maps showing the location of power lines and radio towers for aircraft pilots' use, property ownership, jurisdiction, and more. The jurisdiction map is used for cost apportionment among the responsible agencies.

An incident where GIS was used extensively was the recovery of *Columbia* space shuttle parts. When returning to Earth in 2003, the shuttle broke up as it entered the atmosphere. To recover as many parts as possible and try to determine the cause of the disaster, the National Aeronautics and Space Administration (NASA) engaged with the Federal Emergency Management Agency (FEMA) and the U.S. Forest

Service's incident management teams to manage the recovery process. To ensure that the path of the shuttle was covered as completely as possible by parts recovery teams on the ground, GIS mapping was used. There were four camps set up across Texas and into Louisiana. Each day teams of personnel would go to areas indicated on GIS-created maps and grid out the path of the shuttle. The furthest west camp, located in Corsicana, Texas, alone had 800 people out walking through farmland, pastures, ranches, and along roadways looking for parts that landed on the ground.

Without GIS maps and careful logging of the areas searched, this project could not have been as successful as it was. Approximately 40 percent of the shuttle was recovered.

1. What are some other possible incident-related uses of GIS data?
2. In what ways can GIS data assist fire fighters in the fire stations?

Modified from John F. Kennedy Space Center. "Columbia Recovery Efforts." NASA. Last updated November 22, 2007. http://www.nasa.gov/missions/shuttle/columbia_recovery.html; Howell, Elizabeth. 2013. "Columbia Disaster: What Happened, What NASA Learned." http://www.space.com/19436-columbia-disaster.html#sthash.KtyDi28Z.dpuf

Access Navigate for more resources.

Introduction

Fire department operations can be divided into several areas, which can be further divided into those that are incident focused and those that are not. Not all of the personnel who work for the fire department respond to incidents; many work in supporting roles that aid the incident responders in performing their jobs. Organizations cannot function without support. The modern fire department needs personnel in the fire stations to respond to incidents, but it also needs support and assistance in other areas. If the engines do not mechanically perform up to standards, the fire fighters will not be able to respond. If the radio system is not functioning properly, well-coordinated attacks are not possible. If the self-contained breathing apparatus (SCBA) are not maintained, interior attacks are difficult or untenable. Without this support, the personnel in the fire stations would not have the equipment, training, and facilities they need to perform their jobs. For the purposes of this chapter, those personnel not directly engaged in fighting fires and responding to incidents are considered support personnel.

All of the support functions, under the management direction of the staff, are required to enable the field personnel to conduct emergency and non-emergency

operations. As in other areas of administration and operations, the size and complexity of the support functions are dependent on the needs of the department. In small departments, many of the service and repair functions may be handled through outside contractors. In large organizations, it may be more economically feasible to have internal support personnel to maintain radios and SCBA. For organizations of any size, the support functions must be in place to provide an effective and efficient fire protection system to the community.

Tip

A modern fire department requires many personnel working in the area of support functions to operate. These functions are widely varied in nature and require personnel with expertise in each of the different areas.

Dispatch

The fire department requires the capability of receiving requests for emergency service and sending its units to the scene of the incident. Even in a one-station department, there needs to be someone assigned to answer the phone and alert the station crew that their services are required. Small departments may rely on the police department dispatch system to alert their personnel. Volunteer fire departments issue their personnel pagers so they can be alerted at their homes or jobs. In larger fire departments, the dispatch section is usually a purely fire department function. The personnel assigned may be fire fighters or civilians, depending on the department's preference. Assigning fire fighters to dispatch positions has its positives and negatives. On the positive side, the personnel assigned are familiar with departmental operations and can often anticipate the needs of the units assigned to the incident. During intense activity, a trained fire fighter acting as a dispatcher can be of great help to the incident commander. When a situation is developing rapidly, the incident commander is trying to keep track of resources assigned and responding. He or she is also trying to formulate a plan and give assignments to the equipment that has already arrived or is about to arrive. It is difficult to think of everything when the incident is gaining the upper hand. It takes a while for everyone to arrive and to amass the resources needed to control the incident. Having someone who understands the situation assisting with the tracking of resources is a help in that the incident commander is able to focus on what is happening, and what is about to happen, at the scene.

> **Tip**
>
> Even in a one-station department, there needs to be someone assigned to answer the phone and alert the station crew that their services are required.

On the negative side, having fire fighters as dispatchers is much more expensive in salary and benefits than having civilians in the positions. Another negative is the feeling of helplessness felt by fire fighters when they are sitting at a dispatch console listening to the radio traffic as other fire fighters engage in a tough fight. Fire fighters tend to be, and are trained to be, action-oriented people. When acting as dispatchers they are forced to sit there thinking that their friends and coworkers need help and there is no direct action they can take. There is no release for the adrenaline they build up while listening to what is going on. When dispatching, they have to sit there and pay close attention in case someone needs assistance. In too many incidents, fire fighters have gotten themselves in bad situations and no one heard

them call for help. In some cases, they died because no one came to their rescue. With dispatchers listening to the incident radio traffic, in a controlled environment that is much quieter and less chaotic than the incident scene, the calls for help might have been heard and reported to the incident commander at the scene. Many fire fighters never do adjust to the dispatch position and return to an operations assignment at the first opportunity.

Expanded Dispatch

In certain situations, incidents reach a point where they are beyond the immediate capabilities of the dispatch center. One way that departments handle the communications for incidents that become large or complex is to set up an expanded dispatch, sometimes called the command center, to take some of the load off the dispatchers performing their normal duties. When a large incident occurs, there is an increased volume of radio traffic between the units at the scene and between the command units and the dispatch center. The rest of the department does not shut down when there is a large incident in the jurisdiction. In a jurisdiction with a large geographical area, there may be a major wildland fire occurring, but maybe several structure fires need to be dispatched as well as a hazardous materials incident, medical aid incidents, and normal traffic, such as equipment going in and out of service. It does not take very long for the dispatchers to become overloaded.

The expanded dispatch is set up at a location outside the normal dispatch center. Some departments have a mobile command post and others move into a conference room at headquarters or a classroom at the training center. This allows the personnel assigned to the expanded dispatch to perform their functions without disturbing the regular dispatchers. When the department has several different radio channels available, all of the radio traffic for the large incident is switched to a designated channel. In this way, the regular dispatchers turn that channel's volume down or off and are not disturbed by all of the incident traffic. At the expanded dispatch center, the orders for resources are handled by dispatch recorder personnel. Other personnel are assigned to making the necessary phone calls to get the resources ordered and to gather necessary information, such as estimated time of arrival.

An example of an expanded dispatch center is the geographic area coordination center (GACC) concept used by federal wildland firefighting agencies and some states. In California, the state is divided into two regions represented by the South Operations (South

Ops) dispatch center in Riverside, California, and the North Operations (North Ops) center in Redding. South Ops is staffed by U.S. Forest Service and CalFire personnel. When a major wildland fire occurs in the southern region, and the responsible fire department has to go outside of its jurisdiction for resources, it contacts South Ops. South Ops then dispatches federal and state wildland firefighting resources as necessary to the incident. This allows for the response of numerous resources from various agencies quickly and efficiently.

When coordination is required on a larger scale, resource requests are sent to the National Interagency Fire Center (NIFC) in Boise, Idaho. The National Interagency Coordination Center (NICC) is located at NIFC. The NICC is the focal point for coordinating the mobilization of resources for wildland fire and other incidents throughout the United States. The NICC also provides intelligence and predictive services-related products designed to be used by the internal wildland fire community for wildland fire and incident management decision making (NICC 2018). It is not uncommon for hotshot crews from the Angeles National Forest in Southern California to be unavailable because they are assigned to fires in Montana. When a fire starts in Southern California, a crew may have to be flown in from the East Coast to take their place. This may seem inefficient, but you must fight the fire that is burning. When the fire in Montana starts first, that is where the closest available resources are sent.

South Ops and the NICC track the daily availability of air tankers, helicopters, hand crews, engines, water tenders, shower units, catering services, bulldozers, and just about everything else that could be used for suppressing fires or supporting wildland fire incident operations. By having one central order processing center, the closest available resource can be dispatched when requested. In a region where numerous large-scale incidents occur at the same time on a regular basis, this system is necessary. When South Ops receives a resource request, it checks status information and sends a resource order to the jurisdiction nearest the incident that can fill the request. There have been several incidents in the last few years where the requests were for hundreds of engines for structure protection for one incident. An order of this size requires the request of engines from fire departments, both volunteer and career, local government, state and federal from all over the state and sometimes out of state. Air resources have even come into the United States from Canada to attack fires. At times like this, a well-organized and coordinated system is necessary to track who is available, who is responding, from where, and when they will arrive. All resources ordered must be given order and request numbers, **resource designators**, travel routes, communications frequencies, a rendezvous point, and a reporting location.

Transmission of Alarms

There are several methods in which alarms are transmitted to the fire department. The 911 system has taken the forefront **FIGURE 8-1**. Before the development of the 911 system, the fire alarm box was very common in cities. Attached to a street light or power pole, the alarm box was always available should a fire need reporting. Alarm boxes had several weaknesses as a system, including frequent false alarms in some districts.

The boxes were sometimes vandalized or stolen. Another fault was that they were a one-way communication to the fire department. The dispatcher receiving the alarm had no information from the reporting party as to what the problem was. They were required to send a "full box" to alarms received. This could require the response of a ladder truck, a couple of engines, and a chief to a minor traffic accident. An argument in favor of the alarm box on the corner is that it is easy to understand and use. With the installation of telephones in almost every residence and business, the day of the alarm box is coming to an end. As an added benefit of the 911 system, dispatchers can now ask the reporting party what the problem is and respond with the appropriate resources.

As the fire department has moved into the area of emergency medical services (EMS), two-way communication has gained in importance. In some jurisdictions,

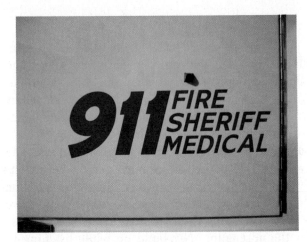

FIGURE 8-1 911 sticker on apparatus compartment door to remind the public of the community's number to call in an emergency.

© Jones & Bartlett Learning

the level of EMS response is determined by this conversation. A heart attack will get a paramedic response, and a cut finger will get an emergency medical technician (EMT) response. Many departments are training their dispatchers in emergency medical dispatch (EMD). Using this system, a trained dispatcher asks questions regarding what is wrong with the victim and can then turn to a page in a pamphlet or computer screen to give instructions to the person at the scene. The instructions include cardiopulmonary resuscitation (CPR), artificial respiration, control of bleeding, imminent childbirth, and the treatment of other medical emergencies. The system is a proven lifesaver. In a medical emergency, when seconds count, the instructions given to the persons at the scene can help preserve the life of the patient until qualified medical help arrives. This system requires one dispatcher to gather the information and type it on a computer screen. Another dispatcher actually dispatches the call and initiates the response of emergency units. The problem occurs when the dispatch center is busy with more than one incident. In critical cases, at least one of the dispatchers must stay on the line with the persons at the scene until units arrive. This ties up a dispatcher and prevents him or her from answering incoming phone calls and dispatching incidents to the stations.

Tip

As the fire department has moved into the area of emergency medical services (EMS), two-way communication has gained in importance.

In many areas, there are people in the community who do not speak English. When these people call in for emergency services, there may not be anyone on duty in the dispatch center who can translate for them. Translation services are now accessible in the form of a conference call initiated by the dispatch center. When the department has an enhanced 911 system that shows the reporting party's address, it helps. Another problem is that many recent immigrants do not understand the 911 system because it does not exist in their native country. They do not speak English well enough to pick up on their own that the system is available and how to use it. They cannot read English or are unable to read and are not able to get the information from a pay phone or the phone book. When there is an emergency, they will drive around looking for a police officer or fire station seeking help. By the time they find help, it is quite possible that someone who could have

been helped has bled to death or died of some other injury that could have easily been handled through quick medical intervention.

There has been an increase in private alarm companies that monitor homes and businesses for burglary and fire. When they receive a fire alarm, they contact the fire department and a response is started. In many cases, the alarms turn out to be false. Police departments across the country require alarm permits to be purchased on an annual basis and charge extra after one or two false alarms. Fire departments are sure to follow as the number of false alarms continues to rise. When a burglar alarm is received, one or two police officers respond. When a structure fire alarm is received, the fire department responds with several pieces of heavy apparatus and numerous personnel. The difference in cost is evident. In areas with volunteer departments, the volunteers leave their jobs and places of business to respond, placing an unnecessary financial burden on themselves and on their businesses. There are also risks associated with responding with red lights and sirens. The initiation of a structure response may leave a first-in area uncovered while the personnel are responding. While they are gone, there may be a legitimate need for their services in their first-in area. All these issues have to be balanced with the need for quick response to a structure fire. Very few fire fighters will call off a response or lessen the amount of equipment responding because it might be a false alarm. If and when it is done, the first-in unit may arrive and report flames through the roof with rescue necessary.

Lookouts

A position that has been used for a long time to report fires in forest areas is the lookout tower. The fire lookout person reports to his or her position in the tower early in the morning and does not come down, except for short periods, until dusk. This can be quite an exciting position in one of the older towers. Some of the older towers are 75 ft high, standing on metal legs. The lookout person sits in a room about 8 ft^2, on a chair that has glass insulators on the ends of the legs. When lightning storms occur, the whole tower can become energized and the lookout person cannot touch anything for fear of being electrocuted. The lookout person is forced to remain in the chair until the storm is over, no matter how much the tower sways in the wind or how many times it is struck by lightning. A modern lookout tower is much lower to the ground and may, in remote areas, include the person's living quarters. These types of lookouts depend

FIGURE 8-2 Fire lookout, Mesa Verde National Park, Colorado.
© Jones & Bartlett Learning

FIGURE 8-3 Osborne fire finder in a lookout tower.
© Jones & Bartlett Learning

on their location on top of a ridge or peak to give them visibility of the surrounding area **FIGURE 8-2**.

In the center of the room is a table with a fire finder, commonly called an Osborne, named after the inventor. Developed in 1915, this is a sighting device mounted on top of a revolving base plate **FIGURE 8-3**. When a fire or smoke is sighted, the fire finder is rotated until the sighting device lines up with the smoke. The degrees of the compass heading are read from the table. When two or three lookouts are able to give a fix on the same smoke, a very accurate location can be determined. Should a triangulation not be possible, the lookout operator provides his or her best speculation to fire dispatch as to the distance the smoke is located from the tower.

GIS/Mapping

Departments in areas of any great size need updated maps on a regular basis. As new streets are added and new subdivisions are built, the maps in all of the stations that respond into the affected area need to be updated. There is also a need for parcel maps to show property ownership. These are used for hazard reduction work, to contact the owners of vacant lots, and to determine ownership of acreage burned in wildland fires.

With the introduction of geographic information systems (GIS), the job of the person responsible for the department's maps has changed **FIGURE 8-4**. The GIS uses maps entered into a computer as a base. The master map can then be overlaid with other information. The map coordinates are tied to different databases that contain the information required. There can be overlays of place names. These include information such as business names, property owner names, and location names in rural areas. In a town it is helpful for the dispatchers to be able to access the address of the local supermarket without having to look it up in the phone book. In rural areas the name may be used to determine the location when a call comes in for a tractor accident at "Yurosek's farm." In wildland areas, the location given may be "the head of Haypress Canyon."

An aircraft sighting given off a global positioning system (GPS) is given in longitude and latitude. These coordinates are not on most maps. The GIS has this information, and the ground location, and its relationship to roads, can be quickly established. This information is also useful when dispatching medical helicopters or firefighting aircraft to an incident scene. Pilots are much better able to plot a course and identify a location by longitude and latitude than by ground-based references, such as the intersection of Tucker Road and Round Mountain Road. Using the GIS, the dispatcher can, with a few keystrokes, find any location (if the information is in the database) and determine in whose first-in area it is located. Use of this information speeds up the dispatch of equipment, thereby reducing response times.

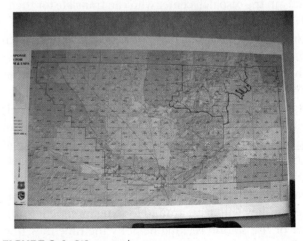

FIGURE 8-4 GIS-created map.
© Jones & Bartlett Learning

Using information from a GPS, a fire can be mapped fairly easily. A ground or air unit, using a GPS, travels the perimeter and records the coordinate points. The information is then fed into the GIS program, and a detailed map of the perimeter can be created as a map layer with other layers showing property ownership, roads, and any other information that is entered into the database that can then be printed out. In wildland areas, where a fire may cover several jurisdictions, this information is very important when it comes to determining each jurisdiction's share of suppression costs. It is also necessary for statistical purposes when justifying budget requirements based on acres burned.

Fire Investigation Unit

All fires should be investigated to determine their cause. As a part of fire prevention, it is important to know how fires started. If they are started by a fault in a piece of machinery or equipment, information gathered from investigations can lead to product recalls or more focused prevention efforts.

The responsibility to conduct a cause determination at a fire scene is that of the officer in charge. If the cause is obvious, such as improper welding; cutting dry grass in the afternoon on a hot, dry day with a power mower; or other fairly easily determined cause, the officer in charge should be able to handle the investigation without assistance. At some fire incidents the cause is not so obvious and requires further expertise and investigation. In these cases, the officer in charge may require the services of the fire investigation unit. Other reasons to notify the investigation unit are fire death(s) or a fire of suspicious origin. If a crime is suspected, the experts need to be called in; they are the arson investigators. In most states if the local fire department does not have the expertise, state investigators may be called upon to assist with cause and origin determination.

Arson Bureau

The arson bureau is either under the fire prevention chief or in its own unit. The personnel in the arson bureau are responsible for determining the cause of suspicious fires and preparing criminal cases against arsonists **FIGURE 8-5**. Arson investigators should also be called any time there is serious injury or death due to a fire. Due to their investigative abilities, the arson investigators may also be utilized as an internal affairs bureau for the fire department, investigating harassment and other complaints, either brought by one employee against another or a member of the public against an employee.

FIGURE 8-5 Fire/arson investigator searching for fire cause and origin.
© Jones & Bartlett Learning. Photographed by Glen Ellman

One of the most heinous crimes committed is that of arson. When a hostile fire is ignited, there is no telling what will burn and who will be injured. Arson fires are started for many different reasons, including spite/revenge, profit, sexual gratification, and to cover up other crimes. The arsonist may be targeting one specific individual, but when gasoline is poured in the stairwell of an apartment building and ignited, often numerous innocent persons are injured or killed. In areas with dry wildland fuels, one arson fire can destroy hundreds of homes and result in losses in the millions of dollars. When these fires are started, the lives of fire fighters and the general public are put at great risk. On January 5, 1995, four fire fighters died and five others were injured fighting an arson fire in Seattle, Washington (Nalder and Wilson 1995). The Esperanza Fire in Southern California in 2006, which resulted in the deaths of five U.S. Forest Service fire fighters, was arson caused (Maclean 2013). There is no doubt that arson will continue and so will the tremendous costs in lives and property. The direct costs are those of the buildings and contents destroyed in the fire. Indirect costs are the suppression effort and the loss of natural resources. Another indirect cost is that when units are fighting an arson-caused fire, their first-in districts may be left uncovered because of lack of manpower. If another fire starts while they are out, it has a chance to gain headway, resulting in higher losses.

Fire Mark

On January 5, 1995, four fire fighters died and five others were injured fighting an arson-caused fire in Seattle, Washington. The fire was deliberately set and the fire fighters were killed when the floor collapsed and they plunged into a burning basement that they were unaware of. As a result of this fire tragedy, the Seattle Fire Department has made it policy that crews responding to a fire be alerted to arson threats and any other known potential hazards. Today, crews around the city routinely conduct inspections on buildings to become familiar with the layouts and potential hazards.

The crime of arson can be determined in many cases, but the conviction of the person responsible is difficult. It is the very nature of fire to destroy evidence of the crime. Much of the material is consumed in the ensuing fire. Fingerprints are destroyed and structural collapse and fire fighters with hose lines disturb the scene. In wildland situations, the **incendiary device** may be thrown from a passing vehicle and the criminal is miles away before the fire is reported. Arson investigators dream of catching the person with the match in their hand, but this is very rare. Often, long hours of investigation and gathering of facts lead to conviction of the person(s) involved for defrauding the insurance company. It is much more difficult to prove arson.

Arson Investigators

Arson investigators should be specially trained in the gathering and preserving of evidence. If the evidence is not properly preserved, the evidence cannot be used in court, the case is dismissed, and the arsonist goes free. Arson investigators are trained in giving testimony as witnesses in a court of law and are trained in the laws of the state regarding arson and collection and preservation of evidence. They are also trained in the art of interviewing suspects and witnesses. In a typical investigation, they interview the fire fighters at the scene, law enforcement officers, building or area occupants, property owners, witnesses, and bystanders. In cases of suspected insurance fraud, they also contact the insurer of the property.

Arson investigators must exert authority at fire scenes and over suspects. A background in firefighting is an asset. When they are on the witness stand, they are often asked for their opinion. Their opinion carries more weight when they come across as trained, experienced professionals.

The qualifications for arson investigators are that they must be knowledgeable in the chemistry and physics of fire. They must have a good understanding of how a fire burns to interpret the scene. They must be systematic and careful in gathering evidence and building the case. The legal system requires the investigator to follow set procedures in order for the evidence collected to be admissible in court. The investigator must have knowledge of the laws that pertain to arson and must be able to gather the facts and put together a complete picture of what happened and when. He or she must also be able to present this information in a professional and understandable manner from the witness stand. When called to testify, the arson investigator is explaining the facts to a jury, not to fire fighters.

The arson unit of the fire department maintains a close relationship with the law enforcement community. By working closely together, their effectiveness is greatly enhanced. In many jurisdictions, the fire and police departments have officers working together on single cases as well as on arson task forces. Arson is sometimes used in an attempt to cover up other crimes, and a united fire and law enforcement effort is required to convict the persons responsible.

The arson investigators also work closely with the district attorney's office in preparation for presentation of the case. As previously stated, the evidence must be properly handled and preserved to be admissible in a court of law.

Cost Recovery

An effort that is becoming more common is **cost recovery**. Through the use of civil suits, the fire department is now starting to pursue the recovery of costs from persons who cause fires. It is not uncommon for the suppression cost of a major wildland fire to run into the millions of dollars. This amount is not totally recoverable in most cases. The recoverable amount is usually limited to the amount of insurance carried by the person who caused the fire. This can be thousands of dollars and is better than nothing. Once the word gets out that cost recovery will happen, it tends to make people more careful about causing fires.

In cases where property is burned for profit or other motive, the fire department may be able to collect as well. The amount expended putting out a car fire may be in the hundreds of dollars and the recovery of these costs frees up department monies for other activities, such as prevention and equipment.

Some jurisdictions have passed extraordinary hazard ordinances, which are used to recover money spent on the mitigation of hazardous materials incidents. Other areas where cost recovery is pursued are criminal negligence and criminal acts. One scenario is where a person has caused a vehicle accident while driving under the influence of drugs or alcohol. Once someone is convicted for driving under the influence of alcohol or drugs, the department can pursue recovery of the costs incurred under the applicable penal code section.

Hazardous Materials Control Unit

In recent years, hazardous materials incidents have received increased public attention. Several major incidents have led the public to demand closer control of hazardous materials stored and used in their towns and cities. States have passed right-to-know laws requiring the producers and users of hazardous materials to file a business plan and inventory of the hazardous materials they are storing on site. To process this information and ensure compliance with legal and code requirements, the agency responsible sets up a hazardous materials unit. This unit is not to be confused with a hazardous materials response unit, which responds to spills and leaks.

Many fire departments want the responsibility of hazardous materials control because they are the ones who would respond to incidents involving hazardous materials. The fire department already has the expertise of inspection technique from fire prevention activities. In other jurisdictions, the responsibility lies with the health department or environmental health department. In some jurisdictions, the responsibility is shared between the fire department and the health department. Often in these situations, the fire department and health department personnel undergo their hazardous materials training together. This aids in building a relationship between the departments from the bottom up. The personnel from both departments are going to respond to incidents together and need to be familiar with each other's operations.

The hazardous materials control unit can provide vital services to the incident response personnel by having the business plans and materials inventories available. The information can be accessed by modem and printed out on a computer printer; a copy can be sent to the field by fax machine or accessed on a mobile computer, such as a tablet or smart phone. Having this information available can save lives and prevent environmental damage at incident scenes. It is preferable to know what is in a chemical warehouse when it is on fire so the right decisions can be made as to how to handle the incident. With some chemicals, a let-burn policy is the safest choice.

The hazardous materials control unit also has personnel with technical expertise to assist in decision making. There are very few fire fighters with a strong background in chemistry. From an inspection standpoint, the control unit personnel are better prepared than fire fighters to identify a hazardous situation. They would more likely recognize the storage of a powerful oxidizer next to a fuel than the average fire fighter would. Identifying these types of hazards may help prevent an incident from occurring.

> **Tip**
>
> The hazardous materials control unit can provide vital services to the incident response personnel by having the business plans and materials inventories available.

Adjutant/Aide

An adjutant, also called an aide, is another support position. The fire fighter assigned to this position assists the chief officer in his or her duties. The job includes running errands, typing memos, and researching information for projects. The aide also acts as a dispatch recorder and message runner at incidents. It is important that the chief keep track of all of the units at the scene, the ones still responding, and their time of arrival. The aide can record this information, freeing the chief to make strategic and tactical decisions for the incident. The aide drives the chief around so he or she can conduct business by cellular phone and review written documents or material on his or her personal computer. When responding to a preplanned incident, the chief can review the plan as the aide drives.

> **Tip**
>
> Being an aide is an opportunity for fire fighters to learn more about the department and how it works.

Basically, the aide does the things requested by the chief. This is an opportunity for fire fighters to learn more about the department and how it works. It is also an opportunity to learn about emergency and non-emergency management. Spending time as a chief's aide can help you prepare for promotion.

Technical Support

The requirements placed on the modern fire department are such that technical assistance is often required in many areas.

Legal Issues

One of these technical areas is legal services. The office of the county counsel or the city attorney reviews contracts and agreements that involve the fire department. Fire districts contract with private attorneys for these services. The fire department may enter into agreements involving mutual aid or automatic aid with adjoining agencies. When this is done, there are legal questions that must be answered about liability and worker's compensation coverage for employees working outside their jurisdictions. If the department decides to lease fire apparatus instead of buying, the department's attorney should review the lease agreement. In many areas, county fire departments contract with cities inside their boundaries for providing fire protection. Another possible contract is a county contracting with the state for protection of state land within the county. The county counsel is consulted to make sure that all of the necessary language is in the contract and that it is legal under the county charter or state law to enter such a contract. Counsel is also necessary any time the department runs into legal problems regarding its relationship with an employee or the public.

Crime Investigation

Investigative technical support can be provided in an agreement in which the police crime lab assists the fire department arson personnel. Very few fire departments are able to afford, or even need, a full crime lab. When the arson personnel have gathered evidence, it is sent to the crime lab for interpretation. The crime lab can determine whether there are traces of flammable liquids or other accelerant in the samples submitted. They can also research fingerprints and other evidence collected at the scene. The police agency is usually the one that administers lie detector tests of suspects, when required.

Weather

For incidents where weather can be a major factor, such as hazardous materials incidents and wildland fires, the National Weather Service can provide assistance. At hazardous materials scenes, the wind direction and relative humidity can affect the size and direction of the evacuation or shelter in place ordered. Weather information is also used at the scene of hazardous materials incidents to predict the speed and movement direction of vapor clouds.

One factor that drives wildland fire behavior is the weather. There have been numerous wildland fire fighters killed or injured because unexpected winds changed a fire's direction or fanned a dormant fire to life. General weather forecasts are available for large areas, much like the ones used by the television news. So-called spot forecasts are available to fire fighters on a local scale. These give a much clearer picture as to what is likely to happen at the incident. Information is gathered at the fire's perimeter and faxed to the National Weather Service. They can then make a local forecast and fax it back to the requesting agency. On long-term incidents (those that last for more than a couple of days), a weather observer can be requested for the incident. The observer sets up a trailer at fire camp to take local observations. The observer uses his or her own observations, those from the personnel on the fire line, and information from the computers of the National Weather Service. By using this information, he or she can make weather predictions specific to the incident area.

SAFETY TIP

Be acutely aware of weather forecasts and shifting winds. There have been numerous wildland fire fighters killed or injured because unexpected winds changed a fire's direction or fanned a dormant fire to life.

Hazardous Materials

In the area of hazardous materials, technical assistance is a must. Computer databases are available to determine the hazards of various materials. Research resources in the form of books are also available. With approximately 2000 new chemicals developed each year, books and computer databases soon become outdated (Fire 2004). By subscribing to a database service, the information is constantly updated.

The health department can be of great assistance in these types of incidents. They have personnel trained in the field of environmental health. These personnel usually have chemistry backgrounds and are trained in the determination of an unknown chemical's properties. It is often necessary to determine at the scene whether a spilled white powder is poisonous, corrosive, or harmless and whether it may require collection and disposal as a hazardous material or just be washed off the highway. Until this determination is made, no real action can be taken because the hazardous properties are unknown. When a major highway has to be shut down at rush hour, a quick and accurate determination is appreciated by everyone involved, especially the public backed up in traffic.

Special Hazards

Other technical sections may be set up in jurisdictions that contain oil refineries or other special hazards. The Los Angeles City Fire Department has a technical section devoted to the Metro Rail. They have already had a major fire in a tunnel under construction and recognized the need for such a section of technical expertise. Many of the requirements in model fire codes came from fire department technical sections recognizing the need for protection systems or engineering and drawing up ordinances to be adopted by the local jurisdiction.

Emergency Medical Services

When departments are involved in emergency medical services on the higher levels, they may have a technical section for this function. The technical section monitors the activities of the paramedics and EMTs in the department by performing quality assurance (QA). QA requires the review of a specified percentage of EMS incidents to ensure that protocols and procedures are followed correctly. They also provide research and development if needed. They ensure that the personnel providing life support to the public are properly trained and equipped. As laws and protocols change, the fire department must stay abreast of the changes. The costs in public image and liability are too high not to. Smaller departments can rely on the county EMS department for their needs.

Information Systems

The fire department always seems to have more information than it can manage. There is information coming from all sides. All expenditures already made and expected need to be recorded for budgeting purposes. Specifications for the new equipment must be generated and the bids sent out to the manufacturers and suppliers. Statistical information from fire and medical aid reports needs to be put into an understandable form. Information that pertains to contracts and agreements must be gathered and

maintained. Training information on new techniques and procedures must be reviewed and applied if it is advantageous. New laws, codes, and ordinances must be reviewed for their effect on the department and its operations. Personnel and payroll information has to be gathered and recorded. Personnel are on vacation, sick leave, or injury, and others are working overtime in their place. Many fire departments are switching to computer-based training and reporting systems and linking the stations through an intranet system. All of these systems require technical support. These are just some examples of the type of information that has to be managed by the fire department.

To manage all of this information, the information systems bureau has been developed. In the smaller departments, much of the information will be managed from outside the department. In larger departments, the information systems bureau can range in number of personnel assigned. Some of the most important people in this bureau are the repair technicians. They make sure the computer system is up and functioning. The people working in the headquarters need to have access to information and to update it as necessary. Even in a small headquarters, with few personnel, when the local area network goes down, the work starts to slow down and become backed up. The problems are compounded when the fire stations are linked through remote terminals to the headquarters. When the department is decentralized, more personnel are required as the technicians have to make site visits outside the headquarters to make repairs.

In the information systems bureau are the personnel who actually handle the information as it comes in. These personnel are assigned to data entry. They may be assigned to entering information from the fire reports received at headquarters or entry of other information into the department's database. An emerging trend is to connect all of the stations to headquarters with wide area networks. The fire reports and other communications can then be entered from remote terminals instead of being written on hard copy and transferred to headquarters through the department mail.

Data management personnel control the flow of information and make sure it is directed appropriately. They may suggest improvements to the system and to how data is generated. They manipulate the information available to present as requested by the department staff in the form of reports. These are used to collect money, justify contract amounts, justify budgets, and track expenditures.

The systems analyst maintains the computer operating system for the department. Depending on how computerized the department is, this person may have a wide array of responsibilities. The analyst helps users with their applications, assisting them in the use of their software programs and automating functions commonly used to save time and increase productivity. This is done through determining user needs and configuring hardware and software. The analyst specs out new equipment for the department so it purchases what it needs.

A job being added to many fire departments is that of Web master and social media consultant. This person or persons maintains the department's Internet presence and keeps social media sites updated with department information. The impact of social media on the modern fire department is not to be taken lightly. Inciweb is a prime example. Inciweb provides regular incident updates. This can go badly if social media is not used responsibly. During the Rim Fire in California in 2013, civilians were posting incorrect evacuation and incident information on sites such as Twitter and Facebook. Their information was nothing more than rumors that spread quickly (Carroll 2013).

Personnel/Human Resources

The fire department personnel clerk is involved with matters regarding the employees of the department. In a department with few employees, the city personnel clerk takes care of personnel matters. Larger departments may have their own personnel clerk assigned.

When personnel are to be hired or promoted, the personnel clerk fills out the necessary paperwork to requisition the position from the agency personnel department. The personnel department sends the required number of names from the list of eligible candidates for interview. Once the person is notified that he or she has been hired, the personnel clerk schedules the medical exam and drug screen. Once cleared, the newly hired employee sees the clerk to complete the required paperwork for setting up payroll deductions, including income tax withholding, and, if the employee chooses, union dues. The personnel clerk handles the enrollment forms for health insurance and other benefits. When personnel are retired or terminated, the personnel clerk ensures that all of the proper forms are completed.

The personnel office issues personnel evaluations and retains them, when they are completed, for record-keeping purposes. When an employee is on work-related injury leave, the personnel clerk makes sure that all of the proper forms are completed and sent to the state and other agencies that are involved in worker's compensation. If there is personal property damage to an employee's property, the personnel clerk

handles the claim forms. In some departments, the personnel clerk completes the estimate of salary and benefit costs for preparation of the budget. This office also maintains a personnel database to record the home addresses and phone numbers of fire department personnel. The personnel clerk works with the payroll clerk to make sure that personnel receive their base salary, overtime, pay raises, and incentives and are paid the correct amount.

In situations where the fire department is not a stand-alone department, outside the fire department is the jurisdiction-wide personnel department. This would be countywide in a county fire department and citywide in a city fire department. The term becoming more common for these departments is *human resources*. The job of human resources personnel is to advertise for applicants when the fire department needs to hire new personnel. They receive and evaluate the applications for employment.

The human resources department administers the entrance and promotional written tests and assists with the oral examinations. The human resources department is given this function to avoid the filing of charges of favoritism. In departments with civil service, the human resources department makes sure that all the rules and laws regarding personnel are followed. It is not uncommon for the human resources department to assign one or more personnel analysts to handle all matters regarding the fire department. In this type of system, the fire department budget is charged for human resources department services.

The human resources department ensures that all federal and state guidelines regarding affirmative action and equal opportunity employment are followed. It also has personnel who are trained in the Americans with Disability Act employment requirements. Were the fire department to want to change any part of an approved entrance or promotional testing process, personnel department staff would be involved to ensure fairness and compliance with the law.

A personnel analyst may be assigned to assist in writing the job requirements for the different job classifications, such as fire fighter and company officer. As the jobs change over the years in terms of responsibilities and requirements, these classifications need to be updated. Not only do they affect the entrance requirements, they also justify pay increases as personnel are required to operate at a higher level of responsibility, skill, and training. These requirements are represented in the job announcement and the testing procedure. An example of this is when federal guidelines regarding driver's licenses were instituted: All job descriptions for fire fighters required to operate heavy vehicles had to be updated to reflect the changes (Commercial Motor Vehicle Safety Act of 1986).

The human resources department is involved with the active recruitment of prospective fire fighters. There are pre-academy tutorial programs to assist persons in preparing to take the fire fighter test. Some of these programs involve only mental skills and others involve weight training and other activities to help recruits pass the physical agility test. It has an affirmative action section that ensures that there are no discriminatory practices and that the targeted groups are reached. Through the use of these types of activities, the department does not have to adjust its standards to meet its recruitment goals.

Business Manager

An important part of the fire department management team is the business manager, who reports directly to the fire chief. This person can be compared to a chief financial officer in a private enterprise. This is usually a civilian position because the person is hired for his or her financial expertise, not firefighting skills. The business manager and the department staff prepare the final budget document to be presented to the jurisdiction for approval. They also participate in the research necessary to justify the budget request. The business manager assists with budget preparation through revenue prediction activities. The department management cannot plan ahead if it does not know how much money will be available. The main budget is a combination of all of the subordinate budgets: the budgets for operations, support services, training, prevention, vehicles, and equipment. When there are separate staff officers in charge of all of the subordinate functions, they submit their budget requests to the business manager, who determines if there is enough money to cover all of the proposals. If not, which is usually the case, negotiation among the staff officers takes place.

The budget is then presented to the jurisdiction's budget analyst for review. The budget analyst works with the business manager and the fire chief on the final proposal that is submitted to the elected officials of the jurisdiction. The elected officials want to make sure their money is well spent and may challenge the fire department's requests in various areas.

A part of the business manager's responsibilities is to keep track of all expenditures. Monthly reports are submitted to the department administration regarding the status of the budget to make sure that the department is staying within its budget and does not run out of money months before the new budget is approved. If one area of expenditure is too high, the other areas will suffer because there is only so much money to cover all of the department's expenses. Special studies are also conducted when a specific project or problem is identified, for example, the cost of maintaining the fleet of department vehicles. This information is used to identify which pieces of apparatus are becoming too expensive to keep in service and need to be replaced.

The business manager must be well versed in all of the federal, state, and local laws that govern the actions of the fire department when it spends or receives money. There are ordinances that govern all aspects of how an agency functions. When equipment is donated, the offer often has to be brought to a vote before the elected officials before being officially accepted. Every penny has to be accounted for. At any time, the jurisdiction can step in and audit the department.

The business manager's staff does all of the day-to-day accounting for the department. They pay the bills and cash the incoming checks. Many departments contract for services from other agencies, such as the personnel department, and budget transfers must be made. When other agencies, in or outside the jurisdiction, contract with the fire department for services, they must be billed. An example of this is when the department responds to a major fire out of jurisdiction on an assistance-for-hire basis. A final bill for the apparatus, personnel, and support services must be prepared and sent. In some states, if the bill is not sent in a timely fashion, it is dismissed.

One area in which the department can recover costs is on motor fuel taxes. When apparatus is operated off road, idled, or pumped for long periods of time, fuel is consumed. The taxes collected for the fuel when it was bought can be recovered from the state if this consumption can be proved. A fire department with many apparatus consumes enough fuel under these circumstances to make it well worth assigning someone to spend the time to recover these funds. In areas with outlying stations, the fire department may be the only department that maintains fuel pumps. Other governmental agencies may access this fuel, and they need to be charged for the amounts used because the fire department is paying the fuel vendor for the deliveries.

The business manager's staff does all of the ordering for the department. The staff keeps track of what has been delivered and when to pay the bill. It may also have responsibility for the warehouse/central stores inventory as a double check to prevent pilferage or misuse of departmental resources.

Incident Business Management

When major incidents occur, such as large wildland fires, floods, and hurricanes, resources from multiple agencies and the private sector are required to mitigate the incident. Prior to their use, agreements and contracts must be in place. The agreements spell out who will pay for what at what rate per hour or per day. Even government agencies cannot pay each other for services provided without a contract in effect. All of the agreements and the system to activate resources are set up prior to fire season. Just the work involved in designing, updating, and administrating the resource ordering system requires thousands of personnel hours annually.

Prior to any privately owned resources being used at an incident, there needs to be a contract signed between the owner and the requesting agency. There are contracts to be written and signed with vendors for every item that may be used on the incident. Persons with equipment—such as helicopters, air tankers, and water tenders—need their equipment inspected on an annual basis before the contract is allowed. At the incident itself, there are equipment time recorders and check-in personnel who keep track of who is at the incident and their status. When equipment is used, the time must be recorded so the owner can be paid for its use. When supplies are delivered, they have to be accepted and signed for. Personnel time must be kept so that crews and other personnel are paid. The personnel filling these positions may never even see the incident they are assigned to. Management teams are brought in on large incidents just for this purpose. The people assigned to them have been trained in keeping records and preparing the necessary forms so the information can be processed properly.

Warehouse/Central Stores

The warehouse/central stores facility is where the items needed on a regular basis by the fire stations and offices are kept **FIGURE 8-6**. This is also the site where much of the department's nonmotorized equipment, such as fire hose, is repaired. The warehouse manager and stock clerks receive, sort, and store items that are supplied to the stations. The warehouse also acts as a central receiving facility for items ordered by the department, making it easier to ascertain whether orders were received than if they were sent directly to the fire stations. When the stations send in supply requisitions, the requested materials are gathered and either delivered by the warehouse personnel or picked up by the station personnel.

FIGURE 8-6 Fire department warehouse.
© Jones & Bartlett Learning

The warehouse may also be capable of making repairs to equipment. Hose that has been damaged can often be recoupled and placed back in service. When the damage is near the end, the coupling is removed, the bad section cut away, and the coupling reattached. The hose is no longer the standard 50 ft in length, but it is long enough to be reused. If the break was near the center, the resultant hose will be too short. It can be reused by cutting it into several shorter hoses to be used as soft suction hoses or tank fillers. Every so often nozzles, gated wyes, and other appliances need repair. They are sent to the warehouse and the personnel either repair them or replace them as needed.

Many departments have an SCBA repair technician. This person makes the necessary repairs to SCBA. When new SCBA are received, the technician tests them to make sure that they are working satisfactorily before they are placed in service. By having this capability in-house, the department can reduce the cost of going outside for service and provide for shorter down time for SCBA. The fire department SCBA will be the technician's number-one priority. The technician's shop may also be equipped with a breathing air compressor for filling SCBA bottles. A stock of filled bottles is kept on hand for exchange when a station needs filled bottles. By keeping filled bottles on hand, the stations can get their equipment back into service more quickly after a major incident that has depleted their supply of filled air bottles. The SCBA for other governmental departments can be repaired here and their bottles filled as time permits. When the department provides emergency medical services, the technician may also have a facility for refilling medical oxygen bottles.

The U.S. Forest Service and Bureau of Land Management have much the same facilities. The U.S. Forest Service has a warehouse for the entire forest and a smaller one located at the headquarters of each ranger district in the forest. Their warehouses supply the needs of all of the personnel, not just fire personnel. These facilities supply materials for suppression, prevention, recreation, timber, and other needs. It is not uncommon for the fire crew assigned to the district headquarters to repaint picnic tables, trail signs, and fire prevention signs in the spring. Usually located at the district headquarters station is a 50–200-person cache of fire tools that must be maintained. Once fire season is over, the firefighting staff that stays on maintains equipment, plants trees, and works on other projects. When not fighting fires or training, the fire crews always have plenty of work to do.

Forest Service and Bureau of Land Management smoke jumpers spend the winter and any other down time between incidents making helicopter rappelling harnesses and other equipment. They are well known for their design and construction of many items of wildland firefighting equipment.

Repair Garage

The fire department may have its own repair garage. The repair garage services the fire apparatus. The mechanics at this facility are civilian employees who work a 40-hour week directly for the fire department. When a fire department is large enough to support its own repair garage, it should have one. Some reasons for this are that the mechanics assigned have specialized knowledge of fire pumps and related equipment and that fire equipment will have the top priority at a fire department facility.

Radio Shop

Fire departments need communications capability. Much communication is carried out through radios installed in the dispatch center, the headquarters, the stations, the vehicles, and with hand-held units. After installation, all of these units will need maintenance. When the department covers a large and topographically diverse area, a system of repeaters is usually installed. These boost the radio signal and provide the capability for any unit in the department to talk to any other. The radio technicians install and service all of the radio equipment used by the department, including pagers **FIGURE 8-7**.

FIGURE 8-7 Radio repair technician's work bench.
© Jones & Bartlett Learning

Wrap-Up

CHAPTER SUMMARY

- Not all of the personnel who work for the fire department respond to incidents; many work in supporting roles that aid the incident responders in performing their jobs.

- Even in a one-station department there needs to be someone assigned to answer the phone and alert the station crew that their services are required.

- One way that departments handle the communications for incidents that become large or complex is to set up an expanded dispatch, sometimes called the command center, to take some of the load off the dispatchers performing their normal duties.

- There are several methods in which alarms are transmitted to the fire department, the most common of which is the 911 system.

- A position that has been used for a long time to report fires in forest areas is the lookout tower. Modern lookouts depend on their location on top of a ridge or peak to give them visibility of the surrounding area.

- Departments in areas of any great size need updated maps on a regular basis.

- The crime of arson can be determined in many cases, but convicting the person responsible is difficult. It is the very nature of fire to destroy evidence of the crime.

- The arson bureau is either under the fire prevention chief or in its own unit. The personnel in the arson bureau are responsible for determining the cause of suspicious fires.

- Arson investigators should be specially trained in the gathering and preserving of evidence. If the evidence is not properly preserved, the case is dismissed, and the arsonist goes free.

- An effort that is becoming more common is cost recovery. Through the use of civil suits, the fire department is now starting to pursue the recovery of costs from persons who cause fires.

- In recent years, hazardous materials incidents have received increased public attention. Several major incidents have led the public to demand closer control of hazardous materials stored and used in their towns and cities.

- An adjutant, or aide, is a fire fighter who assists the chief officer in duties such as running errands, typing memos, and researching information for projects.

- The requirements placed on the modern fire department are such that technical assistance is often required in many areas, including:
  - Legal issues
  - Crime investigation
  - Weather
  - Hazardous materials
  - Special hazards
  - Emergency medical services

- The fire department always seems to have more information than it can manage. To manage all of this information, the information systems bureau has been developed.

- The fire department personnel clerk is involved with matters regarding the employees of the department.

- The business manager, who can be compared to a chief financial officer in a private enterprise, is usually a civilian because the person is hired for his or her financial expertise, not firefighting skills.

- Management teams are brought in on large incidents to keep records and prepare the necessary forms so the incident information and needs can be processed properly.

- The warehouse/central stores facility is where the items needed on a regular basis by the fire stations and offices are kept. This is also the site where much of the department's nonmotorized equipment, such as fire hose, is repaired.

- The fire department may have its own repair garage to service fire apparatus. The mechanics at this facility are civilian employees who work a 40-hour week directly for the fire department.

- Radio technicians install and service all of the radio equipment used by the department, including pagers.

KEY TERMS

Automatic aid Under this system, departments assist each other without regard to jurisdictional boundaries. It is often used in areas where there are county islands within city limits or in interagency areas where a fire starting in one agency's area is a direct threat to another's jurisdiction.

Cause and origin The source of heat that started the fire and exactly where it started.

Cost recovery System on the part of public agencies to recover expenditures incurred in mitigating an incident.

Geographic Information System (GIS) A geographic information system (GIS) is a system designed to capture, store, manipulate, analyze, manage, and present all types of geographical data.

Incendiary device A device used to light a fire. This can be as simple as a lit cigarette folded into a matchbook or more complicated, involving chemical mixtures and a timer or cell phone to activate.

Local area network A system linking the computer terminals of a department on a local scale—for example, at the different desks at headquarters. A wide area network links the computers of several different geographic locations, such as between headquarters and the fire stations.

Mutual aid System in which departments draw up an agreement that they will assist each other upon request.

Quality assurance (QA) Checking work to establish a consistent standard of care provided to patients at incidents by trained medical personnel.

Resource designators An identification system using numbers and letters to identify resources by agency and type.

CASE STUDY

A large wildland fire is burning across local, state, and federal jurisdiction lands. The fire began 3 days ago and is at 10 percent containment. It is currently at 63,000 acres and predicted to burn 150,000 acres before being contained.

The incident commander and her staff are using all available resources. Numerous private resources, in the form of aircraft, bulldozers, water tenders, fire engines, and hand crews have been mobilized. All of these resources are required to be under contract prior to being used at the incident. Also being used are local government fire engines, bulldozers, water tenders, helicopters, and personnel filling overhead positions on the incident management team. The local government resources can be used only if there is an existing contractual agreement specifying such things as reimbursement rates for apparatus and personnel.

The fire camp is set up at a local ranch; this requires a contract for land use between the incident and the landowner is in place. A national contract caterer has been ordered to feed the crews fighting the fire and the large number of personnel supporting them. A shower unit, vehicle fuel, copying service, satellite phones, lunches, and numerous other items are also under contract.

All of this must be arranged annually in preparation for fire season. It is too late to wait until the incident occurs to produce and procure signatures for all of the contracts involved. All apparatus must be preinspected, and showers and caterers must meet national standards.

1. Why can government agencies not commit funds without a contract?

   A. It is unlawful.
   B. It is poor business practice.
   C. It may lead to questions about favoritism.
   D. It may cost too much.

2. To use local government resources on a federal jurisdiction incident requires:

   A. a phone call.
   B. a resource order.
   C. a verbal commitment.
   D. a contract.

3. The contracts for private resource providers are entered into:

   A. when the incident occurs.
   B. when they are needed.
   C. prior to fire season.
   D. on a case-by-case basis.

4. Having a contract in place allows the government to:

   A. take advantage of sale prices.
   B. order resources from private sources.
   C. avoid high per-hour costs.
   D. buy in bulk.

REVIEW QUESTIONS

1. List three reasons why a fire department requires a dispatch center.

2. What is meant by the term *expanded dispatch*?

3. List several ways that alarms can be transmitted to the dispatch center.

4. What valuable information can the hazardous materials control unit provide to operations personnel at the scene of an incident involving hazardous materials?

5. Other than investigating fire cause, what are the functions of the arson unit?

6. Why is it necessary to assign a personnel clerk to the fire department?

7. What functions does the business manager perform when dealing with the department's budget?

8. Why is a business management (finance) section set up on major incidents?

9. Why is a weather specialist important on major incidents?

10. What are the advantages of a fire department having its own repair garage?

11. What do you see your responsibilities being if you were offered the position of chief's aide?

12. Suppose your fire engine has a noise in the transmission and the radio is out of service. Whom do you contact for service?

13. Can the fire department exist without the support services? Justify your answer.

DISCUSSION QUESTIONS

1. Why is there a need for expanded dispatch on major incidents?

2. What are the advantages of the GIS system?

3. As fire departments become more computer dependent, how will the role of information systems expand?

REFERENCES AND ADDITIONAL RESOURCES

Carroll, Rory. 2013. "California Officials Ask Residents to Avoid Social Media for Rim Fire Updates." *The Guardian*, August, 27. https://www.theguardian.com.Commercial Motor Vehicle Safety Act of 1986, 49 U.S.C. § app. § 2701 et seq. (1986)

Fire, Frank L. 2004. *The Common Sense Approach to Hazardous Materials*. New York, NY: Fire Engineering.

Howell, Elizabeth. 2013. "Columbia Disaster: What Happened, What NASA Learned." Tech Media Network. https://www.space.com/19436-columbia-disaster.html

John F. Kennedy Space Center. "Columbia Recovery Efforts." last updated November 22, 2007. NASA. http://www.nasa.gov/missions/shuttle/columbia_recovery.html

Maclean, John. 2013. *The Esperanza Fire*. Berkeley, CA: Counterpoint.

Nalder, Eric, and Duff Wilson. 1995. "The Pang Fire: What Went Wrong—A Disaster Marked by Bad Preparation, Poor Communication and Many Other Mistakes." *The Seattle Times*, June 11.

National Fire Equipment Systems Course. 2014. *S-260 Fire Business Management Principles*. Boise, ID: National Interagency Fire Center.

National Fire Protection Association. *NFPA 1033: Standard for Professional Qualifications for Fire Investigator*. Quincy, MA: National Fire Protection Association.

National Interagency Coordination Center (NICC). 2018. www.nifc.gov/nicc/

Wilson, Carl C., and James C. Sorenson. 1978. *Some Common Denominators of Fire Behavior on Tragedy and Near-Miss Forest Fires*. Broomall, PA: USDA Forest Service.

CHAPTER 9

Training

OBJECTIVES

After studying this chapter, you should be able to:

- Identify the personnel and positions that make up a training bureau.
- Discuss the value of interagency training.
- Describe training facilities used in the fire service.
- Describe the purpose and importance of training in the fire service.
- Explain the difference between technical and manipulative training.
- Describe how an adequate level of training is determined.
- Describe how performance standards are determined.
- Explain how skills are developed.
- Describe the role standard operating procedures play in training.
- Discuss the importance of maintaining training records.
- Describe the relationship of training to incident effectiveness.
- List areas in which fire fighters require training.
- Discuss the importance of safety in training.

Case Study

On February 9, 2007, a 29-year-old female career probationary fire fighter died while participating in a live-fire training **evolutions** at an acquired structure. The victim's class was conducting a live-fire training drill that is required by the department's training protocol for their *NFPA 1001: Standard for Fire Fighter Professional Qualifications, Fire Fighter I*. The victim was part of a four-person engine company, led by an adjunct instructor, that made the initial attack on a training fire in a vacant, condemned, three-story, end-unit townhouse. The scenario called for the victim's crew to enter the front of the townhouse and proceed to the third floor to find and extinguish any fire on the third floor. They were to bypass any fire on the second floor so that the second due engine could practice suppression on that floor. The victim's crew encountered heavy fire on the second floor and third floor stairwell as they proceeded to the third floor. The victim, operating the nozzle, and the adjunct instructor attempted to fight fire on the third floor, but conditions made it untenable. The adjunct instructor was able to exit through a window located on the third floor landing followed by a fire fighter who was backing up the victim on the hose line. However, the victim got stuck attempting to exit the window, which was 41 in. above the floor. The victim became unresponsive as the adjunct instructor and other fire fighters attempted to free her from the window. After she had been freed, she was transported to a local trauma center where she was pronounced dead.

1. One of the requirements of training is that it be as realistic as possible. In this case the training may have been too realistic. What are your thoughts on the level of realism expressed in this training exercise?
2. What are some of the safety measures that could have been put into place to avert this sort of tragedy in a training exercise?
3. If you were a trainee assigned to an instructor and encountered this sort of situation, is there anything you could do to avoid being placed into this hazardous environment? Would you speak up?

NIOSH. 2008. "Career Probationary Fire Fighter Dies While Participating in a Live-Fire Training Evolution at an Acquired Structure–Maryland." NIOSH. http://www.cdc.gov/niosh/fire/reports/face200709.html

 *Access Navigate for more resources.*

Introduction

There are at least two expressions that illustrate the importance of education and training to the fire fighter:

1. Experience is what you have right after you need it.
2. A good fire fighter knows how; an educated fire fighter knows why.

Not every situation you may face in the line of duty can be anticipated or prepared for. It is important that you make yourself open to training and education, no matter how it is presented. Some education is in a formal setting, such as the use of this textbook. Other training and education occurs on the job, either in the academy or in the field. Training and education information can be gathered from magazines and the Internet. The rule here is the Internet is not filtered or approved, so what you see being done may not be the correct method or factually correct. Important, but less formal, training and education is gathered by listening to stories from other fire fighters. These can be co-workers or members of other departments through networking. This is the "lore" of the fire service and is a time-honored tradition that still has great value.

One look at a training and education manual, such as *Fundamentals of Fire Fighter Skills and Hazardous Materials Response 4th Edition* (Jones & Bartlett Learning 2019), with its more than 1300 pages, will tell you that there is much to know to perform the job of fire fighter safely and effectively. Without training, fire fighters are inefficient, possibly ineffective, and exposed to greater risk than necessary. With the general personnel reduction on engine and truck companies in the modern fire service, the need for efficient and effective operational skills is even more crucial. As in most governmental functions, the fire service is being asked to do more with less. Many municipalities have reduced staffing on apparatus and reduced the number of fire stations serving an area. This has resulted in reduced numbers of personnel available at a scene to perform diverse firefighting functions.

There are many challenges in properly training fire fighters. This is especially true in small departments with small budgets, large departments with inadequate training budgets, and volunteer organizations. Volunteer fire fighters are going to have difficulty taking time away from their jobs and businesses to attend the National Fire Academy, for example. Their department may not have dedicated training funds to bring in instructors or send people to courses and seminars. They are truly dedicated individuals doing the best they can under difficult circumstances.

The hazards and risks faced by fire fighters are much the same in any size department. All of them face the challenges of changing technology, such as alternative fueled vehicles and lightweight construction. State and federal regulations also play a part. As more laws and regulations are passed requiring training for fire fighters, it is harder to keep up. Some areas are losing their volunteer members due to the sheer volume of training they must undergo to stay current and safe in today's environment.

Consider this scenario: It is dark, smoky, and hot. You are alone inside a burning structure and running out of air in your self-contained breathing apparatus (SCBA). You cannot even see your hand in front of your face and you are totally disoriented. The heat is starting to increase tremendously. The linoleum under your knees is starting to melt. Your ears feel as if they are on fire. How did you get in this situation? What are you going to do? What should you do? Do you have enough time to make a carefully considered decision? Should you stand up and run? Should you do the low crawl and try to find a doorway? What has happened to the rest of your crew? As panic starts to set in, you have one thing to rely on at that moment, and that is your training. You must not let panic overcome you and reduce you to helplessness. You need some idea of where to go and how to get there. You should be able to conserve your air and activate your personal alarm. You should be able to make the decisions to save your own life. Will you be able to?

Every year, a number of fire fighters are injured or killed in structure fire-related training accidents. Training must be administered in a methodical manner by knowledgeable and experienced professionals. Practical simulations are necessary for realistic training, but they must not be taken to the point of unnecessary risk. As one example, in 2005, an 18-year-old Connecticut fire fighter died after falling from a ladder during training. Another example is *NFPA 1403: Standard on Live Fire Training Evolutions*, which was written after several fire fighters were killed in a training accident.

The same type of scenario can happen on a wildland fire. Suppose the fire has swept around below your position on the hillside. It is now starting to make a run in your direction. When do you deploy your fire shelter and when do you run? If you run, should it be uphill, cross hill, or downhill? Where is your escape route? Where is the safety zone? Can you run through the flames to safety? Are you wearing your proper protective equipment so you have a chance of surviving in your shelter? Have you ever deployed a shelter in wind as strong as it is right now? Where is a good deployment spot? What do you do with your pack and the chain saw you are carrying? Do you surrender them to the fire before you run or try to save them by carrying them with you? Any of these decisions made correctly can save your life. Made incorrectly, they can cost you your life or result in serious injury. You have only seconds to formulate your plan of action and to carry it out. Are you prepared to make these decisions?

The foregoing scenarios could happen to you sometime in your career as a fire fighter. Will you be prepared? Are you going to pay attention in classes and drills? Are you going to practice and maintain your skills? Only you can answer these questions.

After fire fighter fatalities in the line of duty, investigations are made and reports written. These reports are made available on the Internet. Many fatalities, in all types of firefighting, have common denominators. The statement, "those who do not study history are doomed to repeat it" is definitely true. An example of this is the four fire fighter fatalities on the Thirty Mile Fire in the Okanogan-Wenatchee National Forest in 2001 in Washington (U.S. Forest Service 2001). There were violations cited for the 10 standard firefighting orders and the 18 situations that shout, "Watch Out!" (discussed in Chapter 14, *Emergency Operations*). We must gather and examine information available to us so we do not become a case history ourselves. We must also pay strict attention to the material already available. Most of the material was developed after fire fighters lost their lives.

SAFETY TIP

Many firefighting fatalities have common denominators, regardless of the type of firefighting. Pay attention to these cases to avoid becoming one of them.

One of the primary focuses of fire-related conferences and grant proposals is the issue of training. It is repeatedly identified as one of the greatest needs in the fire and emergency services. In the information-rich environment in which we operate today, innovations are developed on an almost daily basis. These innovations, such as the positive pressure ventilation fan (power fan), can make us more efficient and provide a wider margin of safety in a dangerous profession. If fire fighters do not avail themselves of this information, they are the ones at fault. There is no one else to blame. Training is a career-long commitment that all emergency service personnel must make.

Training Bureau

The personnel who devote their time and effort to preparing fire fighters to do their job safely and efficiently, both on the everyday incidents and under extraordinary circumstances, are the personnel of the training bureau. Their job requires them to stay abreast of the latest developments in firefighting technique and safety; they then plan, prepare, and present this information to the personnel in operations.

Small departments may have only one person in charge of training for the entire department. Larger departments require a complete training staff. Regardless of the size and capabilities of the department's training staff, fire fighters are ultimately responsible for their own safety and education. You will be the fire fighter you choose to be. No one can make you better or worse. You have to realize that it is in your own best interest to be the best educated, most physically fit, and best trained fire fighter you can be. It may just save your own life and/or the lives of other fire fighters.

Staff or Operations Function

Some departments look on the training bureau as a staff function. The personnel work a 5-day-a-week schedule and they have offices, usually at headquarters or at the training facility.

The advantage of working the 5-day schedule is that they can get more productive work done in a week than if they were on the operations schedule and working out of a fire station. However, the personnel in operations often see an imaginary dividing line between themselves and the training staff. Like any other staff position, the training personnel are there to support the function of the line (operations) personnel. As previously stated, without the training staff, the operations personnel may not have the required training to perform safely, efficiently, and in compliance with the law. Instead of viewing training as a group of people who just give them more work to do, operations personnel should view the training staff as a valuable asset to their personal welfare and ability to perform their jobs.

In other departments, the training staff is under the command of the operations section. Many in the fire service feel strongly that this is where it belongs. With training officers responding to incidents as safety officers and in other capacities, it stands to reason that they should work under the operations chief. Some departments have battalion training officers. These are usually company officers who have expressed an interest and aptitude for training. An advantage to having training personnel respond to incidents is they can then view the operations taking place and assess training and equipment needs and overall performance.

In large departments with stations spread over large geographic areas and differing greatly in the types of incidents they handle, it is only natural that different areas have different strengths and weaknesses. The needs of each area differ as well. A station in a suburban area performs operations differently than one in the downtown area with high-rise buildings and manufacturing occupancies. The training personnel must develop programs that will maintain the proficiency of the outlying companies so they will be able to operate efficiently when called to a fire in the downtown area. The personnel assigned to training also assist in the preparation of the downtown companies to respond to the outlying areas when a wildland fire is threatening homes.

Training Officer

The overall requirements for a training officer are varied. They can be divided into aptitude and attitude. In terms of aptitude, the training officer must be able to plan effectively. The planning must be flexible enough that a company attending a training session can respond if it is needed to handle an emergency and have the training session resumed when the company returns. The planning also requires the coordination of a wide variety of resources. A wildland class may include instructors from various agencies, such as the local fire department, U.S. Forest Service (USFS), and Bureau of Land Management (BLM). A structural firefighting class could include experts from the building department and the local utility provider. The training officer must make sure that all of the personnel are available and understand what portion of the training each is required to present. Nothing looks worse and destroys the credibility of a training session more than two of the instructors engaging in a heated discussion as to a difference in strategy or tactics in firefighting operations in front of the class. Planning and conducting a drill with numerous agencies and resources—including hand crews, helicopters, dozers, and engines—is understandably difficult. When there is a breakdown in the planning process, the resources are not properly utilized and the whole drill may be more or less a waste of time for all involved. Training officers should go through the training to be certified to the levels specified in *NFPA 1041: Standard for Fire Service Instructor Professional Qualifications.*

Training officers must have the capacity to research and develop instructional outlines (lesson plans) for personal presentation or presentation by others; this requires well-developed communication skills. The training officer must be able to communicate clearly and concisely, orally and in writing. Enough information must be included to get the point across and answer the common questions, but not be overly technical or boring. To be effective as a training officer, one must have **credibility**, which is based on expertise (subject matter knowledge) and relationship (the ability to get along with others). The material must be well organized and build knowledge in a logical sequence.

In addition to preparing material to be presented orally, the training officer must be able to develop instructional packages for individual or group self-study. The ability to take a video presentation from the concept stage to the completed video is a valuable attribute. It used to be that most of the visual aids used in training were in the form of slides, films, or chalk-board presentations. With today's availability of video technology, closed-circuit television, digital cameras, electronic slide programs, computers, computer projectors, and DVD players, training posted to the department's website and video are viable alternatives. Not only do these tools provide for general productions, they also allow trainers to address issues specific to the needs of certain personnel in a department—for example, a video about the inspection of the brakes on fire engines operated by the department or a virtual tour of occupancies within the jurisdiction.

The training officer must be able to present concepts and ideas to groups in a teaching situation **FIGURE 9-1**. Nothing requires you to know more about a subject than to have to present it to others.

FIGURE 9-1 Training officer instructing class on fire nozzles.
© Jones & Bartlett Learning

When you study the material and prepare your lesson plan to make that presentation, you need to know the material in depth. You must not only be able to present it in a logical manner, but you must also be prepared to answer questions. No instructor wants to keep saying "I do not know. I will have to look it up and get back to you on that." When you do have to look into the answer of a question, however, be sure that you do get back to the student in a timely manner.

Training others in concepts is probably the hardest material to present. You cannot hold a concept in your hand and point out its features and parts to the class. Taking the concept of the fire tetrahedron, for example, you cannot see the production of free radicals. You just have to take it on faith that they are there. When instructing a class on the parts of a ladder, you can point them out and show their relation and function to the students. It is much easier to grasp the idea of an item in front of you than to understand a concept or theory.

The position of training officer requires that the person have a positive attitude concerning the importance of training. The officer must understand the importance of training and its relationship to the organization as a whole. A training officer should have experience in as many of the department's areas of activity as possible. Many of the people who apply for a training officer position have one or more areas as a specialty (such as hazardous materials) but have only a general knowledge of the rest of the department's functions. There is no substitute for experience when it comes to having credibility in the eyes of one's peers. The training officer must have demonstrated initiative in past assignments and be a self-starter. The position requires that the person be able to work with minimum supervision on complicated projects. In departments in which the position is considered a specialty position and not awarded on seniority alone, credit may be given in the selection process for having shown an interest in training in the past and having taken courses to function better in the position.

As a whole, the members of the training bureau, with respect to their rank, have the greatest effect on the policies of the department. The training officer has the ability to effect change in the overall direction of the department on a regular basis. When a company officer trains his or her crew to do things a certain way, the only personnel affected are those under his or her direct control. When the training bureau writes a new policy or procedure, it can change operations in every fire station in the department.

Along with the training officer's capability to implement change is the exposure to scrutiny by the whole department. When the training bureau develops a course or program, everyone in the department sees the results. The company officer's work is seen by only the immediate subordinates and supervisor. If the company officer makes a mistake, very few will ever know about it. When the training officer makes a mistake, almost everyone is aware of it. This leads to a greater responsibility, not only to perform top quality work, but also to assist in setting the trends for the future and providing for the safety and efficiency of everyone in the department.

Training Chief

The overall responsibility for the training bureau is usually given to a person of chief's rank. The responsibility and importance of this function warrant a person of this position in the organization. Having the head of the training bureau hold a high-ranking position also lends more authority to his or her decisions when it comes to requiring personnel to assist or participate in training programs. Many times, the training bureau is charged with introducing new procedures and methods to the organization. For them to be accepted requires careful decision making as to their usefulness and method of introduction. One person is often in charge of training overall. This will help establish a coordinated effort to have all of the shifts trained to the same level on the same material and in the same procedures. There needs to be a person who sets the tone for the bureau's activities and leads the planning for a comprehensive program that addresses the department's needs. There also needs to be a central contact person for the higher ranks who can prioritize the requests that come from them and adjust the workloads of the training staff accordingly. Situations will come up that require dropping the normally scheduled program and that demand immediate attention. These are often safety-related problems that must be addressed immediately, such as an increase in vehicle accidents involving department apparatus or an incident-related fatality of a fire fighter.

In situations where the training bureau is closely tied with the local college, the training bureau administrator may serve as the liaison officer. The college's fire science representative will most likely be a person with an extensive education background and a master's degree or higher. The fire department representative needs to be able to operate comfortably on this level.

Company Officers

Departments that have a chief officer in charge of training usually have company officers as the primary training officers. These officers schedule the training and arrange the programs. They also perform much of the actual instruction and needs assessment. This is beneficial for the company officers involved as they are exposed to working closely with chief officers on a daily basis. They improve their planning and evaluation skills through the exercise of their duties. Many company officers who aspire to the chief's rank spend some time working in the training bureau to improve their communication skills and to aid in developing themselves for the staff part of the job that all chiefs perform. In many departments, the training staff prepares and revises the department's operational procedures and administrative procedures. These will surely be included on promotional tests, and there is no better way to prepare yourself for testing on them than to be intimately involved in their preparation and introduction to the line personnel. It is considered a given that the best way to retain the material learned in a subject area is to have to present it to others. Not only does the position of training officer aid in preparing oneself for promotion, but also, in many departments, proven effectiveness in a staff position is a prerequisite for promotion. At the very least, it does not hurt; the interviewers on the promotional oral panel you will face are more than likely training or prevention officers.

On the downside, the training officer does not often have the same measure of control as an operations company officer. The station officer can pretty much plan most of the activities of the crew and schedule program work and other functions as he or she sees fit, as long as they fall within department guidelines. The training officer works on long-term projects when there is time between assisting instructors and handling last-minute changes to schedules. This can be compared to the station officer having the scheduled work interrupted by emergency response. Another factor is that the station functions in the company officer's absence. The program work needs to be completed even if the company officer responsible for that shift is on vacation. When the training officer is at a class or on vacation, the work has a tendency to pile up and be waiting when he or she returns.

In many departments, the training bureau must borrow the apparatus it needs from operations to put on a class, such as driver/operator training or a fire fighter academy. This can lead to problems when the operations personnel do not deliver the equipment on time or other problems crop up. After the equipment is used, it must be returned to the operations team clean and in good working order. If not, the training officer may receive a phone call or memo from an irate station company officer asking why the chain saw was returned out of fuel and dirty from the drill.

Instructors

The instructors are another very important part of any successful training program. Instructors can be from the same department or multiple departments, or from related agencies, such as ambulance companies. Good instructors are made, not born. They can attend courses and seminars on presenting instruction. They can learn from observing and speaking with other instructors. They can access information on training topics and instructional methodology in publications and the Internet. Some people seem to have an easier time addressing a group in a training situation. If you were to ask them, they would tell you that it does not come naturally. They studied, prepared, and practiced to become comfortable with the material and to develop their presentation style.

In a busy department, the personnel of the training bureau do not have the time or the subject matter expertise to teach every class. A part of the training officer's job is to recruit and assist instructors in their presentations. Fire personnel may be the best at presenting course material at the level and in the language that other fire personnel understand. However, some areas of expertise are hard to find among the fire personnel available to a department, so the use of civilians is required. One such area is weather, as it relates to wildland fires. Few fire fighters have the depth of weather knowledge to cover this subject thoroughly. However, the local television meteorologist is likely to go over the students' heads immediately and not present the material effectively. There must be careful coordination with the instructor as to the expectations for the course and the subject matter to be covered, including the knowledge level of the students, to adjust the presentation to meet student needs.

The professional qualifications for the different levels of instructor and what is required at each level is contained in *NFPA 1041: Standard for Fire Service Instructor Professional Qualifications*. In this document, instructors are divided into three levels, with Level I being the lowest and Level III being the highest. Level I instructors are able to utilize prepared material to set up and deliver courses; Level II instructors are able to develop lesson plans and evaluate and coordinate the activities of other instructors; and Level III instructors are able to develop comprehensive training curricula and programs, conduct organizational needs analysis, and develop training goals and strategies.

In departments that are tied closely with the local college, credit is given for the courses. The college can collect money from the students or the state and use the money to pay the instructors. The National Fire Academy (NFA) uses federal funds to pay its instructors. NFA instructors are from all over the country, fire fighters and civilians included, and are required to apply for acceptance into an instructor course for the subject they wish to teach. On successful completion of the instructor course, usually held at the NFA in Maryland, the instructors are certified. The instructors are then notified as to where and when the courses are to be held, and they submit a bid for the amount they wish to receive to instruct the course. The instructors are not allowed to bid on courses taught in their home state; this gives the courses a more national perspective.

Tip

Should you choose to seek training on the Internet, make sure that it is from a credible and reliable source. There are many videos posted on such sites as *YouTube* or *Ask.com* that show fire fighters performing. Some of them are portrayed as training videos. They may or may not show best or even good and safe practices. Any fire fighter who attempts the operations portrayed in these videos should have the appropriate safety measures and supervision in place prior to attempting an operation or skill. Other sites, such as those operated by the National Wildfire Coordinating Group (NWCG), International Association of Fire Chiefs (IAFC), and International Association of Fire Fighters (IAFF) are very credible and can be highly valuable.

Electronic Media Technician

With the increased use of video and other electronic media in today's training environment, an electronic media technician is a valuable resource for any training bureau. One well-planned, 10-minute video, podcast, online course, or other method of electronic transmission can be distributed to every station in the department and instruct all of the shifts very economically. To have a single person perform this training in person in a 50-station, 3-shift department spread over a large geographic area would be just about impossible. With electronic media, the students can turn off

---

Tip

With the increased use of electronic media in today's training environment, an electronic media technician is a valuable resource for any training bureau.

the video when they are called out for an emergency and resume when they return. Members who were absent when the rest of the crew saw the video can view it upon their return.

The purchase of electronic media equipment is expensive, but it can return its initial cost in a short time. The return on investment can be justified by the cost in mileage, fuel, and fire loss experienced by having uncovered first-in areas when having companies report to a central location for training purposes. There are numerous sources of fire- and medical-related videos on the market. These videos can be used by themselves or supplemented with department-specific information.

The acquisition of electronic media staff and equipment gives the fire department the opportunity to perform this function for other intra- and inter-agency departments and related groups. In addition to buying goodwill, the department can arrange for reimbursement through exchange. This avoids one department having to transfer funds to another. Preparing a short video for another department may get you the instructor you wanted for a particular course, at no charge. It also gets the name of your department in front of others, in and out of the firefighting profession. When done properly, through the development of professional materials, your department's image can be enhanced.

Light Duty

Numerous departments are requiring that fire fighters who are restricted from active duty because of injury go on what is often called light duty and assist the training and other bureaus. They may be given data entry, equipment transfer, or other assignments. Light duty gives the department a way to employ these personnel while they are on injury leave. It also gives the personnel assigned to light duty the benefit of exposure to what is required of the training and other staff positions.

Interagency Training

One of the finest examples of training cooperation is when the fire department jointly trains with other agencies. These agencies can be federal, state, or local. Joint training creates an atmosphere of mutual respect among all concerned. When training with the USFS and the BLM, the fire department gets a national perspective on its problems with wildland fires. Through the National Wildfire Coordinating Group, the federal personnel have training resources not normally available on a local level. The National Wildfire

Coordinating Group is an association of agencies that develop training and reference materials for wildland firefighting. They also have a qualification system that can prepare local department fire fighters to participate in firefighting on a national scale. The USFS and BLM personnel, in turn, gain a new respect for the problems faced by local fire fighters in the areas of structural firefighting and hazardous materials. By sharing their expertise, all of the involved agencies gain knowledge and increase effectiveness.

On the local level, fire departments that fight fires together should train together. Through the use of multiagency drills, they become familiar with each other's procedures and capabilities. When all of downtown is going up in flames is no time to find out that your SCBA bottles do not fit their breathing apparatus. History has proven that when things are at their worst, we rely on our neighbors the most to help bail us out of a bad situation.

Tip

One training practice that cannot be emphasized enough is that of personnel who respond together training together. This practice builds mutual respect and increases the efficiency at emergency scenes, when time is of the essence. It is also a good way to discover incompatibilities in equipment and operations. By discovering them before the emergency occurs, they can be addressed, and contingency plans can be made.

Some departments have taken this concept to the point that they have joint recruit training academies. When people go through the academy together, they get to know each other from the very beginning. They are well versed in the procedures of all of the departments involved. It also helps to avoid the bad feelings that sometimes arise between departments operating in proximity to one another. If there is a move to consolidation in the future, the joint training of personnel from their first day will pay dividends in the long run.

Joint training also occurs between the fire department and the private sector. Industrial fire brigades train with the fire department regularly. This gives the fire brigade the benefit of the firefighting expertise of the fire department. The fire department gains information on the processes protected by the private fire brigade. In some cases, private industry sponsors fire department personnel's attendance at special schools; otherwise many fire departments do not have the training budget to send their personnel. This partnership can extend as far as the private sector purchasing equipment and donating it to the local fire department. The fire department could not afford it and industry does not have to provide personnel time to train in using, operating, and maintaining the equipment.

Ambulance Companies

A group that sometimes is overlooked when joint training takes place is the local ambulance company **FIGURE 9-2**. Fire fighters who are EMTs are not allowed to start intravenous lines (IVs) but should at least be prepared to set up the equipment so it can be started by the on-scene paramedic. Ambulance personnel require access to the patient when fire personnel are performing extrication. They need to be taught the dangers of being unprotected from flammable liquids, broken glass, and sharp metal edges at an accident scene. How many times have you seen a picture in the news of fire fighters in full turnouts working a vehicle accident, and right there with them are ambulance company personnel in shirt sleeves without helmets? Everyone is trying to do his or her job to the best of their ability to save a life, but personnel safety must be the number-one priority.

Another example of the training cooperation between fire departments and private ambulance companies is fire fighters going through paramedic school. They then work with ambulance company personnel to gain the hours required to finish their paramedic clinical training. One way to gain advanced medical training and to exchange information and ideas with the local paramedics is to attend base meetings. These meetings are conducted at the local hospitals that serve as the informational contact bases for the paramedics when treating and transporting patients. Base meetings are conducted because paramedics are required to keep their certification current by completing a prescribed number of continuing education units (CEUs) between recertification testings. These classes are presented regularly, and the presentations are made by doctors or specialists in the subject area. The subject areas can include burn treatment, snake bite, head injury, and others. Because medical providers are required to work closely with the fire department at incident scenes, it benefits the fire department to present material on their operations and incident command system to the medical professionals at least annually.

An area that requires special training and coordination is working around medical helicopters, commonly referred to as air ambulances. These craft have come into common use in many areas. Fire fighters not used to working around helicopters are at great risk of injury without the required training. When the local department does not have its own helicopters, it is necessary to coordinate with the air ambulance provider to receive the necessary training. With the high cost of insurance and helicopter parts, the provider will be willing to assist you in training your personnel.

Training Facilities

Training can take place anywhere: The fire department does not require expensive props or facilities to train its personnel. In a search for realism, departments may choose to contact local building owners when buildings are scheduled for demolition to perform ventilation, fire fighter rescue, and other drill scenarios. The most important thing is that the drill be organized and conducted in a safe manner. Realism is good, but being too realistic, such as setting a building on fire with personnel on an upper floor, is unnecessarily dangerous. There should always be a qualified safety officer designated and a plan of operations in place prior to any training that may result in injuries being experienced. It is safer and more efficient if training has been conducted before an emergency situation arises. It is often not possible for fire fighters to be trained in every type of situation they will face in their careers. They must take their training and past experience and relate it to new situations as they arise. Fire fighters must be able to learn from their mistakes. If you burn your hand because you forgot to wear your gloves, you will

FIGURE 9-2 Fire fighters training with ambulance personnel at a disaster drill.
Courtesy of Battalion Chief Steve Pendergrass

SAFETY TIP

There should always be a safety officer designated and a plan of operations in place prior to any training that has any potential for injuries.

be more careful in the future. If you are not, you may need to reevaluate your ability to receive training.

A valuable asset to the fire department is a training or drill tower for ladder training, both ground and truck mounted. Drill towers also allow practice for operations in multistory structures (see Figure 6-4). When the department does not have a drill tower, a multilevel parking structure can serve the same purpose.

A burn building is nice to have when actual live fire training in structures to be demolished is not available (see Figure 6-5). The burn building provides a controlled environment where operations can be performed at little risk to personnel. Through the use of smoke machines and small fires, the burn building can last for years and provide much valuable training.

A drafting pit is necessary for pumper testing and can double as a training prop for pump operator skills testing and certification (see Figure 6-7). By varying the level of the water in the pit, the difficulty of the drafting procedure can be adjusted. By assigning a person to oversee the annual pump testing and sending the operators in with their equipment, the operation can serve as a refresher course on drafting as well as completing the pumper tests.

As the fire department moves into new areas of responsibility, the training facilities must be included. Hazardous materials props are expensive to construct and take up large amounts of room. These types of facilities are usually provided on a regional basis. The required props include leaking pipes, flanges, and chlorine cylinders. Railroad tank cars are another necessary prop. The props are plumbed with compressed air and water to simulate pressure leaks. The props need to be as realistic as possible while providing a measure of safety for personnel **FIGURE 9-3**. For example, it is extremely difficult to do anything in a fully encapsulated suit, especially climbing ladders and working on top of tank cars. It is a long way to

the ground if you slip. Therefore, practice under controlled conditions is necessary to prepare personnel to respond safely in high-hazard situations.

Classrooms are also a necessity for many types of training. When training a large number of personnel or dealing with technical subjects, a classroom is a must. When people are comfortable and distractions are limited, they learn better. When dealing with technical material, dry erase boards or other means of display are often necessary. Many of today's training programs are available on video and the required equipment is necessary for presenting the material.

The inclusion of an electronic media studio allows editing of videos taken during training sessions. Once edited and prepared, the videos can be reproduced and sent out to the stations so that every engine company can gain at least some value from the training, even if they could not attend the actual drill (refer to Figure 6-62).

Off-Site Training

In just about any area, off-site training can be conducted in numerous locations. Every time you complete a fire prevention inspection, you are performing training. When an alarm comes in at 2 AM and you enter the office you inspected last week, you will stand a better chance of finding your way around in the dark. Even the simple operation of driving around in your first-in area serves as training. As you become more familiar with the streets and traffic patterns in your area, your response times can be lowered.

A prime example of off-site use is wildland fire training. By going into the wildland area and burning of areas during drills, several purposes can be served. The most important one is training personnel in wildland firefighting. This does not mean just lighting of a couple of acres and watching them burn. That may be informative as to fire behavior, but it gives little training value. The suppression of training fires in a wildland setting can serve to prepare personnel for the upcoming summer fire season. As always, this must be done as safely as possible with provisions made to keep the fire within the drill site, and adequate personnel, equipment, and water supply must be available. The site must be surveyed for safety hazards and prepared before any fires are lit. Topographical features that pose a hazard should be identified and brought to the attention of all participants. In areas with oil field operations, the grassy area should be searched for rocks, pipes, holes, and pumping unit foundations that may not be readily visible. These areas may contain overhead electrical wires and, because electricity can arc through smoke, they should

FIGURE 9-3 Hazardous materials training.
Courtesy of Kern County Fire Department

FIGURE 9-4 Removing fuel from the base of a power pole prior to vegetation-fire hot drill to prevent fire from damaging the pole.
© Jones & Bartlett Learning

be avoided. All power poles should have the grass removed around the bases so the fire does not spread to them **FIGURE 9-4.** Fence lines with wooden posts must be protected to prevent damage. High heat will weaken barbed wire fences, and fuels should be reduced or removed underneath them. Basically, anything that you do not want to burn should have the immediate area surrounding it adequately cleared before any fires are lit. Piles of tumbleweeds and other high-heat-producing fuels must be considered in planning the drill.

SAFETY TIP

Wildland fire training must be done as safely as possible with provisions made to keep the fire within the drill site; adequate personnel, equipment, and water supply must be available.

Another advantage to wildland fire drills is the reduction of existing hazards. An example is areas where certain roadsides suffer numerous fires during the dry season. They may be able to be burned off to provide a buffer zone so that vehicle fires do not spread to the wildland or grazing lands.

Bear in mind when performing vegetation fire-fighting operations in agricultural areas that the aerial and ground spraying of pesticides may have coated a certain amount of the vegetation. Vegetation firefighting is often done without breathing apparatus, and the presence of pesticides can be an unseen hazard.

In areas with high-rise structures under construction, it may be possible to secure permission to drill in the building on a weekend or holiday. Before the floor

and wall finishes and furnishings are installed, hoses can be dragged around with little chance of damage to the structure. The drill can also acquaint engine and truck company personnel with the construction methods used in the building.

As with any other type of drill, proper preplanning is necessary to provide maximum training during the time allotted and to provide for personnel safety. The drill area should be checked for electrical and other hazards. Vertical shafts must be secured, including any hole large enough for someone to fall into. Construction workers fall into open elevator shafts on a regular basis. There is no reason to think that it could not happen to a fire fighter on a drill.

Tip

Proper preplanning is necessary to provide maximum training during the time allotted and to provide for personnel safety.

An opportunity to engage in training is during preplan inspections of the high-hazard occupancies in your district. These walk-throughs usually require the participation of several engine companies and serve as an opportunity to put some hose on the ground. With several companies at the drill site, one engine can remove its hose load and all hands can help it get back into service. During the preplan, building construction, hazards, strategy, and tactics can all be discussed and reviewed **FIGURE 9-5.** This is also a good time to test communications from inside the building to the outside. In some structures, hand-held radios are unable to transmit or receive. It is better to know this before an incident happens and to plan accordingly.

FIGURE 9-5 Incident preplanning for shopping center using plot plan.
© Jones & Bartlett Learning

With the use of a smoke generator, search and rescue drills can be conducted in different types of buildings. When the proper type of smoke machine is used, there is no residue and the drill can be very realistic from a visibility standpoint. The only thing lacking is the heat. This type of drill can also be performed without smoke by placing a piece of waxed paper over the mask of the SCBA to simulate low visibility. The advantage of this method is that the instructors can easily observe the actions of the personnel performing the drill.

In areas with harbors, ships are a hazard that must be addressed. Shipboard firefighting has many hazards not present in any other type of structure. With the possibility of a fire below deck, the ship turns into a multilevel basement fire that is particularly hard to ventilate. There is always the possibility of falling down one of the ladders that serve as stairs between the decks. The drill should also include a close inspection of the various types of ships one is likely to encounter. Although locations will vary, somewhere on the ship there should be a fire control plan. With many ships having non-English-speaking crews, communications can be a problem. The metal structure of the hull tends to prevent communication by hand-held radio as well.

Aircraft firefighting has special techniques and hazards. The equipment used at major airports is specially designed for the suppression of fires in aircraft **FIGURE 9-6**. Personnel have to be trained in attack, rescue, extinguishment, and overhaul in an airport setting. Just learning to operate the radios and how to deal with the control tower to receive clearance to cross active runways requires extensive training. If the operator of the crash rescue truck were to get excited and pull out in front of a landing jetliner, the results would be catastrophic. The personnel must be trained to operate safely around both propeller and jet aircraft. They must learn how to approach with the engines running. In a military setting, the aircraft may well be loaded with different types of ordnance that pose a great hazard to firefighting personnel.

Training in oil firefighting is a requirement for fire fighters working in areas that have oil refining or production. Due to the large volumes of smoke produced by these types of live fire training facilities, they must be placed far from any residential areas.

Purpose and Importance of Training

The primary purpose of instruction is to change behavior. The first-day-on-the-job fire fighter may not know one end of a fire engine from the other. He or she certainly does not know how to use the wide variety of tools and equipment available on the apparatus effectively. It would be a poor idea to expect this person to respond to a fire and take action without training in what to do, how to do it, and when to do it. Just telling the person "we will show you what to do when we get there" is not effective training either. One of the most important jobs in any department is the thorough training of personnel. The personnel have the right to demand good training, and the department has the obligation to provide it.

FIGURE 9-6 Aircraft rescue fire fighter (ARFF) training.

Courtesy of Jeff Riechmann

> **Tip**
>
> The primary purpose of instruction is to change behavior. The training you receive is intended to mold your behavior so skills become second nature.

One of the most important aspects of any training program is safety. It is not enough to just show someone how to perform an operation; they must be instructed in how to perform it safely. If you were just told to advance the hose line at every fire and "put the wet stuff on the red stuff" you could soon find yourself in a life-threatening situation without a clue as to what to do. As has already been discussed, firefighting is a carefully coordinated operation in which a great many people play a part. When safety is not stressed from the very first training that you receive, when can you be expected to learn it? There are many ways to perform most operations, but only a very few safe ways. Safety must be included and stressed in every training session. It is extremely important that fire fighters develop a safety attitude because it helps to protect them and the other fire fighters working with them.

> ### SAFETY TIP
>
> There are many ways to perform most operations, but only a very few safe ways. Always perform operations the safest way possible.

Some parts of the fire fighter's job are practiced to the point that they become habit. The act of donning an SCBA over and over until you perform all of the steps to department specification is one of these. The idea is that people will do as they are trained, even under times of extreme stress. Being confronted with a situation that requires a rescue is no time to forget how to don your SCBA. It should be so familiar that you can do it almost without thinking. Your mind will be occupied with the thoughts of how to rescue the trapped person, not whether you remembered to test your face mask seal when donning the SCBA.

Another area that must become so ingrained that it is habit is size-up. When faced with devastation, it is important that you are able to size up the situation and determine the course of action to take that will have a positive effect on the situation and not make you one of the victims. Rushing blindly into a bad situation is just going to make it worse for all involved. As you move up in the rank structure and are expected to come up with a plan and assign resources, this training becomes even more important because it is not only your life that you are endangering by your decisions.

Along with the development of habits in the simple tasks comes the development of the thought processes that are required to perform under extreme pressure. By training yourself from the very beginning to think in a logical and orderly fashion in emergency situations, you develop a **command presence**. Persons with command presence seem very cool and collected no matter how bad things have gotten. This does not mean that they do not feel the stress and pressure of the situation. They just know how to deal with it. They do not yell and use profanity on the fireground. They keep their wits about them and perform in an orderly and decisive fashion. One method of dealing with extremely bad situations is to turn your back on the incident for a moment, while you talk on the radio. This does not mean that you should ignore what is going on for long periods. It means that in the face of destruction and devastation, you should turn around and collect your thoughts so you can present a calm demeanor when speaking. People who are calm and collected tend to have that same effect on those around them. Conversely, the opposite is true; if you lose your cool, it is likely that others will too.

As you become better trained and experienced in the performance of your job, you will develop the necessary knowledge to make changes to your plan if it is not working. The knowledge of what other methods and tools are available to handle the situation is critical. At first the knowledge level that using a sledge hammer instead of an axe is more effective on a tile or slate roof is where you will be. As you progress you should be capable of determining different strategies and tactics to handle situations. At that point you will be able to determine whether an offensive or defensive attack is the best strategy for the fire you are in charge of. You should also develop the ability to determine which areas of the plan need to be changed and how to implement the changes. To become a well-rounded fire fighter and incident commander does not happen overnight. It takes many hours of study and years of experience.

Technical Training

Not all fire-related training is manipulative **FIGURE 9-7**. The world is changing at a rapid pace, and so are the hazards and situations that fire fighters must face in the line of duty; fire fighters must be trained to meet the new challenges.

To perform fire prevention functions, fire fighters must be able to read and interpret codes and ordinances. They must be able to speak and write well.

FIGURE 9-7 Fire fighter performing technical training.
© Jones & Bartlett Learning

They are required to come across as professional and knowledgeable when dealing with the public. The reputation of the department as a whole rests on your shoulders when you make contact with business people in the community. If you come across as well trained and competent, that is how the department will be perceived. If you do not, you may be the subject of discussion at the business association meetings. When this is the case, a business owner may possibly even call the chief or local politician to report your shortcomings.

> ### Tip
>
> The world is changing at a rapid pace, and so are the hazards and situations that fire fighters must face in the line of duty. Fire fighters must be trained to meet the new challenges.

Much hazardous materials training is technical. Learning chemistry, even at a basic level, is difficult for many people. It requires study and effort. Learning how to select the correct personal protective equipment (PPE) is not only technical, but is also a life-saving concern. Hazardous materials team personnel must be able to perform tests on samples to categorize chemicals **FIGURE 9-8**. This requires observing the results of the tests and making decisions. A careless or incompetent person may be placed in a dangerous situation when he or she handles the spilled material incorrectly.

In the provision of EMS, much of the learning is technical. You must learn about the human body and its systems. Paramedics learn how to read and interpret electrocardiograms (EKGs) and what information to pass on to the base hospital so the proper drugs can be administered in the field. On the manipulative side, you will be required to learn how to immobilize spinal injuries and properly package victims for removal from the accident scene. The information must be remembered because you may deliver only one baby in your entire career, but the penalty for assisting incorrectly can cost the newborn its life.

Building construction is one of the areas of technical instruction. Seeing photos and visiting construction sites can provide you with the required knowledge about the relative strengths, weaknesses, and problems encountered with roof construction of varying types. It is not necessary to have spent years working in the construction trades to understand truss roof construction and its reaction to fire. This information is essential when determining whether it is safe to get on the roof to perform ventilation and where to place the hole.

Extinguishing agents are based on both chemical and physical reactions. It is not enough to know to open the nozzle and spray water on the fire. You should understand what is happening and why. As extinguishing agents, such as foam, become more complex, you must be able to choose the right agent for the type of fire and determine whether there is enough on hand to start the attack or if you should wait for more to arrive. The situation in which you use up the foam on site and do not extinguish the fire could easily arise. By the time more foam arrives, the fire is back to full intensity and you still do not have enough foam for extinguishment. If you had waited a few minutes for the reinforcements to arrive, the first try would have been successful. These types of decisions are directly connected to technical training in application rates and coverage areas.

When inspecting or supporting extinguishing systems, it is necessary to know how they are designed. It is not enough to know that pipe A connects to pipe B and then to pipe C. You must understand how the valve system works and be able to figure out whether you can adequately support the system based on how many sprinkler heads are open **FIGURE 9-9**. If your water source is inadequate, you must be capable of working through the command structure to request more units to assist you.

A thorough knowledge of the incident command system used by your department is critical. You not only need to know how to initiate the system as the initial attack incident commander, but also whom to contact to get the things you need.

The most important area of technical training is safety. You must be knowledgeable about the design limitations of all of your equipment. This includes

FIGURE 9-8 Hazardous materials technicians testing an unknown substance to determine hazard category (haz cat).

FIGURE 9-9 Sprinkler system riser cutaways showing internal parts of valves.
© Jones & Bartlett Learning

> **Tip**
>
> A thorough knowledge of the incident command and communication systems used by your department is critical.

PPE, SCBA, and everything else. Your knowledge of flashover and backdraft can save your life. A common reason for the failure of aerial ladders is operator error in exceeding the limitations of the equipment. This information is technical. Yes, the ladder might have been extended in the proper sequence with all of the steps followed. The locks might have been set and the controls locked. What might have gone wrong was that it was placed at too low an angle while fully extended, resulting in exceeding the design limitations, causing the ladder to fail.

> **SAFETY TIP**
>
> You must be knowledgeable about the design limitations of all your equipment.

All apparatus has limitations as to what it can do. A good operator knows the design limitations of the apparatus thoroughly. When you try to climb too steep a hill and exceed the apparatus's climbing ability, you have made a technical error in judgment. When you are the pumper operator and the company officer calls for another line to be stretched to attack the fire, you must be able to determine whether you have a sufficient water source to supply the additional line.

Fire equipment is furnished with all manner of tools. The use and application of these tools is an important part of your training. When dealing with forcible entry tools, the study of locks, their construction, and their removal methods are technical information. It is difficult to gain access to locks of all of the different types that you may encounter in your career. When it comes to lifting a rolled-over car off of a victim, you must know how much weight your rescue air bags will lift. If they are insufficient, you will need to call for more resources to accomplish the task. The car is crushing the life out of the victim and your knowledge of your equipment and quick decision making may just save him or her.

A thorough knowledge of the communication system is necessary to perform efficiently. In large departments, numerous channels may be used. It is necessary to know what channel to switch to when performing on an emergency. It is very frustrating to have equipment assigned to the same incident on different channels and to have to hunt them down when you are trying to coordinate resources. In departments with large geographical areas, the radio system may have **repeaters**. It may be necessary to turn off one repeater and turn on another to have clear communications when operating in certain areas. Clear communications are vital to a coordinated attack and personnel safety at an incident. Your ability to use the system correctly and effectively can have a definite effect on the outcome of incident operations.

Another area of communications that requires technical training is written communications within the department. Most departments have forms made up for different purposes. Supplies from central stores or an oil change for the engine often require different forms. One of these is the fire report form. If it is not filled out correctly, the information needed will not be available for planning purposes and for developing statistical information.

Manipulative Training

The area of training with which you will become intimately familiar as a new fire fighter is manipulative training. Every time you advance a hose line or don an SCBA during the initial training, you are undergoing manipulative training. Manipulative training differs from technical training in that you are performing hands-on operations of equipment and tools **FIGURE 9-10**. When you learn how much air is contained in your SCBA cylinder and at which pressure the alarm goes off, that is technical training.

FIGURE 9-10 Fire fighter performing forcible entry training.
Courtesy of Kern County Fire Department

One of the rules of manipulative training is that perfect practice leads to perfect performance. You will be required to perform simple operations again and again under the watchful eye of your instructor according to department guidelines as to the steps to be followed. Learning by trial and error is not conducive to good performance. You will be watched closely and corrected at any misstep to make sure that you do things correctly.

Tip

Perfect practice leads to perfect performance; be sure to use the right techniques when practicing.

Performing drills with hose and other equipment is called *performing evolutions*. They start as the apparatus comes to a stop and you exit your seat. At first they will be very simple and involve performing only one task, such as pulling off the supply hose. As you progress, they will include combinations of tasks, for example, exiting the apparatus, donning your SCBA, and advancing a hose line to the door of the burn building. As you become proficient at the steps of the evolution, they will become more and more complex. By the end of your initial training and assignment to operations, you should be able to show up in the right location with the right equipment, as specified by your company officer, and be ready to perform.

Much of your time will be spent learning to operate the wide variety of tools carried on your assigned apparatus. You must become proficient to the point that you can operate these tools in the dark with limited visibility under stressful conditions. You should also be capable of minor repairs because equipment has a way of malfunctioning when it is most needed. One way you are taught to do this is by properly cleaning the equipment after it is used. There is hardly a better way to become intimately familiar with the workings of a chain saw than to clean it and return it to service after each use.

When you become an engine operator, you must be able to drive safely and perform all operations required at the pump panel. These operations can include hydrant hookups, drafting, relay pumping, supplying attack lines, and supplying master streams. You must be able to perform all of these operations under normal circumstances and troubleshoot when something malfunctions. This position requires a wide array of manipulative and technical knowledge. You must not only be able to hook up everything correctly, but also be able to pump at the proper pressure. Ladder truck operators have many of the same requirements. They must be able to place the apparatus properly so the ladder is kept within design specifications when it is used. Operators of wildland firefighting equipment must be able to operate in four-wheel drive and "read" the terrain so they do not get on a side hill and roll the apparatus. They must be able to traverse rough terrain without tearing up the equipment or getting stuck. Having your apparatus stuck in a hole with the fire rapidly advancing in your direction is no place to be.

One thing that is sought during manipulative training is realistic conditions. Drills should not be conducted only during good weather and under optimum circumstances. Drilling at night is performed to acquaint fire fighters with the actual conditions under which they will commonly work. Performing drills in darkness stresses the need to be intimately familiar with your equipment. When crews drill only during the day, they tend to forget about setting up adequate lighting during night operations. When you are operating at a fire scene at night and you put out the fire, there goes your light. Other advantages are that some of the locations, such as commercial areas that would not be available for practice during the day, are available when the stores are closed and traffic is at a minimum.

When dealing with people who have already mastered the basics, **stress drills** may be performed. When performing stress drills, the drill site is selected to be as close as possible to realistic conditions. The purpose of stress drills is to practice skills under extreme conditions, not to get anyone hurt.

A type of stress drill is the SCBA drill in a pitch-black, smoky room or maze. It is amazing how realistic this can be, even when you know that there is no fire and you are just on a drill using a smoke machine or a covering over the face piece of the SCBA. The claustrophobia that sets in when confronted with total darkness with an SCBA mask on your face can be very frightening. The way to keep yourself from panicking when faced with this situation is to practice. The advantage to performing this type of drill under controlled conditions is that if you do panic, you can tear off your mask and stand up to run for the door, and you will not be harmed by deadly fire gases and heat produced in a fire. Familiarity with this type of situation may just save your life when confronted with it for real.

Another situation faced by fire fighters that is very claustrophobic is performing rescue in a **confined space**. This type of situation must be practiced when wearing SCBA or a supplied air respirator (SAR) as well **FIGURE 9-11**. The fire fighter must get used to the limited mobility of working in tight spaces.

FIGURE 9-11 Confined-space rescue training.
Courtesy of Kern County Fire Department

These spaces may include, but are not limited to, underground vaults, tanks, storage bins, pits and diked areas, vessels, sewers, and silos. To further complicate the situation, the instructor may tell one of you that your SCBA has just failed and you must perform **buddy breathing** with one of the other members of the crew. This skill should be learned and well rehearsed before you try to perform it under stress drill conditions. You should know how to do it before you are tested.

When performing SCBA drills, a common session is to have trainees search for and rescue a downed fire fighter. Dragging someone in full turnout gear who is wearing an SCBA is very hard to do. The victim must wear full PPE to avoid injury as you drag him or her out.

Trying to find someone in smoke and darkness should reinforce that you need to activate your personal alarm device whenever entering a building. You may think that you can find the other fire fighters by the sound of their breathing through their SCBA. The problem is that your SCBA and your partner's are both making the same noise. When the person you are searching for has his personal alarm activated, you may just be able to locate him, disentangle him from whatever he is caught in, and drag him out before either you or he runs out of breathing air.

Another type of stress drill is to have crews perform wildland fire line construction and tell them that they are about to be overrun by the fire. When this happens, they must proceed to the safety zone or deployment site and deploy their fire shelters. This drill should be performed only without fire present. The drill must include shedding tools, packs, fuel bottles, and **fusees**. The personnel then deploy and take up position in their shelters. The hard run to the deployment site aids in the realism, and the instructors yelling at them to hurry adds to the stress. When this drill is performed in a location with high winds that simulate an actual overrun situation, the realism is increased. When winds are light or nonexistent, several power fans can provide sufficient wind to simulate real conditions. The drill should be performed in as realistic conditions as possible. Performing the drill on the station lawn does not adequately prepare the personnel for realistic conditions. The practice of having the crew deploy and lighting a fire that actually overruns their position is dangerous and unnecessary.

Your level of physical fitness will become evident during your performance of stress-type drills. When making the run uphill to the safety zone or dragging a fire fighter around, your overall fitness is tested. It is much better to find out beforehand that you need to work out more than it is to find out under life-threatening conditions.

Even in stress-type drills, safety is still the number-one priority. To perform search and rescue drills, you do not need to set the building on fire for realism. The skills to be practiced are search and rescue, not putting out the fire. It would be extremely dangerous to place someone in a burning building, leave him alone, and tell him not to come out until he is found. That would violate just about every safety rule there is. The instructor must remember the purpose of the training and determine the safest way to perform the drill without unnecessarily endangering the safety of the attendees. It is better to focus on one skill at a time than to construct a situation in which things can get out of hand.

Determining Adequate Levels of Training

When determining the adequate level of training for any job, several criteria must be considered. The first criterion for determining the adequate level of training for any job is whether it is being performed safely. All personnel who perform the job must be able to perform it in the safest manner possible. If the safe way to perform the job requires it to be accomplished while wearing gloves, then you must be able to perform with your gloves on. This is not as easy as it sounds. The design of many pieces of equipment adapted from other disciplines did not take into account a fire fighter wearing thick gloves.

The complexity of the job has an effect on the adequate level of training. A simple job may require only minimal training, whereas a complex job with many steps, such as raising an aerial ladder or drafting from a static water source, will require more extensive training. Not only do these jobs have many steps that must be performed in the proper sequence (the second criterion) to accomplish them, but there are also safety concerns as well. If certain steps are left out, the results could be catastrophic.

The third criterion is with what frequency the job is to be performed. Some jobs are performed on a regular basis, such as donning SCBA and advancing attack lines. In the initial training period, these will be focused on heavily. When you are to the point that you can perform them almost automatically, within department guidelines and in a safe manner, the focus will shift to more complex operations.

As you gain experience and skill, you will be required to learn jobs that are performed less frequently. Although fire fighters drill in deploying their shelters, many fire fighters are never required to perform this operation under real conditions.

That is no excuse not to master the skill. Fire-fighting has never been a routine job; every incident presents a different situation. When you become complacent and think that nothing is ever going to go wrong, something surely will. When it comes to any job that affects your safety and survival, being caught up in the situation is no time to realize that you should have prepared yourself better to deal with it. Remember, the job you perform the least may just be the one you need to be the most familiar with when the situation arises and you are in trouble.

Department policies, and in many cases the law, require that personnel be trained to a minimum prescribed standard. To achieve maximum effectiveness, personnel should be trained beyond the minimum level for efficiency and safety reasons. The minimum acceptable level for wildland fire fighters is 32 hours of training (National Fire Education System 2003). This is enough to put them out on the line in relatively nonhazardous locations. It is surely not enough to place them on the head of an active wildland fire that is exhibiting erratic behavior.

Performance Standards

The basic underlying factor that determines the performance standard for any job is whether it was performed correctly, which means that it was performed within department guidelines as to time and safety and with a minimum of errors.

One of the most commonly used criteria for setting performance standards is time. The department specifies that evolutions that require a single skill or combination of skills are to be performed within a time limit. An example of this is the almost universal standard that you'll be able to don your breathing apparatus in under 1 minute. Time is one of the hardest criteria to justify. It is better to use the criterion that the operation be performed smoothly with no major mistakes and zero safety violations. When too much focus is placed on time, competition develops between crew members and companies, and safety is sacrificed for speed. An area that makes this very evident is ladder training. One of the primary points of raising ladders is to look up before raising the ladder. When personnel compete against each other at the drill tower for time, with no overhead wires, the emphasis becomes one of speed, not the proper steps. When these people go to work in the field, they may very well be in a hurry and raise their ladder without looking up first. The one time that the ladder comes into contact with overhead electrical wires is probably the last time that person will ever raise a ladder, or do anything else.

There is much discussion among training officers as to what percentage of errors is acceptable. Some will be satisfied with a passing grade of 70 percent; others think it should be 80 percent. Realistically, the percentage will vary with the importance of the material being presented. Knowing the numbers of all of the forms used in the department may have an acceptable rate of error of 30 percent. Forms are used in the office and have titles on the top. They are not usually used in emergency situations where speed and accuracy are of great importance. If you make an error on a fire report, it is not going to cost anyone his or her life or property.

When it comes to safety, the acceptable error level is 0 percent, which is referred to as "zero tolerance." If the acceptable level of error were 20 percent, that would mean that on the average, for every 10 responses, personnel could be expected to be injured, injure someone else, or damage equipment 2 times. Morally and financially, this rate is unacceptable. Part of the professional attitude of modern fire fighters is that they approach their jobs with due consideration for the consequences of their actions. When you do not secure the axe you are using on the roof and it falls on someone below, it is your fault, not his or hers, even if the reason he or she ended up in the hospital was that he or she was not wearing a helmet.

Skill Development, Maintenance, and Assessment

Development

Skill is the ability to use one's knowledge effectively and readily in execution or performance. It is a developed aptitude or ability. When you come to work for the fire department, you are not expected to be skilled. You are expected to be able to become skilled through instruction, study, and practice. You are also expected to be physically fit to the point where you will be able to perform the skills you are taught.

A master is someone whose work serves as a model or ideal. Your goal as a student of your job should be to achieve mastery of the tasks expected of you when performing the required jobs. In training a person to the level of mastery, each individual task is required to be performed to a set standard. The scores achieved on the individual parts of the testing are not averaged. In this type of training, the emphasis is on performing the tasks correctly, not on time.

> **Tip**
>
> Your goal as a student should be to achieve mastery of the tasks expected of you when performing the required jobs.

As a fire fighter, you are expected to be able to perform at a consistently high level after a very short time. The job requires that you exhibit mastery of at least the basic tasks of properly donning your PPE and performing suppression actions. Your supervisor does not have the time at an emergency scene to watch you closely. In a normal fire situation, everyone is very busy and the price paid if you make a mistake may be very high.

As you master the basic sequence of skills, you will soon be expected to be able to deal with more complicated situations. This requires the application of both manipulative and technical skills. Firefighting is a dynamic process and the situation is subject to change rapidly. You will be expected to adapt your operations accordingly. Some examples of this are a wildland fire spreading to the roof of a structure, a fire with burn victims, or a vehicle accident with hazardous materials and rescue of trapped persons required. In these situations, you cannot limit yourself to one aspect of the situation. You must be able to tie together all of your training and experience and act appropriately. All of the mentioned scenarios require decision making. The major burden of the decisions falls on the supervisor, but they are not always standing next to you or available. You must become proficient to the point where you can at least begin to take the appropriate action while they are being notified. At the very least, you should be capable of recognizing the hazards present and not unnecessarily endanger your safety or that of others at the scene.

Maintenance

After you have mastered your basic skills and moved on to more complicated operations, you must maintain your proficiency at every skill, even the basics. As you are learning to operate the pumper, you must still be able to don your SCBA in the required time. The only way to maintain skill level is to practice. Once you have mastered something, the tendency is to take it for granted that you are still capable of performing at a high level. This does not hold true for most people. Only through constant review and practice can you maintain your skill level.

Assessment

Once skills are learned, the ability of the student to perform them must be assessed. This is done through testing performed on both technical and manipulative skills. Many different types of testing are used. One of these is the pencil and paper test of technical information; another way is through verbal testing. An exam about ladders could require you to name all of the parts when given a drawing or you could be asked to name all of the parts when looking at one lying on the ground. At the station level, a form of technical testing is the company officer observing how proficient you are at filling out reports and doing other required paperwork.

In manipulative testing, you will be expected to perform operations and evolutions in front of an evaluator. This goes on all of the time at the station level. When the company officer has the crew perform a drill, he or she is evaluating the performance of the personnel and assessing training needs. If the crew is slow or hesitant in performing the required operations, more time should be spent practicing to increase proficiency. The company officer has the added incentive that when the chief wants to see a drill performed, the company officer is going to receive guidance if the crew is inefficient or sloppy, especially if the drill is conducted in an unsafe manner.

At the end of your initial training, you may receive a comprehensive manipulative and technical examination. The manipulative test requires you to show a high level of proficiency in the skills you are expected to perform. The technical test is to determine whether you learned the policies, procedures, rules, and regulations of the department. There may also be specialized testing in the areas of hazardous materials and emergency medicine.

In departments that have structured post-academy training programs for new fire fighters, you can expect a manipulative and technical test at the end of your probationary period. In many cases, the successful completion of this test is required to complete probation and retain your position with the fire department.

A reason for getting companies together and performing testing is the one-department concept. When companies are brought together from different areas of the department's jurisdiction, they are expected to be able to perform together as they would with companies that they drill with on a regular basis. With the standardization of apparatus, any fire fighter should be able to approach any station's apparatus and find the SCBA or other tool on the first try. With standardized hose loads, any combination of personnel should be able to advance hose lines off any of the apparatus. At a typical drill, several engine companies are brought together at one location and told to perform a coordinated attack. As they do this, the drill requires that they support each other's operations with personnel and equipment. When there are vast differences in the way operations are performed, it is immediately evident in the inefficiency and time it takes to complete the required operations. If the companies spend a lot of time talking over what they are going to do before they do it, it is evident that they are not trained to perform the operations the same way.

Standard Operating Procedures

The way to achieve unity and coordination is through the use of standard operating procedures (SOPs) or guidelines. SOPs are written procedures that specify what companies will do when they arrive at the emergency scene. SOPs are developed to apply to the areas of command, communications, safety, tactical priorities, and utilization of companies. With the inclusion of SOPs, the department can function more smoothly because the incident commander knows what to expect from the arriving companies. It also reduces the need for fireground communications because the incident commander does not have to spell out everything required of the companies at the scene. When the first-in truck takes care of turning off the utilities upon its arrival at the scene, the incident commander does not have to remember to request it or specify

Standard Operating Procedure—Horizontal Ventilation with Power Fan

Purpose: To adequately ventilate a structure in a rapid manner with due regard to fire fighter safety, the following procedure will be utilized whenever possible and applicable.

The first engine at the scene will give a size-up and determine the attack mode appropriate to the situation.

When offensive mode interior attack is chosen, ventilation will be performed in the following manner:

1. Personnel with SCBA, charged attack lines, and forcible entry tools will advance to the entry point and prepare for operation.
2. A fire fighter will be detailed to determine the location of the seat of the fire and create a ventilation exit hole, without entering the structure. When this is done, the fire fighter will advise the entry team.
3. The power fan will be placed into operation and the entry point opened.
4. As the smoke is cleared from the path of attack, the entry team will enter and proceed to the seat of the fire.

FIGURE 9-12 Sample SOP.
© Jones & Bartlett Learning

what is to be done. **FIGURE 9-12** is an example of a simple SOP. It can also be used as a standard drill to build and test company proficiency.

Standard Operating Procedure— Horizontal Ventilation with Power Fan

Purpose: To adequately ventilate a structure in a rapid manner with due regard to fire fighter safety, the following procedure will be utilized whenever possible and applicable. The first engine at the scene will give a size-up and determine the attack mode appropriate to the situation. When offensive mode interior attack is chosen, ventilation will be performed in the following manner:

1. Personnel with SCBA, charged attack lines, and forcible entry tools will advance to the entry point and prepare for operation.
2. A fire fighter will be detailed to determine the location of the seat of the fire and create a ventilation exit hole, without entering the structure. When this is done, the fire fighter will advise the entry team.
3. The power fan will be placed into operation and the entry point opened.
4. As the smoke is cleared from the path of attack, the entry team will enter and proceed to the seat of the fire.

Training Records

A necessary step in the training process is maintaining training records **FIGURE 9-13**. Records are necessary to document that training was received by the personnel involved and how they performed. Over a period of time these records can be reviewed, and areas of greater need can be assessed. The review of these records can also identify areas in which training was lacking or missed altogether. When some operation goes especially right or especially wrong, the training records can be accessed to help determine the reason. Also, numerous government regulations and laws require that fire fighters receive training in specified areas. Training records are necessary to verify that the required subjects were covered and to what extent.

When a situation arises where the department has to go to court to justify its actions in a case, the training records of the personnel involved will be requested as evidence. It is imperative that the records be factual, up to date, and easy to understand.

Some governmental programs, such as apprenticeship programs, provide funding to reimburse departments for training costs. These can be accessed only through the maintenance of records of hours spent on training. When the department is linked to the local college and the college receives funding from the state for student hours, the training records are necessary to receive the funds. These funds can then be used to pay instructors and purchase training packages for the department.

Tip

A necessary step in the training process is maintaining verifiable training records. Training records are used to verify that training was completed and may be subpoenaed into a court of law as evidence in a civil or criminal suit.

Relationship of Training to Incident Effectiveness

The overall purpose of training is incident effectiveness. Under the concept of incident effectiveness, operations are performed efficiently and safely. The incident is controlled with minimum loss, and everyone goes back to his station without injury. In some areas, staffing has been reduced to the point where a three-person engine company is expected to perform

Tri-State Fire Protection District
Training Record Report
Attendance Report

Date:_____ Start Time:_____
Station(s):_____ Category: Company – Single or Multi-Company
Instructor(s): _____

Description: Pumper Operator Refresher Mod 1 & 2
Single Engine / Single and Multiple Line Operations
Method: PR **Course** #: Pump Operator Refresher

Credit Hours (Total Time): 3 **Type**: FAE

Print Name(s)/Signature(s) ID # Apparatus Used Line Size Tip Type & Size Net PDP Pass/Fail

Objectives: Pumper operator refresher Mod 1 & 2. Single engine, single and multiple handline operations. Charge lines to proper net PDP and nozzle flow requirement.
Description of Training (Notes) Performed annual pump operator refresher module 1 and 2. Single and multiple line evolutions completed and correct nozzle flows and pressures achieved as indicated above.

Type of Training
Company Multi-Comp. Officer Mutual Aid Night
Tower Burn Classroom Practical Combo. Driver

Instructor Signature
Drillmaster Approval

Equipment Used in Training Session
Feet of 1 3/4" Hose Used
Feet of 2 1/2" Hose Used
Supply Hose Used/Ft.
Feet of Ladders
Number of Engines
Number of Trucks
Gallons of Water
Number of SCBA
Total Number of Fire fighters

FIGURE 9-13 Monthly training record.
Courtesy of Tri-State Fire Protection District

the work traditionally done by a four- or five-person engine company. Fire fighters have been given better tools and procedures; now it is up to the fire fighters to learn how to use them effectively and efficiently.

The best trained engine company in the world will not be able to put out every fire upon its arrival, nor will it be able to save every life. It will, by being properly trained, be able to perform to a high level and make as much of an impact as possible with the available resources at the incidents to which they respond.

Required Training

The modern fire fighter is required to be skilled in many areas of emergency operations. The public expects fire fighters to be able to handle just about any situation that arises. When people dial 911, they want help and they want it fast. An example of this is the training requirements placed on the fire department by the federal government.

Federal requirements for hazardous materials training are referred to in the Super-fund Amendments and Reauthorization Act (SARA) of 1986. The requirements are found in 29 *Code of Federal Regulations* (CFR) 1910.120. Part 1910.120 requires fire fighters and other first responders to hazardous materials incidents to have first responder awareness training. This training is required for personnel who are likely to witness or discover a hazardous substance release and who have been trained to initiate an emergency response sequence by notifying the proper authorities of the release. Their only action would be to notify the proper authorities of the release. They are required to have:

- An understanding of what hazardous materials are and the risks associated with them in an incident.
- An understanding of the potential outcomes associated with an emergency created when a hazardous substance is released.

- The ability to recognize the presence of hazardous substances in an emergency.
- The ability to identify the hazardous substances, if possible.
- An understanding of the role of the first responder awareness individual in the employer's emergency response plan.
- The ability to use the DOT *Emergency Response Guidebook*.
- The ability to realize the need for additional resources and to make appropriate notifications to the communication center.*

First responder operations-level (FRO) personnel are those who respond to releases or potential releases as part of the initial response to the site for the purpose of protecting persons, property, or the environment. They are trained to act in the defensive mode without actually trying to stop the release. They are to contain the release from a safe distance, keep it from spreading, and prevent exposures (basically to isolate, identify, and deny entry). In addition to the requirements of the awareness level, they are required to know the basic hazard and risk assessment techniques; know how to select and use proper PPE provided to the FRO level; have an understanding of basic hazardous materials terms; know how to perform control, containment, and/or confinement operations and how to rescue injured or contaminated persons within the capabilities of the resources and PPE available with their unit; know how to implement basic equipment, victim, and rescue personnel decontamination procedures; and understand the relevant operating procedures and termination procedures.

Title 29 of the *Code of Federal Regulations* goes on to specify the training requirements for the more highly trained positions of hazardous materials technician, specialist, and incident commander/on-scene manager. The average fire fighters should be trained to at least the FRO level, as they are expected to take action when they respond to the incident.

The Federal Aviation Administration (FAA) specifies the training requirements for aircraft firefighting personnel in the Federal Aviation Regulations Part 139. The personnel involved in this type of operation are required to be initially and recurrently trained in the following:

- Airport familiarization, aircraft familiarization, rescue and firefighting personnel safety, and emergency communications systems in the airport, including fire alarms

- Equipment use, including use of the fire hoses, nozzles, turrets, and other appliances required for compliance with Part 139
- Application of the types of extinguishing agents required for compliance with Part 139
- Emergency aircraft evacuation assistance
- Firefighting operations
- Adapting and using structural rescue and firefighting equipment for aircraft rescue and firefighting
- Familiarization with fire fighters' duties under the airport emergency plan**

All rescue and firefighting personnel must participate in at least one live fire drill every 12 months. At least one of the personnel on duty must be trained and current in basic emergency medical care. The training must include at least 40 hours covering bleeding; cardiopulmonary resuscitation; shock; primary patient survey; injuries to the skull, spine, chest, and extremities; internal injuries; moving patients; burns; and triage.

In the area of emergency medical services, the requirements are set on the state or local level. As the level of medical expertise increases from basic first aid through emergency medical technician (EMT) and on through paramedic, the training hours and requirements to maintain proficiency increase.

There is no set time to be spent on firefighting skills maintenance. This can be considered to include all of the manipulative and technical training and practice the department requires. A standard rule of thumb in many departments is 2 hours of drill per day. This is not 2 hours of standing around talking about it, but actual application of skills. The skills to be practiced are hose evolutions, use of PPE, tools, and apparatus. A standard combination drill would be to don PPE, including SCBA; advance an attack line and forcible entry tools to a doorway; enter; and conduct a primary search. All of this is to be done within department SOPs, adhering to safety guidelines, and within a reasonable time. The drill would also be evaluated for its smoothness, coordination of personnel, and whether all of the required tools and equipment were used in the proper manner.

A valuable part of your training will be preparing to perform the duties of the position above you. This requirement is specified in many job descriptions. In many departments, the fire fighter is able to act as the driver/operator and the driver/operator as the

* 29 Code of Federal Regulations (CFR) 1910.120
** Modified from *The Federal Aviation Administration (FAA)*

company officer. In this way, if one of the personnel is missing for any reason, the others can fill in and perform the required operations.

It would be an ineffective crew that could not charge their hose lines at the scene because the driver/operator had fallen and twisted his ankle.

Training Safety

In training, as in any other operation, safety is of the utmost importance. Fire fighters must pay attention to the instructors, as well as to details, when training. It only takes a momentary lapse to end a career or a life through injury when performing realistic training on the drill grounds or elsewhere. As illustrated in the beginning Case Study, training accidents have occurred, and when instructors and participants are not constantly vigilant, they will continue to occur. Training injuries can be grouped in several main areas:

- *Falls*—Typically from ladders or roofs.
- *Being struck*—By apparatus, heavy equipment, or falling objects.
- *Overexertion*—Leading to heart attack, thermal stress, and dehydration. Strains and sprains are often the result of overexertion.
- *Burns*—During live fire training.
- *Carbon monoxide (CO) poisoning*.

Injuries from training accidents have ranged from minor to fatal. Through the proper use of PPE, planning, and vigilance, most if not all of these types of injuries should be avoidable.

You must be able to size up a situation and decide how to approach it without endangering your life or health any more than absolutely necessary. Part of this comes from developing a safety attitude and having zero tolerance for safety violations. When you exhibit this safety attitude and perform on this level, you set the example for those around you. By paying attention to detail and learning all you can about your job, you can be the best fire fighter you are capable of being.

Tip

With promotion comes the added responsibility of training those under your supervision. As a manager, making sure that your subordinates are trained to the highest level possible is a big part of your responsibilities. Their safety as well as yours depends on the training of the whole crew. Firefighting is a team effort and teams are only as strong as their weakest member.

Wrap-Up

CHAPTER SUMMARY

- Without training, fire fighters are inefficient, possibly ineffective, and exposed to greater risk than necessary.
- Regardless of the size and capabilities of the department's training staff, fire fighters are ultimately responsible for their own safety and education.
- Without the training staff, the operations personnel may not have the required training to perform safely, efficiently, and in compliance with the law.
- The overall requirements for a training officer can be divided into aptitude and attitude. In terms of aptitude, the training officer must be able to plan effectively. The training officer should also have a positive attitude concerning the importance of training.
- Having the head of the training bureau hold a high-ranking position (e.g., chief) lends more authority to his or her decisions when it comes to requiring personnel to assist or participate in training programs.
- Departments that have a chief officer in charge of training usually have company officers as the primary training officers who schedule the training, arrange the programs, and perform much of the actual instruction and needs assessment.
- Instructors in a training bureau can be from the same department or multiple departments, or from related agencies, such as ambulance companies.
- With the increased use of video and other electronic media in today's training environment, an electronic media technician is a valuable resource for any training bureau.

- Light duty gives the department a way to employ personnel while they are on injury leave and gives the personnel exposure to what is required of the training and other staff positions.

- One of the finest examples of training cooperation is when the fire department jointly trains with other agencies. These agencies can be federal, state, or local. Joint training creates an atmosphere of mutual respect among all concerned.

- Ambulance personnel require access to the patient when fire personnel are performing extrication. They need to be taught the dangers of being unprotected from flammable liquids, broken glass, and sharp metal edges at an accident scene.

- There should always be a qualified safety officer designated and a plan of operations in place prior to any training that may result in injuries being experienced.

- In just about any area, off-site training can be conducted in numerous locations, for example, high-rise buildings and wildland.

- One of the most important jobs in any department is the thorough training of personnel. The personnel have the right to demand good training, and the department has the obligation to provide it.

- The world is changing at a rapid pace, and so are the hazards and situations that fire fighters must face in the line of duty; fire fighters must be trained to meet the new challenges.

- Manipulative training differs from technical training in that you are performing hands-on operations of equipment and tools.

- The first criterion for determining the adequate level of training for any job is whether it is being performed safely. All personnel who perform the job must be able to perform it in the safest manner possible.

- The basic underlying factor that determines the performance standard for any job is whether it was performed correctly, which means that it was performed within department guidelines as to time and safety and with a minimum of errors.

- When you come to work for the fire department, you are not expected to be skilled. You are expected to be able to become skilled through instruction, study, and practice.

- After you have mastered your basic skills and moved on to more complicated operations, you must maintain your proficiency at every skill, even the basics.

- Once skills are learned, the ability of the student to perform them must be assessed. This is done through testing performed on both technical and manipulative skills.

- The way to achieve unity and coordination is through the use of SOPs or guidelines.

- Training records are necessary for documenting that training was received by the personnel involved and how they performed. Over time, these records can be reviewed, and areas of greater need can be assessed.

- The overall purpose of training is incident effectiveness. Under the concept of incident effectiveness, operations are performed efficiently and safely.

- The modern fire fighter is required to be skilled in many areas of emergency operations. The public expects them to be able to handle just about any situation that arises.

- Injuries from training accidents have ranged from minor to fatal. Through the proper use of PPE, planning, and vigilance, most, if not all, of these types of injuries should be avoidable.

KEY TERMS

Administrative procedures Written procedures for performing staff-related functions, such as reports and other paperwork.

Buddy breathing A technique in which two people share the same SCBA air supply to avoid breathing smoke or toxic fumes.

Command presence The ability to maintain composure in situations that are stressful.

Confined space An area a person can enter to do work, but which has limited means of entry or exit and is not designed for continuous occupancy by a person.

Credibility A state established between persons based on expertise (subject matter knowledge) and relationship (the ability to get along with others). It is especially important between instructors and students.

Evolutions Combination of skills to perform a task. Example: performing all the skills required to don structural PPE and advance a 1¾-in. hose line from an engine up a ladder and into a second-story window.

Extrication The act of removing trapped victims. This term is usually used in reference to vehicle accidents.

Fully encapsulated suit A suit that includes total body protection. When worn with gloves, it gives head-to-toe protection from certain chemicals. The interior is sealed from the outside air.

Fusees Road flares, sometimes called railroad flares.

Liaison officer A contact person for outside agencies.

Operational procedures Written procedures for performing operational functions.

Repeaters A device that receives radio transmissions, boosts the signal, and retransmits the signal. It is used in areas where topography or tall buildings interrupt clear communications.

Stress drills Drills conducted under realistic conditions to develop and test the fire fighters' ability to perform in stressful situations.

CASE STUDY

It has been said that fire fighters do not lie, cheat, steal, or tolerate those who do. As has been proven in the past, this is not always the case.

In 2013, amid a suspected cheating scandal, a group of 14 Las Vegas Fire & Rescue academy members were not allowed to graduate from the fire academy due to allegations of cheating on an exam. The city decided it was not in their best interest to allow the candidates to graduate and assume positions as fire fighters with the department. The recruits have since hired an attorney contesting the reported cheating. They are denying they cheated.

1. If only one person cheated on the exam, what should the others have done about it?

   A. Nothing; it is none of their business.
   B. Put peer pressure on the cheater not to do it again.
   C. Reported the cheater to the instructor.
   D. Reported the cheater to the news media.

2. What are the chances these 14 candidates will successfully pass a full-disclosure background exam in another department with this on their record?

   A. It is not a problem.
   B. They do not have to mention it.
   C. They can just say they are sorry and won't do it again.
   D. It very well may preclude them from being hired.

3. What would be the city's concerns about the integrity of the candidates?

   A. If you'll cheat now, you'll cheat later.
   B. There is no real reason for concern.
   C. It could come up later in another incident.
   D. Fire fighters are not held accountable more than anyone else.

4. From this case study, what is the most important message?

   A. Fire fighters are just people.
   B. People have low expectations of the integrity of fire fighters.
   C. Fire fighters have low expectations of each other.
   D. Being a fire fighter requires you to have high ethical standards.

REVIEW QUESTIONS

1. Why is the training officer position given to a high-ranking officer?

2. Why do the members of the training bureau have such an effect on the direction of the fire department as a whole?

3. Which level of government sets the requirements for first responder operations-level training?

4. Which level of government sets the requirements for emergency medical service training?

5. Why is time alone not a good indicator of the performance level of an engine company?

6. Why is it important for fire fighters to train with the local ambulance company?

7. When there are multiple jurisdictions in the area, what are the benefits of joint training?

8. What are the two basic types of training a fire fighter receives?

9. Which type of training is based on the operation of various tools?

10. Which type of training is learning firefighting-related chemistry?

11. What is the importance of maintaining training records?

12. List two factors in determining an adequate level of training.

13. Why is it important to be cautious when taking training directly from the Internet?

14. What are some precautions you should take prior to attempting to perform training skills portrayed on the Internet?

DISCUSSION QUESTIONS

1. Is the training bureau more of an operations or staff function?

2. What is the one most important factor in any training?

3. Why are stress drills performed?

4. What is the importance of standard operating procedures?

5. Suppose you are the fire chief and have been asked to cut the training budget. How will you defend not making the cut?

REFERENCES AND ADDITIONAL RESOURCES

Brunacini, Alan V. 1985. *Fire Command.* Quincy, MA: National Fire Protection Association.

Dodson, David. 2005. *Safety in Training for the Fire Instructor.* Albany, NY: Delmar Publishers.

Jones & Bartlett Learning. 2019. *Fire Fighter Skills and Hazardous Materials Response.* Burlington, MA: Jones & Bartlett Learning.

National Fire Education System. 2003. *S-130, Firefighter Training* and *S-190, Introduction to Fire Behavior.* Boise, ID: National Interagency Fire Center.

National Fire Protection Association. *NFPA 11: Low-, Medium-, and High-Expansion Foam.* Quincy, MA: National Fire Protection Association.

National Fire Protection Association. *NFPA 402: Guide for Aircraft Rescue and Fire-Fighting Operations.* Quincy, MA: National Fire Protection Association.

National Fire Protection Association. *NFPA 414: Standard for Aircraft Rescue and Fire-Fighting Vehicles.* Quincy, MA: National Fire Protection Association.

National Fire Protection Association. *NFPA 1041: Standard for Fire Service Instructor Professional Qualifications.* Quincy, MA: National Fire Protection Association.

National Wildfire Coordinating Group. 2017. *310-1: Wildland Fire Qualification System Guide.* Boise, ID: USDA Forest Service.

U.S. Forest Service. 2001. *Thirty Mile Fire Investigation.* Washington, DC: U.S. Department of Agriculture.

CHAPTER 10

Fire Prevention

OBJECTIVES

After studying this chapter, you should be able to:

- Describe the activities performed by a fire prevention bureau.
- Identify fire prevention activities and how to implement them.
- Identify methods of fire prevention.
- Discuss the importance of fire information reporting.

Case Study

On February 20, 2003, the fourth deadliest nightclub fire in U.S. history occurred at the Station nightclub in West Warwick, Rhode Island. One hundred patrons died, and approximately 200 were injured. The club was overcrowded with 450 people listening to a rock band. The pyrotechnics used for special effects ignited highly flammable polyurethane foam insulation lining the walls and ceiling area of the stage where the band was performing.

The direct contributors to the large loss were found to be (1) the proximity of the highly flammable polyurethane foam to the pyrotechnics that started the deadly fire; (2) the Station nightclub was not equipped with a sprinkler system, which resulted in the inability to suppress the fire during its early stages of growth;

and (3) the inability of the exits to handle all of the occupants in the short time available to escape the fast-growing fire.

The investigation also concluded that strict adherence to the 2003 model codes available at the time of the fire could have helped prevent the tragedy. Changes made to the code subsequent to the fire made them more effective.

1. What could have been done prior to this tragedy to prevent it?
2. How could prior training of the fire prevention inspector have changed the outcome of this incident?
3. What are some elements of fire prevention that could have prevented this tragedy?

Retrieved from https://www.nfpa.org/Public-Education/Staying-safe/Safety-in-living-and-entertainment-spaces/Nightclubs-assembly-occupancies/The-Station-nightclub-fire

JONES & BARTLETT LEARNING
NAVIGATE 2   *Access Navigate for more resources.*

Introduction

The United States has one of the highest fire death rates per capita in the world (USFA 2015). To prevent these deaths, we must reduce the number of hostile fires that start. Research conducted by the Department of Homeland Security and other groups has shown that for every $1 spent on prevention, $4–$7 can be saved in responding to incidents (ICMA 2009). This does not take into account the toll in lives of the public and fire fighters who are lost combating hostile fires.

Fire prevention codes tend to be reactive, rather than proactive. Whether it is ignorance, cost, competing political agendas, or other factors, we are still not as prepared as we could be as a country to prevent destruction and death due to hostile fires. A look through our history illustrates this point. Fire prevention codes requiring nonflammable interior finishes, built-in fire suppression systems, exits, and building design have primarily arisen from fatality fires. Incidents such as the Triangle Shirtwaist Factory, Iroquois Theater, Our Lady of Angels, and the Beverly Hills Supper Club fires received national attention and action in their aftermath. Keep in mind that the resulting systems were installed for the safety of occupants, not fire fighters.

It is of utmost importance that we as fire fighters try our best to reduce injuries, deaths, and property loss due to fires. One of the best methods is through fire prevention education. There are several programs available, such as *Learn Not to Burn* and *Risk Watch*

from the National Fire Protection Association (NFPA), designed to prevent injuries and deaths through educating children. The primary method to reduce fire loss is through prevention. Responding after the fire has started will be too late for some victims.

Examples of nightclub fires in fairly recent United States history, ranked in order of lives lost, are the Cocoanut Grove fire of 1942 (492 deaths), the Beverly Hills Supper Club fire of 1977 (165 deaths), the Station Nightclub fire of 2003 (100 deaths, mentioned in the Case Study), the Happyland fire of 1990 (87 deaths), and the Ghost Shop fire of 2016 (36 deaths, discussed below). These incidents illustrate that without constant vigilance and effort, we are doomed to repeat the errors of the past. One of the most important and least recognized jobs the fire department performs is that of fire prevention. Prevention does not make the headlines when it is successful. When it is unsuccessful, however, the community suffers fire-related deaths and property losses. Not all fire-related deaths and property loss can be prevented, but through prevention efforts, they can be reduced.

Tip

One of the most important and least recognized jobs the fire department performs is that of fire prevention. Once the fire starts, damage is being done. Even with quick response lives and property are at risk.

A fire-related tragedy occurred in Oakland, CA, on December 2, 2016. Thirty-six people were killed in a non-approved and uninspected warehouse during a concert. This is in spite of the fact that the nearest fire station is one block away. The building was an artist's colony known as the "Ghost Ship." The building was not designed or permitted for human occupation or assembly-type events. The fire department was initially faulted for not making fire inspections on the property.

One of the true measures of a fire department's effectiveness is the amount of loss experienced in the community or jurisdiction. If hazards and unsafe acts can be reduced, there will be a reduction in the area's fire experience, which can be documented through proper record keeping. When fire fighters spend their time sitting around the station waiting for the next emergency, they are not fulfilling their duties as charged by the fire department's mission statement to "protect life and preserve property." To reduce the losses due to fires, effective, focused fire prevention efforts must take place.

Fire Prevention Bureau

Fire prevention is performed by people. In technical or high-risk areas, these people are often fire department employees assigned to the fire prevention bureau. They use technical processes to inspect occupancies and buildings with high life-loss risk, such as factories, schools, and hospitals. They are professionals with advanced training in fire prevention. Many are specialists in the field due to the amount of specialized knowledge required to perform their jobs. It is the responsibility of every fire fighter to have at least a minimum amount of knowledge in the field of fire prevention and to act to correct hazards when they are encountered. Various fire prevention activities are performed by a fire department. As our environment and surroundings become more complicated, many of the fire prevention activities also become more complicated. Typically, the activities that require a higher level of experience or training are handled by the fire prevention bureau.

Staff Function

A fire inspector is specially trained in the science of fire, fire prevention inspection, and enforcement. This person is responsible for performing inspections to identify and reduce hazards and risks. A fire inspector is also responsible for maintaining written records and reports on the inspections made and the findings of those inspections. The inspector is required to

have enough knowledge of fire chemistry, electricity, building construction, safety practices, and codes and ordinances so that a hazard is recognized when he or she sees it. It is the inspector's responsibility not only to identify the **hazards** and **risks** but also to be able to work with the property owner to mitigate them. This may require education of the owner as to what is wrong, suggestions on how to correct the problem, and, if necessary, forcing compliance through legal action.

The position of fire inspector requires a certain amount of salesmanship to gain the property owner's cooperation to reduce the hazards and risks. It is counterproductive and time consuming to have to gain compliance through legal action. When performing fire prevention work, public relations are very important. In many cases, the fire inspector is the only contact the owner will ever have with the fire department. The professional demeanor and capabilities of the inspector are under scrutiny any time an inspection is made. The fire inspector must know his or her job and be professional in appearance and presentation style.

Operations Function

As a fire fighter assigned to an engine or truck company, you may very well be called on to perform fire prevention inspections of businesses in your first-in area. Your knowledge of the related codes and ordinances and prevention techniques will be tested in these situations. If you require an owner to purchase an expensive system or make a building change that is not required by law, you may find yourself in serious trouble when the owner learns the truth. If you overlook a hazard and there is a resultant fire, it reflects poorly on your ability and knowledge and in some cases may result in legal action being taken against the department. You will need at least a minimum of training in fire prevention codes and ordinances to perform your job correctly.

Personnel

Staffing of the fire prevention bureau differs depending on the size and resources of the department. In large departments, the prevention bureau is managed by the chief of prevention, often called the fire marshal. In smaller departments, it may be run by a person of a lower rank. Whatever their rank or position, the personnel assigned to the prevention bureau must understand the importance of the bureau in achieving the overall mission of the fire department.

Fire prevention bureaus are often staffed with a mixture of uniformed (fire fighters) and civilian personnel. The uniformed personnel are necessary because they are charged with enforcing the fire prevention codes and ordinances. Uniformed personnel are given the legal authority to write citations and stop unsafe acts, whereas personnel considered civilians are not. In some areas, civilian personnel or private companies are given this authority as well.

Personnel are selected to work in the prevention bureau based on aptitude, attitude, and experience. The personnel must be able to understand codes and ordinances and to make a reasonable interpretation and enforcement application of them. They must possess a positive attitude about the role fire prevention plays in the fire service. They must be able to maintain their composure in stressful situations. Last, but not least, they must be able to sell fire prevention. Their experience should include active participation in, and a thorough understanding of, fire prevention programs.

The assignment of personnel on limited, restricted, or light duty and fire fighters retired for medical reasons to these positions does not always work out so well. Although they are not fit for active duty, they usually miss their friends at the station and tend to spend a lot of time trying to find out what is happening in operations. The light duty personnel are assigned to the prevention unit only until they are well enough to return to active duty and often lack the training and commitment to performing well as prevention personnel.

Fire Prevention Chief

On the state level, the head of fire prevention is typically the state fire marshal, so called because the job requires more than just fire prevention work. The state fire marshal office often includes research, arson investigation, training, and fire prevention divisions. The office of state fire marshal is often charged with responsibility for facilities deemed above the level of local jurisdiction, such as prisons, hospitals, and pipelines.

The local department level fire prevention bureau is usually headed by a person of chief officer's rank. They may have the title of fire marshal as well. Due to the complex nature of the bureau and its wide range of responsibilities, a person of this rank is required. In large fire departments, prevention bureaus have numerous employees. The fire marshal must often come up with recommendations to the fire chief as to issues that have a wide-ranging effect on the businesses and residences of the jurisdiction. On this level, costs and politics play a large part in making changes to existing policies and requirements. It is a position of major responsibility and visibility. People in this position must be politically astute as well as able to present themselves well in a public forum.

Inspection Officers

Usually working directly for the prevention chief are inspectors who are equivalent in rank to company officers. They are in charge of the major subdivisions of the prevention bureau. These personnel may be in charge of subfunctions, such as **petrochemical** facilities, public assembly, plans review, public education, institutional occupancies, fire investigation, or other areas of special expertise. The personnel in these positions make decisions that can be very costly to businesses in the jurisdiction. They need to be well trained so that they do not make mistakes. They also need to be well spoken because they are trying to sell their recommendations to the business owner.

Inspectors

When officers are in charge of divisions of the fire prevention bureau, fire fighters may act as the inspectors. These personnel have received more training than the average fire fighter on the engine company. Their job is to make the actual inspections. If you were to look at the requirements for a fire prevention inspector, it would be obvious that an assignment to the fire prevention bureau is a good career move to prepare yourself for promotion.

Civilians

In the fire prevention bureau, there may be technical specialists, called water engineers, plans check specialists, and other titles. These personnel are hired for their ability to perform these jobs, not for their firefighting skills. An advantage to hiring civilian personnel is that they are selected for their expertise in this one area. Their job descriptions usually contain requirements for advanced education in the field for which they were hired. They are typically not paid as much as fire fighters, nor do they get the pension benefits paid to fire fighters; therefore, it is easier for the department to get a highly qualified person for less money.

Professional Standards

The training requirements for a skilled fire prevention officer are contained in NFPA 1031: *Standard for Professional Qualifications for Fire Inspector and Plans Examiner*. An inspector of any level may seek training in fire prevention from numerous sources. State fire training programs are available in many states. Courses are available from colleges of all levels that have fire science-related programs. In most cases, courses in fire prevention are required for an associate's degree in fire science. Each of the agencies that publish model building and fire codes provide training and certification to accompany the codes. The National Fire Academy (NFA) presents fire prevention courses on a national level. The NFA courses include Fire Inspection Principles, Fire Prevention Specialist, Code Management, Plans Review for Inspectors, and Fire/Arson Investigation.

Another way to receive training in inspection technique is to accompany fire prevention inspectors from the local department on an inspection. Observing how they perform the inspection and interact with the public can be of great benefit. Knowledge of the code and how to enforce it is not enough. The abilities to prioritize hazards, develop the plan of correction, and write reports are necessary qualifications for the inspector. The NFA has a series of courses offered both on-site and through states' training on fire prevention and public education. These courses, coupled with others on course development and instructional methodology, can help you become an effective communicator in fire safety. Other courses are available from state-level organizations in some areas. The courses in instructional technique are of value because they give you the techniques to assess needs and tailor the program to your audience. You must be able to prepare and present instruction that is clear and understandable, yet motivational as well. The purpose of the program is to get people to change their behavior and become more fire conscious and therefore more fire safe.

A way to find out the latest techniques and methods is by joining fire prevention organizations and subscribing to fire prevention publications. The major fire-related magazines, such as *Fire Engineering* and *Fire Journal*, regularly have articles on code enforcement and other fire prevention–related subjects. In some areas, conferences and forums are held for people working in fire prevention.

These gatherings focus around presentations made by those preeminent in their field of expertise. They also give the person attending the opportunity to share their concerns and network with others in the field. It is not necessary to reinvent the wheel every time you run into a problem. By attending these gatherings, you can become acquainted with others in your field, and a phone call may be all it takes to find the solution to your problem. Websites can be a source of help by allowing you to access the ideas and methods of other people working in prevention.

Fire Prevention Activities

The purpose of fire prevention activities is to prevent the loss of life and property due to fire. This is done to prevent hostile fires from starting, to provide for

life safety in case of fire, and to prevent the spread of fire from one area to another. As we have seen previously, the loss of just one major facility can have a severe negative financial impact on a community or area. When a business that is a major employer burns, many people are out of work and the jobs may never be available again. Even worse, there are numerous documented cases of fires in structures that have led to major loss of life due to correctable hazards, such as locked or inadequate exits. An example of this happening was the fire at Magna International Inc. in Michigan. When the plant was closed due to a fire in March, 2011, the disruption in the parts supply chain caused the closure of plants or slowdown at both General Motors and Mazda Motors. Magna is also a supplier and caused disruption to the manufacture of automobiles at Ford, Chrysler, and Nissan (Runk 2011).

> **Tip**
>
> The purpose of fire prevention is clear: to prevent hostile fires from starting, to provide for life safety in case of fire, and to prevent the spread of fire from one area to another.

When the fire is a major forest fire, large amounts of natural resources are destroyed. This has a direct negative effect on the local timber industry as well as tourism. When Yellowstone National Park burned in 1988, the park was severely impacted. From the tourism standpoint, the businesses in the affected area lost revenue because few visitors wanted to see vast amounts of scorched acreage. Much of the timber that burned was in the range of 150–300 years of age and is considered lost forever. With air pollution and other factors, the replacement trees may never grow to the previous trees' stature.

Fire prevention activities can be divided into four areas: engineering, education, enforcement, and fire cause determination. The areas are interrelated and are all necessary parts of a fire prevention system. Through the use of all four, in an integrated manner, the fire department's objectives of fire and life safety can be accomplished.

The activities performed include:

- The design of fire-safe assemblies and systems
- Review of plans prior to buildings being built or remodeled
- Inspection of fire safety equipment and devices once installed

- Inspections to ensure that the devices are kept in working order and not prevented from proper operation by occupant error, intention, or modification
- Enforcement of codes and ordinances related to fire prevention
- Public education in the methods and benefits of fire prevention and fire safety
- Education of the legislative body on the need for the enactment of fire safety–related legislation
- Investigation to determine fire cause and prosecution of arson when applicable

Fire prevention does not take place only in urban settings. In rural and wildland areas, fire prevention is conducted as well. In these areas the focus is not so much on devices and installations as it is on clearance between combustible vegetation and improvements, such as structures. In wildland areas, public education is extremely important as many of the people in any given area are campers and others who are not permanent residents.

The overall goal of fire prevention activities is to keep people and property fire safe. It is not to issue citations and fines. The most effective and successful fire prevention programs are those that gain compliance through voluntary means. If the program is adversarial, the public is likely to go right back to doing what they were doing before you arrived as soon as you leave.

Tip

The most effective and successful fire prevention programs are those that gain compliance through voluntary means.

Fire Prevention Terms

The term *fire prevention* encompasses the use of inspections, engineering, and education to reduce or eliminate the causes of fires. A fire prevention inspection is the act of making a systematic and thorough examination of a premises or process to ensure compliance with fire codes and ordinances.

In terms of fire prevention, anything that can burn is a potential hazard. Hazards can also be defined as anything that can cause harm to people or equipment. Risks are the activities undertaken in relation to the hazard. A person washing an automotive part in soap and water is not exposed to much of a hazard and is taking relatively little risk. However, that same person

washing parts in an open pan of gasoline by a gas water heater is creating a hazard and operating in a risky manner. In a wildland environment, the hazard is the fuel in the form of dry vegetation and the risk is human action, such as operating gasoline-powered equipment (lawn mowers, chain saws) without proper spark arrestors, on a dry, windy day close to the fuel.

Hazards also occur in the design of buildings. Allowing structures to be built or modified in ways that prevent the orderly and safe exit of the occupants in the event of a fire presents a hazard. When persons enter that structure in such numbers that the exiting capacities are exceeded, their life safety is potentially at risk.

An occupancy is defined as the use or intended use of a building, floor, or other part of a building. Occupancies are divided into general types. One type is a place of public assembly, such as restaurants, night clubs, or churches. Another is educational, which includes schools and day care centers. Hazardous occupancies are those that pose more than an ordinary hazard, such as vehicle repair facilities with welding or spray painting, aircraft hangars, and cabinet shops. Institutional occupancies include facilities, such as jails, hospitals, and group homes, where occupants are hindered or incapable of self-preservation.

The activities performed in a structure determine its occupancy classification. A restaurant would be one kind of occupancy and an aircraft repair facility would be another. Fire prevention revolves around occupancy classifications in many settings. Different occupancies present different hazards and are treated differently in the code and in inspection procedures.

Methods of Fire Prevention

Fire prevention does not begin only after a hazard is in place. Fire prevention activities start before the building is even built. Zoning regulations determine what kind of occupancies can be built and where. A high-hazard occupancy, such as an oil refinery, would not be allowed to be built next to a school and vice versa. In planning for the construction of a building, there are required setbacks from property lines. Setbacks are not totally due to fire safety concerns, but they are a consideration. Another consideration is the accessibility for placement of fire apparatus. If ladder trucks are going to be necessary for rescue and suppression, many jurisdictions require special pads to be poured for this purpose. The fire lane required around shopping malls and other facilities provides for fire department access **FIGURE 10-1**.

Typically, when building plans are submitted, the required fire flow is determined to ensure that

FIGURE 10-1 Fire lane sign with applicable ordinance identified.
© Jones & Bartlett Learning

adequate water is available to fight a fire if one does start. When subdivisions are constructed, one of the first considerations is a hydrant system. It is important that the system be installed and in service before the first foundation is poured, to provide a water supply in case of fire when the structures are in the framing or later stages of construction. In rural areas, water supply is one of the most important considerations from a firefighting standpoint. Hydrant systems are not often available, and fire water tanks may be substituted (see Figure 12-12).

> ### Tip
>
> In the design and construction of buildings, one of the first considerations is fire protection.

In the design and construction of buildings, one of the first considerations is fire protection. These concerns are addressed through engineering. Engineering activities are evident in the design and installation of fire protection and suppression devices and fire-rated assemblies. These devices include fire sprinkler systems, standpipes, detection and notification systems, and fire-rated construction used in the design of the building itself. Through adopted codes, the fire and building departments require the architect to design certain features that can prevent the spread of a fire if one does start. By enclosing vertical shafts, such as laundry and mail chutes, in fire-resistive construction, the upward spread of fire through the structure can be prevented. Areas where hazardous functions are performed are separated in much the same manner. In high-rise buildings, the structural steel is covered with fire-resistive materials, and holes for cables and pipes that poke through floors are sealed with material to resist the spread of smoke and heat through such

FIGURE 10-2 Truss roof construction.
© David Huntley Creative/Shutterstock

penetrations. Something as simple as an intact ceiling assembly can prevent the spread of fire to upper floors by concentrating heat that allows for the proper operation of sprinklers or blocking fire spread.

One type of construction that does not take fire into consideration and causes much concern and danger to fire fighters is lightweight construction. In this form of construction, the floor and roof supporting systems are made from the least amount of material possible **FIGURE 10-2**. As with any system, the assembly is only as strong as its weakest member. When attacked by fire, these systems are weakened and have a tendency to fail and fail early in the fire. If you are inside or on the roof, it is very likely that you could become trapped or injured due to this failure. Extreme caution must be exercised when operating at incidents where this type of construction is used. The simplest way to identify this type of construction is to observe the construction of the building. However, when conducting inspections, you will come across buildings that have been built and occupied for several years. In these cases, a thorough inspection can also reveal lightweight construction. As always, a thorough knowledge of the hazards and where they are located in your response area is essential to your safety.

Hazard Evaluation and Control

The purpose of hazard evaluation is to identify possible accidents and estimate their frequency and consequences. In hazard evaluation, an accident is defined as a specific unplanned sequence of events that has an undesirable consequence. The first event of the sequence is the initiating event. There are usually one or more events between the initiating event and the consequence. The intermediate events are the result of the response of the system and the operators to the initiating event. Different responses to the initiating event often lead to different accident consequences. In theory, it is possible to reduce the consequences of the accident by reacting appropriately to each of the events between the initiating event and the ultimate consequence.

There are two basic methods of hazard evaluation and control in use: (1) adherence to good practice and (2) predictive hazard evaluation.

Adherence to good practice includes observing rules and regulations, meeting the requirements of the accepted standards, and following the practices that have proved best from years of experience with the same processes, the same plant designs and requirements, and the same operating and maintenance procedures. Hazard evaluation procedures, such as checklists and safety reviews, are used to identify deviations from accepted standards and good practices.

Good practices evolve from experience. Experience has been documented in the form of standards or recommended procedures. The standards summarize today's accepted good practice. The NFPA has numerous standards directed toward establishing proper safeguards against loss of life and property by fire. Many of these standards and good practices are the result of lessons learned from analysis of accidents and near accidents. Two of the methods used to identify deviation from good practice are checklists and safety reviews. These methods are used to ensure that design specifications are met, that previously recognized hazards can be identified, and that operating and maintenance procedures conform to principles and practices that have evolved from experience. From a fire fighter's standpoint, a safety review can be equated with a fire prevention inspection. When you conduct a plans check, preapproval for certificate of occupancy, or after-occupancy inspection, you are conducting a safety review. This is usually done using some sort of checklist. The checklist can be as simple as a fire prevention inspection form that lists common hazards, or it may be as complicated as the checklists often used by plans checkers to ensure compliance with code, design, and installation requirements.

Predictive hazard evaluation procedures have been developed for analysis of processes, procedures, systems, and operations that are sufficiently different from previous experience that adhering to good practice may not be adequate. They may also be used when evaluating a very low probability accident with very high consequences for which there is little or no experience. The concept addresses both the probability of an accident and the magnitude and type of the undesirable consequences of that accident.

Once a hazard is identified, it is necessary to evaluate it in terms of the risk it presents to the employees, the public, and the property involved. It is necessary to identify the initiating event, the intermediate events, and the consequences of each accident involving the hazard.

The first step is to identify the hazards that are an inherent feature of the process and/or plant and then focus the evaluation on events that could be associated with the hazards. Several recognized procedures are used to identify the hazards, including the what-if method, hazard and operability studies, failure modes, effects and criticality analysis, and human error analysis. To identify intermediate events, procedures such as fault tree analysis, event tree analysis, and cause-consequence analysis can be utilized. A complete description of these processes and their uses is beyond the scope of this text. Further information is available in texts on the subject.

Devices

Some activities are hazardous to the point where all fuel and ignition sources cannot be controlled. A simple example is a commercial deep fryer. In these types of installations, devices and systems are installed either to control the fire or to prevent its spread **FIGURE 10-3**.

FIGURE 10-3 Class K fire-extinguishing equipment is installed in range hoods.

© Jones & Bartlett Learning

These devices could include automatic and/or manual fire extinguishing systems, interlocks to shut off the gas or electric supply to the fryer, interlocks to control the ventilation system in the exhaust hood, and portable fire extinguishers. Fire prevention devices can be as simple as a lid that automatically closes on a parts washer when there is a fire. They can be as complicated as a smoke removal system that increases the likelihood that a safe exit route for occupants is available in case of fire in a high-rise building.

The most common types of devices you will encounter are portable fire extinguishers and automatic fire sprinkler systems. In every case, the type and number of devices installed will depend on the hazard to be protected against.

Assemblies

Assemblies come in many forms. A fire-resistive door with fire-resistive walls is considered an assembly. When assemblies are properly designed and installed, they can prevent the spread of fire from one part of the structure to another. They must be inspected on a regular basis because owners and occupants often negate their effectiveness by removing part of the assembly, block their operation, or bypass them altogether. Examples of this would be a self-closing door blocked or wired open or holes cut in a fire wall.

Public Education and Public Information

It is not possible to have a fire inspector on duty in every hazardous location at all times that the occupancy is in use. It is also not possible or good public relations to prosecute every case of violation of the fire codes. It is more cost-effective and more productive—while bettering public relations—to convince people ahead of time of the importance of fire prevention. When you visit a business, the people on-site are concerned with good practices and fire prevention measures while you are inspecting. By educating people as to the importance of fire prevention, they will be concerned about it even when you are not there.

One of the most effective forms of public education is that carried out in the school system. By presenting a fire safety message to an assembly at a school, a large

audience can be reached in a relatively short time. The children targeted must be old enough to understand what you are talking about. In numerous cases, very young children have learned to dial 911 and have done so in emergencies. Another program that has worked well with even very young children is Stop, Drop, and Roll. It is simple in concept and easy to understand. If the children are old enough, they can be instructed in more advanced concepts, such as electrical safety, and given fire prevention check sheets to take home and inspect their own houses. Not only are you reaching the children at school, but they are also old enough to go home and talk to their parents about the importance of having a smoke detector and a home exit plan. It is very important that you target your message to your intended audience.

An opportunity to reach children with safety responsibilities is through programs such as the YWCA Super Sitter. This program brings together children who will be babysitting with representatives of the fire and police departments in an educational setting.

Civic groups present an opportunity to spread the fire prevention message as well as serving as a potential source of funds for purchasing fire safety education materials. Often fire departments do not have the funds to purchase handout materials and fire safety education videos and props. By making fire safety presentations and appeals for monetary assistance to civic groups, often these funds can be secured. Some groups have contributed enough money to purchase such expensive props as fire safety trailers **FIGURE 10-4**.

> **Tip**
>
> Civic groups present an opportunity to spread the fire prevention message as well as serving as a potential source of funds for purchasing fire safety education materials.

> **Tip**
>
> It is more cost-effective and more productive to convince people, before a fire occurs, of the importance of fire prevention.

FIGURE 10-4 Fire safety education trailer.
© Jones & Bartlett Learning

These cost approximately $25,000 and are very effective, but beyond many departments' budgets.

Another source of funding is industry. Many companies can use discretionary funds from their advertising budget to purchase equipment for the fire department to use in presenting the fire safety message to the community. This has the twofold effect of helping out the fire department and promoting the public image of the company involved. A fire safety trailer set up at a shopping center is a rolling billboard for the company that has its name painted on the side.

Another way to get the fire prevention message out is to have fire station tours. Groups from the community, such as the Boy Scouts and Girl Scouts and school groups, visit the station. During the visit they are shown the station and the equipment. During these tours and other programs, it is more effective to introduce yourself as Fire fighter John than Fire fighter Jones. It puts the children more at ease and they are more likely to ask questions and have a valuable experience. Young children are more interested in the size of the apparatus and a demonstration of the siren and red lights than they are in a compartment-by-compartment description of the tools and equipment carried and their uses. This is also a good time to cover the proper procedure for children to follow when they are crossing the street or riding their bicycles when a vehicle is approaching with red lights and siren in operation. Another activity that children enjoy is being allowed to spray water from the hose. This must be done with great care as there has been a case where a hose ruptured—knocking a school teacher to the ground—and the fire department was subsequently sued.

One thing that is traditionally done is to have the children try on articles of turnout gear. Should you choose to do this, make sure the gear is free from contamination and that the children are old enough to support the helmet without injuring their necks. During this demonstration, you can discuss how well fire fighters are protected before entering a burning building, and that the children should never reenter a burning building to find a pet or toys, because this has cost children their lives. It is possible to contract a case of head lice from a child putting on your helmet, so the station should have a helmet without a liner for demonstrations. The helmet will be too big for the children anyway and they do not notice the lack of the liner. It is not an acceptable alternative to have the children try on the equipment of the personnel who are off duty that day. Possibly the best alternative is to have a lightweight plastic helmet to use for this purpose.

An effective demonstration is to have a fire fighter don full turnout gear and SCBA. Having the fire fighter crawl into the area as if they are conducting a search gives the children a feel for how fire fighters would appear when looking for them in an emergency. This is done to show the children that it is only a person in the suit. The appearance of a fire fighter in full turnout gear and the sound of the breathing apparatus in action are frightening to small children, and they may be afraid to call out to you when you are searching for them in smoky conditions. At this time, you can cover the point that, in a fire, children should not hide under beds or in closets as this makes them extremely hard to find. An aid for teaching young children is robots **FIGURE 10-5**. Their size is close to a child's and they tend to hold a child's attention while the safety message is presented.

Public education activities that can reach a large number of people in the community in a short time are public service announcements (PSAs). With the help of the local television and radio stations, PSAs can be produced to spread the fire prevention message. Television spots are especially effective in that they include both audio and visual information. Just hearing about something is not nearly as effective as seeing it at the same time. The television station may be willing to use some of their previously shot raw footage and edit it for you to do a **voice over** of your message. A 30-second spot with 3 seconds of each image you want to be presented can show 10 fires or other incidents. When done correctly, this message can have a tremendous impact.

Television and radio can also be used effectively through the news conference format. When there is an issue deemed newsworthy, such as the opening of brush fire season or the high incidence of fires caused by heating equipment, the news departments are often more than willing to give you free air time. This is mutually beneficial because you get your message across and they are always looking for a good story.

The media always seems to be interested in doing a fire department story at Christmas. An effective

FIGURE 10-5 Teaching preschoolers fire safety.
© Jones & Bartlett Learning

way to handle this is to have the media attend a news conference where a dried out Christmas tree is set up with packages around the bottom. The tree is ignited and the rapid spread of the fire is evident. A setup like this has great visual impact and instills the need to be especially careful when celebrating the holidays with a live tree. Public education needs to continue year-round, with Fire Prevention Week being an especially good time to get the media involved. Every year there is a new theme for the week. Some themes used in the past have been Operation EDITH (Exit Drills In The Home) and "I'm alarmed; smoke detectors save lives." These phrases give the media a good hook from which to build their story. Combined with the pertinent statistics and a fire officer in dress uniform or full personal protective equipment (PPE), the message can be presented very effectively.

One method of building a good working relationship with the media is setting up a media day. On media day, representatives from television, print, and radio are invited to the training center. They are equipped with turnout gear and equipment and, accompanied by regular fire fighters, they are allowed to put out small fires in a burn building or other scenario. This gives them the firsthand experience of heat and smoke as well as the weight and difficulty of working in full turnout gear. If done properly and well planned, the experience is enjoyable for all involved and gives the media personnel a new respect for the job fire fighters do and the conditions under which they are required to perform. That night you can just about be guaranteed that every station is going to run footage of the activities.

The positive public relations reaped from this are worth the small amount of money spent to present the activity.

Another way of spreading the fire prevention message is to post signs **FIGURE 10-6**. These can come in the form of billboards, roadside signs, and bumper stickers. Some billboard owners will donate space for fire prevention messages when the sign is not otherwise used.

One of the most important groups that need to be reached with the fire prevention message is elected officials. These are the people who enact the laws. The only way the fire service can enforce fire safety requirements and regulations is through these people making them law. If the required measures are not law, fire fighters can seek only voluntary compliance. An example of this is the controversy over wood shake roofs **FIGURE 10-7**.

In areas with dry vegetation, wildland fires often spread to structures by flying firebrands: particles of burning material carried aloft in the convection column that land on dry wood shake roofs. The roof catches fire and spreads its own firebrands as the fire increases in intensity. When the roof of one house catches fire, the fire is spread through convection of burning material, radiated heat, and direct flame impingement to adjacent structures. If the wind is blowing, the spread can rapidly exceed control efforts and capabilities. This leads to a conflagration. It is not uncommon for large numbers of homes to be lost in these types of fires.

For years fire department spokespeople have tried to get wood shake roofs outlawed, at least in wildland areas. But the wood shake manufacturers have lobbied against restrictive legislation. Fire retardant treatments have been developed, but until recently these products have only lasted for a few years and then become ineffective. A wood shake roof normally lasts for 25 years. New technology and improvements have now created fire retardant treatment for wood shakes that will last throughout the life of the roof. The ability to install aesthetically pleasing wood shakes that are also fire resistant will eventually have a tremendous impact on the spread of fire from rooftop to rooftop.

FIGURE 10-6 Fire safety message on roadside sign.

FIGURE 10-7 Wood shake roof after fire.

Fireworks are another hotly debated issue in many jurisdictions. The fire department would like them restricted or outlawed due to the number of fires and injuries that they cause annually. However, community groups and businesses raise money by having fireworks booths around the Fourth of July, making it politically difficult to outlaw fireworks in many areas. Part of the problem is that one jurisdiction may outlaw them, for example an incorporated city, while being surrounded by a county in which they are legal. Another example is fireworks being outlawed in one state and legal in an adjoining state, with fireworks sales very close to the state line, making them readily accessible to areas where they are outlawed **FIGURE 10-8**. In some states, the state fire marshal's office limits what can be sold to so-called safe and sane fireworks. They do not explode like firecrackers, but they do emit showers of sparks. The idea is that they will be used safely, but this does not stop people from throwing them on shake roofs or into dry grass.

Another subject with this kind of competing interest is sprinkler legislation. In most cases, if the sprinklers were not required when the building was built, they are not required to be retroactively installed. These buildings are called an existing nonconforming use. The best that many fire departments have been able to do is get legislation passed stating that if the building is remodeled to a certain extent, it is required to be brought up to current requirements. Many designers and developers will also fight against sprinkler installation in new buildings due to the additional construction cost. This opposing view of sprinklers is usually a result of misinformation about the operation of sprinkler systems. Many people believe that when one sprinkler operates, they all operate. This is a myth that has been created by the movie industry, because that is how sprinklers are portrayed in the movies and on television. A strong public education campaign is needed to overcome this belief.

Smoke detectors are an area in which many jurisdictions have been successful getting legislation passed because of their proven life-saving effectiveness. They are required in rental property, motels, and hotels and may be required to be installed in existing private homes. Carbon monoxide detectors are now becoming more commonly required as well. Carbon monoxide (CO) is a colorless, odorless, and tasteless gas that kills people through asphyxiation.

The only way that the fire department is going to get the needed codes and ordinances enacted is to continue to gather information on the causes of fires and to bring these statistics to the attention of the local lawmakers. Allies in this fight are the insurance companies that underwrite fire insurance policies. They, too, have an interest in keeping fire losses to a minimum and can help in lobbying the legislators and providing technical assistance.

National organizations that can provide materials for fire education include the U.S. Fire Administration (USFA), National Fire Protection Association (NFPA), International Association of Fire Chiefs (IAFC), International Society of Fire Service Instructors (ISFSI), National Bureau of Standards, National Safety Council, U.S. Forest Service (USFS), Bureau of Land Management (BLM), Underwriters' Laboratories (UL), and others.

State and local level organizations that can be sources of public fire education materials are the office of the state fire marshal, state forester, state fire chiefs' association, safety film producers, electric and gas utility companies, hospital burn units, state fire fighter associations, and fire departments with video production capabilities.

Company-Level Fire Prevention Activities

Fire prevention inspections on the company level serve several purposes. They are useful in that the company members walk through the businesses in their district, which acquaints them with the business owners, the layout and occupancy of the building, and any special hazards. This is also a good time to take a look in the closets, basement, and attic, checking these areas for hidden hazards and construction features. During the visit, the location of electrical, water, and gas shutoff controls should be noted. The business owner can be consulted about where fire would do the most damage and salvage is a priority, usually in areas where records are stored. By having firsthand knowledge of a building and its contents, firefighting is safer and more efficient.

FIGURE 10-8 Fireworks may be for sale in one state, with possession and use punishable by fine and imprisonment (felony) in other states.

© Jones & Bartlett Learning

Performing these inspections gives the company a chance to carry out some public relations work as well. Forming a working relationship with the business owners in your district will encourage them to notify you of any major changes they make in their building or occupancy and encourage them to consult you regarding the fire-related implications of the changes. The public likes to see fire fighters out taking proactive measures to ensure its safety.

A company-level inspection performed in departments with natural hazards includes weed abatement/hazard reduction. During these inspections the company tours its area and notes dry brush and grass that may threaten improvements (structures). By doing so, the company reduces hazards and makes the public more aware of the dangers. Contact is made either in person or by mail with the parties that need to remove combustible vegetation from around their structures. These inspections are an opportunity for the fire fighters to emphasize the need for persons to take care with ignition sources in dry grass. Just the sight of the fire engine cruising the area and inspecting tends to make people more aware of the fire danger.

The intent of this program is not to get people to ruin their natural surroundings by removing every blade of grass and bit of vegetation on their property. It is to break the fuel ladder by preventing a fire from spreading from ground fuels to the structure itself. The usual requirements are a 30-ft break to mineral soil around all structures and within 10 ft of liquid petroleum gas (LPG) tanks, fences, and wood piles; removal of limbs 10 ft from stovepipes and chimneys; and removal of combustible materials like pine needles and leaves from roofs **FIGURE 10-9**. These requirements are often contained in adopted ordinances. Other ways to comply with the intent of these regulations are to landscape with plant species with low combustibility around the structure and to keep them irrigated.

Another form of hazard reduction inspection is checking the area for accumulations of tires. In the last few years there have been fires with thousands of discarded tires involved. These fires can lead to tremendous control problems for the fire department. They burn with acrid black smoke and are usually very time consuming and expensive to extinguish. The water

FIGURE 10-9 Defensible space—an area surrounding a structure cleared of highly flammable vegetation.
© Jones & Bartlett Learning

used to extinguish the fire must be contained to keep the runoff from polluting streams.

Another type of business that has a large accumulation of combustibles is wooden pallet manufacturers and recyclers. These facilities can be located just about anywhere, and once a fire gets started it is usually beyond control in a short time. The problem of stacks of combustibles and their arrangement combined with the design of a pallet make them highly combustible once ignited. In installations with large accumulations of combustible materials, the best the fire department can do is to keep a hostile fire from spreading to surrounding structures.

Fire Prevention Inspection

To conduct a fire prevention inspection of a business or facility properly, preliminary work must be done. Previous inspections should be reviewed to see what violations and corrections were made in the past. If the occupancy or facility is unusual, some preliminary research may be necessary to make sure that you are capable of noticing a violation when you see it. It is good to know which fire protection features are required as this will make you appear more professional because you are familiar with the code. When in doubt it is better not to give wrong information. You should advise the owner that you will conduct the necessary research in the code and get back to him or her.

Tip

The fire department must have the legal authority to enforce fire-related codes and ordinances. This issue is discussed in Chapter 11, *Codes and Ordinances*.

FIGURE 10-10 Fire prevention inspector performing a check of building plans.

© Jones & Bartlett Learning

The inspector needs to be equipped with the tools of the trade **FIGURE 10-10**. These will aid in making the inspection and documentation more thorough as well as making the inspector appear more professional. One essential is a proper uniform or other form of identification to let the business owner know who you are. Protective clothing, such as a hard hat and coveralls, may be required when inspecting dirty or hazardous locations. Writing materials and clipboard, including graph paper, notebook, inspection forms, and violation notices are a necessity **FIGURE 10-11**. A measuring device for determining distances and clearances is important; a 25-ft measuring tape is usually sufficient.

Township of Upper Moreland
Office of the Fire Marshal

117 Park Avenue • Willow Grove, Montgomery County, Pennsylvania 19090
Phone: 215-659-3100 • Fax: 215-659-1364 • Email: uppermoreland.org

FIRE CODE INSPECTION REPORT

OCCUPANCY: CONTROL:

ADDRESS: OWNER/MGR/PRINCIPAL:

ADDRESS:

PHONE:

ICC USE: INSPECTOR: DATE: DAY: TIME:

✓VIOLATION
SEE REMARKS
☐ PERMIT INSPECTION
☐ BLOCK INSPECTION
☐ RESIDENTIAL INSPECTION
☐ TANK INSTALL/REMOVAL
☐ FOLLOW-UP

✓VIOLATION
SEE REMARKS
☐ TEST _____
☐ COMPLAINT _____
☐ OCCUPANCY _____
☐ MAR _____
☐ OTHER _____

1. Fire Extinguishers due to be inspected.
2. Fire Extinguishers not provided/installed properly.
3. Keep EXIT signs illuminated and visible.
4. Provide additional EXIT signs (high and low).
5. Maintain adequate aisle width.
6. Keep fire exits unlocked and free of obstructions.
7. Repair/install panic hardware.
8. Removed obstructions from fire towers or escapes.
9. Maintain egress lighting.
10. Provide/maintain emergency lighting.
11. Maintain 24" clearance at ceiling with combustibles.
12. Repair voids in the ceiling: holes, tiles, etc.
13. Remove combustible storage from boiler room.
14. Maintain proper housekeeping
15. Remove grease from ranges, hoods, duct, fans, etc.
16. Keep electrical appliances clear of combustibles.
17. Repair improper wiring, fuses, grounding, etc.
18. Provide proper storage of compressed gas cylinders.
19. Oxidizers stored separately from flammable liquids, corrosive liquids, combustible materials.
20. Provide proper containers/storage of flammable liquids.

21. Bond wires shall be used when dispensing Class I or II liquids from metal to metal containers.
22. Discontinue smoking and post proper signs.
23. NFPA signs posted at entrance to buildings where hazardous materials are used or stored.
24. Provide/maintain fire zone signs.
25. Install temporary fire zone signs.
26. Discontinue Improper/Illegal burning.
27. Dumpster/combustibles too close to building.
28. Post address on building/sign.
29. Street address posted on rear of strip occupancy.
30. Occupancy limit sign(s) missing/Improperly posted.
31. Failure to obtain required permit.
32. Required test or drill logs missing/incomplete
33. Required emergency plan missing or out of date.
34. Fire alarm system annual test report.
35. Fire sprinkler system annual test report.
36. Sprinklers obstructed.
37. Other suppression system test report. (List in remarks).
38. Knox Box Key Check
39. Other (List in remarks)

REMARKS: _____

_____

_____

_____

_____

_____

_____

☐ Haz Mat/Non-Permit/Req'd. Site   ☐ Sprinkler   ☐ Standpipe   ☐ Cooking System   ☐ AFA   ☐ UST/AGT

IN THE INTEREST OF FIRE SAFETY AND TO COMPLY WITH THE UPPER MORELAND TOWNSHIP FIRE CODE, THE ABOVE VIOLATIONS MUST BE CORRECTED IMMEDIATELY. FAILURE TO COMPLY WILL RESULT IN PENALTIES AS SET FORTH IN THE FIRE CODE OF UPPER MORELAND TOWNSHIP. A REINSPECTION WILL BE ON OR ABOUT 30 DAYS. RECEIPT OF NOTICE ACKNOWLEDGED:

SIGNATURE:_____ INSPECTOR:_____ BADGE:_____

PRINT NAME: _____ FIRE MARSHAL:_____ DATE:_____

THIS REPORT IS BASED UPON OBSERVATIONS AT THE TIME OF SURVEY WHICH MAY NOT DISCOVER ALL HAZARDS.

FIGURE 10-11 Fire code inspection form.

Courtesy of Township of Upper Moreland/Office of the Fire Marshall

A camera with flash to document hazards and photograph evidence can come in handy. Code books for easy reference during the inspection can be used when you are confronted with a situation that you are not sure about, when the code section needs to be shown to the owner, or when you are writing a citation or inspection report. A citation book and red tags for taking equipment out of service should also be available.

Fire codes typically allow entry for routine inspections or when the fire official has reason to believe an unsafe condition exists. Entry cannot usually be made by personnel without compliance with the following conditions: You must show proper credentials, request entry, and make a reasonable effort to locate the owner of a vacant building prior to entry. When an inspector is refused entry, an inspection warrant or administrative warrant may be obtained from a judge under civil code procedures. An inspection warrant is not the same as a criminal search warrant. They are not as specific as a search warrant and are designed to allow you to investigate any possible fire code violations. Legal precedent allows you to obtain an inspection warrant to enter the property if you can prove that a logical sequence of inspections is occurring in the geographical area (Bahme 1976). The inspection warrant is used for the period of inspection but typically cannot exceed 14 days. An inspection warrant normally does not allow forced entry into the building, and it must be served during normal business hours. In some jurisdictions, part of the licensing procedure for businesses is that the owner agrees to periodic fire prevention inspections. You should know if this is the case in your area before you attempt to make the inspection.

In most situations, when you approach the business owner and explain your visit as the standard fire prevention inspection, they usually invite you in and offer to show you around. Some businesses may request that you call and make arrangements before your visit. This is not always done to hide violations but to make sure that someone is able to assist you with your inspection. Remember that the goal of fire prevention is to educate and gain voluntary compliance. It is not always convenient for a business owner to drop what he or she is doing to show you around. Some businesses have peak business hours during the day and it is usually best to avoid these times. For example, a restaurant owner may be reluctant to accompany you on an inspection during the lunch hour but may be happy to join you in mid-afternoon. In seasonal businesses, such as agricultural packing facilities, it is best for you to visit when they are in operation, but not best for the owner as they are very busy. In these occupancies, an appointment is highly recommended. As you perform the inspection,

approach it in a consistent manner. After you have introduced yourself and received permission to proceed, start around the outside and work your way inside. While outside you can check for accumulations of combustibles, location of utilities shutoffs, presence of fire hydrants and automatic extinguishing systems, posted and visible address, and other important data.

When you reenter the business, let them know that you are ready to proceed with the interior inspection. When performing the inspection, it is not critical whether you start at the top floor and work down or start at the bottom floor and work up. However, it is critical that you inspect the entire facility. You will develop your own style after you have completed several inspections. In a production or assembly facility, it is usually a good idea to start at the beginning of the process, with the raw materials, and go all the way through to the final product. As you tour the premises, take note and document any violations and ask that they be corrected. It may be necessary to explain to the owner or his representative exactly what you want and why as part of the education process.

If you encounter a situation that you are unsure of, it is better to take the time to look it up, or at least document it so you can research it later, rather than just make a mental note and possibly forget or ignore it because you are not sure. Do not attempt to bluff the business owner regarding an issue that you are unsure of. There is nothing wrong with letting the business owner know that you will need to conduct some research about the topic and then get back to him or her with an answer. It is much better to conduct yourself in this fashion; you will give the business owner the impression that you are taking an interest in the business and are trying to work with him or her. Do not be bashful about asking to look in closets and behind closed doors. Often combustibles are stored in closets right next to water heaters.

During an inspection is a good time to talk to the owner about any special concerns he or she may have about what he or she is doing or planning to do. It is also a good time to determine whether any remodeling has been done that may affect fire control systems or firefighting operations. When hazards and violations are found, they need to be prioritized as to their importance. The most important are those that jeopardize life or safety and have the potential for excessive property damage. As you seek correction of these violations, some are going to be more time critical than others. A blocked fire exit needs to be corrected in your presence because it is a direct threat to the life safety of the building's occupants. A fire extinguisher overdue for maintenance or inspection can be given a reasonable abatement period.

All hazards and violations should be documented, even if they were corrected in your presence.

This will provide a record of your findings. It can also provide a listing of the hazards detected at the facility for your records. The written documentation serves as a notice to the owner that the hazards need correction. In many instances, you will not be talking to the owner of the business or the building. You will be speaking with a manager. When you provide this person with written documentation, he or she can present it to the owner and get the violations corrected.

Upon completion of the inspection, be sure to thank the business representative for his or her time and assistance. A little effort on your part goes a long way toward developing a working relationship between the fire department and local business owners. It also makes it easier for the person who inspects the business next time. Always remember that the business owner is your customer and treat him or her as such.

In some situations, you may be required to develop a plan of correction. This situation occurs when there is a long list of violations or chronic hazard situations. By prioritizing the hazards and working with the business owner, you can develop a plan of periodic reinspections to ensure progress. Through this process you can avoid having the business shut down while corrections are made. The plan of correction can be as simple as advising the owner, in writing, that you will be back in 14 days to reinspect the premises, and if he or she is not in compliance at that time, a citation will be issued. In other circumstances, the reinspection and progress check process may take much longer, depending on the amount of work to be done.

When you reinspect and find the occupancy is still not in compliance, several courses of action are available. One is to provide a verbal warning by reminding the occupants of their legal obligation and that civil and criminal action can be taken against them. The other is to write a citation and send the matter to court or before a civil enforcement board. Depending on the immunity laws in your state, a failure to follow up on violations found can open you and the jurisdiction to a lawsuit. When you fail to act, you could be assuming a certain amount of liability. Long-term noncompliance generates poor attitudes on everyone's part. Inspectors tend to lose interest because of frustration. The fire department's credibility can be lost. Once the business owners see that they can ignore your requests, they can develop the attitude that nothing you require is of any importance. When there is a disagreement between the fire prevention bureau and the owner, several courses of action are available. A board of appeals may hear an appeal from the owner or rule on the application of alternate materials and methods to correct a situation that is called for in many fire and building codes. Many fire codes also provide that the fire official may modify provisions of the code upon written application by the building owner when it is impractical to carry out the strict letter of the code.

Determination of Fire Cause

Typically, it is the responsibility of the fire official to determine the cause of every fire that occurs in his or her jurisdiction. As with other responsibilities, the determination may be delegated to the other members of the fire department. Scene responsibility lies with the highest ranking officer at the fire. The person who usually performs the primary investigation is the company officer. It is the responsibility of every member of the fire department at the scene to be aware that something had to cause the fire and to protect the area of origin as much as possible.

Cause determination is important from a prevention standpoint. If we do not know what is causing the fires, we will have a hard time targeting our prevention efforts. If the cause of numerous fires is children playing with matches, we may need to be more aggressive with the public education program in the schools. Cause determination has also uncovered problems with equipment and processes that can be handled through design changes or product recalls. Fire cause and fire statistics are often used to modify existing codes and standards. These measures can greatly reduce the fire loss in the community.

The first people inside the burning structure are the fire fighters on the nozzle. Their actions are critically important to the determination of the cause. A cause investigation starts with the observations made by the personnel at the scene. Many indicators lead to the site of the fire's origin. Some of the questions you should be prepared to answer at a structure fire are:

- Was the door locked?
- What were the positions of interior doors?
- Were the windows broken when you arrived?
- Was the fire particularly hard to extinguish?
- Was the electricity on when you arrived?
- Did you notice anything out of the ordinary?
- What portion of the building was involved in fire when you arrived?

When the fire is extinguished, it is important not to disturb the suspected site of origin any more than necessary. Some portion of the overhaul may have to be delayed until cause determination is made. Taking a shovel and removing all burned material from the structure may be the usual procedure, but it can destroy evidence if done in the area of origin. One of the worst mistakes is to find a gas can and run to your officer with it in your hand, shouting "Look what I found!" You have in essence just compromised that piece of evidence. This ranks right up there with telling the news media that the fire was started by electricity, then finding out that it was arson and the electricity was off. The defense attorney may very well show the film clip of your statement in court, during the trial, to discredit your testimony and credibility in the eyes of the jury.

Tip

When the fire is extinguished, it is important not to disturb the suspected site of origin any more than necessary.

At all fire scenes, the area of origin must be protected. In a wildland fire scenario, it is important to take note of the general area where the fire was burning on your arrival. As the fire burns, it can spread in all directions, making it hard to pinpoint the area of origin. Your initial observation can narrow this area considerably. At wildland fires it is important to protect the area of origin by keeping vehicles and foot traffic out. Causes have been determined from items as small as a single match or piece of carbon from an exhaust. One person walking through the origin area could easily destroy this fragile evidence. Another recommendation in remote areas is to take note of any vehicles leaving the area. Write down a description and license number whenever possible.

The next step in the cause determination is the reconstruction. By working back from the farthest points of fire extension (i.e., area of least damage), the trained investigator can use numerous clues to narrow the way to the point of origin. Some clues are less obvious than others. Training in the science of cause determination is important to all fire fighters so they do not inadvertently destroy evidence.

Once in the area of origin, the tedious work begins. Material is removed a little at a time until the cause of the fire is uncovered. A part of the reconstruction is witness statements. Determining what people at the scene saw and an examination of the structure can create a chronology of the fire's development.

After the facts are determined from the first steps of the investigation, they are evaluated. Numerous factors can influence the behavior of a fire. The investigator should never approach the fire with any preconceived notions as to what happened. The facts should be evaluated as they are. If you are sure that it was electrical, you may overlook the gasoline burn pattern on the floor.

Once the facts have all been examined and evaluated, a conclusion may be drawn from the determination of the fuel source, the heat source, and the act or omission that caused the two to come together. Not all fires require a long, drawn-out investigation to determine their cause. In some cases, the cause is obvious. When the homeowner says the pan of grease on the stove caught fire and set the kitchen cabinets ablaze, the cause may be evident. In other cases, the investigation may take weeks to conduct. In any case, once the fire department releases the scene and all fire fighters leave, any evidence found after that may be worthless in court.

The types and complexity of investigations can be divided into three basic categories. The first and most commonly conducted is the basic investigation. In this investigation, only general information, such as time, place, and type of material burned, is necessary. This is the minimum information needed for the fire report. When someone is working on his fuel system and accidentally sets his car on fire, a basic investigation is usually conducted.

A more involved investigation is the technical investigation. These investigations are more in-depth to determine a more complicated cause of origin. When a certain process or device is suspected to be the fire cause, the investigation can get more technical.

In these cases, evidence must often be collected and sent to the crime lab for further investigation to determine its chemical makeup. In some cases, personnel from the Bureau of Alcohol, Tobacco, Firearms and Explosives (BATF) or other specialists are called in to assist with cause determination.

The third type of investigation is called for when the fire is incendiary or suspicious. In these investigations, cause determination experts are called in to assist with gathering and processing evidence. The cause of the fire may have already been determined by engine company personnel. A strong smell of gasoline in a dwelling after a fire is a good indication that a crime was committed. In most jurisdictions where there has been a death related to the fire, arson investigators are automatically called for cause determination. The cause may be as simple as smoking in bed, but no chances are taken with an incorrect determination in these cases.

In cases where the ranking officer is unsure or unable to determine the fire cause, the arson investigators may be called for. It is better to err on the side of caution than to leave the scene and then find out the cause was arson.

Fire Information Reporting

Fire reports are generated for several reasons. The statistical data generated from the reports can be used for budget justification, trend analysis, future planning, needs assessment, the identification of faulty equipment, code changes, enhanced safety devices, and legislation. Fire reports are used to generate statistical data on a local, statewide, and national scale. By keeping track of what types of fires are experienced, where they are occurring, why they are occurring, who is having them, and how they are starting, the fire service can identify focus areas for fire prevention efforts. Any statistics collected are only as good as the input. If the input or fact gathering that generates the statistics is faulty, the conclusions drawn from them will be faulty as well. For example, if an expensive fire loss occurs and the amount is not properly recorded in the fire reporting system, the dollar loss for the whole year will be incorrect.

In the management of any program, there must be a focus. There must be goals and objectives. To set the goals and objectives we must first determine the problem to be addressed. Without a plan of action, personnel can work for years and never really get anywhere in reducing the fire problem. Many personnel hours expended on fire prevention without a reduction in the fire loss should lead the department to question whether they are addressing the root problem. If most of the fires are started by faulty wiring, and prevention efforts are only talking to people about combustibles stored by water heaters, the prevention job is not getting done.

The information that is required on a standard fire report and the coding for the information it contains are specified in NFPA 901: *Standard Classifications for Incident Reporting and Fire Protection Data*. This guideline indicates how to tabulate the data collected so they can be used to compile statistics on a regional and nation-wide basis. Most states require reporting of all fire-related civilian and fire fighter casualties. The information is provided through the National Fire Incident Reporting System. Reports are transmitted electronically to the states and then to the National Data Center.

Wrap-Up

CHAPTER SUMMARY

- Research conducted by the Department of Homeland Security and other groups has shown that for every $1 spent on prevention, $4–$7 can be saved in responding to incidents.
- In technical or high-risk areas, fire prevention is typically performed by fire department employees assigned to the fire prevention bureau.
- A fire inspector is specially trained in the science of fire, fire prevention inspection, and enforcement and is responsible for performing inspections to identify and reduce hazards and risks.
- A fire fighter assigned to an engine or truck company may very well be called on to perform fire prevention inspections of businesses in his or her first-in area.
- Fire prevention bureaus are often staffed with a mixture of uniformed (fire fighters) and civilian personnel.
- An inspector of any level may seek training in fire prevention from numerous sources.
- The purpose of fire prevention activities is to prevent the loss of life and property due to fire.
- A fire prevention inspection is the act of making a systematic and thorough examination of a premises or process to ensure compliance with fire codes and ordinances.

- Fire prevention activities start before the building is even built.
- The purpose of hazard evaluation is to identify possible accidents and estimate their frequency and consequences.
- By educating people as to the importance of fire prevention, they will be concerned about it even when you are not there.
- Fire prevention inspections on the company level allow company members to walk through the businesses in their district, which acquaints them with the business owners, the layout and occupancy of the building, and any special hazards.
- If the occupancy or facility is unusual, some preliminary research may be necessary to make sure that you are capable of noticing a violation when you see it.
- Typically, it is the responsibility of the fire official to determine the cause of every fire that occurs in his or her jurisdiction. Scene responsibility lies with the highest ranking officer at the fire.
- The statistical data generated from fire reports can be used for budget justification, trend analysis, future planning, needs assessment, the identification of faulty equipment, code changes, enhanced safety devices, and legislation.

KEY TERMS

Fire flow The total volume of water required to control the fire incident expressed in gallons per minute.

Hazards Things within the environment that can cause harm to people or equipment.

Incendiary Deliberately set.

Occupancy The use or intended use of a building.

Petrochemical Related to oil refining and production facilities.

Risks The chance that humans take in relationship to the hazard(s).

Voice over A presentation technique, often used in television, where the audience sees video and hears the voice of the narrator but does not see the narrator.

CASE STUDY

According to the USFA, home fires spike in winter months. Cooking and home heating are the leading causes of residential building fires during the winter. The risk of fires also increases with the use of supplemental heating, such as space heaters.

The Consumer Product Safety Commission (CPSC) estimates that home heating was associated with an average of 33,300 fires and 180 fire deaths per year from 2005 to 2007.

CO is also a serious threat in the winter months. Any fuel-burning appliances in the home, including furnaces and fireplaces, are a potential CO source. CO is called the invisible killer because it is an odorless, colorless, and poisonous gas.

The following is a short list of incidents involving fire-related fatalities:

- In Citra, Florida, a fire killed five children on November 8, 2010. Their home did not have smoke alarms.

- In Penfield, New York, a 54-year-old man died of CO poisoning in November, 2012. Prior to his death, the home's CO alarms reportedly beeped and were removed from the house.

1. What would be the most effective way for the fire department to prevent these tragedies?
   A. Quick response by fire fighters
   B. Effective water application at the scene
   C. Use of the proper firefighting resources
   D. Fire prevention education provided to the public

2. In anticipation of home heating-related incidents, related fire prevention messages should begin:
   A. during the summer.
   B. during the fall.
   C. during the winter.
   D. during the spring.

3. From a loss prevention standpoint, what is the best strategy for the local fire department to anticipate which types of incidents are likely to occur at any specific time of the year and adjust to address them?
   A. Adjust fire prevention messages and efforts by time of year.

B. Provide a large number of suppression resources all year because fires occur in all seasons.

C. Increase fire suppression resources during peak seasons.

D. Quick response and suppression is better than prevention.

4. The most effective fire prevention program that keeps people safe at work and home requires:

A. enforcement of fire codes.

B. continual inspection of occupancies.

C. application of the fire code during construction.

D. directed public education.

REVIEW QUESTIONS

1. List the four areas of fire prevention and give an example of each.

2. Give an example of a fire prevention device.

3. List two methods of presenting fire prevention education to the public.

4. Suppose you have been given the responsibility of preparing a fire prevention presentation for a third grade class. What are some of the things you can do to get your point across?

5. List two advantages to a fire company performing fire prevention inspections in its first-in district.

6. Why does the fire prevention bureau inspect the more complicated occupancies?

7. Explain the difference between a hazard and a risk.

8. What is the reason for weed abatement in an area with a dry climate?

9. Illustrate the chain of authority that allows you to require compliance with the fire code.

10. Which fire code is adopted in your area?

11. Illustrate the organization of a standard fire prevention bureau.

12. Where can a fire fighter seek fire prevention training?

13. When faced with an owner who fails to comply with the required corrections, what actions should you take?

14. List three reasons why fire statistics are collected.

DISCUSSION QUESTIONS

1. How valuable is fire prevention in providing comprehensive fire protection to a community?

2. Why can't the fire protection problems in a community be solved solely by building more fire stations and hiring more emergency response personnel?

3. You are the fire chief and have been asked to cut the fire prevention budget. How will you defend not making the cut?

REFERENCES AND ADDITIONAL RESOURCES

Bahme, Charles W. 1976. *Fire Service and the Law*. Quincy, MA: National Fire Protection Association.

The Center for Chemical Process Safety. 1985. *Guidelines for Hazard Evaluation Procedures*. New York, NY: American Institute of Chemical Engineers.

International City/County Management Association (ICMA). 2009. "Fire Prevention Saves Lives and Budgets." ICMA .http://icma.org

National Fire Protection Association. 2018. *Deadliest Fires and Explosions in U.S. History*. Quincy, MA: National Fire Protection Association.

National Fire Protection Association. 2019. *The Station Nightclub Fire*. Quincy, MA: National Fire Protection Association.

https://www.nfpa.org/Public-Education/Staying-safe /Safety-in-living-and-entertainment-spaces/Nightclubs -assembly-occupancies/The-Station-nightclub-fire

National Fire Protection Association. *NFPA 901: Standard Classifications for Incident Reporting and Fire Protection Data*. Quincy, MA: National Fire Protection Association.

National Fire Protection Association. *NFPA 921: Guide for Fire and Explosion Investigations*. Quincy, MA: National Fire Protection Association.

National Fire Protection Association. *NFPA 1031: Standard for Professional Qualifications for Fire Inspector and Plan Examiner*. Quincy, MA: National Fire Protection Association.

Oakland Fire Department. 2017. Origin and Cause Report Incident #2016-085231. Oakland, CA. Oakland Fire Department.

Runk, D. 2011. "GM Cancels Shifts Due to Magna Factory Fire." *MSN News*, March 3. https://www.deseretnews.com /article/700115281/GM-cancels-shifts-due-to-Mich-parts -supplier-fire.html

United States Fire Administration (USFA). 2015. *Facts about Fire in the U.S.* Emmitsburg, MD: United States Fire Administration.

CHAPTER 11

Laws Affecting Fire Fighters

OBJECTIVES

After studying this chapter, you should be able to:

- Give an overview of the types of laws in the United States.
- Discuss the court system.
- Describe how to handle personnel complaints.
- Discuss the legal components of fire prevention activities.
- Describe how codes are developed.
- Discuss the legal considerations at emergency incidents.

Case Study

The requirement that emergency vehicle drivers wear seat belts is a point of contention among many fire service personnel. Annually, many fire fighters, EMS personnel, and police officers are killed or seriously injured after being ejected from vehicles where their seat belts are left hanging.

On July 8, 2008, a 25-year-old volunteer fire fighter was fatally injured after being ejected in a fire truck rollover. The driver lost control of the fire truck, swerved off the left side of the road, returned to the pavement, and overturned on the right side of the road. The victim, another fire fighter riding in the fire truck, was not wearing a seat belt and was ejected out of the driver's side window. The pumper's 725-gallon water tank detached from the truck body and landed on top of the victim in the street. The victim was pinned underneath the water tank and died from injuries sustained in the crash.

On July 16, 2012, a 30-year-old volunteer fire fighter died after being ejected from a fire engine.

The victim, who was riding in the right front seat, was responding with one other fire fighter (the driver) to a reported motor vehicle crash. The fire engine traveled approximately 1.3 mi from the station when the driver lost control of the engine in a curve. The engine left the paved road and crashed into trees on the right side of the roadway. The victim was ejected from the engine and landed in a wooded area. A contributing factor to this death was the nonuse of a seat belt.

1. Considering that seat belts are mandatory for the public, why would emergency responders want to be exempted from such a law?
2. Even if seat belt use for responders were exempted, should seat belt use still be fire department policy? Why or why not?
3. In 2010, approximately 15 percent of fire fighter fatalities were driving related. What can be done to lower this number?

Modified from the following sources: Nicol, Susan. 2013. "RI Responders Balk at Seat Belt Repeal Measure." Firehouse. http://www.firehouse.com/news/10978791/ri-responders-balk-at-seat-belt-repeal-measure
NIOSH. 2013. "Volunteer Fire Fighter Dies After Being Ejected From Front Seat of Engine—Virginia." NIOSH. http://www.cdc.gov/niosh/fire/reports/face201223.html
NIOSH. 2008. "Fire Fighter Dies after Being Ejected from a Pumper in a Single Vehicle Rollover Crash—New York." NIOSH. http://www.cdc.gov/niosh/fire/reports/face200825.html

 Access Navigate for more resources.

Introduction

There are many laws, regulations, and standards that have an impact on fire departments and fire fighters. They fall into broad categories, such as employment and workplace law, building construction, fire prevention, medical regulations, and emergency operations. As has been mentioned in previous chapter policies, procedures and guidelines are published within the department. Part of the reason for these is to ensure employees stay "within the law" when performing their official functions. The fire department and fire fighters are not exempt from the law, even in extreme cases. Failure to know and follow applicable laws and standards can end up in civil and criminal court cases and termination of employment. The fire department, as well as individual members, can be found liable if applicable laws and standards are not followed.

Definitions of Laws

Laws are pieces of enacted legislation. There are different types of laws in the United States, led by the supreme law, the U.S. Constitution, with which no other laws may conflict. All legal authority for governmental action in the United States comes from the Constitution. The U.S. Congress passes legislation within the confines of the Constitution.

Statutory laws are those that are adopted by Congress (federal statutes) and those that have been passed by state legislatures (state statutes). Statutory law also includes local laws, usually called ordinances.

Codes are systematically arranged, comprehensive collections of laws. **Regulations** are rules designed to implement a statute based on an agency's interpretation of that statute. Such rules provide procedural requirements for program operations. Typically, they are published through an official process that allows for public comment. Regulations have the same effect as law and must be complied with once they are published in final form.

The federal government organizes its statutes applying to a certain subject by placing them in the *Code of Federal Regulations* (CFR). States do much the same thing. The laws regarding public safety may be in the health and safety code. Criminal laws are organized into the penal code. Examples of codes that are adopted by local government ordinance are the building and fire codes. The hierarchy of legal enforcement is designed so that the state laws can augment or increase provisions in federal law, but they cannot weaken the federal law. The same relationship exists between local ordinances and state or federal laws. The types of codes are not to be confused. Federal regulations and state legislation are not the same as the adoption, by ordinance, of a building or fire code.

The constitutionality of laws is determined by the judicial system. When a law or the enforcement of a law is questioned, it may be taken before the courts. Once these issues are decided, the rulings are referred to as precedents. These precedents are used as the basis for future interpretation and enforcement of the laws.

Laws are applicable only if they have been properly adopted and not been reversed by a court decision. This is known as judicial review. Through court decisions, laws are changed, and in some cases, suspended until a final decision is reached. In modern times, initiatives passed by voters may create a law whose enforcement is held up in the court system for years. Appeals may be made at successive levels until the Supreme Court agrees to hear the suit or refuses to hear it, thereby leaving the decision of the next lower level of court standing.

When the law is not specific in a matter, the laws and previous court decisions are reviewed to determine what is most closely the intent of the law. Laws are not created overnight, and many new processes and situations occur that have not been addressed specifically in the law or in a code that has been adopted. In some states, in lieu of an existing law, a legislative counsel or attorney general's opinion governs conduct, action, or procedure.

It is important to have a good idea of what the applicable laws are in any situation. Fire fighters are charged with the authority and the responsibility to enforce only laws that exist, not just what they think is a good idea. When you, acting under the authority of your position, tell a business or property owner to make a correction, such as reduce a hazard or provide a fire extinguisher, they have two choices: to go ahead and do it or to challenge you. They can say that they want you to cite the provision of the code that requires them to have fire extinguishers serviced and maintained periodically. If you are not familiar with the code, and you are wrong, they have every reason to contact your supervisor and report your inadequacies. When you know the fire extinguisher regulations, you can produce the code section that requires compliance with your orders. A third alternative is for them to let you go ahead and issue the citation and take the matter before a judge. What

they will ultimately have to do depends on the decision rendered by the court.

In some circumstances, the word of the fire department employee is not the final word, no matter what the law says. A business owner may be granted a variance to avoid complying fully with the letter of the law. A variance can be granted for a number of reasons. The code allows a variance provided that the intent of the code provisions is met. A variance allows the business owner an alternative to meeting the strict letter of the code as long as life safety and fire safety are provided. At other times, the variance may be granted for political or economic reasons.

The Court System

Before looking at the court system and its organization, it is necessary to understand the concept of jurisdiction. The cases that a court can hear are considered to be within its jurisdiction. When a case can be first heard in a certain court, it is considered to have original jurisdiction. When the case must first be heard in a lower court and then the lower court's decision is appealed to the higher court, the higher court is an appellate jurisdiction. In the case of fire departments, the meaning of jurisdiction is the limits of territory within which its authority may be exercised. An example is that the city fire chief cannot go into a county fire department station and give orders to the fire fighters, because he is outside his jurisdiction. Jurisdiction is a very important concept to understand whenever talking about who can legally do what, where, and when. If you are acting outside your jurisdiction, you are not legally taking action, no matter how good your intentions.

Tip

Be sure to stay within your jurisdiction. In the case of fire departments, the meaning of jurisdiction is the limits of the territory within which its authority may be exercised.

The court system is divided into different levels. As you go up the levels, the matters brought before the court tend to be more complex and the rulings are more widely applicable. On the federal level, the highest court is the U.S. Supreme Court **FIGURE 11-1**. Most cases heard before this court are questions about the constitutionality of a law. The next lower level is the circuit court of appeals, which hears appeals from the federal district courts. The federal district courts deal with matters of federal law. When a crime—such as

FIGURE 11-1 Unites States Supreme Court Building Washington, DC.
© VisualField/Shutterstock

building an illegal campfire—is committed on federal property, the accused person must appear in federal district court. The illegal act occurring on federally managed land gives the federal court jurisdiction.

On the state level, the highest court is typically the state supreme court. This court usually hears appeals from the district courts of appeal. The district courts of appeal typically hear appeals from the district or superior courts. The district or superior courts are where most of the state trials are heard. When someone is charged with arson, the case is first heard in the district or superior court. In certain cities and counties, there are municipal or county courts where misdemeanor offenses are heard, with the exception of those committed by juveniles, which are heard in district or superior court. If you wrote someone a citation for illegal open burning of trash, he or she would be required to appear in municipal court. If someone were arrested for possession or discharge of illegal fireworks, a felony in some states, he or she would appear in superior court. Both situations may start with the issuing of a citation, but the offenses are very different in the eyes of the law.

Liability

There are two kinds of action that may be taken in a court of law in regards to liability. One is criminal, where accountability is sought under criminal law. Criminal cases are prosecuted by an attorney general or district attorney in an attempt to prove that a crime has been committed. A crime is defined as an unlawful act as described in the criminal codes.

The other kind of liability is civil liability. The purpose of civil liability is to recover any losses incurred by injured parties and to create behavioral changes in defendants. Civil courts can award compensation for damages to an injured party (plaintiff) from the person causing injury (defendant) and divide and assign a degree of responsibility for the damages incurred in cases involving multiple defendants. A person or organization does not need to be found criminally liable to be held civilly liable. As an example, a defendant may be acquitted of murder in a criminal trial but be found civilly liable in an unlawful death civil lawsuit.

Lawsuits

We live in a society where lawsuits are becoming increasingly common. When someone's house burns down, for example, they may file a claim against the fire department for not saving it. This claim can be litigated as a civil suit. The jurisdiction may decide to settle out of court.

The way to avoid lawsuits is to do your job correctly every time. Should your actions result in someone filing a claim for damage or a lawsuit, you will most likely be questioned regarding the industry standards, department standards (policies and procedures), your knowledge of them, your training, and whether you were acting within your scope of employment. You will be judged on whether you were exercising reasonable and prudent judgment. You must also take the time to document, through the use of reports, what you did. When lawyers approach the jurisdiction to start a suit, the evidence of a job done correctly and properly documented can often stop the proceeding in its tracks.

When a civil suit is filed, it is because of a tort—a wrongful act. The term tort is used in civil, not criminal actions. Civil actions are brought with the intent of seeking monetary compensation. Torts can result from either nonfeasance, misfeasance, or malfeasance. Nonfeasance is a failure to act. If you were to respond to a medical aid incident and failed to splint a broken leg before the victim was moved, thereby further injuring the victim, that is a failure to act. Misfeasance is doing something wrong that you are lawfully allowed to do. Responding with red lights and siren is legal. It releases you from obeying the traffic laws to a certain point. It does not allow you to run stop signs at a high rate of speed without due regard to public safety, as is specified in many departments' standard operating procedures and guidelines. Malfeasance is wrongdoing or misconduct. Driving the engine to a fire while intoxicated is an example of malfeasance. Yes, the engine needs to get to the fire, but when you are under the influence of alcohol in this situation, you are violating the laws against driving under the influence. Using the red lights and siren does not release you from your obligation to obey the driving under the influence laws. The greater public good is served by having a short delay while another person shows up to drive the engine than for you to try to operate it yourself. In addition, if a fire fighter is impaired in some way, he or she should not respond at all. His or her response to the incident may endanger himself/herself or others due to an impaired condition.

Tip

The way to avoid lawsuits is to do your job correctly every time, actively exercising reasonable and prudent judgment while working within your scope of employment.

The defense that you were just doing the best you could under the circumstances is often not good enough. When things do go wrong, it is our responsibility to find out what went wrong and take measures to prevent a recurrence. The ability to perform properly at incidents comes from experience and training. In many cases, you may lack the experience. If you paid attention to the training and practiced, you should be able to perform as expected. If not, your performance will be below that which is expected and you will have a hard time defending your actions in a court of law. The training must also be documented. Judges and juries are suspicious when documentation of training completed is not available when requested.

One of the ways to address these issues is through policies. Policies clarify or provide direction for specific situations. The policies must be agency specific and regularly reviewed to ensure that they are still valid. The policies must be written and delivered to the members of the department so that they are understood. The department must be able to verify that they have been received and all personnel have not only seen them, but understand them as well. The only way to verify understanding is through regular testing. All personnel in a supervisory capacity must follow the policies and set the example if they expect others to do the same.

There are several simple policies that you can use to limit your liability and help to protect yourself against a lawsuit. They are:

- Develop a reminder process that will send up a notice when it is time to go back to a business to reinspect for previous problems.
- Conduct as thorough an inspection as you are capable of. In other words, do your best.

- Maintain accurate records of your inspections, variances, complaints, and so forth. These records can be used to justify the reason you approached a business to conduct an inspection.
- In court, testify to those things you actually know.
- Have someone with you when making your inspections, preferably an employee or owner of the business. However, this action will not eliminate the possibility that someone could accuse you of misconduct.
- If you are refused entry for an inspection, obtain an inspection warrant. Do not force your way in.
- Finally, treat all people fairly and honestly to eliminate the potential of a business owner claiming that he or she was being discriminated against or picked on.

The overriding issue is that the public expects you to do a job and do it well. It is not just done to avoid lawsuits. It is the responsibility of every fire fighter to perform to the best of his or her ability and to perform correctly. To you, it may be just another incident; to the people involved it may be the most important event in their lives.

Personnel Complaints

Most departments have a standard procedure for complaints brought against members by the public or other members of the department. An example of a standard procedure is as follows:

1. The person wishing to file the complaint should speak to the accused's supervisor or the fire chief.
2. The complaint is discussed with the appropriate officer. The officer is to explain the options available to the complainant. Should the complainant decide to pursue the matter, the personnel complaint form should be completed.
3. The form is forwarded to the fire chief or a designated representative, and a decision is made as to the correct investigative procedure based on the nature of the complaint.
4. At the conclusion of the investigation, a determination is made as to the disposition of the complaint and what action, if any, is warranted against the employee.
5. The person complaining is notified by mail, in writing, of the results of the investigation, as is the accused.

The person making the complaint should be assured that the complaint will be properly investigated. Also, the department should make sure that no adverse consequences occur to any person or witness as a result of having brought a truthful complaint or provided truthful information in any investigation of a complaint.

It is standard policy that any employee who attempts to discourage, delay, or cover up any complaint received by a citizen shall incur disciplinary action.

It is the complainant's responsibility to be truthful and as accurate as possible in presenting information that he or she believes should be investigated. The complainant should understand that malicious or false complaints against department personnel could subject him or her to possible criminal and/or civil action. The department may assist in pursuing such action. The complaint form should state that the person signing the complaint does affirm under penalty of perjury that the facts contained therein are truthful.

Harassment-Free Workplace

The federal government and the courts hold management responsible for harassment in the workplace. Consequently, a clear understanding of what constitutes harassment is essential to the development of a harassment-free work environment.

Harassment is defined as coercive or repeated, unsolicited, and unwelcome verbal comments, gestures, or physical contacts, including retaliation for confrontation or reporting harassment. Harassment includes:

- *Physical conduct.* Unwelcome touching; standing too close; inappropriate or threatening staring or glaring; and obscene, threatening, or offensive gestures.
- *Verbal or written conduct.* Inappropriate references to body parts, derogatory or demeaning comments, jokes, or personal questions; sexual innuendoes; offensive remarks about race, gender, religion, age, ethnicity, sexual orientation, political beliefs, marital status, or disability; obscene letters or telephone calls, catcalls, or whistles; sexually suggestive sounds; loud, aggressive, inappropriate comments, or other vocal abuse.
- *Visual or symbolic conduct.* Display of nude pictures; scantily clad or offensively clad people; display of intimidating or offensive religious, political, or other symbols; display of offensive, threatening, demeaning, or derogatory drawings, cartoons, or other graphics; offensive T-shirts, coffee mugs, bumper stickers, calendars, or other articles.
- *Work environment.* Any area where employees work or where work-related activities occur, including field sites, fire stations, buildings, and facilities. Also included are vehicles or other conveyances used for travel.
- *Responsibility.* Managers, supervisors, and employees, as well as contractors, cooperating agency personnel, and volunteers are responsible for creating and sustaining a harassment-free environment by their individual conduct, through job supervision, coaching, training, and other behavior and means. All employees, contractor personnel, and visitors must take personal responsibility for maintaining conduct that is professional and supportive of this environment. Employees who witness harassment are instructed to report it to the proper authority.

Individuals who believe they are being harassed or retaliated against should exercise any one or more of the following options as soon as possible:

- Tell the harasser to stop the offensive conduct.
- Tell an officer or supervisor about the conduct.
- Contact the fire chief, or any other individual who would take action; for example, a union representative or the agency equal employment opportunity representative.

Fire Prevention

Fire prevention enforcement is based on codes and ordinances. Codes and ordinances fall under the broad description of laws. Laws are written and adopted on all three of the levels of government: federal, state, and local. There are clearly laid-out relationships among the different levels of government and their influence on each other.

One of the greatest challenges to those who develop fire-related codes and ordinances is gaining support for the restrictions and additional costs that the codes will require. Politics plays an important part in the code creation and adoption process. At many points in our history, the need for a code restriction or feature was not recognized until tremendous losses in lives and property had taken place.

Codes often impose limitations and increased costs on builders and developers. Limitations may come in the form of building placement on property, with setback and clearance restrictions to allow for ladder truck access and space between buildings. Valuable land is taken up to comply with these requirements. Codes also require built-in fire protection in buildings over certain square footages, based on local ordinances adopting the code. Construction costs increase due to these restrictions. In wildland-urban interface areas, fuel modification may be required to be as much as 300 ft from the structure to prevent the spread of fire from the natural vegetation to the structure. In addition, the installation of adequate water systems and fire hydrants is expensive. Many jurisdictions also charge permit and plan check fees. All of these requirements involve increased construction costs and use of available space.

Remember that if we do not study history, we are doomed to repeat it. The need for codes was identified out of many disastrous fires that struck cities and wildland areas and exacted huge tolls in lives and property. However, common sense does not always prevail. The Station Nightclub fire of 2003 and the Ghost Ship Fire in 2016 illustrate that even with modern codes, there still exists risk for major loss of life due to fires when occupancies are not in compliance. The Camp Fire in California of 2018 demonstrated that when fuels, adverse weather, and structures are all in alignment, the losses will be dramatic.

In spite of the problems with enacting and enforcing codes and ordinances, the impacts of lack of codes or weak enforcement can be catastrophic. Examples of this occur in other countries where codes are lacking or enforcement is weak, for example, the nightclub fire in Brazil in January 2013, where 233 young people perished in a fire. This was the deadliest nightclub fire since one in December 2000, when a welding accident reportedly set off a fire at a club in Luoyang, China, killing 309 (CNN Library 2013).

The fire prevention bureau has the legal responsibility and authority to enforce fire-related codes and ordinances. This authority does not just come about by the fire department deciding what they want to do. The fire department and its personnel cannot demand that action be taken by a property owner unless there is authority to inspect.

The most often cited case in relation to fire prevention is *See v. City of Seattle* (1967). In this case the U.S. Supreme Court has held that administrative entry, without consent, of the portions of commercial premises that are not open to the public, may be compelled only through prosecution or physical force within the framework of a warrant procedure.

The U.S. Supreme Court has also set forth guidelines for inspection agencies. The following recommendations are given to assist fire inspectors in operating within these guidelines:

1. Inspectors must be adequately identified. It is recommended that inspectors wear some type of uniform. Commonly recommended is a blazer, which looks professional but is not as threatening as wearing a badge.

2. Inspectors must state the reason for the inspection. Many times people at the site will joke that you are there to shut them down. This is not always taken as funny and you must be careful about what you say.

3. Inspectors must request permission for the inspection. There are times that are inconvenient for the occupants to undergo an inspection. This does not always mean that they are trying to hide something. It might just mean that they are extremely busy. Just make arrangements to come back at a more convenient time.

4. Inspectors should invite the person in charge to walk along during the inspection. When the manager or his or her designee accompanies you, it is a good chance to point out problems, ask questions pertinent to the inspection, and discuss voluntary compliance with the required changes. You must remain professional at all times and complete the inspection; the persons at the business have work to accomplish as well.

5. Inspectors should carry and follow a written inspection procedure, making it less likely to overlook something important. If you are called away or otherwise interrupted during the inspection, it also is easier to pick up where you left off.

6. Inspectors should request an inspection or administrative warrant if entry is denied. This will rarely be necessary. If all other methods of gaining voluntary submission to an inspection fail, then an inspection or administrative warrant will be necessary.

7. Inspectors may issue stop orders for extremely hazardous conditions, even if entry is denied, while warrants are being issued.

8. Inspectors should develop a reliable record-keeping system of inspections. By keeping records of past inspections, the inspector can update information such as phone numbers and contact persons, owner and business name information, and record of past violations. If the system of voluntary compliance is used, it is important to know when

reinspecting whether the last violations were corrected. If they were not, stronger measures may be required to ensure future compliance.

9. Inspectors should have guidelines available that define conditions whereby they may stop operations without a search warrant or obtaining permission to enter. These guidelines are necessary in that if you shut down an operation, you are costing the business money and it may very well end up in a lawsuit. It is important that these procedures be applied uniformly and impartially.

10. Inspectors should be sure that all licenses and permits indicate that compliance inspections can be made throughout the duration of the permit or license. Before any inspection is made, the inspector must be sure that he or she has the right to inspect the premises. After you have required expensive modifications or shut down a portion of a business is not the time to find out that you were acting outside of your authority.

11. Inspectors must be trained in fire hazard recognition and in applicable laws and ordinances. When you enter a business, it is your responsibility to make a thorough, professional inspection. Ignorance of the law is no excuse for either party—you or the owner. When an inspection is made, you may be held liable for fire, injury, or death if there is a fire resulting from something you overlooked or failed to enforce. It is just about a sure bet that you will be named in the lawsuit. For this reason, personnel assigned to the fire prevention bureau receive special training and inspect the more complex businesses and processes. Personnel on engine companies are not as highly trained and, in most cases, perform the less complex inspections. As an inspector, you need to possess the technical expertise to identify physical illegalities, such as substandard electrical installations, excessive amounts of flammable liquids stored, and lack of or inoperative fire extinguishers. Procedural illegalities include failure to obtain required permits for hazardous operations, unsafe welding activities, improper smoking areas, and unrestrained flammable gas cylinders that may tip over. Structural deficiencies include inadequate or breached fire walls, improper fire exit design, and improper installation of fire doors.

The state fire prevention laws and codes are based primarily on model national codes. In some states the codes are divided by type, such as a building code, health and safety code, welfare and institutions code, and administrative code. In most states, the state fire marshal interprets and recommends state-level legislation as it relates to fire and life safety. The state fire marshal may also be charged with the responsibility of enforcing regulations and codes in state buildings and in areas with no established fire prevention bureau, such as rural areas. The state fire marshal may delegate authority to the local jurisdiction when there is an established fire prevention bureau, usually in county, city, or district fire departments. The state fire marshal can delegate the authority to perform these inspections but still retains the responsibility to see that they are completed.

On the local level are the local fire departments and local codes and ordinances. Most jurisdictions adopt model codes through ordinance. These codes may be adopted in part or in whole, as the local jurisdiction wishes and finds politically acceptable. Adoption of the fire prevention code usually designates the fire chief as the primary enforcement authority. The chief then delegates that authority to the fire marshal and inspectors.

Adopting a model code does not mean that the jurisdiction cannot establish ordinances that are not addressed in the model code to deal with particular local problems. An example of this would be a weed ordinance in areas where dry vegetation is a summertime fire hazard. Another example would be regulations regarding high-rise buildings as there may not be language in the adopted model fire regulations that refer to this type of building. One area that is gaining in acceptance is a residential fire sprinkler ordinance. Another common local ordinance is one that specifies hydrant spacing in the jurisdiction. It is the responsibility of the local governing body to enact the required legislation, through ordinance, to make these requirements law.

Relationship of Federal, State, and Local Regulations

When discussing the relationship of federal, state, and local regulations, the subject of jurisdiction comes into play. The jurisdiction is the limits of territory in which authority is exercised. Territory is not strictly geographical, however; it can also apply to situations.

When several levels of laws are encountered, usually the most stringent law is enforced. This is not always the case, especially when different levels of government jurisdiction are involved. A fire prevention person for the forest service would enforce federal regulations on federal lands. He or she would not perform a prevention inspection in a business in an incorporated city. Conversely, a local fire department would not have jurisdiction in a federally owned post office or other federal building. This should not prevent the local fire department from performing preincident planning at federal buildings. If there is a fire, it is the responsibility of the local fire department to respond.

The jurisdictional lines get somewhat blurred in that the local fire department does have jurisdiction over a contract post office operated from a privately owned building, which is common in suburban or rural areas. Another example is a locally protected housing area, on private land, in the middle of a national forest. The fire department and the forest agency should work together on hazard reduction to prevent a structure fire from spreading to the forest and a forest fire from burning structures. Through cooperation, the fire prevention mission can be accomplished within jurisdictional boundaries to satisfy the needs of both parties.

The state fire marshal usually has jurisdiction in state-owned buildings, such as prisons and government office buildings. The state fire marshal may make an agreement that allows the local fire department to enforce state regulations in certain occupancies.

Other governmental agencies are involved as well. The local zoning commission regulates what types of occupancies are allowed and where. It is uncommon for a high-hazard occupancy to be allowed in a residential area. This may be as simple as stopping someone from having a woodworking business in his garage. The local building department is responsible for the enforcement of building codes, plan checks, and the inspection of buildings during construction, whether it is new construction or remodeling. The building department also determines occupancy types allowed in buildings, dependent on construction type. A building that was originally a restaurant may not be allowed to be used as a cabinet shop without major modification. In the average jurisdiction, the building department may be aware of remodeling only when a building permit is requested, whereas the fire department makes periodic inspections of businesses and is likely to discover evidence of major work done or a change of occupancy. When the fire department discovers alterations or changes in occupancy, it should notify the building department and work with that staff to ensure that the facility complies with construction, fire, and life safety regulations. There may be times when the business owner is reluctant or hesitant to comply with the correction notice issued by the fire department. At that point law enforcement personnel may be needed to serve an inspection or administrative warrant or to arrest persons who willfully disregard action required to correct fire code violations.

Some problems are beyond the legal authority of the fire department alone to handle. It is important for fire department personnel to know where their jurisdictional authority begins and ends. The violations found may not be of the fire code, but the building code or other ordinances. In these circumstances, personnel must know how and to which department to make a referral. An example is an abandoned house. It is a health hazard if left open and may be a life hazard if used as a sleeping area by transients. The provisions of the fire code may not provide jurisdiction for fire department action. A referral can be made to the building department and the hazard abated. Another common scenario is that an inspector is out in the station's responsibility area performing weed abatement and notices an accumulation of garbage behind a house or business. If it is not an accumulation of combustibles, it is not within the inspector's jurisdiction to correct it, but by giving a referral to the health department, the nuisance can be cleaned up.

> ### Tip
>
> Some problems are beyond the legal authority of the fire department alone to handle. In these circumstances, personnel must know how and to which department to make a referral.

It is good public relations to have a working knowledge of the responsibilities and jurisdictions of the local public agencies. That way, when someone calls to complain about the neighbor's accumulation of trash in their yard, you can assist in resolving the problem. In cases where the accumulation is an eyesore and not a fire hazard, no direct action can be taken by the fire department. If you can make a referral that handles the problem, the person who complained is going to remember that you, the fire department, were of assistance. You did not just say that there was nothing you could do about it because it was not your jurisdiction. By working together, the fire department and the other public agencies can assist each other and perform their functions to the public's satisfaction.

Model Fire Codes

The Building Officials and Code Administrators (BOCA), the International Conference of Building Officials (ICBO), and the Southern Building Code Congress International (SBCCI) have all developed model fire prevention codes. ICBO publishes the *Uniform Fire Code*, BOCA publishes the *Basic Fire Prevention Code*, and SBCCI publishes the *Standard Fire Prevention Code*. Additionally, the National Fire Protection Association (NFPA) publishes a series of codes and standards known as the National Fire Codes; one of these is *NFPA 1: Uniform Fire Code*.

The three organizations—ICBO, BOCA, and SBCCI—recently teamed together to create a nationwide fire code together with a nationwide building code to go hand in hand with the fire code. This task was undertaken in an effort to create uniformity and ease of use of the fire codes across the nation. These two documents will replace the individual building and fire codes that each of the three model code organizations previously published. The benefit of creating a set of national codes is that building designers and architects can use the same set of regulations in Tennessee, Alaska, or California. In the year 2000, the first editions of the *International Fire Code* and the *International Building Code* were published by the newly formed International Code Council. This new set of codes is quickly gaining in popularity and has been adopted in all or part of 46 states across the country.

The use of a nationally recognized model fire prevention code is usually more desirable than a locally written code because the model codes have been nationally developed and represent a broad spectrum of fire prevention experience. The nationally recognized codes also give building experts, such as architects and engineers, a familiar base from which they can design a structure's built-in fire protection features. The national codes also offer a means to gain formal interpretations of the code's intent if a local inspector does not clearly understand a particular code requirement. Additionally, nationally recognized codes undergo a constant review process with a new, updated edition published every 3 years, which makes them more in line with current fire protection theories and technology.

> ### Tip
>
> Using a nationally recognized model fire prevention code is usually more desirable than using a locally written code.

Perhaps one of the most important factors in adopting one of the model codes is that they are companion codes to the publishing organization's building codes. This fact is very important in that it minimizes the likelihood that there will be conflicting code requirements.

The model fire prevention codes are typically divided into sections or chapters that deal with certain topics of fire protection. Typically, the first chapters deal with administrative items, such as a model ordinance that officials can use to adopt the code. The codes typically then define the authority of the fire official and define the responsibilities of property owners to maintain their premises in a fire-safe condition. One of the more important code sections usually follows next, and this section defines the fire protection terminology used in the code. The following chapters typically deal with the proper installation and maintenance of fire protection features, and lastly they offer specific requirements for the use and protection of equipment, processes, occupancies, and hazardous materials.

One important point is the recognition of the NFPA's National Fire Codes. Many jurisdictions have, by ordinance, adopted the whole set of codes and standards, which has the positive effect of having almost any fire protection problem covered by a code or standard. It has the disadvantage of making the inspector responsible for the knowledge and research capabilities to understand many separate fire protection regulations.

At this point, it is important that you understand the difference between a code and a standard. A code is written in language that mandates what should be done and is written such that it can be enforced as a law. A **standard**, on the other hand, expresses how something should be accomplished and is typically written as a set of recommendations. In other words, the codes stipulate when to install a fire sprinkler system in a particular building; the standards address how to design and install the sprinkler system. When standards are adopted as part of an ordinance, these recommendations have the force of law and may be enforced as such.

When a certain model code is adopted, it is imperative that a certain edition (year) be adopted. It is illegal for the ordinance simply to state, "Adopt the most recently published edition" of a code, as this denies the public its right to due process. The specific edition must be identified in the ordinance. This will give the public the opportunity to comment on, or protest, the enforcement requirements that are identified.

One thing that must be kept in mind is that a building built under an older edition of the adopted code is subject to that code and not necessarily to the latest edition of the adopted code. This often changes when a major renovation is done on the structure. The amount of renovation varies among jurisdictions. As part of that process, the building must be brought up to code to meet the new code. A building that was constructed prior to the adoption of the new code, but is compliant under the code existing at the time, is referred to as "existing nonconforming."

Occupancy Classification

When the building or fire code is to be applied, the first thing that must be determined is the occupancy classification of the building. It is important to select the classification that most accurately fits the use of the building. Most requirements of the code come from this classification. Several examples of occupancies are listed in **TABLE 11-1**.

Many of the occupancy examples have subcategories. These subcategories vary from code to code. An assembly occupancy may be subdivided based on the size of the occupant load or type of use, such as restaurant or theater. Educational occupancies may be subdivided into regular schools and day-care centers. Institutional occupancies may be subdivided into restrained or unrestrained occupants. Residential occupancies may be subdivided based on the number of units or the type of residential setting (i.e., dormitory, board and care, hotel/motel, apartments). Finally, storage occupancies may be subdivided based on the combustibility of contents (fire load) stored in the building. The classification of the occupancy is important for many reasons. The building code requires limits as to the height and area of a building

TABLE 11-1 Sample Occupancy Codes

Occupancy	Letter Designation
Assembly	A
Business	B
Educational	E
Institutional	I
Mercantile	M
Residential	R
Storage	S

Courtesy of International Code Council (2012)

depending on the occupancy classification and the types of construction materials. Generally, the more fire resistive the structure, the larger it is permitted to be. Additional area may also be added if a building is protected throughout by automatic fire sprinklers. It is also important, from a building code standpoint, to determine the occupancy of a building so that the mixed-occupancy fire separations can be determined. For example, a model building code may require only a one-hour fire-rated separation between a business (such as an office) and a mercantile (such as a clothing store) use, whereas it may require a three-hour fire-rated separation between a moderate-hazard storage occupancy and a business occupancy.

Tip

When the building or fire code is to be applied, the first thing that must be determined is the occupancy classification of the building.

Construction Types

Both model building codes and *NFPA 220: Standard on Types of Building Construction*, can be used to determine the type of construction used in a building. This type of construction is typically denoted by a shorthand notation such as Type I, II, III, IV, and V. It may also be followed by a number or letters such as Type IV 2 HR, or Type IV Unprotected **FIGURE 11-2**. These notations and numbers refer to what the building is constructed of and what the hourly ratings of its structural members are (i.e., exterior and interior bearing walls, columns, beams and trusses, floors, and roofs).

The model building codes determine how close buildings of a certain construction type can be from another building or property line. The model building codes and *NFPA 101: Life Safety Code*, determine other important factors, such as the occupant load of a structure, the size of the means of egress, the number of exits, and the travel distance to those exits. For multistory buildings, the codes require certain hourly ratings around stairwells and building shafts, such as trash chutes and elevator shafts. The codes at times mandate the installation of fire sprinkler systems, fire alarm systems, or standpipe systems. They can also be used to determine the need for emergency lighting and exit signs. Additionally, these codes require the adherence to requirements on the types of interior wall and ceiling finish that is permitted in areas of the structure. These items are only a brief overview of complex codes and interpretations that can be up to 1000 pages long.

FIGURE 11-2 A. Unprotected steel construction, which may fail quickly when exposed to fire. B. Unprotected steel construction after fire.

© Jones & Bartlett Learning
Courtesy of Kern County Fire Department

Code Development

Codes are most often developed in response to a disaster. After a major disaster that could have been prevented, there is a public outcry as to why it was not prevented. The fire codes requiring that exits open outward and be kept unlocked were developed after numerous multiple-life-loss fires. Building codes requiring enclosing shafts in **fire-resistive construction** were the result of rapid fire spread in buildings caused by unprotected shafts. Examples of 20th- and 21st-century multiple-life-loss incidents are listed by occupancy type in **TABLE 11-2**.

Codes are developed by recognizing the need for increased public safety. Legislation that sets the groundwork for the process to begin is sponsored and passed. Public input is sought and ordinances are written in answer to the needs and the input. Examples of this process are public right-to-know laws. Right-to-know laws were enacted after it was discovered that many businesses were storing large amounts of hazardous materials on their premises. Many times these businesses were operating close to residential areas and schools. After a few major incidents, the public came to realize the need for disclosure of just what was being stored, manufactured, and used in their neighborhoods. These laws require businesses to disclose, through a business or management plan and hazardous materials inventory, the materials they have on-site, their use, method of storage, and location.

Committees of business personnel, technical experts, and public safety agency personnel were formed to create model legislation. The legislation was designed so that the businesses were required to disclose their materials without requiring them to give away any trade secrets.

TABLE 11-2 Major U.S. Fires Causing Loss of Life (More than 25 Deaths)		
Date	**Location and Occupancy**	**Lives Lost**
December 5, 1876	New York, NY—theater	300
June 30, 1900	Hoboken, NJ—steamship piers	326
January 12, 1903	Boyertown, PA—opera house	170
December 30, 1903	Chicago, IL—theater	602
June 15, 1904	East River, NY—steamship	1030
March 4, 1908	Collinwood, OH—grammar school	175

March 25, 1911	New York, NY—clothing factory	145
April 10, 1917[a]	Eddystone, PA—ammunition company	133
May 15, 1929	Cleveland, OH—clinic	125
April 21, 1930	Columbus, OH—state penitentiary	320
September 8, 1934	New Jersey coast—S.S. Morro Castle	125
March 18, 1937[a]	New London, TX—school	294
April 23, 1940	Natchez, MS—nightclub	207
November 28, 1942	Boston, MA—nightclub	492
July 6, 1944	Hartford, CT—circus tent	168
July 17, 1944[a]	Port Chicago, CA—munitions depot	300
October 20, 1944[a]	Cleveland, OH—gas company	130
June 5, 1946	Chicago, IL—hotel	61
December 7, 1946	Atlanta, GA—hotel	119
March 25, 1947	Centralia, IL—coal company	111
April 16, 1947[a]	Texas City, TX—S.S. Grand Camp	468
December 21, 1951	W. Frankfort, IL—coal company	119
December 1, 1958	Chicago, IL—grade school	95
October 10, 1963	Indianapolis, IN—fairgrounds	74
November 23, 1963	Fitchville Township, OH—nursing home	63
February 7, 1967	Montgomery, AL—restaurant	25
April 5, 1968[a]	Richmond, IN—sporting goods store	41
January 9, 1970	Marietta, OH—convalescent home	32
December 20, 1970	Tucson, AZ—hotel	28
December 30, 1970[a]	Hyden, KY—coal mine	38
February 3, 1971	Woodbine, GA—chemical plant	25
February 10, 1973	Staten Island, NY—gas storage tank	40
June 24, 1973	New Orleans, LA—nightclub	32

(continues)

TABLE 11-2 Major U.S. Fires Causing Loss of Life (More than 25 Deaths)   (continued)

Date	Location and Occupancy	Lives Lost
November 15, 1973	Los Angeles, CA—apartment home	25
June 30, 1974	Port Chester, NY—apartment home	34
March 9, 11, 1976[a]	Oven Fork, KY—coal mine	26
October 24, 1976	Bronx, NY—social club	25
May 28, 1977	Southgate, KY—supper club	165
June 26, 1977	Columbia, TN—county jail	42
December 22, 1977	Westwego, LA—grain elevator	36
April 2, 1979	Farmington, MO—boarding home	25
November 21, 1980	Las Vegas, NV—hotel	85
April 19, 1995[b]	Oklahoma City, OK—federal office building	168
September 11, 2001[b]	Arlington Co., VA—the Pentagon	189
September 11, 2001[b]	New York, NY—World Trade Center	3000+
February 20, 2003	Providence, RI—nightclub	100
December 2, 2016	Oakland, CA - illegal residential/entertainment space	36

[a]Indicates explosion.
[b]Indicates terrorist attack.
Data from National Fire Protection Association. 2003. *Fire Protection Handbook* (pp. 2–13). Quincy, MA: National Fire Protection Association.

Along with the model legislation, model regulations were developed stipulating the storage and handling of the hazardous materials. These regulations were then added to all of the model codes and later adopted by local ordinance across the nation. The development of codes and ordinances is an ongoing process that has shifted with a more proactive stance taken by the parties involved as needs change.

Relationship of Codes to Standards

Codes are written as bodies of regulations that can be adopted in whole or in part by ordinance. Standards are recommendations on how things should be designed or done. They are usually adopted by policy or as part of a memorandum of understanding. The NFPA has many standards that are used in the design of fire apparatus and other equipment. Most fire departments specify that the equipment they are ordering from the manufacturer meets the current applicable standard requirements. Manufacturers, in turn, design their equipment to meet the latest standard and mention this fact when they advertise in magazines and other literature.

When a standard is adopted as part of a memorandum of understanding between a jurisdiction and its employees, the items referred to must meet the standard. An example of this is the adoption of *NFPA 1500: Fire Department Occupational Safety and Health Program*. When the standard is adopted in whole, all equipment bought by the department will meet the standard. It may go as far as to state that all equipment that does not meet the standard will be replaced by a specified date.

Standards are usually adopted as a matter of policy instead of by ordinance. This adoption method recognizes their use without their having the force of law. In the case of many model codes, a body of standards that illustrates the points of the code accompanies them.

Legal Considerations at Emergency Incidents

Scene Management

In many jurisdictions the agency in charge of the emergency scene is determined by law. Some states recognize the public agency with primary investigative authority as the scene manager. In the absence of a representative of the agency with primary investigative authority, the highest ranking member of the public safety agency at scene will be the scene manager. Once the agency with primary investigative authority arrives, the individual from that agency assumes scene management authority.

In the case of traffic accidents in a city, the agency with primary investigative authority is the local police department. On state and federal highways in rural areas, this may be the state highway patrol. When the fire department arrives before law enforcement, they assume scene management responsibility. Responsibility for scene management would then be transferred to the agency with primary investigative authority when a member of that agency arrives on the scene. As a matter of practicality, when the incident is primarily based on functions provided by the fire department, such as a vehicle fire, hazardous materials incident, or vehicle rescue, the law enforcement agency may relinquish command of the incident to the fire department. Another option is for the fire and law enforcement persons in charge (incident commanders) to participate in unified command of the incident as outlined in Chapter 13, *Emergency Incident Management*.

It is important to know and establish who has scene management authority, because that agency is the lead agency and ultimately responsible for the incident. In the case of fires, when the fire department has an arson unit, it is clearly the agency with primary investigative authority.

> **Tip**
>
> It is important to know and establish who has scene management authority because that agency is the lead agency ultimately responsible for the incident.

Personnel Safety

29 CFR 1910 Occupational Safety and Health Standards: Operating in Immediately Dangerous to Life and Health (IDLH) Atmospheres

On May 1, 1995, the Occupational Safety and Health Administration (OSHA) issued compliance instructions for states recognizing OSHA standards, to ensure uniform enforcement standards for all workers operating in atmospheres that are immediately dangerous to life and health (IDLH). This instruction further addressed the regulations requiring respiratory protection first issued in 1971. Title 29 CFR Part 1910 requires that fire fighters involved in interior structural firefighting operations be provided with and use self-contained breathing apparatus (SCBA). The regulation goes on to specify that all fire fighters involved in interior firefighting and utilizing SCBA shall operate in a buddy system with two or more personnel, commonly referred to as "two-in, two-out" **FIGURE 11-3**. The fire fighters operating in the buddy system shall be in direct voice or visual contact or tethered with a signal line. Radios and other means of electronic contact shall not be substituted for direct visual contact for employees within the individual team in the danger area. Identically equipped and trained fire fighters are required to be present outside the hazard area prior to a team entering and during the team's work in the hazard area in order to account for and be available to assist or rescue members of the team working in the hazard area.

A minimum of four individuals are required: two individuals working as a team in the hazard area and two individuals present outside the hazard area for assistance or rescue at emergency operations where entry into the danger area is required. One of these outside

FIGURE 11-3 Fire fighters operating in an IDLH atmosphere.
Courtesy of Kern County Fire Department

fire fighters may be assigned to other duties such as incident command; the other can have no responsibilities other than to account for the team inside.

What this means to you as a fire fighter is that you are not advised to make an interior attack on a structure fire with only two or three trained fire fighters at the scene. Every reasonable attempt must be made to provide a minimum of four personnel at the scene. The two who remain outside must be identically equipped; this means with full personal protective equipment (PPE) and SCBA. One of those remaining outside must stay in direct contact with the entry team. The other may have other duties but must stay available to assist if necessary. The total number of personnel required at the scene increases to five when one is assigned to operate the pumper, one is assigned to incident command, one is the contact person, and two are inside. This is not a problem for personnel working for larger departments with four-person engine companies and a command officer in close proximity. In rural areas or on volunteer fire companies, it is entirely possible to have two or three personnel at the scene with additional assistance 15 minutes or more away.

The regulation does allow entry with fewer than four personnel at scene if a rescue is needed. Another situation would be that the fire is small and you feel that you could extinguish it quickly with an interior attack. In these situations, you are faced with a dilemma. Do you obey the regulation as written and make an indirect attack or do you go in and attack the seat of the fire? The only way that you can literally stay within the regulation and still do the job that needs to be done is to change the atmosphere prior to entry. Your first operation may be to open the window to the fire area, make an indirect attack, and knock down the body of the fire. Then ventilate the structure to remove the "atmosphere immediately hazardous to life and health." This could be done by placing a power fan in the doorway and proceeding as soon as the smoke clears.

In many instances laws that do not consider every possible scenario are enacted. The 29 CFR 1910 regulation was not issued with the intent to limit firefighting operations, but it still directly affects fire fighters.

23 CFR Part 634: Fire Fighter High-Visibility Safety Apparel

In an effort to protect the safety of fire fighters, emergency medical responders, and other workers who work in the right-of-way of federally funded highways, the Federal Highway Administration has initiated title 23 *Code of Federal Regulations* part 634. The regulation states that:

All workers within the right-of-way of a Federal-aid highway who are exposed either to traffic (vehicles using the highway for purposes of travel) or to construction equipment within the work area shall wear high-visibility safety apparel. Fire fighters or other emergency responders working within the right-of-way of a Federal-aid highway and engaged in emergency operations that directly expose them to flame, fire, heat, and/or hazardous materials may wear retroreflective turn-out gear that is specified and regulated by other organizations, such as the National Fire Protection Association. Fire fighters or other emergency responders working within the right-of-way of a Federal-aid highway and engaged in any other types of operations shall wear high-visibility safety apparel*.

Definitions (634.2) within Part 634 cover what is meant by *workers* and *high-visibility safety apparel* **FIGURE 11-4**. Any exceptions for emergency responders are incorporated in the definition of *workers*:

Workers means people on foot whose duties place them within the right-of-way of a Federal-aid highway, such as highway construction and

FIGURE 11-4 Fire fighter wearing reflective vest over firefighting PPE to increase visibility when working a roadside incident.
© Jones & Bartlett Learning

* Federal Register/23 CFR section 634.3.

maintenance forces; survey crews; utility crews; responders to incidents within the highway right-of-way; Fire fighters and other emergency responders when they are not directly exposed to flame, fire, heat, and/or hazardous materials*.

Structural firefighting PPE is not considered to be highly visible safety apparel as it does not meet the standard as specified in ANSI/ISEA 207, *Standard for High-Visibility Public Safety Vests.*

> ### SAFETY TIP
>
> Traffic safety vests should not be worn over fire fighter PPE when there is a danger of flame, fire, heat, and/or hazardous materials contact.

NFPA 1901: Standard for Automotive Fire Apparatus applies to all fire apparatus contracted for on or after January 1, 2009. The standard requires "one traffic safety vest for each seating position, each vest to comply with ANSI/ISEA 207, *Standard for High-Visibility Public Safety Vests*, and have a five-point breakaway feature that includes two at the shoulders, two at the sides and one at the front" (2009, § 5.8.3)†.

> ### SAFETY TIP
>
> Traffic safety vests must be equipped with a five-point break-away feature to prevent fire fighters becoming entangled in passing vehicles or other hazards.

Vehicle Markings

NFPA 1901 requires the marking of the rear of apparatus with **retroreflective** lettering and striping. The striping is to be displayed in a chevron pattern with 6-in.-wide stripes, sloping downward from the center point and covering at least 50 percent of the rear of the vehicle **FIGURE 11-5**. The colors for the striping must be contrasting and are allowed to be red, fluorescent lime, fluorescent yellow, or yellow in color. Additional retroreflective material may be used on the tail flap of the hose bed cover to enhance the visibility of the vehicle.

FIGURE 11-5 Fire engine with approved retroreflective striping on the rear panels.
© Jones & Bartlett Learning. Photographed by Glen E. Ellman

Operation of Emergency Vehicles

In 1988, the federal government passed legislation that required all operators of vehicles over 26,001 pounds gross vehicle weight (GVW) or towing trailers over 10,000 pounds to have a Class B driver's license (Commercial Motor Vehicle Safety Act of 1986). A normal automobile operator's license is a Class C. This legislation affected fire departments in that many fire vehicles weigh over 26,001 pounds. The Class B license requires a valid medical card that must be renewed every 2 years. A medical card requires a physical examination by a medical doctor. The cost of having to send all personnel for a medical exam every 2 years was seen as being prohibitively expensive for fire departments. After public agency lobbying of state legislators, an exception was granted in many states. The exception relieves fire fighters of having to see a doctor every 2 years for the driver's license medical exam. In some states, fire fighters can complete a medical questionnaire instead. Other states exempt fire fighters from all Class B requirements.

When members of the fire department are operating emergency vehicles on a nonemergency basis, they are subject to all of the same traffic laws as the general public. When the vehicles are being operated as authorized emergency vehicles and are responding to emergencies, the laws specify certain exemptions. Usually, to be considered an authorized emergency vehicle, the vehicle must be equipped with a minimum of a siren and one steady burning red light to the front.

---

* 23 CFR 634.2.

† Reproduced with permission from NFPA 1901–2009, Automotive Fire Apparatus, Copyright© 2008, *National Fire Protection Association*. This reprinted material is not the complete and official position of the NFPA on the referenced subject, which is represented only by the standard in its entirety.

Most state vehicle codes state that upon the approach of an authorized emergency vehicle displaying the required emergency warning devices, the driver of every other vehicle shall yield the right-of-way and drive to the right-hand edge or curb of the highway clear of any intersection and shall stop and remain stopped until the authorized emergency vehicle has passed. When approaching an emergency vehicle parked at the side of the road with emergency lights flashing, drivers must slow down and move over to the outside lane, proceeding with caution until safely past the emergency scene. All pedestrians shall proceed to the nearest curb or place of safety and remain there until the authorized emergency vehicle has passed.

One important fact is that the vehicle must be an authorized emergency vehicle. The fact that the operator is a fire fighter does not create the assumption that the vehicle is authorized. This comes into question when volunteer fire fighters respond to the station or the incident scene from home or work in their personal vehicles. They are only authorized to respond as emergency vehicles when properly equipped and doing so is allowed under the state vehicle code.

No state's vehicle laws relieve persons operating authorized emergency vehicles from the duty to drive with due regard for the safety of all persons and property. In effect, no matter how many red lights you display and how loud your siren, you are still required, by law, to drive carefully and defensively. In many departments, the policy is that you never exceed the speed limit by more than 5 mph and that you come to a complete stop before proceeding for all red lights and stop signs. Many times at red lights at a cross street, the car in the lane nearest you will see you and stop, but the car in the farther lane will not see you and will not stop. When caution is not exercised by the fire fighter driving the fire apparatus, they may be involved in a collision.

> ### Tip
> No state's vehicle laws relieve persons operating authorized emergency vehicles from the duty to drive with due regard for the safety of all persons and property.

Some fire fighters assume that the display of red lights and siren gives them the right to drive as fast as they want at any time they please. This leads to what is called "sirencide." If there is a traffic accident, they are going to have to prove that they were operating the vehicle in a safe manner and that they were responding to a true emergency. Responding to get a cat out of a tree is not considered an emergency.

Good Samaritan Laws

Numerous states have so-called Good Samaritan laws, which state that a person who voluntarily assists an injured person is not chargeable with responsibility for any errors or omissions in the care provided. This applies only if one is acting within the scope of one's training. If you were to perform an emergency appendectomy on someone and you were only trained to the level of an EMT, you would be way outside the scope of your training and could be held liable for injury to the victim. The line becomes more finely drawn if you were to give CPR to a victim and you did not clear her airway properly before you started. Suppose you tried to ventilate her and forced a piece of foreign material down into her throat that prevented her from receiving any air. This would render the CPR ineffective and prove you did not perform it properly. You would have been acting within the scope of your training by trying to perform CPR but would not have performed it correctly. You could be held liable to some extent.

You could also run into problems if you ask a person stopped at the scene of the traffic accident to assist you. When you do this without ascertaining his level of medical training, you should be very careful about what you ask him to do. You could be held liable if he were to take some action that injured the victim or himself because he would have been acting at your request. It pays to be careful and think before you act.

Infectious Disease

An issue that has become more prominent in the workplace is HIV/AIDS and other infectious diseases. This issue involves the people fire fighters encounter on emergency incidents, as well as fire fighters themselves. The Federal Rehabilitation Act of 1973 prohibits discrimination because of a handicap. The act applies to federal agencies and any organization receiving federal contracts in employment or the provision of services. In addition to covering people who are actually disabled, the act prohibits discrimination against those who have a history of a handicap and those perceived as having a handicap (*School Board of Nassau County v. Arline*).

A series of cases have held that persons with HIV are protected by the act. In 1987 the United States Supreme Court held that persons with contagious diseases were handicapped within the meaning of federal physical handicap statutes. In 1988, the Ninth Circuit

Court of Appeals (*Chalk v. U.S. District Court*), found AIDS to be a contagious disease within the meaning of *Arline* and therefore a handicap within the meaning of the Federal Rehabilitation Act.

Law prohibits discrimination in employment because of race, religion, color, national origin, ancestry, physical handicap, some medical conditions, marital status, or sex of any person.

Agencies have issued opinions that emergency medical care personnel have a responsibility to provide emergency medical care to a victim when responding to emergency and rescue incidents. Emergency care personnel cannot wait for mechanical breathing devices to start CPR, as this may further endanger the life of the victim; therefore many departments have issued each fire fighter a small shield device to be used when performing mouth-to-mouth resuscitation on victims. These devices need to be carried at all times while you are on duty. When you are in the grocery store, in your uniform, buying food for the shift, and someone has a heart attack is no time to realize that you are in violation of department policy by not having your shield with you.

In general, if you feel you have been exposed to an infectious disease, you can report it to the receiving hospital and it will notify the health officer. One problem lies in the fact that the person you were treating does not have to undergo a test for HIV/AIDS or other communicable disease if she or he does not want to. Should the person choose to be tested, under confidentiality laws, the hospital may not be able to advise you of the results unless the victim agrees.

In many cases it is against the law to disclose to others the information that someone is HIV positive or has AIDS. If you worked at a medical aid incident and the victim told you she was HIV positive, that information is to be kept confidential. Writing the information on the board at the fire station or in any way advising the other shifts is ill advised. The absolute best way to avoid these problems is to wear your PPE every time you deal with victims, with no exceptions. Assume that every victim contact is a possible exposure and protect yourself appropriately.

Being able to get test results back from a victim from whom you suffered an exposure is no consolation. If he or she is HIV positive or has AIDS, finding out after exposure is probably too late to do you much good.

Local jurisdictions have enacted AIDS-specific discrimination laws. These laws vary in their content and scope, and it is important that you become aware of them.

SAFETY TIP

The absolute best way to avoid exposure to an infectious disease is to wear your PPE every time you deal with victims, with no exceptions.

Health Insurance Portability and Accountability Act

The Health Insurance Portability and Accountability Act (HIPAA) affects fire fighters in their jobs due to their response to rescues and medical aid incidents. HIPAA states that health information regarding a victim can be given only to someone directly involved in the treatment of the victim. In a practical sense, if you are at the scene of a medical emergency, you are allowed to provide information regarding the victim's status only to the ambulance personnel or other medical care provider directly involved. This precludes you from providing medical information to the news media, law enforcement, other personnel at scene not directly involved with the treatment of the victim, or bystanders. It also covers any victim care report or station logbook information that you may gather; these records must be kept confidential.

Wrap-Up

CHAPTER SUMMARY

- Codes and ordinances fall under the broad description of laws. Laws are written and adopted on all three of the levels of government: federal, state, and local.
- All legal authority for governmental action in the United States comes from the Constitution.
- In the case of fire departments, the meaning of jurisdiction is the limits of territory within which their authority may be exercised.
- The way to avoid lawsuits is to do your job correctly every time. You must also take the time to document, through the use of reports, what you did.

- The department should make sure that no adverse consequences occur to any person or witness as a result of having brought a truthful complaint or provided truthful information in any investigation of a complaint.

- The federal government and the courts hold management responsible for harassment in the workplace. Consequently, a clear understanding of what constitutes harassment is essential to the development of a harassment-free work environment.

- The fire prevention bureau has the legal responsibility and authority to enforce fire-related codes and ordinances.

- When discussing the relationship of federal, state, and local regulations, the subject of jurisdiction comes into play.

- The benefit of creating a set of national codes is that building designers and architects can use the same set of regulations anywhere in the country.

- Codes are most often developed in response to a disaster. After a major disaster that could have been prevented, there is a public outcry as to why it was not prevented.

- Codes are written as bodies of regulations that can be adopted in whole or in part by ordinance. Standards are recommendations on how things should be designed or done.

- In many jurisdictions the agency in charge of the emergency scene is determined by law. Some states recognize the public agency with primary investigative authority as the scene manager.

- Standards have been put in place to promote personnel safety at incidents. Among those are the following:
  - 29 CFR 1910: Operating in IDLH Atmospheres
  - 23 CFR 634: Fire Fighter High-Visibility Safety Apparel
  - *NFPA 1901: Standard for Automotive Fire Apparatus*

- When members of the fire department are operating emergency vehicles on a nonemergency basis, they are subject to all of the same traffic laws as the general public. When the vehicles are being operated as authorized emergency vehicles and are responding to emergencies, the laws specify certain exemptions.

- Numerous states have so-called Good Samaritan laws, which state that a person who voluntarily assists an injured person is not chargeable with responsibility for any errors or omissions in the care provided.

- An issue that has become more prominent in the workplace is infectious diseases. This issue involves the people fire fighters encounter on emergency incidents, as well as fire fighters themselves.

KEY TERMS

Civil liability The accountability of an individual under civil law.

Codes A law that can be established by legislative action, but is most commonly created by an administrative agency or a local entity.

Crime An unlawful act as defined in the criminal codes.

Felony A serious crime, such as murder, arson, or rape, for which the punishment is either imprisonment in a state prison for more than 1 year or death.

Fire-resistive construction Construction that has been designed to resist the effects of heat from fire.

Good Samaritan laws Laws stating that a person who voluntarily assists an injured person is not chargeable with responsibility for any errors or omissions in the care provided.

Harassment Coercive or repeated, unsolicited, and unwelcome verbal comments, gestures, or physical contacts, including retaliation for confrontation or reporting harassment.

Malfeasance Dishonest, intentionally illegal, or immoral action.

Misdemeanor A crime punishable by up to 1 year in a county jail or by a fine usually not to exceed $1000, or both.

Misfeasance Mistaken, careless, or inadvertent action that results in a violation of law.

Nonfeasance A failure to act when action is required.

One-hour fire-rated separation A fire-rated assembly that should resist breakthrough for a period of 1 hour. An example of this type of construction is the use of 5/8-in.-thick fire-rated gypsum (15.875 mm) wallboard or a combination of wallboard and plaster. All of the electrical boxes must be metal and not plastic.

Any penetrations through the assembly must be properly protected to prevent the spread of fire.

Perjury False statements in a sworn document or testimony.

Regulations Rules designed to implement a statute based on an agency's interpretation of that statute.

Retroreflective A surface, material, or device (retroreflector) that reflects light or other radiation back to its source; reflective.

Scope of employment The complete range of activities an employee might reasonably be expected to perform while carrying out the business of the employer.

Standard A document, the main text of which contains only mandatory provisions and is in a form suitable for mandatory reference by another standard or code or adoption into law.

Statutory laws Laws adopted by Congress (federal statutes) and those that have been passed by state legislatures (state statutes).

Tort A civil wrong leading to a legal claim for damages.

CASE STUDY

Every year wildland and urban fire fighters are put at tremendous risk and sometimes die, due to wildland fires in the wildland-urban interface. Examples include:

- The Cedar Fire in 2003 in which a fire fighter died protecting a structure
- The Esperanza Fire in 2006 in which five U.S. Forest Service fire fighters died protecting a structure
- The Yarnell Fire in 2013 in which 19 Prescott, Arizona fire fighters died protecting a town from wildland fire

One of the model fire codes that addresses this issue is the wildland–urban interface (WUI) code. The WUI code takes a two-pronged approach to protect buildings from fire: (1) Remove flammable materials from around the building, and (2) construct the building of fire-resistant material. The law requires that homeowners do fuel modification to 100 ft (or the property line) around their buildings to create a defensible space for fire fighters and to protect their homes from wildland fires (Land Use Solutions for Colorado).

New building codes protect buildings from being ignited by flying embers, which can travel as much as a mile away from the wildfire. Ignition-resistant standards are designed to prevent embers from igniting a building.

1. The WUI code is a(n):

   A. demonstration code.
   B. example code.
   C. exhibit code.
   D. model code.

2. The WUI code, directly or indirectly, protects all except:

   A. homes.
   B. fire fighters.
   C. power poles.
   D. homeowners.

3. The WUI code requires fuel modification up to:

   A. 10 ft.
   B. 30 ft.
   C. 50 ft.
   D. 100 ft.

4. The ignition (fire) resistant standards refer to:

   A. grass and brush.
   B. building materials.
   C. roadways.
   D. ingress and egress.

REVIEW QUESTIONS

1. What is the supreme law of the United States with which no other laws must conflict?

2. May a state law be different from a federal law? Explain.

3. If a postal vehicle, fire engine, and private vehicle arrived at an intersection at the same time, who would have the right of way? Use the relationship of the levels of law as your guide.

4. A failure to act is considered which type of failure?

5. What court case determines the responsibility of a fire prevention bureau to enter premises?

6. How are model codes developed? Who may suggest changes to model codes? How are model codes adopted at the local level?

7. What type of occupancy is a school?

8. What is meant by the term *one-hour fire-rated separation*?

9. Why is the location of a building in relation to the property line important from the standpoint of stopping fire spread?

10. What is meant by a building being classified as an A occupancy?

11. What are your responsibilities when responding to an emergency with red lights and siren activated?

12. Is it legal to refuse treatment to a person with HIV who is bleeding when you are not equipped with full medical PPE? Justify your answer.

13. If you act within the scope of your training at an accident scene while you are off duty and the victim dies, can you be held liable? Why, or why not?

14. Under OSHA standards, how many persons must be at a scene before interior firefighting can be considered? What factors affect this decision?

15. What are the PPE requirements for operating on federal-aid highways? What are the exceptions?

DISCUSSION QUESTIONS

1. Why are codes often not enacted until after a major incident?

2. Should fire fighters be exempt from liability?

3. Do you think that the wearing of high-visibility vests and the installation of retroreflective striping on the rear of emergency vehicles will greatly enhance responder safety on the highways?

REFERENCES AND ADDITIONAL RESOURCES

Building Officials and Code Administrators. 2000. *Basic Fire Prevention Code*. Country Club Hills, IL: Building Officials and Code Administrators.

Chalk v. U.S. District Court, 840 F2d. 701 (9th Cir. 1988).

CNN Library. 2013. *Nightclub Fires Fast Facts*. Atlanta, GA: CNN. www.cnn.com/2013/09/12/world/nightclub-fires-fast-facts

Commercial Motor Vehicle Safety Act of 1986, 49 USC Chapter 311 (1986).

Darlington, Shasta, Marilia Brochetto, and Dana Ford. 2013. "Fire Rips through Crowded Brazil Nightclub, Killing 233." *CNN*, January 28. http://www.cnn.com/2013/01/27/world/americas/brazil-nightclub-fire/

Federal Highway Administration docket No. FHWA-2008-0157, 23 C.F.R. 634 (2008).

Federal Rehabilitation Act of 1973, 29 U.S.C. § 701 (1973).

Health Insurance Portability and Accountability Act of 1996, Pub.L. 104-191 (1996).

International Code Council. 2012. *International Fire Code*. Country Club Hills, IL: International Code Council, Inc.

International Code Council. 2012. *International Building Code*. Country Club Hills, IL: International Code Council, Inc.

International Conference of Building Officials. 2012. *Fire Code*. Whittier, CA: International Conference of Building Officials.

Land Use Solutions for Colorado. 2019. *Wildland-Urban Interface Code (WUI code) Planning for Hazards*. https://planningfor-hazards.com/wildland-urban-interface-code-wui-code

National Fire Protection Association. 2003. *Fire Protection Handbook*. Quincy, MA: National Fire Protection Association.

National Fire Protection Association. *NFPA 1: Uniform Fire Code*. Quincy, MA: National Fire Protection Association.

National Fire Protection Association. *NFPA 101: Life Safety Code*. Quincy, MA: National Fire Protection Association.

National Fire Protection Association. *NFPA 220: Standard on Types of Building Construction*. Quincy, MA: National Fire Protection Association.

National Fire Protection Association. *NFPA 1500: Fire Department Occupational Safety and Health Program*. Quincy, MA: National Fire Protection Association.

National Fire Protection Association. *NFPA 1901: Standard for Automotive Fire Apparatus*. Quincy, MA: National Fire Protection Association.

Nicol, Susan. 2013. "RI Responders Balk at Seat Belt Repeal Measure." *Firehouse*. http://www.firehouse.com/news/10978791/ri-responders-balk-at-seat-belt-repeal-measure

NIOSH. 2008. "Fire Fighter Dies after Being Ejected from a Pumper in a Single Vehicle Rollover Crash—New York." *NIOSH*. http://www.cdc.gov/niosh/fire/reports/face200825.html

NIOSH. 2013. "Volunteer Fire Fighter Dies after Being Ejected from Front Seat of Engine—Virginia." *NIOSH*. http://www.cdc.gov/niosh/fire/reports/face201223.html

School Board of Nassau County v. Arline, 480 U.S. 273 (1987).

See v. City of Seattle, 387 vl. 541, 87 S. Ct 1737 (1967).

Southern Building Code Congress International. 1999. *Standard Fire Prevention Code*. Birmingham, AL: Southern Building Code Congress International.

CHAPTER 12

Fire Protection Systems and Equipment

OBJECTIVES

After studying this chapter, you should be able to:

- Describe the purpose and components of public and private water companies.
- Discuss the importance of a dependable water supply system.
- Describe the components and importance of a fire department water supply program.
- Describe fire detection systems and their components.
- Describe the different types of extinguishing agents.
- Describe different types of extinguishing systems and their components.

Case Study

The City of Scottsdale, Arizona, is widely recognized as a leader in built-in automatic sprinkler systems. In 1985, the city passed an ordinance requiring every commercial and multifamily building to be outfitted with a complete fire sprinkler system. The ordinance also requires that single-family residences built after January 1, 1986, be fully outfitted with an approved fire sprinkler system. Sprinkler systems are also required in major remodeling projects. These systems have been credited with saving a number of lives locally and across the nation.

Sprinkler facts:

- Sprinkler systems have been used in industrial protection for more than 100 years.
- Only the heads that have been activated by heat, release water.
- Water damage from sprinkler discharge will be much less than from fire hoses used to control an interior fire.

- Smoke alarms provide early alert of a fire, but do not control the fire.
- The National Fire Protection Association (NFPA) has no record of a multi-fatality fire in a fully sprinkler-equipped occupancy where the system was operating properly.
- Home sprinklers add 1 percent to the cost of new construction in a home.
- Residential sprinklers are designed to fit in with home décor and are barely visible.

1. With fire sprinklers so effective and inexpensive, why are they not required in all homes?
2. Is there an ordinance requiring single-family residential fire sprinklers in your area? Why or why not?
3. If the answer to the previous question is no, at what square footage are fire sprinklers required in a structure?

City of Scottsdale, AZ. 2013. "Fire Sprinkler Systems: Meet Your Personal Firefighter." ScottsdaleAZ.gov. http://www.scottsdaleaz.gov/fire/residentialsprinkler

 JONES & BARTLETT LEARNING NAVIGATE 2 *Access Navigate for more resources.*

Introduction

Almost every large city has experienced a major conflagration fire. Conflagration fires share several common denominators. Delayed alarms were a contributing factor in the great fires of New Orleans (1788) and New York City (1845), in which the night watchman in the city hall bell tower fell asleep. Fire alarm boxes were finally installed in New York City in 1870. In the Great Chicago Fire of 1871, the initial reported location was over a mile from the actual fire location. In addition, the alarm boxes in the fire area were not activated.

Water is the most common extinguishing agent used for combating fires. A lack of water supply and equipment contributed to the size of these fires and many others. This deficiency led to the improvement and installation of water systems with less reliance on surface water, such as rivers that were subject to freezing over. Over the years water systems have been developed to the point where they have become dependable, making water readily available at many fire scenes.

Fire sprinklers have been shown to be 97 percent effective in controlling the spread of fires. Automatic firefighting devices have been developed to aid in the application of water and other firefighting agents.

Additives have been devised for water to make it as effective as possible.

Water is not the only extinguishing agent available to modern fire fighters. In occupancies or applications where water may cause damage or be ineffective, other extinguishing agents have been developed. These agents come in many different forms, and fire fighters must be acquainted with their uses and applications.

Tip

Water is the most common extinguishing agent used for combating fires, but it is not the only extinguishing agent available to fire fighters. Fire fighters should be acquainted with the uses and applications of other extinguishing agents.

Components of fire protection and life safety systems also include exits and alarms. People are much more likely to escape a fire if a working smoke alarm system is installed. A major contributing factor to loss of life in assembly occupancy fires is a lack of properly functioning exits (e.g., the Beverly Hills Supper Club). The keys to built-in fire protection are adequate exits, fire-resistant construction, properly operating fire suppression and detection systems, and maintenance.

Water Companies

Public Water Companies

Water is so fundamental to firefighting that a good water supply is one of the most important factors in municipal fire protection (Insurance Services Office [ISO] 2013). Close cooperation and communication are necessary between the fire department and the water company to ensure that an adequate water supply is available for firefighting operations. The fire department must keep in mind that the primary reason for a water company to exist is to provide for the everyday needs of its customers.

Water companies are set up in several ways, one of which is under public utility laws, allowing it to act as a monopoly. It would be extremely rare for two competing water companies to provide service to an area under parallel systems. The water company usually has an elected board of directors with a president, board members, and a secretary. These people are from the area the water company serves, much like a school board. The water company is allowed to charge for the water provided at rates set by the public utilities commission. The money collected is used to administrate, maintain, and improve the system. An alternative is a water system owned and operated by the local government.

When new structures, such as large buildings or subdivisions, are in the planning stage, the fire department is often involved in specifying the water system requirements from a fire protection standpoint. This is done in cooperation with the builder, the building department, and the water company. By addressing concerns for fire water supply during the planning phase, serious problems with lack of fire flow can be avoided later.

In some smaller systems, a large fire can severely tax the capabilities of the whole water system. By having a good working relationship with the water company personnel and knowledge of the water system, it may be possible to have the pressure in the system boosted to provide for more fire flow. Likewise it is good practice to let the water company personnel know when the fire is under control so pumps and other equipment are not run needlessly. The water company should also be notified when any testing or flushing of the water system is to take place. Testing and flushing often stir up sediment in the water system and may cause customer complaints to the water company office. Testing and flushing can also cause the pressure to drop and demand on the water system to increase, the same as flowing large amounts at a fire would.

The agreement with the water company should also include that it let the fire department know when the system is undergoing major repairs, when water company personnel are flushing the system, or when individual hydrants are out of service. By knowing and respecting each other's needs, the water company and fire department can maintain a good working relationship.

Private Water Companies

Private water companies also exist, usually in industrial and commercial complexes.

They may store their own water in reservoirs or tanks or receive it from the public water company. The private water system maintains all of its own distribution and storage equipment. The fire department should check on the system periodically to ensure that it is in operating condition.

Water Supply Systems

All water supply systems must first have a storage capability. The size of the storage capacity and adequacy of the system are determined by several factors. The frequency and duration of droughts is a major factor. In the late 1980s and early 1990s, the West Coast experienced drought conditions for 7 years in a row. Water supplies were so depleted that residents in some cities were prohibited from watering their lawns and washing their cars. Many fire departments conducted their firefighting drills without charging hose lines to conserve water.

A second factor is the danger to the system from natural disaster. Earthquakes can sever supply lines from reservoirs to pumping plants and individual mains, leaving whole areas of cities without water supply for days at a time. This happened during the earthquakes in San Francisco in 1906 and 1989. Fortunately, in 1989, the fire department was equipped with fire boats that could pump water from the bay to engines on shore. Floods can affect water systems by making pumping plants inoperative. Flood waters can ruin the electric motors on the pumps, destroy electrical distribution systems, or wash out reservoirs and water distribution systems. Tornadoes and other natural disasters that disrupt power supplies and destroy the buildings housing the water company equipment can have the same effect.

FIGURE 12-1 Municipal water pumping facility with vertical turbine pumps.
© Jones & Bartlett Learning

FIGURE 12-2 Municipal water storage tower.
© Jones & Bartlett Learning

Water systems take their supplies in a variety of ways. In the gravity system, the water source is at a higher elevation than the city. The water is collected in reservoirs and gravity-fed through pipes into the system **FIGURE 12-1**. This gravity-forced concept is used in applications where water towers are used to keep a constant pressure on the system.

Direct pumping systems are used where a reservoir or river is a water source. The water is pumped straight from the source into the system. Automatic pressure controls are built into the pumps to maintain a consistent pressure on the system.

Combination systems are used where areas of need are removed from the water source, perhaps because of a higher elevation or remote location. The pumps in the system provide water directly to the system as well as pumping water into storage tanks. By filling the tanks at times of low use, the pumps do not have to provide the total supply to the system at times of high use **FIGURE 12-2**.

In many systems, the storage is underground in water-bearing strata called aquifers. In a wet year, the aquifer will have a high water level under the ground, called the water table. In dry years, the water table will recede as water is pumped out. Not all wells will be able to access the full range of the aquifer at any one time. The shallow wells will run dry as the water table recedes. Over the years, wells tend to silt up and become plugged, at the least reducing their efficiency. In a system with only one or two wells, it is easy to see that an equipment failure could cause total loss of water supply to the system.

To prevent loss of water supply, most systems of any size are equipped with duplication of wells, pumps, tanks, and sources **FIGURE 12-3**. This way, if one part

FIGURE 12-3 Vertical turbine well pumps with water storage tank in a rural area.
© Jones & Bartlett Learning

of the system is out of service, the other components can come online and take up the slack. Other safeguards used are gasoline- or diesel-powered pumps or generators to back up electric equipment. Electrical supply lines are laid underground to protect them from weather and vehicle accidents damaging power poles and other equipment.

In countries where terrorism and political unrest are common, two of the first things to be attacked are the electrical and water supply systems. In the United States, there have been very few instances of this type of activity, and the systems are not protected. A power transformer can be taken offline by a single rifle shot. Surrounded by nothing but chain-link fences, these installations are vulnerable to attack, even by vandals.

The adequacy of a water system is gauged by its ability to meet several criteria. The first is the average daily consumption, figured over the last 12 months. The second is the maximum daily consumption, which is the highest demand in a 24-hour period over the last 3 years. If there has not been a major fire in the last 3 years, the system may be adequate for domestic, commercial, and industrial use, but inadequate in the case of a major fire. The peak hourly consumption is the maximum amount of water used in any given hour of a day. The maximum daily consumption is normally about 1.5 times the average daily consumption. The peak hourly rate varies from 2 to 4 times a normal hourly rate. Both maximum daily consumption and peak hourly consumption should be considered to ensure that water supplies and pressure do not reach dangerously low levels during these periods and that adequate water will be available if there is a fire (NFPA 2008).

For firefighting purposes, the minimum recognized water system is 250 gpm for 2 hours (ISO 2013). This is not much in firefighting terms. A structure fire of any great size would require a much higher flow rate than this and quite possibly for a longer time. It is not uncommon to have the fire department applying 4000 gpm of water on a large fire in a warehouse or manufacturing facility.

Distribution System

Once the water is removed from the storage area, it goes through the treatment plant to make it fit for drinking. When it leaves the treatment facility, the water enters the distribution system. The distribution system is made up of underground piping of various sizes. These pipes are called water mains. The largest of these are the primary feeders. The primary feeders are widely spaced and carry the water to the various areas to be distributed by smaller mains. Like the other parts of the system, duplication is important. The primary feeders are often looped or cross connected so that water enters the system from at least two directions, preventing dead ends and pressure drops throughout the grid. Gridding becomes extremely important during times of high demand, such as large fires. If two high-flow pumpers are operating from the same main and water came in from only one way, the

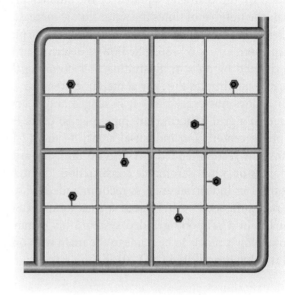

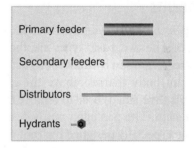

FIGURE 12-4 Gridded water main system.
© Jones & Bartlett Learning

pumper closer to the source would rob most of the water from the main. With gridding, the water flows in from two directions and both pumpers can be supplied to the limits of the system **FIGURE 12-4**.

The water from the primary feeders then flows into intermediate size pipes called secondary feeders. The secondary feeders reinforce the grid within the loops of the primary feeders and concentrate the water supply in high-demand areas.

The piping that serves individual hydrants and blocks of consumers are the distributors. These mains may also be installed in a grid system, which allows them to reinforce each other in times of high demand.

The main sizes most commonly in use in newer systems are 8, 12, and 16 in. It is recommended that main sizes be a minimum of 6 in. to allow for required fire flow (ISO 2013). Should the water mains be smaller in size, the fire department should preplan accordingly to meet the required fire flows of occupancies in the area. An increased main size reduces loss of pressure due to the friction of the water against the inside of the pipe. Increased size also allows for higher flow rates at the same pressure. A simple formula used to illustrate this point is that the diameter squared of the larger pipe divided by the diameter squared of the smaller pipe equals the number of times the water flow is increased. A 12-in. pipe is expressed as $12 \times 12 = 144$. A 6-in. pipe is expressed as $6 \times 6 = 36$. The flow is determined by 144 divided by $36 = 4$, illustrating that doubling the pipe size quadruples the flow at the same pressure.

If it does not exceed 600 ft in length and is connected in a grid pattern, 6-in. pipe may be used. For shopping centers and industrial areas, 8- and 12-in. mains are recommended. In heavily built-up areas of homes or other flammable construction, the main size used in industrial areas is recommended. Valves are recommended to be placed a maximum of 800 ft apart. In a properly gridded system, this distance would allow a repair to be made to the main with only 800 ft being deactivated (ISO 2013).

Fire Hydrants

Several types of fire hydrants are in use today **FIGURE 12-5**. The two basic types are the wet barrel and dry barrel. Common hydrant construction consists of the hydrant body (barrel) above the ground with pentagon nuts (five-sided) used to remove the caps and operate the stem. The pentagon nut is designed to be turned using a specially designed hydrant wrench. The purpose of the special nut is to foil vandals who would turn on hydrants. The nut can be operated with a large pipe wrench if the need arises. The openings, usually 2½, 4, or 4½ in. in size, are situated in a horizontal position. The thread on these outlets is commonly national standard thread. The threads are protected with caps. Most of these caps, when coming from the factory, are brass, making them attractive to persons wanting to sell them for their scrap value. When replacing stolen caps, plastic caps are often used. The only problem with these is that if they are installed too tightly, the nut can twist off when fire fighters try to remove them. Some departments equip their hydrants with quick-connect fittings. It is important for fire fighters to be acquainted with the fittings on any hydrants in their own and neighboring jurisdictions and to make sure their pumpers are equipped with the necessary adapters.

The wet barrel hydrant has water in it at all times. There is sometimes a spring-loaded valve underground that operates and shuts off the hydrant if the aboveground portion is sheared off by accident. When the caps are removed and the valve is opened, the water flows out the opening. With this type of construction, the hydrant can have several openings valved separately, which allows the pumper to be attached to the hydrant with one line. Other lines can be added later without shutting down the hydrant. Not all wet barrel

FIGURE 12-5 A. Wet barrel hydrant common in warm climates. B. Dry barrel hydrant common in climates where subfreezing temperatures are experienced.

© Jones & Bartlett Learning
Courtesy of Jeff Riechmann

hydrants are set up this way. Some types have only one valve. Whichever cap is removed is the opening the water flows from. This can be a problem if the pumper is initially attached to a smaller opening and, as the fire increases in size, it is determined that the larger opening or attaching another line to the hydrant is necessary. One way to deal with this is to always keep the water tank on the pumper full. This will give the operator a period of time, relying on tank water to supply the fire stream, to shut down the hydrant and make the necessary hookups. Depending on the tank size on the pumper and the fire flow being used, the time to perform the hookup may be very short. Pumper operators should practice this operation. This operation can be dangerous for the fire fighters on the nozzles if the lines being used on the fire are supporting an interior attack. Losing the water supply to the hose line while operating on an interior attack is dangerous, to say the least.

The dry barrel hydrant is designed for use in cold climates where freezing is a problem. The water is held back by a valve underground, which is operated by the stem protruding out the top of the hydrant. The hydrant also has a drain several feet below ground level, surrounded by gravel, that allows the water to drain out of the aboveground part of the hydrant once it is turned off. This drain is set up so that water is not forced out of it when the hydrant is being operated. In a dry barrel system it may take a while for the water to rise in the pipe and water to flow out of the hydrant. This is normal but can be disconcerting when the water is needed in a hurry.

Another type of hydrant, not to be confused with the dry barrel type, is the dry hydrant. The dry hydrant is installed at a static water source for ease of setting up drafting operations. The hydrant head is positioned alongside a road at a lake or pond. The pumper hooks up to the dry hydrant and drafts water from the source. The pipe extends out into the water source, under the surface, and has a screen on the end to keep out debris **FIGURE 12-6**. This greatly speeds up the operation because the operator does not need to extend heavy hard suction hoses into the water source (see Figure 6-26). In areas where ice forms on pond surfaces, it is not necessary to chop a hole in the ice to get at the underlying water. A modified version of the dry hydrant is sometimes attached to ground-level water tanks. By drafting from the tank, the operator can increase the flow into the pumper over that provided by gravity.

Hydrants are installed where required and operate off the regular water mains. The piping that extends off the water main to the hydrant is called the **bury** **FIGURE**

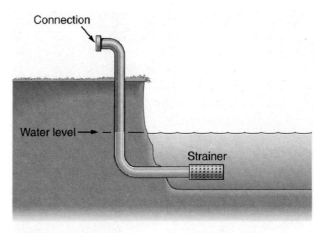

FIGURE 12-6 Dry hydrant suction source for drafting.
© Jones & Bartlett Learning

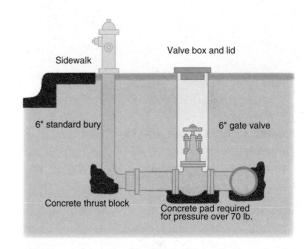

FIGURE 12-7 Schematic of hydrant with underground plumbing.
© Jones & Bartlett Learning

12-7. This pipe is of a specified size determined by the type of hydrant it supplies. A concrete **thrust block** is often installed where the elbow is installed in the bury. A valve installed between the hydrant and the main allows the hydrant to be turned off for removal or maintenance. This also comes in very handy when the hydrant is knocked off in an accident and needs to be shut down. Hydrants should be installed with the openings far enough above the ground to allow easy hookup, leaving enough room to swing the handle of the hydrant wrench when removing the caps or opening valves. The openings should be pointed toward or parallel to the road for ease of hookup. Fences and walls should not be allowed to encroach on the area around the hydrant where they would interfere with its operation. There should be enough room between the hydrant and the street that it is not in danger of being hit by vehicles. A common way of protecting hydrants is to erect vertical pipe barriers (bollards) around them.

FIGURE 12-8 Fire fighter servicing a hydrant.
© Jones & Bartlett Learning. Photographed by Glen E. Ellman

- Operating the stem of the hydrant to check for proper operation.
- Lubricating and cleaning any parts as necessary.
- Clearing weeds and obstructions from the area of the hydrant to provide for ease of location and access under poor visibility conditions, such as fog or darkness.

> **Tip**
>
> Part of a complete prefire program is an annual or semiannual hydrant inspection conducted by fire department personnel in their first-in district.

Many departments paint the hydrant barrels a highly visible color, such as yellow. Another way of identifying hydrant location is to affix reflective roadway markers in the street; blue is often used to distinguish them from lane markers. In areas where snow is a problem, poles are erected by the hydrant to mark their location. Any damage or repairs needed should be brought to the attention of the responsible party to place the hydrant back into serviceable condition.

The situation may arise where it is necessary to flush certain hydrants in a water system. This operation is usually performed by the water company. If the fire department wishes to flush several hydrants, the water company should be notified. When flushing the hydrants, it is important to open and close them slowly to prevent damage to the water main. If at any time a valve through which water is flowing is closed quickly, **water hammer** can occur. This condition is caused by a large volume of water flowing from an opening being shut down too rapidly. The large volume of water moving through the opening has great momentum. If it is suddenly stopped, the momentum of the water will cause it to send a shock wave back through the system, with resultant damage. The same rule applies when operating pumpers at hydrants. All nozzles and discharge valves should be opened and closed slowly to avoid water hammer. This can be demonstrated with a common garden hose and nozzle. Open the nozzle and let the water flow. Let go of the nozzle trigger, stopping the flow, and the hose will jump. On such a small scale no real damage is done. If the flow were 1000 gpm and you did the same thing, the forces at work would be greatly multiplied.

Another problem to watch out for is causing destruction to the roadway or causing a traffic accident. A hydrant flowing 1000 gpm at 40 psi moves

On airport or other special property, hydrants are installed as the situation dictates. At some airports, the hydrants are under the ground. This allows them to be available on the runway aprons without being a hazard to aircraft. Hydrants are also commonly installed on piers extending out into the ocean to provide firefighting water supply.

Hydrant spacing is specified by local ordinance. The purpose is to concentrate availability of fire flow at the blocks or groups of buildings to be protected. Some standard rules of thumb for spacing are 250 ft in compact mercantile and manufacturing districts and 500 ft in residential districts. The ISO requires maximum spacing of 330 ft in commercial and industrial districts and 660 ft in residential areas.

Part of a complete prefire program is an annual or semiannual hydrant inspection conducted by fire department personnel in their first-in district. This program acquaints the fire fighters with hydrant locations and provides required maintenance **FIGURE 12-8**. The maintenance should include:

- Removing all the caps.
- Inspecting the condition of the outlet threads.
- Checking the hydrant barrel for foreign objects.
- Replacing missing caps as necessary.
- Replacing worn or missing cap gaskets.

approximately 4 tons of water a minute with tremendous force. This force can easily undermine a roadway or cave in the door of a passing car, possibly even causing the driver to lose control.

SAFETY TIP

If at any time a valve through which water is flowing is closed quickly, water hammer can occur. Always open and close valves slowly to avoid damage to water systems.

Hydrant testing is done on new systems to test the flow rates **FIGURE 12-9**. It is also done periodically on older systems to ensure the system is still performing well. The same precautions should be taken as when flushing hydrants. The test may consist of opening one hydrant to test its individual flow or opening several at once to test the ability of the water system to perform under high-demand conditions. Standard hydrant testing is performed using two hydrants: a pressure hydrant and a test hydrant. The test process consists of several steps and can be found online.

Hydrant Painting

Hydrants are painted for visibility and because their barrels are often made from cast iron. Keeping them covered with a good coat of paint prevents corrosion of the hydrant. Hydrants are often color-coded to identify their flow capabilities. The NFPA has developed a commonly used color coding system. When the NFPA system is used, hydrants capable of flowing 1000 gpm or more have their caps and bonnet painted green. Hydrants flowing 500–999 gpm have the caps painted orange. Hydrants with a flow capacity of 499 gpm or less have their caps painted red. An additional marking system sometimes used is to color-code hydrants installed on dead-end mains with at least one cap painted black. Hydrants that are out of service may have the bonnet painted white **FIGURE 12-10**.

A word of caution here is that not all fire departments follow this color coding scheme, so local knowledge of the color coding system used is a must. Hydrants that are part of a private system located on a public street may be painted all one color to identify them. Private hydrants on private property are painted the color the owner wishes. The use of color coding allows the pumper operator to make a quick decision as to which hydrant to use if several are present and gives the operator some idea of the capabilities of the water supply at hand. When the color codes are included on the hydrant maps and prefire plans carried in the pumper, it is easier to pick the location of the high-flow hydrants before arriving at the

FIGURE 12-9 Hydrant flow testing equipment, including static pressure gauge mounted on a hydrant cap, a pitot gauge for testing flow pressure, and an adjustable hydrant wrench for opening hydrant caps and hydrant valves.
© Jones & Bartlett Learning

FIGURE 12-10 Nonfunctional hydrant at a roadside rest stop. The caps are painted white to indicate the hydrant is out of service.
© Jones & Bartlett Learning

scene. In 1976, for the bicentennial celebration of the U.S. Constitution, many fire hydrants were painted in patriotic motifs. Although this was attractive artistically, it was not such a good idea from the fire fighter's viewpoint because the color coding was painted over in many cases.

Water Systems Program

Fundamental to any fire department's ability to extinguish fires is its ability to fully utilize the available resources, one of which is water. A thorough and complete knowledge of the water systems available is required. A water system program is implemented to promote cooperation between the fire department and the water companies. Records of each hydrant should be maintained, as should complete and detailed maps of the water systems **FIGURE 12-11**.

The first step in implementing a water system program is to meet with water company officials to establish a working relationship and to execute an agreement on testing and maintaining the system. A letter of working agreement is written and signed by the responsible officials of the water company and the fire department. The fire department's role is traditionally to service and maintain the hydrants as far as minor repairs are concerned. If required, the need for major repairs is reported to the water company.

Water company records contain a description of the water system and its capabilities, including the type of water system and its components, storage capacity, normal pumping capacity, and emergency pumping capacity. The water company record may also contain information on the procedures needed to boost flow during a large-demand fire and emergency phone numbers for contacting water company personnel.

> **Tip**
>
> Fundamental to any fire department's ability to extinguish fires is its ability to fully utilize the available resources.

A grid map is maintained showing the location of hydrants and the size of mains serving them. The hydrants should be numbered on this map so individual

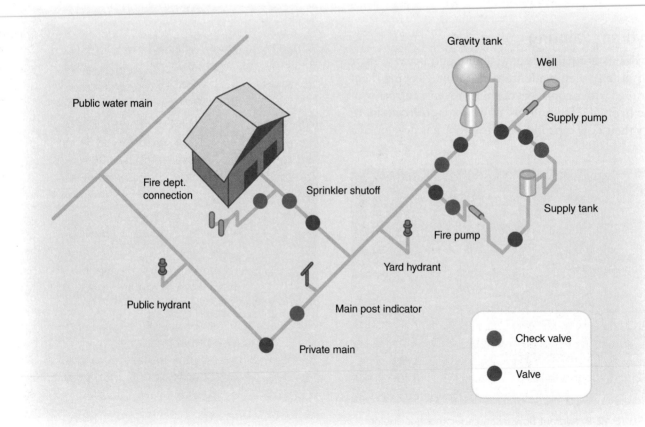

FIGURE 12-11 Water system schematic illustrating supply to a fire sprinkler system in a protected occupancy.
© Jones & Bartlett Learning

hydrants can be referenced for record-keeping purposes. Other maps are maintained and placed in the fire apparatus for reference at the fire scene. Individual hydrants need not be numbered. Their location and color coding should be noted to aid in at-scene decision making. Valves should be noted on these maps to aid in shutting down portions of the system in case of sheared-off hydrants or broken water mains. The map should also include the emergency phone number of the water company.

Hydrant survey and service records that show the size, type, and make of the hydrant should be maintained. The flow test dates and any flushing of the hydrants should be recorded. If successive flow tests over several years show a serious decline in the hydrant's flow capability, the system may need major repair. The date the hydrant was serviced should be recorded so that no hydrants go too long without being checked.

Hydrants require very little maintenance over their life spans; however, they should be checked and serviced annually to ensure that they are in proper working order. By keeping track of their condition and when they were maintained, fire fighters can identify any that are regularly in need of service due to vandalism or being backed into by vehicles. You may not think about them until you need them, but it sure is nice to be able to find them and have them in working order when the need arises. An additional advantage to company-level hydrant service is that it requires you to visit them and helps you remember where they are when they are needed.

In many jurisdictions where there are not built-up residential, commercial, or industrial areas, there may not be a regular water system. In other areas, the water system may not be adequate due to additional construction of structures after the water system was installed. In these situations, it is up to preplanning and the fire fighters' resourcefulness to find adequate water to control fires.

Auxiliary sources of water supply can, and often do, exist **FIGURE 12-12**. The water may be available in holding areas in the form of reservoirs, tanks, cisterns, and swimming pools. It may be available in the flowing state in canals, rivers, or streams. There may be public or private construction equipment in the area with water-carrying capability that may be pressed into service.

By preplanning these auxiliary water supplies, they can be identified. Their locations should be shown on the maps carried in the apparatus. Often, after contacting the owner, a connection can be installed on a tank, or a dry hydrant can be installed at a drafting source. Some owners may even agree to purchase a

FIGURE 12-12 A fire water storage tank with a fire department fitting should be clearly marked. This storage tank has the marking obscured by an advertising sign for the on-site business. The fire pumper connection is circled.

© Jones & Bartlett Learning

small pump to fill apparatus out of their swimming pool. In the case of helicopter bucket operations, the water source needs to be gauged for depth as well.

Private Fire Protection Systems

NFPA 25: *Standard for the Inspection, Testing, and Maintenance of Water-Based Fire Protection Systems* addresses issues with sprinkler systems, underground piping, fire pumps, storage tanks, water spray systems, and foam-water application systems. With the importance and wide applicability of fire sprinklers in built-in fire protection, the NFPA has a series of standards that apply to them. Additional standards that apply to fire sprinkler systems include NFPA 13: *Standard for the Installation of Sprinkler Systems*, NFPA 13D: *Standard for the Installation of Sprinkler Systems in One- and Two-Family Dwellings and Manufactured Homes*, NFPA 13E: *Recommended Practice for Fire Department Operations in Properties Protected by Sprinkler and Standpipe Systems*, and NFPA 13R: *Standard for the Installation of Sprinkler Systems in Low-Rise Residential Occupancies*.

Private fire protection systems are those designed to protect individual occupancies from fire. They may be installed in private homes, businesses, manufacturing plants, or public buildings. One of their main purposes is to alert the building occupants and/or the fire department of an incipient fire, reducing loss of life and property through early detection. Other systems are designed not only to alert, but also to control or extinguish fires.

Detection Devices

Detection devices may be as basic as a smoke detector, sometimes referred to as a smoke alarm, in the home. Different types of smoke detectors are designed for use in various locations **FIGURE 12-13**. The U.S. Fire Administration (USFA 2019), in its *Smoke alarm outreach materials*, states:

- Three out of five home fire deaths result from fires in properties without working smoke alarms.
- More than one-third (38 percent) of home fire deaths result from fires in which no smoke alarms are present.
- The risk of dying in a home fire is cut in half in homes with working smoke alarms*.

A detector in common use is the ionization chamber detector. These detectors may be either wired into the building's electrical system or battery operated. The ionization chamber detector contains a small amount of radioactive material that ionizes the air entering the chamber. As ionized smoke particles enter the chamber, they disrupt the process and the alarm is triggered. The advantage of these types of detectors is they are inexpensive to produce and do not need visible amounts of smoke to activate. A problem with this type of detector is that the battery is sometimes removed when it gets weak and the owner forgets to replace it. They are also subject to false alarms due to dust, steam, burning toast, and so forth.

FIGURE 12-13 Smoke detector. The NFPA public education section prefers to call these smoke alarms because their purpose is to sound an alarm when smoke is detected.
Courtesy of Kidde Residential and Commercial Division

Flame or light detectors come in two basic variations. The ultraviolet detector measures light waves, and when those associated with high-intensity flames are detected, the alarm is triggered. The infrared detector measures either the flame flicker or the total infrared component of the flame.

There are detectors that rely on visible smoke to trigger the alarm. They work on the same principle as the light beams that are used in doors of stores to alert the employees that someone has entered. The light beam is aimed at the receiver, and when the beam is interrupted, the alarm is triggered. Another variation has a photosensitive device that is hit by the light only when it is scattered by smoke particles. Several variations of this type of detector are available for specific applications (NFPA 2018).

Rate of rise detectors are used in some occupancies. These measure the rate of temperature rise in the monitored area. If the rate exceeds specified amounts, the alarm is triggered.

Fixed-temperature detectors are designed to melt at a certain temperature or measure the deflection of a bimetallic strip. Another type has a liquid in a glass bulb that breaks when the liquid expands due to heating. These can be rendered ineffective due to accumulations of dirt or paint acting as insulation.

> **Tip**
>
> The two main purposes of detection devices are to warn building occupants of the fire start and to get the alarm to the fire department as soon as possible. If the alarm is local notification only, it will still require someone to call the fire department.

Carbon monoxide (CO) detectors are becoming more prevalent, especially in residential settings. They are designed to trigger an alarm based on an accumulation of CO over time. Sources of CO are flame-fueled devices such as ovens, furnaces, water heaters, fireplaces, and so on. According to the Centers for Disease Control and Prevention (CDC), during 2010–2015 there were 2244 deaths from unintentional carbon monoxide poisoning. In 2015 there were 393 deaths, with 36 percent of the deaths occurring in the winter months of December, January, or February (CDC 2017).

The manual-pull alarm station that you are probably familiar with from elementary school is another alarm-triggering device **FIGURE 12-14**. When someone

---

* U.S. Fire Administration (USFA 2019), in its *Smoke alarm outreach materials*.

FIGURE 12-14 Manual-pull fire alarm activator.
Courtesy of Jeff Riechmann

pulls down on the handle, a switch is triggered and the alarm sounds. Newer installations include a strobe light as well to alert the hearing impaired.

A water flow switch or excess flow alarm, mounted on the riser of a sprinkler system, is another means of notifying building occupants of a fire. If a sprinkler head opens, the flow of water through the system causes a switch to trigger the alarm. Most of these types of alarms also include electrical connections to an alarm company so that if no one is around to hear the alarm, it still gets transmitted to the fire department (see Figure 12-18).

Attached to the automatic and manual pull stations in an occupancy is the fire alarm panel. This is usually kept in a locked room or at the security desk in a larger occupancy. The panel can tell the fire department responders in which area, or zone, the fire alarm has been activated. One of the problems with the fire at the Cathedral of Notre Dame in Paris in 2019 was there were no detectors in the area of origin and the fire department took 23 minutes to determine exactly where the fire was (Hubbard 2019). This gave it time to gain headway and once hoselines were deployed the fire was in the free burning stage and beyond immediate control.

Many facilities have a combination of these devices to sound the alarm. The system needs to be monitored at some level for the alarm to be transmitted to the fire department. There are several ways of accomplishing this, including security guards, alarm-monitoring companies, and direct tie-ins to the fire department dispatch office or fire station. Sometimes these systems are plagued with frequent false alarms, which leads to two problems. First, after a while the fire department starts to take it for granted that the alarm is false and does not respond in a timely fashion or responds without the full **first-alarm complement** of resources. Second, if the alarm is intercepted at a reception desk at

the facility and not transmitted to the fire department, the fire may have a chance to become well established before someone checks it out.

Extinguishing Agents

Water

Water is the most common fire-extinguishing agent in use today. From a mechanical standpoint, water is easily applied from many types of devices. If we were to compare water to sand as an extinguishing agent, these attributes would become immediately obvious. Water can easily be bent around corners in plumbing or flexible hoses. It can be deflected off surfaces and redirected into inaccessible spaces. By adding pressure, it can be lifted to great heights and carried great distances, either inside a hose or outside as a hose stream. It will assume the shape of any container into which it is placed, making it adaptable to different shaped tanks on apparatus. It will seek the lowest level, making it run off from upper floors instead of having to be shoveled out. Being a natural product, water is readily absorbed and does not cause environmental problems when used on nonhazardous substances. In most places, water is plentiful and inexpensive.

Water has the ability to extinguish fires through cooling and smothering. It has one of the highest values of **specific heat** of any known substance, allowing it to absorb great amounts of energy. One gallon will absorb approximately 1280 BTUs in the process of raising its temperature from 62°F to 212°F. Another quality is its **latent heat of vaporization**. When this gallon of water at 212°F turns to steam, it absorbs an additional 8080 BTUs. It requires 970 BTUs per pound to change water from a liquid to steam at 212°F (Fire 2009). The most effective absorption of heat energy is achieved when the complete volume of water applied is vaporized (converted to steam). This vaporization also produces a volume increase of 1 to over 1700 (NFPA 2008). One gallon of water also produces 223 cubic feet of steam. This has the potential of displacing oxygen, especially in a confined space, such as an attic. At an efficiency rate

> **Tip**
>
> Water has one of the highest specific heats of any known substance, allowing it to absorb great amounts of energy. Water is best applied to the heat source in a spray or droplet form to increase its surface area, allowing it to absorb more heat per unit (gallon) applied.

of 90 percent, 50 gallons of water would produce enough steam to completely fill a one-story, 35-ft by 45-ft building. There are numerous delivery systems available to apply water to fires.

Foam

Although water is the traditional extinguishing agent, the development of foam products that increase its effectiveness over that of plain water are becoming increasingly popular. Foam has three properties that make it capable of extinguishing and/or preventing fires: insulating, cooling, and forming a vapor barrier. It can cover the fuel surface, forming a layer of insulation between the heat source and the fuel supply. This is illustrated when foam is applied to grass and brush or structures prior to an advancing fire. It can cool the fuel surface below its ignition temperature. In the case of liquid fuels, it can reduce vapor production by reducing the evaporation rate through temperature reduction. Foam can also form a barrier between the vapors produced and the ignition source.

Foam Components

The components of foam are water and foam concentrate. When the concentrate is proportioned into the water, at the correct rate for the application, foam solution is produced. When air is introduced into the foam solution, finished foam is produced. Some foams do not require large amounts of bubbles to work well. Others are as thick as shaving cream when applied correctly.

Types of Foam

Two basic classes of foam are available to the fire fighter, both of which have particular strengths as firefighting agents. The traditional purpose of foam agents is to extinguish flammable liquid (Class B) fires by cutting off vapor production at the surface. As you will remember from the Chemistry and Physics of Fire chapter, liquids must produce vapor to burn. If there is not enough vapor above the liquid's surface, the fire will go out. Class B foam forms a layer above the surface of the liquid that cuts off the production of vapor. Water alone is unable to do this because it has a specific gravity greater than that of most commonly encountered flammable liquids (refer to Table 4-1). If you were to spray water on the surface of the flammable liquid, it would sink, increasing the volume of the burning spill, and cause the flammable liquid to spread, creating more of a problem. As the foam is applied to the burning liquid surface, some is sacrificed as steam as it cools the surface of the liquid.

Chemical foams were the first to be developed. They were generated using a hopper into which two chemical powders were dumped. The powders were introduced into the hose stream, the chemical reaction took place, and finished foam was produced. The problems arose when the powders would cake and not mix properly, leading to inconsistent foam quality. The gas bubbles formed by the chemical reaction would burst the hose if the nozzle were shut off prematurely. Chemical foams are not commonly used today.

Mechanical foam is another type of foam. It is formed by the introduction of foam concentrate into the hose stream. The bubbles are formed by mechanical instead of chemical action. Air is entrained in one of the following three ways to create the characteristic bubbles found in foam when it is applied: (1) The hose stream passes through special aeration devices; (2) compressed air is forced into the hose stream; or (3) as the solution leaves the nozzle, air is entrained. There are several types of mechanical foams available and in use; they include protein foam, fluoroprotein foam, alcohol-type protein foam, and aqueous film-forming foam (AFFF).

Protein foam is made of natural protein solids that have been broken down chemically. Protein foam concentrate has a strong resemblance in smell and appearance to blood. Protein foam has excellent fire-fighting capabilities in that it is an extremely water-retentive compound with high strength and elasticity. Protein foam is nontoxic despite its smell. Like most other foams, protein foam is usable in 3 percent and 6 percent concentrations in water. The stiffness of protein foams can be a detriment. The foam does not flow around and seal behind obstructions very well. This property also makes it susceptible to being blown around on the surface of the liquid by the wind.

Fluoroprotein foam is an improved version of protein foam and consists of a protein foam base with fluorinated agents added. This gives the foam produced good fuel-shedding ability, making it suitable for subsurface injection into large fuel storage tanks. Fluoroprotein foam also works well with dry chemical extinguishing agents, making it a good choice for use on three-dimensional flammable liquid fires.

Alcohol-type protein foams were developed for use on the class of materials known as polar solvents. **Polar solvents**, such as alcohol, are miscible in water. Hydrocarbon fuels, such as gasoline, are not miscible in water. When regular foams are used on polar solvents, the water in the foam dissolves into the polar solvent and the foam blanket is destroyed.

The most popular type of synthetic foam is the AFFF, commonly pronounced as "A triple F." This type of foam creates a sudsy blanket of trapped bubbles and then breaks down mechanically into a very thin film of water on the surface of the liquid fuel. It is self-sealing in that it has the capability of sealing itself around pipes and other structures protruding from the surface of the burning liquid. It also reseals itself when the foam blanket is disturbed if there is sufficient foam available on the surface. When produced as AFFF alcohol-type concentrate, the foam can also be used on polar solvents. This type of foam typically comes in a concentrated formulation that is used at 3 percent on hydrocarbon fuels and 6 percent on polar solvents. The problem with this foam is that it is not very sudsy and it is hard to tell if effective foam is being produced. It is also difficult to tell if the foam blanket has broken down, which may endanger fire fighters working in the foamed area. When used to blanket flammable liquid spills for the purpose of vapor prevention, combustible gas indicators should be employed **FIGURE 12-15**. To be effective, foam must be periodically reapplied to maintain an effective vapor barrier.

The best plan of action is not to walk across the flammable liquid spill, even when foam is in place. If you must, be sure to drag your feet as you walk slowly. Do not lift them for each step. This reduces the chances of breaking the foam blanket and creating a cloud of flammable vapor around you. If the foam blanket is broken and does ignite momentarily, it should reseal and the fire should go out. The natural inclination is to run, but this will only disturb the foam blanket more, increasing the intensity of the fire around you.

High-expansion foams are produced by running the foam solution over specially designed netting while forcing air through the netting with a powerful fan. This creates a foam with an expansion ratio as high as 1000 to 1 (National Foam, 2019). The foam is introduced into a space (a basement is a good example), displacing the oxygen and insulating the unburned materials from heat. Care must be taken to use fresh air in generating the foam. Combustion products cause the foam to break down at a faster rate. It could also be possible to suck flammable vapors into the foam generator, causing a foam with flammable vapor inside the bubbles. High-expansion foam will reduce visibility to nearly zero, and it would be easy to become lost or trip over objects in a foam-filled area.

> **Tip**
>
> Foam must be periodically reapplied to maintain an effective vapor barrier.

Class A Foam

Foam designed for Class A fires is different than that designed for Class B fires. The foams are not interchangeable. Class A foam is designed to be used at much lower concentrations than Class B agents. The percentages are in the 0.1–1 percent range. This is adjustable to produce foam with the desired qualities. The lower the proportion of foam concentrate, the wetter the foam will be. When used in conjunction with a compressed air foam system, Class A foam that will stick to vertical surfaces and shield them from the heat of the fire can be created. This quality allows fire fighters to pretreat areas in advance of the fire. Ordinary water would just run down and end up on the ground or evaporate. A type of foaming agent coming into more common use is fire-blocking gel. This agent forms a protective barrier that holds water suspended better than regular Class A foams, even on vertical surfaces. Gels are used in wildland firefighting to pretreat structures, trees, and other flammables before the fire approaches. "Gelling" increases fire fighter safety as the fire fighters are able to pretreat the structure and leave the area prior to the fire arriving. The gels may be rewetted after they are applied to increase their effectiveness once they have dried.

Wetting Agents

Wetting agents act much the same as soap: They reduce the surface tension of the water and allow it to soak into the fuel at a faster rate. In this way, more of the water remains on the fuel, and as it soaks in it tends not to evaporate as fast as surface water would. To demonstrate this, take a piece of cloth and pour some plain tap water on it. Observe how long it takes to soak in. Then take some water with dish soap in it and perform

FIGURE 12-15 Hazardous materials detection instruments, including combustible gas indicator.
© Jones & Bartlett Learning

the same experiment again. The water with the soap will soak in much faster because it has less surface tension. When performing fire overhaul operations, wetting agents are especially helpful because less water is needed to accomplish the job and the deep-seated embers are extinguished more quickly.

Many of these additives add no particular color or odor to the water. If you had ever thought about drinking water from a hose stream, the possibility that there may be wetting agents or other chemicals added should convince you not to. They would give you intestinal problems if you were to take them internally.

Encapsulator Agents

Encapsulator agents can be used on Class A, B, C (solar panels, lithium-ion batteries, and transformers), and D fires. They do not release steam and render hydrocarbon spills nonflammable. They are recommended for car fires, especially lithium-ion battery hybrid and electric cars. They are added as a percentage (0.25–3 percent) solution to water and increase the water's cooling ability as well as containing many of the toxic and carcinogenic fire byproducts (Haddon 2017).

Fire Retardant

Agents applied on wildland fires are divided into two categories: short- and long-term retardants. All are water based. Water, foam, and wetting agents are classified as short-term retardants and fire suppressants. Their effectiveness is directly related to the moisture content they retain when on the fuel. They lose their effectiveness when the water evaporates and the fuel dries out after their application. Suppressants are used in direct attack and mop-up operations. They may be applied from the air or from ground-based units.

Agents that react chemically with the fuel and retain their effectiveness after they have dried out are classified as long-term retardants. Retardants are used for indirect attack. Long-term retardants are almost always applied from the air. They can be applied by air tankers or helicopters. They contain pigments so they remain visible from the air. New types of pigments have been developed. In areas where there are rock outcroppings and structures, fugitive pigments are used. These pigments lose their color and basically disappear after a while from exposure to the sun. Regular pigments are iron oxide (rust)-based and tend to need rain to wash them off the fuel and into the soil (NFES 2018).

Carbon Dioxide

Carbon dioxide gas (CO_2) extinguishes fires by smothering. Carbon dioxide is an inert gas with the ability to dilute the oxygen in the fire area to a level where the fire will be extinguished. CO_2 systems work best in installations where air flow can be controlled, preventing premature dilution of the product. The CO_2 comes out of the extinguishing system as a mixture of vapor and dry ice particles, which is very cold, and it has a cooling effect when applied directly to the burning fuel surface. CO_2 systems are installed in areas where water is not the extinguishing agent of choice. CO_2 is selected because it can extinguish fires without leaving a residue, does not promote rust, does not create water runoff, and does not conduct electricity (NFPA 2008). Some of the areas where CO_2 is preferred are telephone switching rooms, fur storage, and computer installations. In occupancies where chemicals are stored and using water would create runoff that would have to be treated as hazardous waste requiring expensive cleanup, the cost of a CO_2 system is justified (see Figure 12-25).

Halogenated Agents

Halogenated agents extinguish the fire by breaking the chemical chain reaction, whereas CO_2 extinguishers operate by smothering. Halogenated agent systems are more effective, but there is concern about their effect on the ozone layer because they are chlorofluorocarbons. The concentrations of halogenated agents used in fire extinguishment are not considered hazardous. However, the chemical by-products of their use can be harmful, and self-contained breathing apparatus (SCBA) should be worn when coming into contact with these agents. An additional concern when utilizing halogenated agents is that they may not extinguish deep-seated combustion in cellulosic materials (wood, paper). Examples of these fire-extinguishing compounds are brand names such as Halon, followed by the Army Corps of Engineers numbering system based on their chemical makeup: 1211, 1301, and so forth (NFPA 2008).

Halogenated agent systems are installed and operate much like the carbon dioxide systems. Their use is being phased out because of environmental concerns.

Clean Agents

So-called clean agents have been developed to replace halogenated agents. The requirements of these agents are that they have the same cleanliness (lack of residue), extinguishing capability, and low toxicity of halogenated agents but do not deplete the Earth's ozone layer. These systems come in various configurations; one is hydrofluorocarbon (HFC) based and works through heat absorption. Another uses inert gases: nitrogen, argon and carbon dioxide. Additionally, there is one that uses a special liquid that vaporizes and suppresses fire through cooling (Koorsen Fire & Security 2017).

Dry Chemical

Dry chemical extinguishing systems use a mixture of finely divided powders. These powders are treated to resist caking and are water repellant. Their effectiveness lies in their ability to break the chemical chain reaction. They also absorb some of the radiated heat from the fire and displace oxygen in a limited way. Effective on flammable liquid (Class B) and electrical fires (Class C), some formulations can also be used on ordinary combustibles (Class A). When used on ordinary combustibles, their application should always be followed up with water to extinguish deep-seated embers.

> **Tip**
>
> When used on ordinary combustibles, the application of dry chemicals should always be followed up with water to extinguish deep-seated embers.

Dry Powder

The extinguishing agents used on combustible metals (Class D fires) are dry powder agents. The agent may come in a bucket, pail, or extinguisher. These agents control the fire by forming a coating on the burning surface and excluding oxygen. Some are nothing more than dry sand, and others are graphite or special powders. Some contain plastic beads, which melt and help to form a coating.

Water is not commonly used on combustible metals as it may react violently, especially with magnesium and sodium, causing explosions. These explosions can spray burning material onto fire fighters, leading to burn injuries. The bright light generated by the explosion can also cause eye injuries. When water is to be used, it must be applied in flooding amounts.

> **SAFETY TIP**
>
> Water is not commonly used on combustible metals because it may react violently, especially with magnesium and sodium, causing explosions.

Extinguishing Systems

Sprinkler Systems

Automatic sprinkler systems have been in service for more than 100 years. They have an excellent record of fire suppression. Sprinklers have been statistically shown to control 97 percent of the fires where they were activated. The remaining fires were not controlled due to improper maintenance, inadequate or shut-off water supply, incorrect installation or design, and obstructions. A common misconception about fire sprinklers is that if one head is activated, they all activate. This is regularly shown in television shows and movies but is incorrect. With the exception of deluge systems, only the heads that are affected by heat activate. The activation of a minimum number of heads to contain the fire reduces water damage.

One of the main causes of fires getting out of control is the lack of immediate detection. The best time to attack a fire is in its incipient stage. When there is a delay between the fire's ignition and discovery, the fire has a chance to gain headway and enter the free-burning stage. The primary asset of a sprinkler system is its ability to act on a fire without human intervention. Modern sprinkler systems are very reliable, are always on duty, and are not busy elsewhere when a fire starts. They activate regardless of whether anyone has discovered the fire. There have been instances where the sprinkler system activated and extinguished a fire that was not even discovered until after it was out.

Residential Sprinklers

Home fire sprinklers significantly reduce the dangers posed by lightweight construction. Fitting a home with sprinklers reduces the chance of death by fire by 80 percent and reduces property loss by 71 percent according to the NFPA and its *Fire Sprinkler Initiative: Bringing Safety Home* (NFPA 2018). A Centers for Disease Control and Prevention/NIOSH alert states that over 60 percent of the homes in the United States are constructed with lightweight wood truss construction techniques (CDC/NIOSH 2005).

The efficiency of these systems has led to their adaptation for use in residences. Installed in a very simplified form, the system does not have the extensive valves required on systems installed in commercial or industrial occupancies. They also have a much-reduced pipe size in their plumbing and therefore reduced water supply and plastic piping are allowed. There is a shutoff valve installed on the supply side of the system for ease of turning off the flow in case of operation. There should be a water flow alarm installed so if the system activates, the fire department can be notified. The heads are designed to sit in a recess in the ceiling and drop below ceiling level when activated.

As with any other sprinkler system, the owner or occupant needs to understand how the system works and its limitations. If the system is activated, the owner or occupant should notify the fire department. A built-in fire sprinkler system will not protect from a fire that is burning on the roof, which is a particular problem where shake (wood shingle) roofs are in use. The local requirements may not specify that sprinklers need to be installed in the attic, where fires occur as well. The system should not be shut down until the local fire department has responded and ensured that the fire that activated the system is completely extinguished.

Commercial and Industrial Sprinkler Systems

A sprinkler system consists of several basic components. Sprinklers are distributed throughout the occupancy **FIGURE 12-16**. There is a pipe bringing water in from the water system. The water then passes through an **open screw and yoke valve (OS and Y)** or **post indicator valve (PI)**. The purpose of these valves is to make it obvious under quick inspection whether they are open or closed **FIGURE 12-17**. Both of these types of valves are usually kept locked in the open position to prevent tampering. The water then enters the main control valve. Main control valves differ depending on whether the system is dry pipe or wet pipe.

Wet Pipe System

The wet pipe system has water under pressure behind the sprinkler heads at all times. This allows the system

FIGURE 12-17 OS and Y valves chained and locked in the open position. The fire department connection is on the left end of the plumbing.
© Jones & Bartlett Learning

to operate rapidly when a head is opened. These systems are suitable for use in installations where freezing is not a problem. The control valve on a wet pipe system is a check valve to keep water from the sprinkler system from reentering the domestic supply **FIGURE 12-18**. The clapper in the check valve is usually in the closed position. It opens fully only when a sprinkler head opens. There are pressure gauges on each side of the clapper. In a properly operating system, the pressure gauge on the sprinkler side of the valve should read the higher pressure, because as pressure surges enter the system, the gauges raise the clapper momentarily and the higher pressure ends up trapped above the valve, keeping it in the closed position. There is also an alarm valve that activates when excess water flow is detected. A **retard chamber** that allows pressure surges to dissipate without triggering the alarm will be attached.

There is a main system drain mounted above the clapper on the main valve body to allow draining of the system for repairs.

Dry Pipe System

The dry pipe system is somewhat more complicated in design. This system is used in areas where freezing temperatures may occur, such as outside storage or unheated buildings. Because water expands as it freezes, if the system is allowed to become filled with water, the expansion will rupture the plumbing. As the temperature warms, the parts that were held in place by the ice fall to the floor. In a room with a high ceiling, falling chunks of cast iron fittings can be hazardous. The dry pipe system has compressed air in the lines behind the sprinkler heads. The compressed air keeps

FIGURE 12-16 Assortment of fire sprinkler heads mounted for classroom display.
© Jones & Bartlett Learning

FIGURE 12-18 Sprinkler riser, valves, and water flow alarms located outside of the building where freezing temperatures do not occur.
© Jones & Bartlett Learning

FIGURE 12-19 Fire department connection and post indicator valve.
© Jones & Bartlett Learning

the clapper in the main valve body from opening and admitting water. There is water on the underside of the clapper valve, and as soon as it opens, the water is admitted to the system. In especially large systems, if one head were to open, there can be a 1-minute delay before enough air is released to let water come out of the open sprinkler.

In areas where large amounts of water are needed immediately, deluge systems are used. They are set up with sprinkler heads that are open all the time. Fire detection devices open the valves and allow water into the system when necessary.

Once the water leaves the main valve, it enters the riser. There is a **fire department connection** attached to the riser **FIGURE 12-19**. This connection is used to boost the pressure in the system by attaching a hose from a pumper. The standard operating pressure for this pumper is 150 psi, considerably more pressure than most water supply systems. The fire department connection is on the discharge side of the main valve so water from the pumper is not forced back into the domestic water supply. It is important when attaching to the sprinkler system that the pumper is not connected to a hydrant that directly supplies the water flow to the sprinkler system.

When the water leaves the riser, it enters the feed mains. These pipes are then connected to the cross mains, which are attached to the branch lines. The branch lines are where the individual sprinkler heads are attached. At the highest and furthest point in the system is the inspector test drain. This is used to verify that the required pressure is available to every sprinkler head **FIGURE 12-20**.

Variations of the dry pipe system are the deluge and preaction systems. Deluge sprinkler systems are used in high fire hazard occupancies, such as plywood manufacturing. The sprinkler heads in this type of system are not heat activated and are referred to as open sprinklers. The system is activated by a supplemental fire detection system. When the deluge valve opens, water enters the plumbing and is immediately discharged through the open heads.

Preaction sprinkler systems are designed with a preaction valve holding the water back and have closed sprinkler heads. The sprinkler system is connected to a detection system. When the detection system activates, the preaction valve opens, allowing water into the piping. When heat becomes sufficient to open the sprinkler head, water is discharged.

Sprinkler Heads

Sprinkler heads come in numerous styles and are divided into three main types. The pendant type head is designed to be installed in the hanging down position **FIGURE 12-21**. The upright head is installed with the head above the pipe. The sidewall sprinkler head is used in a horizontal position. Each head is equipped with a deflector that divides the water flow into a fine spray. The deflector also directs the spray. The head types are not interchangeable: If pendant heads are

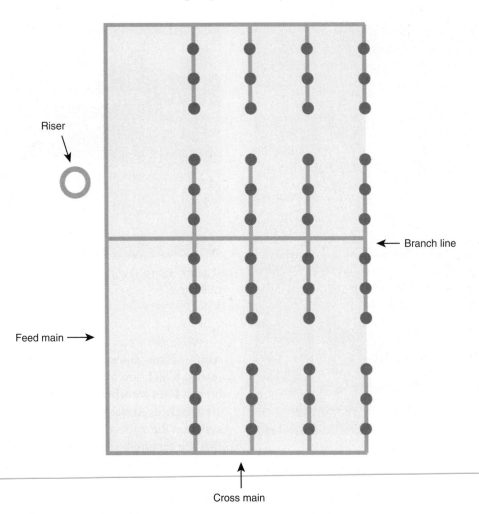

Riser

Branch line

Feed main

Cross main

FIGURE 12-20 Sprinkler piping schematic.
© Jones & Bartlett Learning

FIGURE 12-21 Pendant sprinkler head.
Courtesy of Tyco Fire Protection Products

used in the upright position, the heads will not function correctly.

Sprinkler heads are designed to operate at different temperatures. Either the frame of the head is painted or the glass bulb contains a colored liquid that is color coded for their operating temperature. The maximum

ceiling temperatures range from 100°F to 625°F (NFPA 2008). Should any painting of the ceiling or walls be done, the sprinkler heads themselves must be masked off from paint. Once the painting is completed the masking material must be removed or it may interfere with the proper activation of the head(s) during a fire. Sprinkler heads can be rendered ineffective when stock is piled too close to them, limiting their spray pattern. Another factor affecting their effectiveness is the use of solid shelving. The spray can be kept from penetrating to the seat of the fire by deflection. To avoid the problem experienced with solid shelving, in some instances heads are installed between the shelves. This does, however lead to a greater risk of heads being knocked off or damaged during shelf stocking operations, especially where forklifts are used to move stock.

A new method of delivery for water through sprinkler heads has been developed by the Victaulic Corporation. It is called the Vortex system and delivers water as very fine droplets propelled through special sprinkler heads by compressed nitrogen. The water comes out as a mist to cool the fuel and the nitrogen dilutes the oxygen. This system can be used in electrical installations, leaves no residue, and is not toxic to the environment.

Standpipe Systems

A fire protection system that provides a fire hose attachment station on each floor is the standpipe system. By having a plumbing system installed in the building with fire department hose connections, it is not necessary to take the time to lay hose to each floor. The Class I standpipe is designed for fire department use and has a 2½-in. hose connection **FIGURE 12-22**. The Class II standpipe is designed for use by building occupants until the fire department arrives and has a 1½-in. hose connection with hose attached. The Class III system has both a 2½-in. hose connection for fire department use and a 1½-in. connection with hose for occupant use or a 2½-in. connection on a 1½-in. reducer and hose that is easily removable to allow access to the 2½-in. connection.

A caution to fire fighters is the inclusion of pressure-reducing valves (PRV) installed in standpipe connections in **high-rise buildings**. The inclusion

FIGURE 12-22 Standpipe outlet connection for 2½-in. fire hose.
© Stephen Coburn/Shutterstock

of PRVs was considered a contributing factor in the deaths of three fire fighters at the One Meridian Plaza fire in Philadelphia, Pennsylvania, on February 23, 1991 (Naum 2011). These devices reduce the pressure at the outlets in the standpipe system due to the high pressure required by the built-in protection system to boost water to the highest floors. This causes the pressure to be too high on the lower floors. The PRVs are installed to reduce the pressure coming out of the standpipe connection. This may cause the outlet pressure to be too low for modern automatic fog nozzles. The solution to this problem is either to use straight bore nozzles or to remove the PRV when a fire attack is mounted from a standpipe connection.

Foam Systems

Foam systems can be stationary or vehicle mounted. There are four ways that foam solution is created from foam concentrate and water. The first of these is through eduction using a device called an eductor **FIGURE 12-23**. Eductors operate on the Venturi principle. The water is forced through a small opening that expands after it passes over the orifice where the foam concentrate is admitted to the fire stream. On portable eductors, the water inlet is attached to the hose coming from the pumper. The outlet is attached to the hose going to the nozzle. There is a plastic hose attached to the eductor that is inserted into the foam concentrate container, usually a 5-gallon bucket or 55-gallon drum. Where the plastic hose attaches to the eductor body, there is a plastic ball on the inside, which is designed to act as a check valve, preventing water from being forced into the foam concentrate container. After use of the eductor, the plastic ball should be removed and cleaned. If it is not, dried concentrate can cause the eductor to malfunction. These devices are also called in-line eductors. On the top there may be a dial to set the percentage of the foam solution produced, usually either 3 percent or 6 percent depending on the concentrate and application requirements.

FIGURE 12-23 In-line foam eductor.
© Glen Ellman

FIGURE 12-24 Class A and B foam concentrate containers.
© Jones & Bartlett Learning

The second way foam solution is produced is through injection. In this type of system, the foam concentrate is directly injected into the water. These systems may be set up to perform the injection before or after the pump.

The third method is called batch mixing. In batch mixing, the foam concentrate is introduced directly into the water reservoir. In wildland firefighting operations, the operators often carry premeasured amounts of Class A concentrate on the apparatus. When the alarm is received, they pour the concentrate into the tank and start the pump to circulate the solution so it is well mixed when they arrive at the scene. Wetting agents are usually added in this way. This method is also used on helicopters. The helicopter has an on-board tank of foam concentrate.

When the load of water is picked up, in either a bucket or tank, the required amount of foam concentrate is introduced into the water. The drawback to batch mixing is that to get a consistent concentration of the foam solution, the water tank must be completely emptied and refilled every time the solution is recreated. With Class A foam, careful proportioning

is not that critical; with Class B foam, the concentration is very important and batch mixing is not recommended **FIGURE 12-24**.

The fourth method is premixing. In premixing, the foam concentrate and water are mixed to the proper concentration and stored in this way. Foam extinguishers are designed this way. The problem with premixing is that foam solution and wetting agent solutions are corrosive on steel and cast iron. It is not a good idea to leave premixed foam in an apparatus tank for long periods of time. As a general rule, anytime foam or wetting agents are used, the system must be drained and flushed to avoid corrosion of either the tank or plumbing. If the apparatus is equipped with a plastic tank, carrying premixed foam may seem like a good idea. However, the plumbing, such as the pump casing, is cast iron or steel and still presents a problem.

An addition to Class A foam systems that is gaining in popularity is the compressed air foam system. The system is equipped with an air compressor that pumps air into the hose stream after the pump. The use of this type of system does not rely on the nozzle or other device to aerate the foam. It also makes the hose lines lighter because they contain air as well as foam solution. By adjusting the level of foam concentrate and air in the solution, the foam can be made for different purposes. A very dry foam will stick to vertical surfaces. A wet foam will not stick but provides better water penetration into the fuel surface.

Class B foam is very wet and does not stick to vertical surfaces. It works well on two-dimensional fires—those that have width and length, such as a pool of liquid. On three-dimensional fires—those that have vertical surfaces as well—it does not work well. An example would be flaming liquid spraying from a flange or piping leak. In this scenario, the Class B foam would be able to extinguish the pool fire, but not the fire at the source of the leak. In this situation, a dry chemical extinguisher can be used to extinguish the fire at the site of the leak.

> **Tip**
>
> Any time foam concentrate is used, the manufacturer's specifications as to the concentration of the product must be closely followed. Foam concentrates from different manufacturers should not be mixed in the foam concentrate tank on fire apparatus, as this may clog the foam concentrate injection system.

In the case of an aircraft hangar or other special occupancy, the foam system is preplumbed and discharges out of sprinkler heads or special nozzles. The system is designed to rain the foam down on the flaming material. As the foam reaches the floor, it spreads out and can seal over the surface of the burning material beneath the aircraft. Whenever Class B foam is used on flammable liquids, it is to be applied gently to the surface. Plunging the foam under the liquid surface will just disrupt the foam blanket that you are trying to create. When Class B foam is being used, plain water hose streams are not to be used on the surface that is being foamed. They will only dilute and decrease the effectiveness of the foam blanket.

Any time foam concentrate is used, the manufacturer's specifications as to the concentration of the product must be closely followed. Care must also be taken to determine the type of fuel involved. Regular AFFF will not work on polar solvents, and the proper concentration of AFFF alcohol-type concentrate must be used when dealing with polar solvents. Class A foam is not designed to be used on Class B fires, and vice versa.

When using foam eductors and other in-line devices, the manufacturer's specifications must be followed as pertains to flow rate, inlet pressure, length of line after the eductor, and elevation above the eductor. If the flow rate on the nozzle is set incorrectly or the nozzle is not all the way open, the foam you are expecting is not what you will get. Class B foam does not give very good visual clues as to its concentration. You may think you are forming a good sealing blanket over a product when in fact you are not.

Gas Extinguishing Systems

Other fire suppression systems have been developed to address specific problem areas. Where water will cause excessive damage to stock or electrical installations,

> **Tip**
>
> When using foam eductors and other in-line devices, the manufacturer's specifications must be followed as pertains to flow rate, inlet pressure, length of line after the eductor, and elevation above the eductor.

gas-type systems are installed. The advantage of the gas systems is that, unlike a dry chemical system, they do not leave a residue. Also, unlike a fire sprinkler system, they do not cause water damage or short out electrical equipment.

Carbon Dioxide

In carbon dioxide installations, the product is stored in large cylinders or a tank **FIGURE 12-25**. When a fire occurs, the gas is released into a piping system and expelled from nozzles in the area to be protected. CO_2 is not poisonous but can be harmful to humans because of its ability to dilute the oxygen content of the room, leading to asphyxiation. The system is to be installed with a warning system to evacuate occupants prior to discharge. In any operation where a CO_2 system has been used to attack the fire, personnel must wear SCBA to avoid the danger of low oxygen concentration.

Carbon dioxide will extinguish most types of fires but dissipates rapidly enough that immediate follow-up is necessary. When used on burning liquid fires, it will extinguish the flame but not cool metal parts of the liquid's container. If the metal parts are at a temperature above the ignition temperature of the liquid, reignition can occur. In ordinary combustible fires, CO_2 will not penetrate and

FIGURE 12-25 Carbon dioxide fire extinguishing system at a farm chemical storage facility.

© Jones & Bartlett Learning

extinguish deep-seated smoldering. In this type of fire, follow-up with water is necessary to ensure that the fire will not reignite. The concentration of the CO_2 in the atmosphere must be sufficient to lower the oxygen content to a point where the fire is extinguished. In large areas, this would require prohibitive amounts of product.

SAFETY TIP

In any operation where a CO_2 system has been used to attack the fire, personnel must wear SCBA to avoid the danger of low oxygen concentration.

Dry Chemical Systems

Stored in a container, the powder may or may not be under direct pressure. In a pressure extinguisher, the powder is forced out a tube that extends into the bottom of the container by the pressure of the expellant gas above it. In this type of extinguisher, the gas is usually nitrogen, which is nonreactive. In an extinguisher with the expellant gas stored in a remote reservoir, the reservoir must be punctured to release the gas into the container expelling the contents. The expellant gas in this type of extinguisher is usually carbon dioxide. This type of extinguisher is commonly carried on fire equipment because it is easy to refill the powder and attach a new pressure cartridge after use **FIGURE 12-26**.

Relatively inexpensive and easy to store, dry chemical systems are very common. They are installed in range hoods in restaurant kitchens and other areas where flammable liquids need to be extinguished. These systems are mounted in heavy machinery and racing vehicles due to their dependability under rough treatment and their extinguishing capability. Their ability to perform well in the presence of water is a definite plus. Foam does not perform well on three-dimensional fires. The dry chemical can be discharged into the fire stream at the nozzle and extinguish the fire when foam or water alone will not do the job. These systems are installed on aircraft rescue firefighting apparatus in a twinned system with AFFF. The water in the foam cools the metal parts, the foam extinguishes the pool fire, and the dry chemical can extinguish the liquid as it runs down the fuselage. This same evolution can be performed with plain water and a dry chemical extinguisher on vehicle fires.

FIGURE 12-26 Various types of fire extinguishers.
Courtesy of NFPA National Fire Protection Association

Wet Chemical Extinguishing Systems

Wet chemical extinguishing systems—Class K—were developed to extinguish fires in commercial cooking operations, such as deep fryers. Previously, dry chemical and wet chemical Class B systems performed this function. The oils used now are vegetable based, not animal based. They have a lowered ignition temperature, and the cooking equipment retains heat better than the old systems. The purpose of the wet chemical system is to reduce the temperature of the burning liquid as well as to apply the extinguishing agent.

Fire Extinguishers

Over the years, portable fire extinguishers have been developed using almost every type of extinguishing agent. Multipurpose dry chemical extinguishers are designed to be used on ordinary combustible (Class A), flammable liquid (Class B), and energized electrical (Class C) fires. Plain water extinguishers are suitable for use on Class A fires. There are CO_2 and halogenated agent extinguishers for Class A, B, and C fires as well. A flammable metal (Class D fire) extinguisher may be as simple as a bucket of dry sand with a scoop or shovel used to apply the sand. It is important that the sand be dry because burning magnesium can break the water molecule into hydrogen and oxygen and react explosively when exposed to water.

Others will have powdered graphite and plastic pellets that melt and form a coating over the burning metal. Fire extinguishers come in all sizes from 2 to 250 pounds of extinguishing agent **FIGURE 12-27** and **TABLE 12-1**.

Obsolete Extinguishers

It is quite possible that you may find obsolete types of fire extinguishers still in use. Some of these include soda acid and carbon tetrachloride. If you do come across these types of extinguishers, they should be removed from service. The soda acid extinguisher has a small amount of strong acid inside and must be handled carefully. The carbon tetrachloride–type extinguisher contains a toxic substance and must be disposed of properly. If you come across an extinguisher and are not sure of its type, contact your local extinguisher service company, and it can assist you.

Fire Pumps

In some occupancies, fire pumps are installed either to boost pressure or to ensure a water supply to the firefighting system. Two types of fire pumps are commonly used for this purpose. One is basically the same type of pump that is installed in fire engines, a centrifugal pump. It is driven by either a gas or diesel engine or electric motors **FIGURE 12-28**. For safety purposes, one of each (diesel and electric driven) should be installed to provide service in case of a power failure. The other type is the vertical turbine pump (see Figure 12-1). Designed to operate automatically under demand, these systems also have a manual start. Fire fighters should be familiar with the location and operation of these pumps in their district.

NFPA 13E requires that one of the first fire pumpers at the scene connects to and supports the built-in fire protection system. This is to ensure that if the fire pump does not operate or is insufficient the system can still be adequately supported.

Pressure-Reducing Devices

In high-rise buildings, pressure-reducing devices are installed because the amount of pressure needed to

SAFETY TIP

Fire departments need to be equipped with the proper combination of hose diameter and nozzles for operating in high-rise structures.

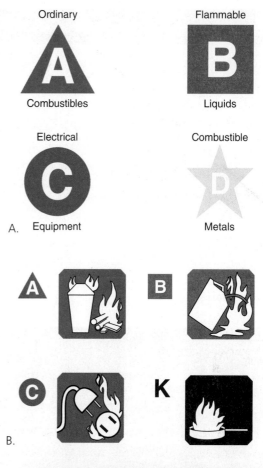

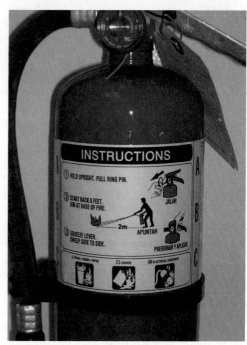

FIGURE 12-27 A. Older versions of fire extinguishers are labeled with colored geometrical shapes with letter designations. B. Newer fire extinguishers are labeled with a picture label system. C. Many fire extinguishers can be used to fight more than one class of fire.

TABLE 12-1 Extinguisher Use Based on Fire Class

Extinguisher Type	Class A Fire	Class B Fire	Class C Fire	Class D Fire	Class K Fire
Type A	X				
Type BC		X	X		X
Type ABC	X	X	X		X
Type D/dry sand				X	
Type K		X	X		X
CO_2	X	X	X		

© Jones & Bartlett Learning

FIGURE 12-28 Diesel- and electric-powered centrifugal fire pumps used to boost pressure in sprinkler/standpipe system.
© Jones & Bartlett Learning

pump water to the 100th floor would cause entirely too much pressure on the 50th floor. These devices can seriously influence the effectiveness of a fire stream. As with any of the other factors that affect firefighting, it is extremely important that fire fighters preplan their area, making sure that they understand the limitations and operation of any systems that will be used in firefighting. Most importantly, fire departments need to be equipped with the proper combination of hose diameter and nozzles for operating in high-rise structures. Without these, hose streams have very little reach and are rendered ineffective, creating a safety hazard for firefighting personnel.

Wrap-Up

CHAPTER SUMMARY

- Water is the most common extinguishing agent used for combating fires. Over the years, water systems have been developed to the point where they have become dependable, making water readily available at many fire scenes.
- Water is so fundamental to firefighting that a good water supply is one of the most important factors in municipal fire protection.
- One way water companies are set up is under public utility laws. Private water companies also exist, usually in industrial and commercial complexes.
- All water supply systems must first have a storage capability. The size of the storage capacity and adequacy of the system are determined by several factors.
- Once the water is removed from the storage area, it goes through the treatment plant to make it fit for drinking. When it leaves the treatment facility, the water enters the distribution system.

- A water system program is implemented to promote cooperation between the fire department and the water companies. Records of each hydrant should be maintained, as should complete and detailed maps of the water systems.
- Private fire protection systems are those designed to protect individual occupancies from fire. They may be installed in private homes, businesses, manufacturing plants, or public buildings.
- Detection devices may be as basic as a smoke detector, sometimes referred to as a smoke alarm, in the home. Different types of smoke detectors are designed for use in various locations.
- Water has the ability to extinguish fires through cooling and smothering. It has one of the highest specific heat values of any known substance, allowing it to absorb great amounts of energy.
- Although water is the traditional extinguishing agent, the development of foam products that increase its effectiveness over that of plain water are becoming increasingly popular.
- Wetting agents act much the same as soap: They reduce the surface tension of the water and allow it to soak into the fuel at a faster rate.
- Agents applied on wildland fires are divided into two categories: short- and long-term retardants. All are water based.
- Carbon dioxide gas (CO_2) extinguishes fires by smothering. CO_2 systems work best in installations where air flow can be controlled, preventing premature dilution of the product.
- Halogenated agents extinguish the fire by breaking the chemical chain reaction, but there is concern about their effect on the ozone layer because they are chlorofluorocarbons.
- So-called clean agents have been developed to replace halogenated agents.
- Dry chemical extinguishing systems use a mixture of finely divided powders. These powders are treated to resist caking and are water repellant.
- The extinguishing agents used on combustible metals (Class D) are dry powder agents. They control the fire by forming a coating on the burning surface and excluding oxygen.
- Automatic sprinkler systems have been in service for more than 100 years. They have an excellent record of fire suppression.
- The wet pipe system has water under pressure behind the sprinkler heads at all times. This allows the system to operate rapidly when a head is opened.
- The dry pipe system is somewhat more complicated in design. This system is used in areas where freezing temperatures may occur, such as outside storage or unheated buildings.
- Sprinkler heads come in numerous styles and are divided into three main types: pendant, upright, and sidewall.
- A fire protection system that provides a fire hose attachment station on each floor is the standpipe system.
- Foam systems can be stationary or vehicle mounted. There are four ways that foam solution is created from foam concentrate and water: eduction, injection, batch mixing, and premixing.
- Other fire suppression systems have been developed to address specific problem areas.
- In carbon dioxide installations, the product is stored in large cylinders or a tank. When a fire occurs, the gas is released into a piping system and expelled from nozzles in the area to be protected.
- In a pressure extinguisher, the dry chemical powder is forced out a tube that extends into the bottom of the container by the pressure of the expellant gas above it.
- Wet chemical extinguishing systems were developed to extinguish fires in commercial cooking operations, such as deep fryers.
- Over the years, portable fire extinguishers have been developed using almost every type of extinguishing agent.
- It is quite possible that you will find obsolete types of fire extinguishers still in use. If you do come across these types of extinguishers, they should be removed from service.
- In some occupancies, fire pumps are installed either to boost pressure or to ensure a water supply to the fire-fighting system.
- In high-rise buildings, pressure-reducing devices are installed because the amount of pressure needed to pump water to a higher floor would cause entirely too much pressure on the lowest floors.

KEY TERMS

Aquifer The underground layer of water-bearing permeable rock or unconsolidated materials (sand, gravel, etc.).

Average daily consumption The amount of water used daily by water system customers; computed by dividing the total water used by the number of customers over a period of a year by 365. Expressed in gallons or liters.

Bonnet The top of a hydrant.

Bury The piping that extends from the water main to the hydrant.

Combustible gas indicator A device that measures the percentage of lower explosive limit concentration of gas in the atmosphere. This device must be used by trained personnel for proper interpretation of the readings.

Dead-end mains Water mains that are not gridded into the system. Water flows into them from only one way.

Fire department connection Fittings connected to the fire protection system used by the fire department to boost the pressure and/or add water to the system.

Fire flow The total volume of water required to control the fire incident expressed in gallons per minute.

First-alarm complement The equipment normally dispatched when a fire is first reported.

Fugitive pigments A coloring agent added to fire retardant that is dropped from aircraft.

Halogenated agents Fire-extinguishing agents containing the elements from Group 7 on the periodic table of the elements (halogens).

High-rise buildings A multi-storied building that is over 75 ft in height; commonly encountered as an office building, hotel, or condominiums.

Inert A substance that will not react with other substances.

Latent heat of vaporization The amount of heat a material must absorb when it changes from a liquid to a vapor or gas.

Maximum daily consumption The highest amount of water used in a 1-day period by the customers of a water system; computed by finding the highest amount of water used in 1 day out of 365 days. Expressed in gallons or liters.

Open screw and yoke valve (OS and Y) A valve with a hand wheel that exposes a threaded rod when in the open position. The hand wheel looks much like a steering wheel and can be locked with a chain and padlock so it cannot turn.

Peak hourly consumption The highest amount of water used in 1 hour of 1 day; determined by finding the highest use per hour in a 24-hour period, and expressed in gallons or liters.

Polar solvents Liquids that will mix readily with water because water is a polar substance. The common polar solvents are alcohols, aldehydes, esters, ketones, and organic acids.

Post indicator valve (PI) A valve with an indicator body that sticks up out of the ground. The body has a small window that says either "shut" or "open" depending on the position of the valve.

Retard chamber A small tank attached to sprinkler systems that allows pressure surges to dissipate their energy before they enter the system and set off the water flow alarm.

Specific heat The ratio between the amount of heat necessary to raise the temperature of a substance compared to the amount of heat required to raise the same weight of water the same number of degrees. Water has a specific heat value of 1.

Thrust block A mass of concrete poured on the outside of an angle fitting and extending back to native soil. The purpose is to prevent surges in flow through a pipe from flexing the fitting and wiggling it in the ground, which would, over time, form a larger and larger underground space, possibly allowing the pipe fitting to pull apart.

Water hammer A pressure surge or wave caused by water in motion when it is stopped suddenly.

Water mains A pipe that carries water in a water system.

Water table The underground depth at which point the ground is totally saturated by water.

CASE STUDY

Fire sprinklers have proven to be 97 percent effective in controlling fires. They control the fire before it gets the chance to grow and produce smoke and toxic byproducts—the things that kill most people in dwelling fires. They are almost like having a fire fighter stationed inside your home, ready to respond 24 hours a day and 7 days a week. They are never busy on another assignment or delayed in their response. They are not understaffed or suffering a rolling brownout due to budgetary shortfalls.

A common misperception, promoted in television shows and movies, is that sprinklers all activate at once. This is not true. The only sprinkler heads activated are those affected by the heat of the fire. Fire sprinklers add approximately 1 percent to the cost of a home.

Sprinklers are the most effective fire safety device ever invented. The National Fire Protection Association reports that people with working smoke alarms in their home have a 50 percent better chance of surviving a fire. Adding sprinklers and smoke alarms increases your chances of surviving a fire by over 97 percent.

1. How effective are sprinklers in controlling fires?

   A. 25 percent
   B. 50 percent
   C. 75 percent
   D. 97 percent

2. How are residential sprinkler heads activated?

   A. Switch
   B. Heat
   C. Smoke alarm
   D. Manually

3. What kills most people in residential fires?

   A. Fire control efforts
   B. Structural collapse
   C. Burning to death
   D. Smoke and toxic byproducts

4. What is the most effective fire safety device ever invented?

   A. Fire sprinklers
   B. Fire fighters
   C. Fire engines
   D. Smoke detectors

REVIEW QUESTIONS

1. What is the most common extinguishing agent used by fire fighters?

2. List three components of a water supply system.

3. What are the differences between direct pumping and gravity-fed water systems?

4. List the names of the mains in a distribution system and show their relationship to each other.

5. Why is it necessary for a supply system to be gridded?

6. What is the difference between a wet barrel and dry barrel hydrant?

7. List the steps to be performed in hydrant maintenance.

8. How are hydrants flow tested?

9. What are the component parts of a water systems program?

10. List two types of detection systems and explain their operation.

11. Why are fire sprinklers so important in new home construction?

12. What are the differences between dry pipe and wet pipe sprinkler systems?

13. List two fire suppression systems that do not use water and explain how they work.

14. In terms of wildland use, list short-term and long-term retardants and explain how they are used in firefighting.

15. What is a foam eductor, and how does it work?

DISCUSSION QUESTIONS

1. From an engineering standpoint, what can be done to reduce the risk of damage from hostile fires in structures in your community?

2. With fire sprinklers rated as 97 percent effective in stopping hostile fires, why are they not included in more structures? What can be done to change this?

3. Justify buying a new pumper with a built-in foam system, even when it is more expensive than one without a built-in system.

REFERENCES AND ADDITIONAL RESOURCES

CDC/NIOSH Alert. 2005. *Preventing Injuries and Deaths of Fire Fighters Due to Truss System Failures.* Publication Number 2005-132. May. www.cdc.gov/niosh

Centers for Disease Control and Prevention. 2017. *QuickStats: Number of Deaths Resulting from Unintentional Carbon Monoxide Poisoning, by Month and Year — National Vital Statistics System, United States, 2010–2015.* Morbidity and Mortality Weekly Report (MMWR); 66:234. doi:10.15585/mmwr.mm6608a9

City of Scottsdale, AZ. 2013. "Fire Sprinkler Systems: Meet Your Personal Fire fighter." *ScottsdaleAZ.gov.* https://www.scottsdaleaz.gov/fire/residentialsprinkler

Fire, Frank L. 2009. *The Common Sense Approach to Hazardous Materials.* New York, NY: Fire Engineering.

Friedman, Raymond. 2008. *Principles of Fire Protection Chemistry and Physics.* Quincy, MA: National Fire Protection Association.

Haddon, Carl. August, 2017. "Encapsulator Agent Proves Safe and Effective on Transformer and Solar Panel Fires." *Fire Apparatus and Emergency Equipment.* www.fireapparatusmagazine.com/articles/print/volume-22/issue-8/departments/to-the-rescue/encapsulator-agent-proves-safe-and-effective-on-transformer-and-solar-panel-fires.html

Hubbard, Lauren. April, 2019. "What Caused the Massive Fire at Notre-Dame Cathedral?" *Town & Country Magazine.* www.townandcountrymag.com/leisure/arts-and-culture/a27166214/notre-dame-fire-cause/

Insurance Services Office (ISO). 2013. *Fire Protection Rating Schedule.* New York, NY: Verisk.

Koorsen Fire & Security. August, 2017. *What Are the Best Clean Agent Fire Suppression Systems?* https://blog.koorsen.com/what-are-the-best-clean-agent-fire-suppression-systems

Lamm, Willis. 2001. *Procedures for Flow Testing Hydrants.* Orinda, CA: Moraga-Orinda Fire District. http://www.firehydrant.org/info/ftest1.html

National Fire Equipment System Course. 2018. *Basic Air Operations.* Boise, ID: National Interagency Fire Center.

National Fire Protection Association. 2008. *Fire Protection Handbook,* 20th ed. Quincy, MA: National Fire Protection Association.

National Fire Protection Association. 2018. *Fire Sprinkler Initiative: Bringing Safety Home.* Quincy, MA: National Fire Protection Association. https://www.nfpa.org/Public-Education/Staying-safe/Safety-equipment/Home-fire-sprinklers/Fire-Sprinkler-Initiative

National Fire Protection Association. *NFPA 10: Standard for Portable Fire Extinguishers.* Quincy, MA: National Fire Protection Association.

National Fire Protection Association. *NFPA 11: Standard for Low-, Medium-, and High-Expansion Foam.* Quincy, MA: National Fire Protection Association.

National Fire Protection Association. *NFPA 12: Standard on Carbon Dioxide Extinguishing Systems.* Quincy, MA: National Fire Protection Association.

National Fire Protection Association. *NFPA 13: Standard for the Installation of Sprinkler Systems.* Quincy, MA: National Fire Protection Association.

National Fire Protection Association. *NFPA 13D: Standard for the Installation of Sprinkler Systems in One- and Two- Family Dwellings and Manufactured Homes.* Quincy, MA: National Fire Protection Association.

National Fire Protection Association. *NFPA 13E: Recommended Practice for Fire Department Operations in Properties Protected by Sprinkler and Standpipe Systems.* Quincy, MA: National Fire Protection Association.

National Fire Protection Association. *NFPA 13R: Standard for the Installation of Sprinkler Systems in Low-Rise Residential Occupancies.* Quincy, MA: National Fire Protection Association.

National Fire Protection Association. *NFPA 14: Standard for the Installation of Standpipe and Hose Systems.* Quincy, MA: National Fire Protection Association.

National Fire Protection Association. *NFPA 17A: Standard for Wet Chemical Extinguishing Systems.* Quincy, MA: National Fire Protection Association.

National Fire Protection Association. *NFPA 18: Standard on Wetting Agents.* Quincy, MA: National Fire Protection Association.

National Fire Protection Association. *NFPA 25: Standard for the Inspection, Testing, and Maintenance of Water-Based Fire Protection Systems.* Quincy, MA: National Fire Protection Association.

National Fire Protection Association. *NFPA 291: Recommended Practice for Fire Flow Testing and Marking of Hydrants.* Quincy, MA: National Fire Protection Association.

National Fire Protection Association. *NFPA 1150: Standard on Foam Chemicals for Fires in Class A Fuels.* Quincy, MA: National Fire Protection Association.

National Foam. 2019. *High Expansion Foam.* http://nationalfoam.com/foam-concentrates/high-expansion-foams/high-expansion/

Naum, Christopher. 2011. *One Meridian Plaza High Rise Fire: Twenty Years Ago.* Tulsa, OK: PennWell Publishing. The Company Officer.com. http://thecompanyofficer.com/tag/high-rise-firefighting/

U.S. Fire Administration. 2019. *Smoke Alarm Outreach Materials.* Emmitsburg, MD: U. S. Fire Administration. http://www.usfa.fema.gov/prevention/outreach/smoke_alarms.html

Victaulic Corporation. 2017. *Fire Protection Systems Catalog* (p. 67). Easton, PA: Victaulic Worldwide.

CHAPTER **13**

Emergency Incident Management

OBJECTIVES

After studying this chapter, you should be able to:

- Discuss the management responsibility at an emergency incident.
- Explain the need for a plan at every incident.
- List and describe the five major components of the National Incident Management System (NIMS).
- List and describe the components of the Incident Command System (ICS).
- List the positions in the ICS and their functions.

Case Study

On June 18, 2007, nine career fire fighters (all males, aged 27–56) died when they became disoriented and ran out of air in rapidly deteriorating conditions inside a burning commercial furniture showroom and warehouse facility. The first-arriving engine company found a rapidly growing fire at the enclosed loading dock connecting the showroom to the warehouse. The assistant chief entered the main showroom entrance at the front of the structure but did not find any signs of fire or smoke in the main showroom.

He observed fire inside the structure when a door connecting the rear of the right showroom addition to the loading dock was opened. Within minutes, the fire rapidly spread into and above the main showroom, the right showroom addition, and the warehouse. The burning furniture quickly generated a huge amount of toxic and highly flammable gases along with soot and products of incomplete combustion that added to the fuel load. The fire overwhelmed the interior attack, and the interior crews became disoriented when thick black smoke filled the showrooms from ceiling to floor. The interior fire fighters realized they were in trouble and began to radio for assistance as the heat intensified. One fire fighter activated the emergency button on his radio. The front showroom windows were knocked out and fire fighters, including a crew from a mutual-aid department, were sent inside to search for the missing fire fighters. Soon after, the flammable mixture of combustion by-products ignited, and fire raced through the main showroom. Interior fire fighters were caught in the rapid fire progression, and nine fire fighters from the first-responding fire department died. At least nine other fire fighters, including two mutual-aid fire fighters, barely escaped serious injury.

National Institute for Occupational Safety and Health (NIOSH) investigators concluded that, to minimize the risk of similar occurrences, fire departments should:

- Develop, implement, and enforce a written incident management system to be followed at all emergency incident operations.
- Develop, implement, and enforce written standard operating procedures that identify

incident management training standards and requirements for members expected to serve in command roles.
- Ensure that the incident commander is clearly identified as the only individual with overall authority and responsibility for management of all activities at an incident.
- Ensure that the incident commander conducts an initial size-up and risk assessment of the incident scene before beginning interior firefighting operations.
- Train fire fighters to communicate interior conditions to the incident commander as soon as possible and to provide regular updates.
- Ensure that the incident commander establishes a stationary command post, maintains the role of director of fireground operations, and does not become involved in firefighting efforts.
- Ensure the early implementation of division/group command into the Incident Command System.
- Ensure that the incident commander continuously evaluates the risk versus gain when determining whether the fire suppression operation will be offensive or defensive.
- Ensure that the incident commander maintains close accountability for all personnel operating on the fireground.
- Ensure that a separate incident safety officer, independent from the incident commander, is appointed at each structure fire.
- Ensure that crew integrity is maintained during fire suppression operations.

1. How could the application of an incident management system have affected the outcome of this tragic incident?
2. What is the importance of a designated safety officer at incidents?
3. Why is it important for the incident commander not to become directly involved in fire suppression operations at a fire incident?
4. Why is it so important to constantly evaluate risk versus gain at any incident?

National Institute for Occupational Safety and Health. 2009 . "Nine Career Fire Fighters Die in Rapid Fire Progression at Commercial Furniture Showroom—South Carolina F2007-18." NIOSH website. http://www.cdc.gov/niosh /re/reports/face200718.html

Introduction

Incidents come in all types and sizes. It is the responsibility of every responder at the incident scene to do his or her part to ensure a swift and successful conclusion to the incident. Emergency incident management is one of the main factors in the successful outcome of incidents in which more than one company is assigned. This is true of small incidents with two companies all the way up to disasters requiring a multijurisdictional or national response, such as the Gulf of Mexico oil spill in 2010, Hurricane Sandy in 2012, the Boston Marathon bombing in 2013, and Hurricane Florence in 2018.

To aid in the effective management of incidents, the Incident Command System (ICS) has been developed. The ICS presents a structure that is adaptable to all incident types. The ICS facilitates resources from different agencies, public and private, working together in a coordinated fashion. This is accomplished through training, common terminology, and an established command structure. By learning the ICS prior to an incident, resources from different agencies and disciplines can come together at the scene and operate in an effective, coordinated manner.

At any incident, a coordinated response and proper utilization of resources is vital. The availability of resources is not unlimited, so you must use whatever you have effectively and efficiently. The incident management system discussed in this text is adopted for use by the National Incident Management System (NIMS). The national system was adopted as a result of numerous large incidents in which resources were not able to function properly due to misunderstandings on how the incident should be managed. By assuming one system and exercising it on a day-to-day basis, resources are able to communicate effectively, as well as understand and implement the system on the appropriate scale when necessary.

The ICS, as presented here, has been tested on major incidents of all types utilizing resources from the Federal Emergency Management Agency (FEMA), the United States Forest Service (USFS), the Bureau of Land Management (BLM), the National Aeronautics and Space Administration (NASA), the Environmental Protection Agency (EPA), the Federal Bureau of Investigation (FBI), the Department of Homeland Security (DHS), the Bureau of Alcohol, Tobacco, Firearms and Explosives (BATF), state and local governments, the Red Cross, the Salvation Army, and the military. The ICS has been a proven performer and is constantly adapting and growing to meet incident-driven needs.

Management Responsibility

Emergency incident management is primarily the responsibility of the first-in fire officer because most incidents require the resources of only one company. However, not all incidents can be handled by the resources that initially arrive at the scene. Should an incident escalate, the first-in officer has the opportunity to turn the incident into a major disaster through poor management or, conversely, to have a positive effect on the outcome of the incident. Some situations require more resources and possibly even long periods of time to control. The first-in personnel must have a complete knowledge of the hazards and potential of the incident as well as their own priorities and capabilities. The only way an incident can be brought under control is through the total and coordinated efforts of all personnel at the scene.

It is the responsibility of every fire fighter at the scene to be involved in assisting in the control of an incident. A new fire fighter may feel that he or she is not involved in the decision-making process. This is not true; all incidents require that everyone at the scene be alert and aware of the hazards and needs at the incident as it develops. The standard rule is, "Victims do not arrive at the scene in fire trucks." Your actions at the incident have a direct impact on its outcome. For this reason you need to be aware of the decision-making process and capable of participating in it.

> ## Tip
>
> Should an incident escalate, the first-in officer has the opportunity to turn the incident into a major disaster through poor management or, conversely to have a positive effect on the outcome of the incident through proper management. It is the responsibility of every fire fighter at the scene to be involved in assisting in the control of the incident.

Incident Planning

Every incident must have a plan. The plan must be formulated to achieve effective utilization of resources and to resolve the incident with as little further damage as possible. The first decision to be made in forming an incident action plan is what the objectives are. Objectives are defined as something toward which effort is directed; an aim or end of action; a goal. If there is no clearly defined goal in mind for the incident, it cannot be determined how to achieve the goal, nor can progress toward the goal be measured. Objectives for different kinds of incidents will, of course, differ. The objectives in a hazardous materials incident may be only to control the spread of the material. In a wildland incident, they may be, initially, to protect structures in the fire's path and then to effect control. Objectives must be clearly stated and understood as well as being measurable and attainable.

> **Tip**
>
> Every incident must have a plan. The plan must be based on objectives that are clearly stated, measurable, attainable, and understood.

Strategy

The next step in the formulation of the plan is to determine the strategies necessary to achieve the objectives. Strategy is the art of devising plans to achieve a goal or objective. From a firefighting standpoint, **strategy** may be defined as the method used to coordinate the tactical operations of units to achieve the desired incident objectives. All of the strategies must support the objectives. Strategies are based on the size-up of the incident, objectives, strategic priorities, and mode of attack selected.

Tactics are the actions taken to achieve the strategies. Examples of tactics would be performing an interior search, ventilation, and advancing hose lines to the interior of a structure. In a wildland situation, tactics could include constructing fire control lines with bulldozers and hand crews.

Size-Up

At this point, it is common to ask yourself, "How will I ever be able to take into account all of the factors that are required to come up with a plan of operation?" In his book, *Fire Fighting Tactics*, Lloyd Layman presented the process of size-up (Layman 1972). The first-in officer is to size up the incident. Size-up is a mental process requiring the incident commander (IC) to perform the following steps:

1. *Determine facts when the alarm is received.* The time of day, which will affect whether life safety is likely to be a major factor; the weather, especially important in wildland incidents; the address or location of the incident; what type of

occupancy is involved, such as a school, hospital, or vehicle; what type of incident is reported—fire, traffic accident, or hazardous materials spill. These are just a representative sample of the facts the fire fighter can start to consider as soon as the alarm is received. Size-up starts even before units arrive at the scene.

Preplanning is also a part of the size-up function. Through proper preincident planning and inspection, many facts about the occupancy involved will already be known. A tremendous aid in entering a business to fight a fire after dark is to have been on a preplan or inspection tour under nonemergency conditions. Knowing something of the building layout is a great help when the smoke is thick and the rooms are pitch black.

2. *Anticipate probabilities.* Probabilities are the likelihood of what may be involved in the reported incident. If it is a reported small brush fire on a hot, windy, summer day, it will probably spread quickly. A traffic accident or car fire at rush hour on the freeway will probably cause access problems for fire apparatus. Some probabilities can be anticipated through study and experience; others cannot.

3. *Assess your own situation.* This refers to the actual situation facing the fire fighter. This could mean the ladder truck is in the shop for repair and the building on fire is higher than the longest ladder available at the scene. Perhaps the fire is putting up a large column of smoke, visible as you leave the station, or perhaps there is no smoke showing and building occupants report that they have it under control **FIGURE 13-1**. There is a wide variety of situations that you will be confronted with

FIGURE 13-1 When performing size-up, all related facts must be taken into consideration. No two incidents are exactly the same.

Courtesy of Kern County Fire Department

over the course of your career. You must be prepared to think clearly and act decisively in all of them.

4. *Make a decision.* This is the time when the fire fighter, considering the facts, probabilities, and his or her own situation, determines what is the best course of action under the given circumstances. Are additional resources needed? Should the attack strategy be an offensive or defensive mode of attack? This is when the IC makes or breaks the incident; it is the time to determine the objectives, strategies, and tactics. If the strategy is too conservative, the incident could exceed control measures; if it is too liberal, resources might be wasted and unavailable if another incident occurs.

5. *Plan the operation.* This is when the tactics are put into action. The IC communicates the plan to the resources at the scene and things start to happen. The attack is started. Resources report back with their progress. Are we winning or losing? Are we going to save the structure or create a new parking lot?

Size-up does not stop when the plan is put into operation. When resources report their progress and the IC observes what is happening, the process starts over again with new facts. As the facts, probabilities, and a fire fighter's own situation change, there will have to be new decisions and adjustments to the plan of operation. As previously mentioned, size-up is a process. It starts before resources leave the station and continues throughout the incident.

> **Tip**
>
> Do not stop your size-up when the plan is put into operation. Size-up is an ongoing process that reevaluates incident factors and resource accomplishments in controlling the incident.

The person who wishes to become an effective IC will go back over the size-up process after the incident and evaluate where improvements could be made. A group critique of the operations performed is a good idea. This is called an *after-action review* (AAR). AARs should be examinations of what went right and wrong. When approached positively and not used to place blame or criticism, the people involved can be honest, and the AAR can be used as a training tool.

When you arrive at the scene, you may not have enough resources to have any direct effect on the incident at that time. One thing that can be done is to give a comprehensive size-up that will help your supervisor determine what the incident's resource needs are going to be. It is often better to scout the incident and give a good size-up than to get involved in taking action that will have little effect. By giving a good description, your supervisor can get the additional resources needed started to the scene. Time is critical. If the supervisor is still several minutes out, he or she can at least get the additional resources ordered. Most ICs would rather have to turn some resources around than not have enough and have the incident overwhelm the resources.

> **Tip**
>
> The person who wishes to become an effective IC will go back over the size-up process after the incident and evaluate where improvements could be made.

Wildland Fire Report of Conditions

One of the benefits of a complete report of conditions is that it requires the consideration of many of the elements necessary to perform a size-up. The following list is an example of the requirements for a report of conditions for wildland fires. This information should be given to your supervisor or the dispatch center over the radio. The other units responding should be able to hear your description as well (National Wildfire Coordinating Group 2018).

1. *Correct location.* Often, reports of wildland fires come in from passersby or persons who see the smoke but are some distance from the fire. It is important to make sure to give a corrected location when necessary. This may also require you to give the other responding units directions to the scene. By doing so, responding units can more easily approach the fire and the dispatch center can give the corrected information to any aircraft that are requested.

2. *Size.* The current size of the fire. This is given as the number of acres involved. No one is perfect at this, especially when the fire is in hilly terrain. Give your best estimate **FIGURE 13-2**.

3. *Fuel type.* In determining the type of additional resources that may be necessary, the type of fuel burning is critical. In grass and light fuels, control and extinguishment can usually be accomplished with pumpers and

FIGURE 13-2 It is especially difficult to size up a wildland fire at night.
Courtesy of Kern County Fire Department

dozers assisted by aircraft when necessary. The fuel burns clean and leaves little mop-up work. In heavy fuels, such as brush and timber, the extinguishment and mop-up require hand crews. With these fuels, helicopters are very effective. Also, the heavier fuels are more likely to produce spot fires.

4. *Slope and aspect.* The slope of the hill the fire is on has a lot to do with its rate of spread. A fire out in the flats is not going to spread nearly as rapidly as a fire at the toe of a slope. A fire on the very top of a hill tends to burn downhill slowly. The slope also affects the type of resources needed. A steep slope may not allow pumping equipment to be used without putting in hose lays. This tactic is much slower to place into operation than pump and roll tactics. Slopes that are too steep or rocky for dozers or hand crews to operate will necessitate an indirect attack.

   Aspect is the relationship of the slope to the points of the compass. A fire on a south-facing slope tends to have lighter fuels and have the sun shining on it for most of the day. The fuels will also be drier, causing them to burn more rapidly. North-facing slopes tend to have heavier fuels and do not get as much sun.

5. *Rate of spread.* The rate of spread is given as slow, moderate, or rapid. In planning for resource needs, the rate of spread of the fire directly affects the size of the fire perimeter by the time the additional resources arrive. The higher the rate of spread, the more difficult the fire will be to contain, and more control line will have to be put in due to its increased size.

6. *Exposures.* The exposures should be specified as to the number and type. Exposures consist of structures and other improvements, such as power poles. When structures are endangered, resources must be ordered to protect them because they are of the highest priority.

7. *Weather conditions.* Wind is going to have a direct effect on the spread of wildland fires. They tend to spread rapidly with the wind and slowly against it. A strong downslope wind can drive a fire at extremely high rates of spread in a downhill direction. This is contrary to the common wisdom that fires spread uphill at a faster rate.

8. *Potential of the fire.* Is the fire going to be easy to control or is this going to be a major operation? Will resources be needed for a couple of hours or a couple of days? Should the order be placed for a base/camp to be set up?

9. *Additional resource needs.* In your best estimate, what additional resources are needed to control the fire? Resources from outside your own department can require a 4- to 8-hour lead time. If you need help, ask for it. One of the great truths of firefighting is that if you do not order what you need, it is not coming.

10. *Objectives.* All of the resources responding need to know what the plan is. Clearly stated objectives, as well as a good size-up, can aid them in getting a mental picture of what you have and what you will need them to do when they arrive.

Structure Fire Report of Conditions

In the size-up, or report of conditions, of a structure fire, many of the considerations are the same as for wildland fires. These considerations include:

1. *Correct location.* This is self-explanatory.
2. *Height/stories.* The height of the structure involved. Specify which floor the fire is on.
3. *Size.* The size of the structure. There is a big difference in the resource requirements for a single-family dwelling versus a large commercial building.
4. *Type of structure.* A wood-frame structure is handled differently than a precast concrete tilt-up structure. Factories have different resource requirements than hospitals.
5. *Location and area involved.* Is the fire in the front or the back? The size of the area involved is usually given as a percentage. For example, 10 percent up to (and including) fully involved.
6. *Level of involvement.* Whether the fire is visible. Examples are nothing showing, smoke showing, smoke and flames showing, and through the roof.

7. *Exposures.* List the number and type of exposures. This could be a vehicle parked next to the structure or a separate building in proximity that is threatened. Remember that radiated heat is not affected by wind.

8. *Potential of fire.* Is the fire small or is it coming out of every window and threatening to spread to the adjacent structures?

9. *Additional resource needs.* How many and what type of additional resources are needed? For example, one more engine, three more engines, or a ladder truck.

10. *Objectives.* State your priorities. When the structure is fully involved, all you may be able to do is protect the exposures until sufficient resources arrive and water supply is set up to attack the fire. Specify whether the attack is to be offensive, defensive, or combination.

11. *Obtain an all clear.* Life safety is the primary objective at any incident and the all clear signifies that the primary search is completed and all victims who could be rescued have been. This is not going to be possible in every case but should always be one of the first considerations.

One of the best things the first-arriving unit can do is to give the others responding a clear description of what is occurring. This way they get a mental picture of the incident and start their own size-up process. At this time, the responding higher-level supervisor may wish to pose questions as to what you have and what you need. This helps him or her to get his or her thought process started as well.

Strategic Priorities

Most of the strategic priorities in use today are based on those identified by Layman. Layman divided the strategic priorities into the following seven areas (Layman 1972):

1. Rescue is the first strategic priority. Life safety is the most important consideration in any incident operation. In fire situations, this may require evacuating people from the fire's path, searching structures looking for trapped occupants, and erecting ladders to pull people from imminent danger. Another method of rescuing people from fires is to place hose lines that cut off the fire spread toward the trapped persons into operation. In hazardous materials situations, rescues may involve evacuation, sheltering in place, or actually pulling people out of tanks and contaminated areas. The first

search of the incident is called the primary search. When it is completed and all live persons are rescued, the all clear message is transmitted to the IC. Any search after that is called a secondary search. Some ICs are of the opinion that if the incident is so well involved that no initial search can be conducted, any search conducted at the incident is considered a secondary search.

2. Exposures are those items that are not yet involved in the incident, but soon will be if no action is taken to protect them. As mentioned in Chapter 4, *Chemistry and Physics of Fire*, radiated heat is a very real problem at fire scenes. When hose lines are placed into operation to protect adjoining structures or foam is applied to the exterior of structures to protect them from advancing wildland fires, these tactics are considered exposure protection **FIGURE 13-3**. Exposures can also exist inside a structure. When a fire spreads through a structure, it usually starts in one area, so other areas of the structure can be considered interior exposures. Interior hose lines are placed so spread of the fire is prevented from happening, in effect protecting interior exposures.

3. Confinement is the act of stopping the spread of the incident. Depending on the scale of the incident, confinement can have different meanings. In a hazardous materials incident, confinement may involve diking the spill to keep it out of a storm drain or waterway. In a structure fire incident, it may mean confining the fire to the building or even block of origin. A typical confinement objective on wildland fires is to confine the fire to one canyon or one side of a road.

FIGURE 13-3 Fire fighters performing exposure protection at a structure fire incident.

4. Extinguishment in the case of fires is the act of putting the fire out. In terms of hazardous materials, it could be equated to stopping the source of the leak.

5. Overhaul is performed when the fire is extinguished. Fire fighters must make sure that the fire is totally out, which often requires checking concealed spaces in attics and inside walls. Overhaul also includes placing the structure into a condition in which it can be turned over to the owner. This does not include structural repairs. On smaller incidents, such as a kitchen fire, it usually includes cleaning up and removing fire debris and water and evacuating smoke. This is not only good safety practice, but good public relations as well.

   At wildland fires, overhaul is commonly called mop-up. Mop-up consists of searching out embers that could threaten control lines, dropping dangerous trees, and improving the fire line.

   Although dirty and non-glamorous, overhaul/mop-up is extremely important, often taking longer than extinguishment operations to accomplish. There is nothing like having to return to a fire scene to combat a new fire due to a rekindle caused by insufficient or improper overhaul/mop-up. During overhaul operations is no time to let down your guard. In structure fires, at this point in the fire, carbon monoxide concentrations are at their highest, and self-contained breathing apparatus (SCBA) must be worn until the atmosphere can be changed. Fire may have weakened the structure to the point that it is not safe to enter, and operations may have to be conducted from the exterior. At wildland fires, trees that were damaged by the fire have a tendency to fall down, so a lookout must be maintained when working around them.

6. Salvage operations can be performed at any time during the incident. Salvage is the act of saving the contents of the building from damage due to incident operations. We cannot do anything about the damage that has occurred before we get to the scene, but we can try our best to prevent any unnecessary further damage after we arrive. Firefighting usually involves applying water to control the fire. Salvage is the actions taken to protect building contents from damage caused by water or falling debris. Salvage covers are used to cover furniture and other objects to prevent damage; where applicable, objects are carried outside. In some areas these items must be guarded to prevent them from being stolen. After wildland fires, dozer and hand lines are rehabilitated to prevent erosion when the rain comes. Salvage is not just good practice to reduce dollar loss; like overhaul, it is good public relations and a part of modern firefighting professionalism.

7. Ventilation is another function that can be performed at any time during incident operations. Considered an incident support operation, ventilation aids in quick and efficient fire control as well as enhancing safety. Ventilation reduces the possibility of flashover and backdraft. It also makes the firefighting job easier by increasing visibility and reducing heat and steam felt by the fire fighters performing the attack. Ventilation properly performed and used as part of a coordinated attack can also act as a confinement tool on the fireground. By venting smoke and heat buildup from an attic, the attic can be prevented from reaching its ignition temperature. Some types of ventilation operations can be used to cut off fire spread by redirecting heat to the outside of the structure.

The seven strategic priorities are not necessarily performed in the order in which they are presented. The main idea is to have a list of options to consider and to aid in the mental process of prioritizing operational objectives.

> **Tip**
>
> Rescue is the first strategic priority. Life safety is the first priority at any incident, and any victims that are savable should be dealt with first.

The common abbreviation for Layman's strategic priorities is RECEO SV.

Modes of Attack
Offensive Mode

Strategies are based on differing modes of attack. In the offensive mode, resources are applied directly to controlling the fire or other incident. This method is often used on small fires or incidents where fire fighters can gain access without unreasonable risk to their safety and there is still something to save. There have been instances where rescuers have lost their lives on incidents that were essentially body recovery. The IC must constantly be aware of the risks versus benefits of actions taken at an incident. It is a common saying that fire fighters will risk much to save much, such as the lives of viable victims, and risk little to save little. As a fire fighter, it is your responsibility to be aware of what is going on around you and to evaluate your situation constantly. Direct attack on a wildland fire or interior attack on a structure fire would be considered the offensive mode **FIGURE 13-4**.

Defensive Mode

In situations where the fire is too large or too well established, the defensive mode is used. A defensive fire attack would primarily involve the protection of exposures. In the constant evaluation of risk versus benefit, if it is found that the risk to personnel is too high in relation to the outcome of their actions, the IC should initiate actions in the defensive mode or switch from offensive to defensive mode **FIGURE 13-5**. Structure fires where structural collapse is imminent or has already occurred would require the defensive mode. Attacking a block-sized structure fire with exterior lines and elevated streams from ladder trucks would

FIGURE 13-4 Fire fighters performing offensive interior attack at structure fire incident.
Courtesy of Kern County Fire Department

FIGURE 13-5 Fire fighters performing defensive operation with an elevated stream from a ladder truck. The structure is too compromised by fire damage to enter.
Courtesy of Kern County Fire Department

be considered defensive mode. A wildland fire being pushed through brush with 30-ft flame lengths would also require the defensive mode. Hazardous materials incidents, which require a precautionary approach, are usually handled initially in the defensive mode until the material and its hazards can be identified.

> **Tip**
>
> In the offensive mode, resources are applied directly to controlling the fire or other incident.
>
> In situations where the fire is too large or too well established, the defensive mode is used.
>
> There are situations where both offensive and defensive modes may be used at the same time—in a carefully coordinated fashion—to control an incident.

Combination Mode

There are situations where both offensive and defensive modes may be used in a combined attack mode to control an incident. This must be done in a carefully coordinated attack. The combination mode requires good communications between incident command and the suppression resources, and among the suppression resources themselves, to avoid one group adversely affecting the safety or operations of the other. Combined attacks are often used on large wildland incidents in the form of direct attack on the slow-moving parts of the fire and indirect attack on the rapidly moving parts. Another example is to add structure protection. The flanks of the fire are being attacked directly or indirectly to establish **perimeter control**. Other resources are assigned within the fire perimeter, protecting structures as the fire moves past. Resources assigned to perform **structure protection** do not take action to control the perimeter of the fire; their sole function is to protect the structures in the fire's path.

Combination attacks require careful coordination and should not be used for interior attack structural firefighting. The introduction of hose streams into

> **SAFETY TIP**
>
> Combination attacks require careful coordination and should not be used for interior-attack structural firefighting. The introduction of hose streams into a ventilation hole or other opening may reverse the outflow of smoke and heat, causing injury or death to interior attack forces.

a ventilation hole or other opening may reverse the outflow of smoke and heat, causing injury or death to interior attack forces.

Tactics

Tactics are the direction and employment of resources to achieve the set objectives. Engine companies are usually involved at the tactical level of firefighting. Once the objectives are determined and communicated and the strategies are selected, the individual companies can employ the tactics necessary to achieve the objectives.

Some examples of tactics would be to perform ventilation to control fire spread or putting in a dozer line and then backfiring to slow the head of a wildland fire.

Tasks

Tasks are jobs that should be completed in a specified period of time and are necessary to achieve the tactics required by the incident plan. Most new fire fighters will initially be expected to operate on the task level. An operation such as advancing a hose line into a burning structure requires the execution of several tasks. One of the first would be to don the SCBA. Another would be taking the hose line from the apparatus and stretching it out. As you can see, even a simple tactic is made up of a good number of tasks. It is very important that the fire fighter master these tasks. Every chain is only as strong as its weakest link. If a fire fighter does not know how to don the breathing apparatus quickly, the hose line will not be stretched in a timely manner and the whole operation will be slowed down. In firefighting, efficiency and proficiency lead to incident effectiveness.

To look at this whole process and the relationship of the parts, let's look at a relatively simple, but common, fireground operation at the scene of a structure fire with one bedroom involved. The objectives would be to ensure life safety and confine the fire to the room involved, causing as little property damage as possible. The strategies would be primary search and interior attack, offensive mode. The tactics would be to ventilate the roof, advance a hose line through the front door attacking the fire at its seat, conduct the search, and perform salvage. Some of the tasks involved would be to don SCBA, ladder the roof, cut the ventilation hole, advance the hose line, search the structure, remove any occupants, attack the seat of the fire, and so on. It is easy to see that the incident could require the effort of numerous personnel to bring it to a swift and successful conclusion. This incident could easily utilize two engine companies and a truck company to perform all of the operations.

Even an incident of this small size requires a plan with closely coordinated management and effective communications among all of the personnel involved. When the companies start the attack, a lack of coordination may mean that one company directs its hose stream into a window while the other company is advancing its hose through the interior. The interior attack personnel are going to get steam burns if this happens. If things do not go as planned, and the incident switches to the defensive mode, it will become more complex still. Exposures will have to be protected, water supplies will need to be developed further, crews will need rehabilitation, and so forth.

Without a well-thought-out plan of operations, the incident we just looked at can easily turn into a minor disaster. All of the fire fighters at the scene must be aware of how the operation is proceeding. If the plan is not working, the objectives, strategy, and tactics will have to be changed to take into account the new situation.

One of the key elements here is communication. The IC needs to communicate the plan to the company commanders, and the company commanders must communicate their portion of the plan to their companies **FIGURE 13-6**. Operations should not proceed until all participants have a clear understanding of their responsibilities. This does not mean that they stand around for 30 minutes and discuss it. The personnel, through their knowledge of strategic priorities and tactics, should not need lengthy instructions. The IC tells the company commander to make an interior attack and primary search. The company commander tells his or her crew which size hose line to use and that they are going to make an interior attack and primary search. The fire fighters know that SCBA will be required, as well as what other tools to bring in case forcible entry is necessary. Beforehand, the fire

FIGURE 13-6 IC and Operations Chief at a command post discussing strategy and tactics in preparation to communicate them to the companies at the incident by radio.

Courtesy of Kern County Fire Department

fighters should know whose responsibility it is to pull the hose and who will get the tools.

Communication should be two-way. The IC not only needs to communicate decisions to the company commanders, but must also receive information from them in return. At an incident of any size much larger than a vehicle fire, the IC cannot see all of the incident. In a structure fire, the IC cannot see through the walls. At a wildland incident, the fire may be spread over many acres, even extending over the back side of a hill or into a canyon. On large wildland incidents, the incident command post may be miles from the actual incident. When an objective is met, the IC may have other work for the involved company and needs to know the company is available. If some part of the plan is or is not working, it must be communicated back to the IC. When the operation is going wrong, the plan must be amended to correct the deficiencies. The IC cannot make these decisions if not informed of the necessity for change.

An example of this would be from the previously mentioned bedroom fire. The truck company is told to ventilate the roof. The engine company is waiting to perform interior attack once the hole is completed. If the truck company officer does not notify the IC that the ventilation operation is completed, the engine company will be waiting at the door for the order to advance its line. Once the hole is completed, the IC may want the truck company to start salvage operations. Without two-way communication, the operations are slowed down and inefficient use is made of resources. At worst, the fire develops to the point where it is beyond the capabilities of the resources at the scene or someone is seriously injured.

National Incident Management System

In response to attacks on September 11, 2001, then-President George W. Bush issued Homeland Security Presidential Directive 5 (HSPD-5) in February 2003. HSPD-5 called for the creation of NIMS; identified steps for improved coordination of federal, state, local, and private industry response to incidents; and described the way these agencies will prepare for such a response.

The secretary of the DHS announced the establishment of NIMS in March 2004. One of the key features of NIMS is the ICS.

The Incident Command System

An incident is an occurrence, caused either by humans or by natural phenomena, that requires response actions to prevent or minimize loss of life or damage to property and/or the environment. The ICS is applicable to all incident types.

Examples of incidents include:

- Fire, both structural and wildland
- Natural disasters, such as tornadoes, floods, ice storms, or earthquakes
- Human and animal disease outbreaks
- Search and rescue missions
- Hazardous materials incidents
- Criminal acts and crime-scene investigations
- Terrorist incidents, including the use of weapons of mass destruction
- National special security events, such as presidential visits or the Super Bowl
- Other planned events, such as parades or demonstrations

Given the magnitude of these types of events, it is not always possible for any one agency alone to handle all the management and resource needs. Partnerships are often required among local, state, tribal, and federal agencies. These partners must work together in a smooth, coordinated effort under the same management system.

The ICS is a standardized, on-scene, all-hazard incident management concept. The ICS allows its users to adopt an integrated organizational structure to match the complexities and demands of single or multiple incidents without being hindered by jurisdictional boundaries.

The ICS has considerable internal flexibility. It can grow or shrink to meet different needs. This flexibility makes it a very cost-effective and efficient management approach for both small and large situations.

History of the Incident Command System

The ICS was developed in the 1970s by a still-active group known as FIRESCOPE (**FI**refighting **RES**ources of **C**alifornia **O**rganized for **P**otential **E**mergencies). The ICS was developed following a series of catastrophic fires in California's urban interface. Property damage ran into the millions of dollars, and many people died or were injured. The personnel assigned to determine the causes of this disaster studied the case histories and discovered that response problems could rarely be attributed to lack of resources or failure of tactics. What were the lessons learned?

Surprisingly, studies found that response problems were far more likely to result from inadequate management than from any other single reason. Weaknesses in incident management were often due to the following:

- Lack of accountability, including unclear chains of command and supervision
- Poor communication due to both inefficient uses of available communications systems and conflicting codes and terminology
- Lack of an orderly, systematic planning process
- No common, flexible, predesigned management structure that enables commanders to delegate responsibilities and manage workloads efficiently
- No predefined methods to integrate interagency requirements into the management structure and planning process effectively

A poorly managed incident response can be devastating to our economy and our health and safety. With so much at stake, we must effectively manage our response efforts. The ICS allows us to do so. It is a proven management system based on successful business practices. This text introduces you to basic ICS concepts and terminology.

ICS Built on Best Practices

The ICS is a proven management system based on successful business practices. It is the result of decades of lessons learned in the organization and management of emergency incidents.

The ICS has been tested in more than 30 years of emergency and nonemergency applications, by all levels of government and in the private sector. It represents organizational best practices, and—as a component of NIMS—has become the standard for emergency management across the country.

NIMS requires the use of the ICS for all domestic responses. NIMS also requires that all levels of government, including territories and tribal organizations, adopt the ICS as a condition of receiving federal preparedness funding.

What ICS Is Designed to Do

Designers of the system recognized early that the ICS must be interdisciplinary and organizationally flexible to meet the following management challenges:

- Meet the needs of incidents of any kind or size
- Allow personnel from a variety of agencies to meld rapidly into a common management structure
- Provide logistical and administrative support to operational staff
- Be cost effective by avoiding duplication of efforts

The ICS consists of procedures for controlling personnel, facilities, equipment, and communications. It is a system designed to be used or applied from the time an incident occurs until the requirement for management and operations no longer exists.

ICS Features

The ICS is based on proven management principles, which contribute to the strength and efficiency of the overall system. ICS principles are implemented through a wide range of management features, including the use of common terminology and clear text, and a modular organizational structure. The ICS emphasizes effective planning, including management by objectives and reliance on an incident action plan (IAP).

The ICS helps ensure full utilization of all incident resources by:

- Maintaining a manageable span of control.
- Establishing predesignated incident locations and facilities.
- Implementing resource management practices.
- Ensuring integrated communications.

The ICS features related to command structure include chain of command and unity of command, as well as unified command and transfer of command. Formal transfer of command occurs whenever leadership changes.

Through accountability and mobilization, the ICS helps ensure that resources are on hand and ready to deploy. The ICS also supports responders and decision makers by providing the data they need through effective information and intelligence management.

Common Terminology and Clear Text

The ability to communicate within the ICS is absolutely critical. An essential method to ensure the ability to communicate is using common terminology and clear text.

Tip

Clear text must be used for communications at incidents. Do not use radio codes, agency-specific codes, or jargon.

A critical part of an effective multiagency incident management system is for all communications to be in plain English. That is, use clear text. Do not use radio codes, agency-specific codes, or jargon.

The ICS establishes common terminology allowing diverse incident management and support entities to work together. Common terminology helps to define:

- *Organizational functions.* Major functions and functional units with incident management responsibilities are named and defined. Terminology for the organizational elements involved is standard and consistent.

- *Resource descriptions.* Major resources (personnel, facilities, and equipment/supply items) are given common names and are typed or categorized by their capabilities. This helps to avoid confusion and to enhance interoperability.

- *Incident facilities.* Common terminology is used to designate incident facilities.

- *Position titles.* ICS management or supervisory positions are referred to by titles, such as officer, chief, director, supervisor, or leader.

Each of the aforementioned areas will be covered in more detail in the remaining sections.

Modular Organization

The ICS organizational structure develops in a top-down, modular fashion that is based on the size and complexity of the incident, as well as the specifics of the hazard environment created by the incident. As incident complexity increases, the organization expands from the top down as functional responsibilities are delegated. The structure of ICS at an incident is "built to fit the incident."

The ICS organizational structure is flexible. When needed, separate functional elements can be established and subdivided to enhance internal organizational management and external coordination. As the ICS organizational structure expands, the number of management positions also expands to address the requirements of the incident adequately.

In the ICS, only those functions or positions necessary for a particular incident will be filled. On a small incident, such as a car crash into a tree with one victim, if one person can manage all of the major functional areas, no further organization needs to be implemented. Positions that are not assigned are assumed by the next higher level of management. Basically, if you are the IC and you do not assign an Operations Chief, then you are also the Operations Chief. In simple terms, regarding management responsibility, "what you do not assign, you assume." If one or more of the major functional areas reaches the point where it requires independent management, a person is assigned the responsibility of that function.

Tip

The typing of resources is expressed through *kind* and *type*. An engine would be a kind of resource (e.g., an engine versus a water tender). The type would be determined by the resource's capability. A Type 1 resource has a greater capability than a Type 2, and so on.

Management by Objectives

All levels of a growing ICS organization must have a clear understanding of the functional actions required to manage the incident. Management by objectives is an approach used to communicate functional actions throughout the entire ICS organization. It can be accomplished through the incident action planning process, which includes the following steps:

1. Understand agency policy and direction.
2. Assess incident situation.
3. Establish incident objectives.
4. Select appropriate strategy or strategies to achieve objectives.
5. Perform tactical direction (applying tactics appropriate to the strategy, assigning the right resources, and monitoring their performance).
6. Provide necessary follow-up (changing strategy or tactics, adding or subtracting resources, etc.).

The first objective for all incidents is to provide for responder and public safety. The term *responder* should be used instead of *fire fighter*, because there are usually more than just fire fighters responding to and taking action at the incident. There may be responders from law enforcement, emergency medical services (EMS), and possibly others. The objective must take into account *all* responders. This objective is implied in unwritten IAPs and written in formal IAPs. The chosen strategy and tactics must not disregard this objective.

Reliance on an IAP

In the ICS, considerable emphasis is placed on developing effective IAPs. An IAP is an oral or written plan containing general objectives reflecting the overall strategy for managing an incident. An IAP includes the identification of operational resources and assignments and may include attachments that provide additional direction.

Every incident must have a verbal or written IAP. The purpose of this plan is to provide all incident supervisory personnel with direction for actions to be implemented during the operational period identified in the plan. IAPs include the measurable strategic operations to be achieved and are prepared around a time frame called an operational period.

IAPs provide a coherent means of communicating the overall incident objectives in the context of both operational and support activities. The plan may be oral or written except for hazardous materials incidents, which require a written IAP.

At the simplest level, all IAPs must have the following four elements:

1. What do we want to do?
2. Who is responsible for doing it?
3. How do we communicate with each other?
4. What is the procedure if someone is injured?

Manageable Span of Control

Another basic ICS feature concerns the supervisory structure of the organization. *Span of control* pertains to the number of individuals or resources that one supervisor can manage effectively during emergency response incidents or special events. Maintaining an effective span of control is particularly important on incidents where safety and accountability are a top priority.

Span of control is the key to effective and efficient incident management. The type of incident, nature of the task, hazards and safety factors, and distances between personnel and resources all influence span of control considerations. Maintaining adequate span of control throughout the ICS organization is very important.

The number of reporting elements one can effectively manage on incidents may vary from three to seven, and a ratio of one supervisor to five reporting elements is recommended. If the number of reporting elements falls outside of these ranges, expansion or consolidation of the organization may be necessary. There may be exceptions, usually in lower-risk assignments or where resources work close to each other.

Predesignated Incident Locations and Facilities

Incident activities may be accomplished from a variety of operational locations and support facilities. Facilities will be identified and established by the IC depending on the requirements and complexity of the incident or event. It is important to know and understand the names and functions of the principal ICS facilities.

Incident Facilities

The incident command post (ICP) is the location from which the IC oversees all incident operations. There is generally only one ICP for each incident or event, but it may change locations during the event. Every incident or event must have some form of an ICP. The ICP may be located in a vehicle, a trailer, a tent, or within a building. The ICP will be positioned outside the present and potential hazard zone but close enough to the incident to maintain command. The ICP will be designated by the name of the incident (e.g., Trail Creek ICP, Gulf ICP).

Staging areas are temporary locations at an incident where personnel and equipment are kept while waiting for tactical assignments. The resources in the staging area are always in available status. Staging areas should be located close enough to the incident for a timely response, but far enough away to be out of the immediate impact zone. There may be more than one staging area at an incident. Staging areas can be co-located with the ICP, bases, camps, helibases, or helispots. Resources in the staging area are to be ready to respond within 3 minutes **FIGURE 13-7**.

A base is the location from which primary logistics and administrative functions are coordinated and administered. The base may be co-located with the ICP. There is only one base per incident, and it is designated by the incident name (e.g., Trail Creek base). The base is established and managed by the logistics section.

A camp is the location where resources may be kept to support incident operations if a base is not accessible to all resources or the incident is of a large enough scale as to require extended transportation times from the base to the tactical work assignments. Camps are temporary locations within the general incident area, which are equipped and staffed to provide food, water, sleeping areas, and sanitary services. Camps are designated by geographic location or number. Multiple camps may be used due to incident size, but not all incidents will have camps.

A helibase is the location from which helicopter-centered air operations are conducted **FIGURE 13-8**. Helibases are generally used on a more long-term basis and include such services as fueling and maintenance.

FIGURE 13-7 Engines in the staging area are ready to respond in 3 minutes or less when activated.
Courtesy of Kern County Fire Department

FIGURE 13-8 Multiple helicopters operating from a helibase at an incident.
Courtesy of Kern County Fire Department

The helibase is usually designated by the name of the incident (e.g., Trail Creek helibase).

Helispots are more temporary locations at the incident, where helicopters can safely land and take off. Multiple helispots may be used and are referred to by number.

Resource Management

ICS resources can be factored into the following two categories:

1. *Tactical resources.* Personnel and major equipment items on assignment to incidents that are available or potentially available to the operations function are called tactical resources. Tactical resources are always classified as being in one of the following statuses:
   - *Assigned* resources are working on an assignment under the direction of a supervisor.
   - *Available* resources are assembled, have been issued their equipment, and are ready for immediate assignment.
   - *Out-of-service* resources are not ready for available or assigned status.
2. *Support resources.* All other resources required to support the incident. Food, communications equipment, portable toilets, supplies, and fleet vehicles are examples of support resources.

Maintaining an accurate and up-to-date picture of resource utilization is a critical component of resource management. Resource management includes processes for:

- Categorizing resources
- Ordering resources
- Dispatching resources
- Tracking resources
- Recovering resources

It also includes processes for reimbursement for resources, as appropriate.

Integrated Communications

The use of a common communications plan is essential for ensuring that responders can communicate with one another during an incident. Communication equipment, procedures, and systems must operate across jurisdictions (interoperability).

Developing an integrated voice and data communications system, including equipment, systems, and protocols, must occur prior to an incident.

Effective ICS communications include the following three elements:

1. *Modes.* The hardware systems that transfer information.
2. *Planning.* Planning for the use of all available communications resources.
3. *Networks.* The procedures and processes for transferring information internally and externally.

Chain of Command and Unity of Command

In the ICS, **chain of command** means that there is an orderly line of authority within the ranks of the organization, with lower levels subordinate to, and connected to, higher levels. **Unity of command** means that every individual is accountable to only one designated supervisor to whom he or she reports at the scene of an incident.

These principles clarify reporting relationships and eliminate the confusion caused by multiple, conflicting directives. Incident managers at all levels must be able to control the actions of all personnel under their supervision. These principles do not apply to the exchange of information. Although orders must flow through the chain of command, members of the organization may directly communicate with each other to ask for or share information.

The command function may be carried out in one of the following two ways:

1. As a *single command*, in which the IC will have complete responsibility for incident management. A single command may be simple, involving an IC and single resources, or it may be a complex organizational structure with an incident management team.
2. As a *unified command*, in which responding agencies and/or jurisdictions with responsibility for the incident share incident management.

Unified Command

A unified command may be needed for incidents involving:

- Multiple jurisdictions
- A single jurisdiction with multiple agencies sharing responsibility
- Multiple jurisdictions with multiagency involvement

If a unified command is needed, ICs representing agencies or jurisdictions that share responsibility for the incident manage the response from a single ICP **FIGURE 13-9**.

A unified command allows agencies with different legal, geographic, and functional authorities and responsibilities to work together effectively without affecting individual agency authority, responsibility, or accountability.

Under a unified command, a single, coordinated IAP will direct all activities. The ICs will supervise a single command and general staff organization and speak with one voice.

Transfer of Command

The process of moving the responsibility for incident command from one IC to another is called **transfer of command**. Transfer of command may take place when:

- A more qualified person assumes command. The emphasis here is on qualification, not higher rank.
- The incident situation changes over time, resulting in a legal requirement to change command.
- Changing command makes good sense (e.g., an incident management team takes command of an incident from a local jurisdictional unit due to increased incident complexity).

FIGURE 13-9 This incident requires unified command among fire, EMS, law enforcement, and the health department. This fire, resulting from a vehicle accident (tanker truck of jet fuel), required rescue, treatment, and transportation of the driver; fire suppression for both fuel and wildland fires; hazardous materials response; and traffic control.

Courtesy of Kern County Fire Department

- There is normal turnover of personnel on long or extended incidents (i.e., to accommodate work/rest requirements).
- The incident response is concluded and incident responsibility is transferred back to the home agency.

The transfer of command process always includes a transfer of command briefing, which may be oral, written, or a combination of both. Once the predetermined time for transfer occurs, a formal transfer of command takes place, usually during an incident shift change briefing, and the transfer is announced on all incident radio frequencies. The method often used by federal incident management teams is to shadow the present command for a period of time and negotiate a time when formal transfer of command will occur. This process is referred to as a *transition of command*.

Accountability

Effective accountability during incident operations is essential at all jurisdictional levels and within individual functional areas. Individuals must abide by their agency policies and guidelines and any applicable local, tribal, state, or federal rules and regulations. The following guidelines must be adhered to:

- *Check-in*: All responders, regardless of agency affiliation, must report in to receive an assignment in accordance with the procedures established by the IC.
- *IAP*: Response operations must be directed and coordinated as outlined in the IAP.
- *Unity of command*: Each individual involved in incident operations will be assigned to only one supervisor.
- *Span of control*: Supervisors must be able to supervise and control their subordinates adequately, as well as communicate with and manage all resources under their supervision.
- *Resource tracking*: Supervisors must record and report resource status changes as they occur.

Mobilization

At any incident or event, the situation must be assessed and response planned. Resources must be organized, assigned, and directed to accomplish the incident objectives. As they work, resources must be managed to adjust to changing conditions.

Managing resources safely and effectively is the most important consideration at an incident. Therefore, personnel and equipment should respond only

SAFETY TIP

To provide for accountability of resources, no resource should self-dispatch to an incident. Resources should respond only when properly dispatched.

when requested or when dispatched by an appropriate authority.

Information and Intelligence Management

The analysis and sharing of information and intelligence is an important component of the ICS. The incident management organization must establish a process for gathering, sharing, and managing incident-related information and intelligence.

Intelligence includes not only national security or other types of classified information, but also other operational information that may come from a variety of different sources, such as:

- Risk assessments
- Medical intelligence (e.g., surveillance of hospitals and other medical providers for occurrences of certain types of reportable diseases, such as smallpox)
- Weather information
- Geospatial data
- Structural designs
- Toxic contaminant levels
- Utilities and public works data

ICS Organization

The ICS organization is unique but easy to understand. There is no correlation between the ICS organization and the administrative structure of any single agency or jurisdiction. This is deliberate, because confusion over different position titles and organizational structures has been a significant stumbling block to effective incident management in the past.

For example, someone who serves as a chief every day may not hold that title when deployed under an ICS structure. In departments that require incident qualification training and experience be documented and tracked, an advantage of the ICS is that all positions and the personnel trained to staff them are identified before an incident occurs.

Performance of Management Functions

Every incident or event requires that certain management functions be performed. The problem must be identified and assessed, a plan to deal with it developed and implemented, and the necessary resources procured and paid for.

Regardless of the size of the incident, these management functions will still apply.

There are five major management functions that are the foundation upon which the ICS organization is developed **FIGURE 13-10**. These functions apply whether you are handling a routine emergency, organizing for a major nonemergency event, or managing a response to a major disaster. They are:

1. *Incident command* sets the incident objectives, strategies, and priorities and has overall responsibility for the incident.
2. *Operations* conducts operations to reach the incident objectives. It establishes the tactics and directs all operational resources.
3. *Planning* supports the incident action planning process by tracking resources; collecting, analyzing, and disseminating information; and maintaining documentation.
4. *Logistics* provides resources and needed services to support the achievement of the incident objectives.
5. *Finance/administration* monitors costs related to the incident. It provides accounting, procurement, time recording, and cost analyses.

Incident Command

The IC has overall responsibility for managing the incident by establishing objectives, planning strategies, and implementing tactics **FIGURE 13-11**. The IC is the only position that is always staffed in ICS applications. On small incidents and events, one person, the IC, may accomplish all management functions. The IC is responsible for all ICS management functions until he or she delegates the function. The rule that applies here, as previously stated, is, "What you do not assign, you assume."

ICS Sections and Position Titles

Each of the primary ICS sections may be subdivided as needed. The ICS organization has the capability to expand or contract to meet the needs of the incident.

A basic ICS operating guideline is that the person at the top of the organization is responsible until the authority is delegated to another person. Thus, on smaller incidents where these additional persons are not required, the IC will personally accomplish or manage all aspects of the incident organization.

To maintain span of control, the ICS organization can be divided into many levels of supervision. At each level, individuals with primary responsibility positions have distinct titles **TABLE 13-1**.

Using specific ICS position titles serves the following three important purposes:

1. Titles provide a common standard for all users. For example, if one agency uses the title *branch chief*, another *branch manager*, and so forth, this lack of consistency can cause confusion at the incident.
2. The use of distinct titles for ICS positions allows for filling ICS positions with the most qualified individuals rather than by seniority.
3. Standardized position titles are useful when requesting qualified personnel. For example, in deploying personnel, it is important to know if

FIGURE 13-11 An incident commander briefing operations personnel on objectives using GIS-created incident map.
© Jones & Bartlett Learning

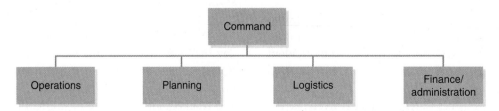

FIGURE 13-10 Five major management functions of the ICS.
FEMA publication National Incident Management System/Department of Homeland Security

TABLE 13-1 Supervisory Position Titles

Organizational Level	Title	Support Position
Incident command	Incident commander	Deputy
Command staff	Officer	Assistant
General staff (section)	Chief	Deputy
Branch	Director	Deputy
Division/group	Supervisor	N/A
Unit	Leader	Manager
Strike team/task force	Leader	Single resource boss

FEMA NIMS IS100.b p. 5.

the positions needed are unit leaders, technical experts, etc.

Incident Commander's Role and Responsibilities

The IC has overall responsibility for managing the incident by establishing objectives, planning strategies, and implementing tactics. The IC must be fully briefed and should have a written delegation of authority. Initially, assigning tactical resources and overseeing operations will be under the direct supervision of the IC.

Personnel assigned by the IC have the authority of their assigned positions, regardless of the rank they hold within their respective agencies.

In addition to having overall responsibility for managing an entire incident, the IC is specifically responsible for:

- Ensuring incident safety.
- Providing information services to internal and external stakeholders.
- Establishing and maintaining liaison with other agencies participating in the incident.

The IC may appoint one or more deputies, if applicable, from the same agency or from other agencies or jurisdictions. Deputy ICs must be as qualified as the IC.

Selecting and Changing Incident Commanders

As incidents expand or contract, change in jurisdiction or discipline, or become more or less complex, command may change to meet the needs of the incident.

Under the NIMS rank, grade, and seniority are not the factors used to select the IC. The IC is always a highly qualified individual trained to lead the incident response.

Expanding the Organization

As incidents grow, the IC may delegate authority for performance of certain activities to the command staff and to the general staff. The IC will add positions only as needed **FIGURE 13-12**.

Command Staff

Depending upon the size and type of incident or event, it may be necessary for the IC to designate personnel to provide information, safety, and liaison services for the entire organization. In the ICS, these personnel make up the command staff and consist of the:

- Public information officer, who serves as the conduit for information to internal and external stakeholders, including the media or other organizations seeking information directly from the incident or event.
- Safety officer, who monitors safety conditions and develops measures for ensuring the safety of all assigned personnel.
- Liaison officer, who serves as the primary contact for supporting agencies assisting at an incident.

The command staff reports directly to the IC.

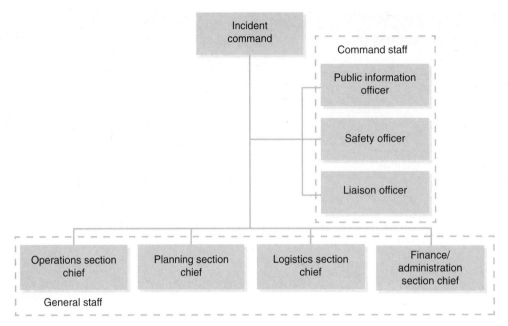

FIGURE 13-12 Expanding the ICS organization.
FEMA publication National Incident Management System/Department of Homeland Security

General Staff

Expansion of an incident may also require the delegation of authority for the performance of the other management functions. The people who perform the other four management functions are designated as the general staff. The general staff is made up of four sections: operations, planning, logistics, and finance/administration; and it reports directly to the IC (see Figure 13-10).

ICS Section Chiefs and Deputies

As mentioned previously, the person in charge of each section is designated as a chief. Section chiefs have the ability to expand their section to meet the needs of the situation. Each of the section chiefs may have a deputy, or more than one, if necessary. The deputy:

- May assume responsibility for a specific portion of the primary position, work as relief, or be assigned other tasks.
- Should always be as proficient as the person for whom he or she works.

At large incidents, especially where multiple disciplines or jurisdictions are involved, the use of deputies from other organizations can greatly increase interagency coordination.

Operations Section

Until operations is established as a separate section, the IC has direct control of tactical resources. The IC will determine the need for a separate operations section at an incident or event. When the IC activates an operations section, he or she will assign an individual as the operations section chief. The operations section chief will develop and manage the operations section to accomplish the incident objectives set by the IC. The operations section chief is normally the person with the greatest technical and tactical expertise in dealing with the problem presented by the incident.

Maintaining Span of Control. The operations function is where the tactical fieldwork is done and the most incident resources are assigned. Often the most hazardous activities are carried out there. The following supervisory levels can be added to help manage span of control:

- Divisions are used to divide an incident geographically.
- Groups are used to describe functional areas of operation.
- Branches are used when the number of divisions or groups exceeds the span of control and can be either geographical or functional **FIGURE 13-13**.

Divisions are used to divide an incident geographically. The person in charge of each division is designated as a supervisor. How the area is divided is determined by the needs of the incident.

The most common way to identify divisions is by using alphabet characters (A, B, C, etc.). Divisions are designated in a clockwise fashion beginning at the front of the incident (Division A). Other identifiers

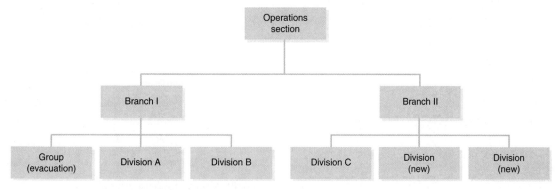

FIGURE 13-13 Operations section: establishing branches.
FEMA publication National Incident Management System/Department of Homeland Security

may be used as long as division identifiers are known by assigned responders. An example of this would be to divide a multistory building into divisions numbered by the floor they are on, such as Division 1 on the ground floor, and so forth.

The important thing to remember about ICS divisions is that they are established to divide an incident into geographical areas of operation.

Groups are used to describe functional areas of operation. The person in charge of each group is designated as a supervisor. The kind of group to be established will also be determined by the needs of an incident. Groups are normally labeled according to the job that they are assigned (e.g., rescue group, ventilation group, structure protection group). Groups will work wherever their assigned task is needed and are not limited geographically, and they may work across division boundaries.

Divisions and groups can be used together on an incident; they are at an equal level in the organization. One does not supervise the other. When a group is working within a division on a special assignment, division and group supervisors must closely coordinate their activities. The evacuation group could be working in Division C assisting the public in evacuating a subdivision.

If the number of divisions or groups exceeds the span of control, it may be necessary to establish another level of organization within the operations section, called branches. The person in charge of each branch is designated as a director. Deputies may also be used at the branch level. Branches can be divided into groups or divisions or a combination of both.

While span of control is a common reason to establish branches, additional considerations may also indicate the need to use these branches, including:

- *Multidiscipline incidents.* Some incidents have multiple disciplines involved (e.g., firefighting, law enforcement, health and medical, hazardous materials, public works and engineering, energy) that may create the need to set up incident operations around a functional branch structure.
- *Multijurisdictional incidents.* In some incidents it may be better to organize the incident around jurisdictional lines. In these situations, branches may be set up to reflect jurisdictional boundaries.
- *Very large incidents.* Very large incidents may be organized using geographic or functional branches.

Managing the Operations Section.

While there are any number of ways to organize field responses, branches and groups may be used to organize resources and maintain span of control.

The IC or operations section chief at an incident may work initially with only a few single resources or staff members.

The operations section usually develops from the bottom up. The organization will expand to include needed levels of supervision as more and more resources are deployed.

Task forces are a combination of mixed resources with common communications operating under the direct supervision of a leader **FIGURE 13-14**. Task forces can be versatile combinations of resources, and their use is encouraged. The combining of resources into task forces allows for several resource elements to be managed under one individual's supervision, thus lessening the span of control of the supervisor. An example of this would be two fire engines and a water tender to support them under one task force leader (FEMA).

Strike teams are a set number of resources of the same kind and type with common communications operating under the direct supervision of a strike team leader. Strike teams are highly effective management units. The foreknowledge that all elements have the same capability and the knowledge of how many will be applied allows for better planning, ordering, utilization, and management.

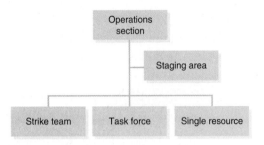

FIGURE 13-14 Operations section: task force, strike team, and single resource.

FEMA publication National Incident Management System/Department of Homeland Security

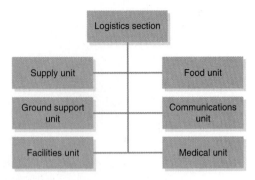

FIGURE 13-15 Planning section.

FEMA publication National Incident Management System/Department of Homeland Security

Unlike strike teams, single resources may be individuals, a piece of equipment and its personnel complement, or a crew or team of individuals with an identified supervisor that can be used at an incident.

It is important to maintain an effective span of control. Maintaining span of control can be made easier by grouping resources into strike teams, task forces, divisions, or groups. Another way to add supervision levels is to create branches within the operations section.

At some point, the operations section and the rest of the ICS organization will contract. The decision to contract will be based on the achievement of tactical objectives. Demobilization planning begins upon activation of the first personnel and continues until the ICS organization ceases operation.

Planning Section

The planning section can be further staffed with four units **FIGURE 13-15**. In addition, technical specialists who provide special expertise useful in incident management and response may also be assigned to work in the planning section. Depending on the needs, technical specialists may also be assigned to other sections in the organization.

The planning section units are:

- *Resources unit.* This unit conducts all check-in activities and maintains the status of all incident resources. The resources unit plays a significant role in preparing the written IAP.
- *Situation unit.* This unit collects and analyzes information on the current situation, prepares situation displays and situation summaries, and develops maps and projections.
- *Documentation unit.* This unit provides duplication services, including the written IAP. It maintains and archives all incident-related documentation.
- *Demobilization unit.* This unit assists in ensuring that resources are released from the incident in an orderly, safe, and cost-effective manner.

Logistics Section

The IC will determine if there is a need for a logistics section at an incident and will designate an individual to fill the position of the logistics section chief. If no logistics section is established, the IC will perform all logistical functions. The size of the incident, complexity of support needs, and the incident length will determine whether a separate logistics section is established. Additional staffing is the responsibility of the logistics section chief.

The logistics section is responsible for all of the services and support needs, including:

- Ordering, obtaining, maintaining, and accounting for essential personnel, equipment, and supplies.
- Providing communication planning and resources.
- Setting up food services.
- Setting up and maintaining incident facilities.
- Providing support transportation.
- Providing medical services to incident personnel.

The logistics section can be further staffed by two branches and six units **FIGURE 13-16**. Not all of the units will be required; they will be established based on need. The titles of the units are descriptive of their responsibilities.

The logistics service branch can be staffed to include:

- *Communication unit.* This unit prepares and implements the incident communication plan, distributes and maintains communications equipment, supervises the incident communications center, and establishes adequate communications over the incident.
- *Medical unit.* This unit develops the medical plan, provides first aid and light medical treatment for

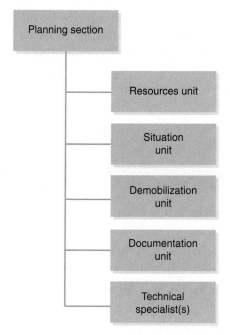

FIGURE 13-16 Logistics section.
FEMA publication National Incident Management System/Department of Homeland Security

personnel assigned to the incident, and prepares procedures for a major medical emergency.

■ *Food unit.* This unit is responsible for providing meals and drinking water for incident personnel and obtains the necessary equipment and supplies to operate food service facilities at bases and camps.

The logistics support branch can be staffed to include:

■ *Supply unit.* This unit determines the type and amount of supplies needed to support the incident. The unit services nonexpendable equipment and orders, receives stores, and distributes supplies. All resource orders are placed through the supply unit. The unit maintains inventory and accountability of supplies and equipment.

■ *Facilities unit.* This unit sets up and maintains incident facilities. It provides managers for the incident base and camps. This unit is also responsible for facility security and facility maintenance services: sanitation, lighting, and cleanup.

■ *Ground support unit.* This unit prepares the transportation plan. It arranges for, activates, and documents the fueling and maintenance of assigned ground transportation. The unit arranges for the transportation of personnel, supplies, food, and equipment.

Finance/Administration Section

The IC will determine if there is a need for a finance/administration section at the incident and if so, will designate an individual to fill the position of the finance/administration section chief. If no finance/administration section is established, the IC will perform all finance functions.

The finance/administration section is set up for any incident that requires incident-specific financial management. The finance/administration section is responsible for:

■ Contract negotiation and monitoring.

■ Timekeeping.

■ Cost analysis.

■ Compensation for injury or damage to property.

More and more, larger incidents are using a finance/administration section to monitor costs. Smaller incidents may also require certain finance/administration support.

For example, the IC may establish one or more units of the finance/administration section for such things as procuring special equipment, contracting with a vendor, or making cost estimates for alternative response strategies.

The finance/administration section may staff four units **FIGURE 13-17**. Not all units may be required; they will be established based on need.

The units in the finance/administration section include:

■ *Procurement unit.* This unit is responsible for administering all financial matters pertaining to vendor contracts, leases, and fiscal agreements.

■ *Time unit.* This unit is responsible for incident personnel time recording.

■ *Cost unit.* This unit collects all cost data, performs cost-effectiveness analyses, provides

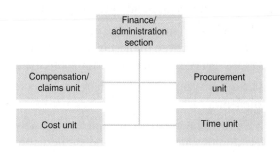

FIGURE 13-17 Finance/administration section.
FEMA publication National Incident Management System/Department of Homeland Security

cost estimates, and makes cost-savings recommendations.

- *Compensation/claims unit.* This unit is responsible for the overall management and direction of all administrative matters pertaining to compensation for injury as well as claims-related activities kept for the incident.

Wrap-Up

CHAPTER SUMMARY

- It is the responsibility of every responder at the incident scene to do his or her part to ensure a swift and successful conclusion to the incident.
- Should an incident escalate, the first-in officer has the opportunity to turn the incident into a major disaster through poor management or, conversely, to have a positive effect on the outcome of the incident.
- If there is no clearly defined goal in mind for the incident, it cannot be determined how to achieve the goal, nor can progress toward the goal be measured.
- Strategies are based on the size-up of the incident, objectives, strategic priorities, and mode of attack selected.
- Size-up is a mental process requiring the IC to perform the following steps:
  - Determining facts when the alarm is received
  - Anticipating probabilities
  - Assessing one's own situation
  - Making a decision
  - Planning the operation
- Rescue is the first strategic priority. Life safety is the most important consideration in any incident operation.
- Fire attack may be performed in offensive mode, where resources are applied directly to controlling the fire or other incident, or defensive mode, where the risk versus gain is too great and the decision is made to protect exposures and not to address the source of risk directly.
- Once the objectives are determined and communicated, and the strategies are selected, the individual companies can employ the tactics necessary to achieve the objectives.
- Most new fire fighters will initially be expected to operate on the task level.
- Homeland Security Presidential Directive 5 (HSPD-5) called for the National Incident Management System (NIMS); identified steps for improved coordination of federal, state, local, and private industry response to incidents; and described the way these agencies will prepare for such a response.
- An incident is an occurrence that requires response actions to prevent or minimize loss of life or damage to property and/or the environment.
- The Incident Command System (ICS) was developed following a series of catastrophic fires in California's urban interface.
- The ICS is a proven management system based on successful business practices. It is the result of decades of lessons learned in the organization and management of emergency incidents.
- The ICS may be used for small or large events. It can grow or shrink to meet the changing needs of an incident or event.
- ICS principles are implemented through a wide range of management features, including the use of common terminology and clear text, and a modular organizational structure.
- There is no correlation between the ICS organization and the administrative structure of any single agency or jurisdiction.

- Every incident or event requires that certain management functions be performed. The problem must be identified and assessed, a plan to deal with it developed and implemented, and the necessary resources procured and paid for.
- The IC has overall responsibility for managing the incident by establishing objectives, planning strategies, and implementing tactics.
- As incidents grow, the IC may delegate authority for performance of certain activities to the command staff and to the general staff.

KEY TERMS

All clear The short descriptive phrase indicating that a primary search of the structure for victims has been completed.

Base The location from which primary logistics and administrative functions are coordinated and administered. The ICP may be co-located with the base.

Camp The location where resources may be kept to support incident operations if a base is not accessible to all resources or the incident is of a large enough scale as to require extended transportation times from the base to the tactical work assignments.

Chain of command An orderly line of authority within the ranks of the organization, with lower levels subordinate to, and connected to, higher levels. The organization of management at the incident starts with the IC and develops downward.

Delegation of authority The provision of authority to an outside entity to manage an incident within another's jurisdiction.

Helibase The location from which helicopter-centered air operations are conducted.

Helispots A natural or improved takeoff and landing area intended for temporary or occasional helicopter use.

Incident An occurrence, either caused by humans or natural phenomena, that requires response actions to prevent or minimize loss of life or damage to property and/or the environment.

Operational period The period of time scheduled for execution of a given set of tactical actions as specified in the IAP. It may be as long as 24 hours for a wildland incident or as short as an hour for a hazardous materials incident.

Perimeter control Controlling the edges of a wildland fire.

Primary search A rapid search of all involved and exposed areas that are affected by the fire but can be entered, to verify removal and/or safety of all occupants. Should this not be possible, a secondary search is conducted as soon as it is safe to do so.

Pump and roll A tactic used in grass fires utilizing pumpers that can drive while the pump is operating. Hose lines are connected to the apparatus, and water is sprayed to extinguish the fire edge.

Rehabilitate (1) To rehabilitate personnel means that they rest, cool off, and replenish body fluids. (2) To rehabilitate a fire line means to construct water bars to direct water runoff and prevent erosion. Under the federal "Minimum Impact Suppression Tactics" policy, rehabilitation may mean the erasure of fire lines as much as possible. In other words, to cause as little damage as possible controlling the fire as fire is a part of the natural environment.

Rekindle To reignite after a fire was thought to be extinguished. This commonly happens in attics, basements, and walls of structure fires and in logs on wildland fires. This is usually due to incomplete overhaul/mop-up.

Salvage A firefighting procedure for protecting building contents from damage due to water or falling debris.

Spot fires In heavier fuels, flying fire brands can land outside the fire perimeter and start new fires.

Staging areas Temporary locations at an incident where personnel and equipment are kept while waiting for tactical assignments.

Strategy The method used to coordinate the tactical operations of units to achieve the desired incident objectives.

Structure protection Protecting structures in danger of being consumed by an advancing wildland fire.

Tactics Actions taken to achieve strategies.

Tasks Specific things that must be done to bring a job to completion.

Transfer of command The transfer of the role of incident commander from one person to another.

Unity of command The organizational principle in which every individual is accountable to only one designated supervisor to whom he or she reports at the scene of an incident.

CASE STUDY

A large, weather-related incident has struck a major city and its surrounding area. The incident requires the response of numerous city departments such as health, fire, law enforcement, public works, and animal control. In addition, outside resources have been ordered from the surrounding county and the state. A disaster has been proclaimed on both the local and state level (the only level of government that can "declare" a disaster is the federal government).

The incident requires sheltering of the displaced; rescue, treatment, and transport of the injured; security for evacuated areas; control of power, water, and sewer systems; public transportation to move personnel and victims; coordination of hospitals receiving the wounded; and documentation of efforts, planning, and preparation for recovery. Resources need to be sent where they are most needed when they are received. Personnel need to be cared for, housed, and fed. The public needs to be kept informed as to what is happening, both with the incident and with control efforts.

All of this requires coordination on the street level and the management level. Without proper coordination and management, the incident can turn into a free-for-all of competing demands and freelancing to get the job done. A unified command is set up involving members of law enforcement, fire, public works, and other agencies with major involvement. Their responsibility is to prepare and manage an IAP that clearly specifies the objectives for the incident and provides for incident needs to accomplish the stated objectives.

1. Why was Unified Command chosen to manage the incident?

   A. complexity of the incident
   B. nature of the incident
   C. time of year
   D. incident size

2. Why would the city need to activate NIMS to manage the incident?

   A. The incident is limited to one jurisdiction.
   B. The incident is limited to one function.
   C. There is multiagency involvement.
   D. The incident is limited to one level of government.

3. Using the positions outlined in this chapter, what is the first position that would need to be established to manage the incident?

   A. Safety officer
   B. Information officer
   C. Liaison officer
   D. Incident commander

4. What facility will be required by the incident to support the responders, who will be there for several days?

   A. Base
   B. Helibase
   C. Staging area
   D. Evacuation shelter

REVIEW QUESTIONS

1. What ICS position is in charge at an incident?

2. What are the three attack modes?

3. Interior structural attack is performed in which mode?

4. List the seven strategic priorities.

5. List the steps in a size-up. Give a listing of the information necessary for structure and wildland size-ups.

6. What are the five main components of NIMS?

7. Whose responsibility is it to implement the ICS on incidents?

8. What are the positions of the command staff?

9. What are the positions of the general staff?

10. Which position has control of tactical resources at an incident?

11. What is the difference between a group and a division?

12. What is the difference between a strike team and a task force?

13. What does it mean to be in available status?

14. When operating an engine on an ICS-managed incident, which unit do you see for fuel?

DISCUSSION QUESTIONS

1. Why is unified command important in the case of a multiagency or multi-jurisdiction incident?

2. Why is it important to have a commonly understood incident command system in place before agencies are required to help each other out in times of major emergencies?

3. With the activation of resources on a nation-wide basis (e.g., Oklahoma City bombing, the Pentagon and the New York World Trade Center attacks, Hurricane Katrina, Gulf of Mexico oil spill), do you think the fire service needs more or less standardization in incident command systems?

4. Using examples of various incident types (flood, fire, snow storm, hurricane, wildland fire, etc.), what would be the appropriate ICS organization to manage these incidents? Using these incident types, create an ICS organizational chart for the ones that you are most likely to encounter in your area.

5. Why should "provide for responder and public safety" be the first objective in any incident action plan?

REFERENCES AND ADDITIONAL RESOURCES

Federal Emergency Management Agency. 2010. "IS-100.b—-Incident Command System (ICS) 100 Training." FEMA. www.FEMA.gov

Layman, Lloyd. 1972. *Fire Fighting Tactics*. Quincy, MA: National Fire Protection Association.

National Institute for Occupational Safety and Health. 2009. "Nine Career Fire Fighters Die in Rapid Fire Progression at Commercial Furniture Showroom—South Carolina F2007-18." NIOSH. https://www.cdc.gov/niosh/fire/reports/face200718.html

National Wildfire Coordinating Group. 2018. *Incident Response Pocket Guide*. Boise, ID: National Wildfire Coordinating Group.

CHAPTER 14

Emergency Operations

OBJECTIVES

After studying this chapter, you should be able to:

- Identify the personnel who might be working at an emergency scene.
- List the *16 Firefighter Life Safety Initiatives*.
- Identify important considerations when working at a structure fire.
- Identify important considerations when working at an electrical distribution equipment fire.
- Identify important considerations when working at a wildland fire.
- Identify important considerations when working at a wildland–urban interface fire.
- Identify important considerations when working at a petroleum fire.
- Identify important considerations when working at hazardous materials incidents.
- Identify important considerations when working at weapons of mass destruction incidents.
- Identify important considerations when working with emergency medical service operations.
- Identify important considerations for highway incident safety.
- Identify important considerations when working at vehicle accidents.
- Identify important considerations when working at an aircraft fire.
- Identify important considerations when working at a technical rescue incident.
- Discuss the importance of decision-making skills and how they relate to incident safety.

Case Study

On June 30, 2013, 19 fire fighters lost their lives on the Yarnell Hill Fire in Arizona. The fire fighters were all part of the Granite Mountain Hot Shot crew based out of the Prescott, Arizona, Fire Department. The crew was constructing line with the fire being pushed away from them by the wind. A thunderstorm over the fire in the dissipating stage caused gusts of up to 40 mph in all directions, pushing the fire back toward and over the crew, trapping them. At 5 PM, the temperature was 95°F and the relative humidity was 17 percent. That, coupled with sustained winds of 26 mph with gusts over 40 mph, could have caused the rate of spread to increase to the point where it would have been impossible for any firefighting resources, in the air or on the ground, to implement any kind of effective fire suppression action. Some reports put the wind-driven fire as moving at 15 mph. That is much faster than most people can run (a 4-minute mile), and the people who are running that fast are doing it on a track with running shorts and running shoes, not in varying terrain with obstacles, wildland personal protective equipment (PPE), and boots.

In this chapter, the importance of lookouts, communications, escape routes, and safety zones (LCES) and the 10 standard firefighting orders are emphasized. LCES is considered to be the minimum safety standard to be in place prior to engaging a fire. In this case, part of the LCES was in place with communication between the lookout and the crew leader. The only Hot Shot who survived the burn-over was the person serving as the lookout.

1. Hot Shots are some of the most well-trained wildland fire fighters in the world. In regard to decision making, what put them where they were?
2. Do you think their LCES was adequate for the situation? Why or why not?
3. What could they have done differently in regard to LCES that may have avoided this disaster?
4. Which of the 10 Standard Firefighting Orders were violated?

Modified from Gabbert, Bill. 2013. "Weather Conditions during the Tragedy at Yarnell Hill, and Where Do We Go from Here." Wildfire Today, July 4. http://wildfiretoday.com/2013/07/04/weather-conditions-during-the-tragedy-at-yarnell-hill-and-where-do-we-go-from-here/

JONES & BARTLETT LEARNING
NAVIGATE 2   *Access Navigate for more resources.*

Introduction

The personnel resources of the fire department can be divided into two general areas: operations (also known as line) and support (also known as staff). The operational personnel resources of the fire department, not to be confused with the operations section in the Incident Command System (ICS), consist of the people in the fire stations and their supervisors. Operations/line personnel are the personnel charged with response to and the mitigation of incidents.

One of the fundamental roles of the fire department is to respond to emergencies. Not every incident is an emergency, and not every emergency is the responsibility of the fire department to manage. In many types of incidents, the fire department does what it can in light of its legal and resource limitations.

The job and types of incidents responded to by fire fighters are constantly changing. In the beginnings of firefighting fire fighters responded to local fires. Nowadays the fire service has taken on a much expanded role. Fire fighters assist in all incident types at local, state, and national levels. Incident types include, but are not limited to, structure, vehicle, and wildland fires and search and rescue due to blizzards, flooding, tornadoes, hurricanes, bomb cyclones (winter hurricanes), earthquakes and structural collapse, airplane crashes, vehicle crashes, mineshaft rescues, confined space rescues, hazardous materials incidents, active shooter incidents, volcanic eruptions, and, most commonly, medical aid for ill and injured persons. The days of fires being the only responsibility for fire fighters have long since passed. At the minimum, modern fire fighters are required to be trained and proficient in responding to all incident types anticipated in their jurisdiction.

Government, on all levels, is finding that no single fire organization can go it alone. It is too expensive and not justified to have sufficient resources on hand at all times to defend against the worst case scenario. This is why a national ordering system for resources has been established to mobilize resources when needed on regional, state, and national levels (Chapter 8). The country has also adopted a National Incident Management System (Chapter 13) to ensure that responders can be integrated and organized into a management structure that they understand using common terminology. All of this requires training (Chapter 9). Fire fighters may be called into action to respond to regional-, state-, and national-level incidents.

It has always been that fire fighters were ready and willing to assist their neighbors. On several occasions in the past, large cities would have such catastrophic fires that fire apparatus was placed on railroad trains and transported to the scene from neighboring jurisdictions (Chapter 3). More recent examples of large-scale national-level responses with fire fighters drawn from across the country are Hurricane Katrina affecting the Gulf Coast in 2005, the terrorist attacks on the Twin Towers in New York City and the Pentagon in Virginia in 2001, and wide-spread flooding in Houston (the result of Hurricane Harvey) and other parts of Texas in 2017. By far, the worst disasters in terms of life and dollar loss and widespread damage have been hurricanes, floods, tornadoes/tornado outbreaks, bomb cyclones, and earthquakes.

To address large-scale disasters there are several examples of pre-staffed, trained, and equipped national response teams. There are currently 28 Federal Emergency Management Agency trained and equipped Urban Search and Rescue Task Forces (USAR) throughout the county. These teams are primarily staffed with local government fire fighters. The task forces specialize in such incident types as structure collapse and floods. They provide urban search and rescue, disaster recovery, and emergency triage and medicine. They and their equipment are deployed to emergency and disaster sites within 6 hours of notification. An example of their use is the pre-deployment (staging) of 1200 USAR Team personnel to the East Coast in the region of the Carolinas for hurricane Florence in September of 2018.

The U.S. Forest Service has 15 predesignated Type 1 (most complex incidents) and 36 Type 2 (less complex incidents) Interagency Incident Management Teams (IMT). These teams are comprised of federal, state, and local fire department personnel that respond when needed across the country to large-scale incidents, primarily wildland fires, but also flooding, other disasters, and the terrorist incidents in 2001. The teams are on call on a rotational basis both regionally and nationally. When on call, the members of the team must be ready to respond within 2 hours of notification.

In addition to the teams, trained and qualified local fire fighters can be called upon to assist on local and national levels with all incident types. Examples have been logistics personnel (Chapter 13) assigned to assist with large-scale evacuations and live-find search dog handlers assigned to search wreckage at passenger train crashes, etc.

In this chapter, you are expected to take everything you have learned throughout this text and this course and put it all together. That is how it happens in the real world of emergency response. All your skills, knowledge, and abilities get tested at real incidents. Emergency response is not a video game where failure just means you restart and go again. In this profession the price of failure can be injury and even death—yours or someone else's.

After studying this chapter, take a look back at the *16 Firefighter Life Safety Initiatives* and review what you have learned from this course. Then consider the numerous things that could have gone wrong in these examples. In addition, reflect on what has gone wrong in past incidents, costing civilians and fire fighters their lives.

Keeping in mind that fires and other incidents that the fire department responds to occur in so many different situations that not all of them can be covered in this text, here are some examples:

Structure Fire Example

Suppose you arrive at a scene securely buckled in your seat belt in the engine. As you don your self-contained breathing apparatus (SCBA) and exit the cab, you can see flames and smoke coming from a one-story single-family residence. The officer on your engine tells you to take the 1¾-in. attack line and advance it to the front door. The other fire fighter arrives at the door with an axe and a hand light. She opens the door and you advance down the hallway toward the seat of the fire. Once inside the room, you determine the fire is in the closet. You quickly extinguish it with a short burst of water from your nozzle. You feel great about having defeated the dragon and look forward to bragging about the experience to your friends.

However, take a moment and think about what else was taking place while you were in the hallway. The officer was performing a size-up and giving a report on conditions over the radio while assuming command. Your driver/operator was setting up the pump to provide you with water through the hose. The second-in company laid a supply line to the hydrant to ensure that you would not run out of water. The truck company secured the utilities, performed ventilation, and started to search for any victims.

Wildland Fire Example

Suppose your engine is dispatched to a growing wildland fire. The smoke is really starting to come up late in the afternoon. The scene is a beehive of activity with crews unloading, dozers starting up, and helicopters flying overhead. Homes are threatened. The incident commander (IC) needs you and your crew to protect structures directly in the path of the advancing fire. As the fire reaches one structure, the heat is intense and the embers and smoke make it hard to see. You extinguish small ignitions around the structure, and the fire roars onward to threaten more structures. You pick up your hoses, load them up on the engine, and go to do it again. Sometimes this goes on for several days in a row. When the job is done, there is a great feeling of satisfaction that you and your company saved people's belongings. On your way home you drive by a "Thank You Fire Fighters" sign posted by local residents and your chest swells with pride because you know you have performed effectively and safely.

So what was happening while you were focused on your tasks? Everything appeared smooth and well-coordinated. All personnel were trained and ready to do their jobs. The apparatus and equipment were well maintained and in good working order. The officers utilized their training and experience to determine strategy and tactics. The ICS was used and everyone was aware of who to report to and who was in charge. The IC stayed above the fray at a good vantage point to ensure your safety.

FIGURE 14-1 Fire fighters and ambulance personnel treating a victim at the scene of vehicle accident.
© Jones & Bartlett Learning

with special incident types as well as regular fire department functions, such as firefighting and medical aid.

The fire department is not the only agency that shows up at an incident that has a say in determining the objectives, strategy, and tactics to be used **FIGURE 14-1**. Law enforcement and other public agencies assist and sometimes are in command of incidents in which the fire department is involved. In some types of incidents, the number of representatives from other agencies may outnumber the fire fighters who perform the necessary tasks to mitigate the incident. An incident of this type would be managed using unified command.

As discussed in Chapter 13, *Emergency Incident Management*, the first objective of any incident is to provide for responder and public safety. This chapter focuses primarily on fire fighter safety, but the general rules should apply to all responders.

SAFETY TIP

The worst hazard is often the one that is not recognized. You must train yourself to evaluate situations as they arise and anticipate situations that can harm you. When you become complacent and stop paying attention, or think that things are just routine, you just may be in the greatest danger.

Tip

The fire department is not the only agency that shows up at an incident that has a say in determining the objectives, strategy, and tactics to be used. Interagency communications must be considered at all incidents.

Personnel

The type of personnel who can be expected to respond to an incident depends on the incident type. Just as some doctors specialize in different types of medicine and parts of the human body, the fire department has its own specialists. These personnel have been trained in dealing

16 Firefighter Life Safety Initiatives

In 2004, the National Fallen Firefighter Foundation (NFFF) examined the causes of line-of-duty fire fighter deaths. As a result of this meeting they put forth the *16 Firefighter Life Safety Initiatives* to be followed by all fire fighters. These are part of the *Courage to Be*

Safe program, which is a short course available on the Internet and worthy of your completion; it is available through the NFFF's *Everyone Goes Home* program. Successfully completing the course can make you a safer fire fighter and provide you with a certificate of completion, which can be added to your resume when seeking employment. The initiatives are (NFFF 2019):

1. Define and advocate the need for a cultural change within the fire service relating to safety; incorporating leadership, management, supervision, accountability, and personal responsibility.
2. Enhance the personal and organizational accountability for health and safety throughout the fire service.
3. Focus greater attention on the integration of risk management with incident management at all levels, including strategic, tactical, and planning responsibilities.
4. All fire fighters must be empowered to stop unsafe practices.
5. Develop and implement national standards for training, qualifications, and certification (including regular recertification) that are equally applicable to all fire fighters based on the duties they are expected to perform.
6. Develop and implement national medical and physical fitness standards that are equally applicable to all fire fighters, based on the duties they are expected to perform.
7. Create a national research agenda and data collection system that relates to the initiatives.
8. Utilize available technology wherever it can produce higher levels of health and safety.
9. Thoroughly investigate all fire fighter fatalities, injuries, and near misses.
10. Grant programs should support the implementation of safe practices and/or mandate safe practices as an eligibility requirement.
11. National standards for emergency response policies and procedures should be developed and championed.
12. National protocols for response to violent incidents should be developed and championed.
13. Fire fighters and their families must have access to counseling and psychological support.
14. Public education must receive more resources and be championed as a critical fire and life safety program.
15. Advocacy must be strengthened for the enforcement of codes and the installation of home fire sprinklers.
16. Safety must be a primary consideration in the design of apparatus and equipment.

In terms of practical application of the *16 Firefighter Life Safety Initiatives*, every fire fighter should adhere to the following:

- *Duty and responsibility.* Make every day a training day so everyone goes home.
- *Fire fighter maintenance program.* Receive regular medical checkups, get regular exercise, and eat healthily.
- *Rehab guidelines.* Stop before you drop, stay hydrated, and monitor vital signs.
- *Passengers when responding to incidents.* Wear full PPE, get in the apparatus, sit down, fasten

Fire Mark

The Nine Questions list was developed as NFFF's Firefighter Life Safety Initiative 12, which states: "National protocols for response to violent incidents should be developed and championed." (NFFF 2019) This is especially timely in light of recent incidents where two fire fighters in western New York were gunned down responding to an incident on Christmas Eve, 2012. A more recent event is the four fire fighters in Gwinnett County, Georgia who were taken hostage by a deranged man while responding to an incident. The nine questions are:

1. Do you use risk–benefit analysis for every call?
2. Do you have an effective relationship at all levels with the law enforcement agencies in your community?
3. How good is the information you get from your dispatcher?
4. Do you allow members to first respond directly to the scene?
5. Does your law enforcement agency use an incident management system?
6. When responding to a potentially violent incident, do you seek out a law enforcement officer when you arrive?
7. Have you told your fire officers/personnel that it is OK to leave the scene if things start to turn bad?
8. Is there a point where you don't respond or limit your response to violent incidents?
9. Is your uniform easily mistaken for that of law enforcement?

your seat belt, and ride with drivers who will get you there in one piece.

- *Drivers when responding to incidents.* It is not a race; safe is more important than fast. Stop at all red lights and stop signs, and if others do not pull over—don't run them over.
- *Interior firefighting.* Work as a team, stay together, stay oriented, manage your air supply, and take the proper tools with you for any interior operation. Every member should have a radio, provide constant updates, and constantly assess for risk versus benefit.

Rapid Intervention Teams

The assignment of one or more companies as rapid intervention crews (RICs, also referred to as rapid intervention companies or teams [RITs]) at working incidents provides the ability to initiate a rescue effort immediately to locate, rescue, or assist fire fighters who are in trouble at the scene of an incident. RIC members should be standing by wearing their full protective clothing with SCBA ready for immediate use. They should have forcible entry tools, rescue rope, and any other equipment that could be needed quickly. At hazardous materials incidents or other situations where special protective equipment is required, the RIC should be ready with the same level of protective clothing and equipment that the entry team requires. Some departments have established fire fighter assist and search teams to accomplish rapid intervention tasks.

Two In, Two Out

The Occupational Safety and Health Administration (OSHA) has created a regulation commonly referred to as "two in, two out." The regulation specifies that whenever personnel are operating in atmospheres that require SCBA, especially atmospheres that are described as **immediately dangerous to life and health (IDLH)**, the procedure of two in, two out must be used. OSHA recognizes that "conditions present during an advanced interior structural fire create an IDLH atmosphere." OSHA states that this applies to hazardous materials incidents and fire-related incidents that have gone "beyond the ignition stage"—in other words, fires in the free burning, fully developed, or decay stages (OSHA 1986, 29 CFR 1910.120).

In some departments one, two, or three personnel will be first at the scene. If they wish to initiate interior fire attack when an IDLH atmosphere is present or likely to be present, engine company personnel must first alter the IDLH atmosphere, with the following exception:

If, upon arrival at the scene, fire fighters find an imminent life-threatening situation where immediate action may prevent the loss of life or serious injury, such action shall be permitted with less than four fire fighters on the scene, when actions are based on appropriate concepts of risk assessment and management. Such action is intended to apply only to those rare and extraordinary circumstances when, in the fire fighter's professional judgment, the specific instance requires immediate action to prevent the loss of life or serious injury and four fire fighters have not yet arrived on the fireground.

In essence, an interior attack with fewer than the OSHA-required four personnel at the scene can be made if there is a high probability of effecting a rescue or stopping the fire from threatening persons in imminent danger. Prior to entry by the full at-scene complement of personnel, there must be an announcement over the radio and a transfer of command to the next incoming company or chief officer.

Personnel operating in hazardous areas at incidents must operate in two-person teams, called the buddy system. They are required to be in communication with each other at all times, either visually or by voice contact. Being in contact by radio is not considered to meet the requirement of direct contact. They are to remain close enough together to render aid if one of the team members requires assistance. Backup team members are also required to maintain either visual or voice contact as well. Of the four personnel at a scene when operating in IDLH or potentially IDLH atmospheres, the entry team must consist of two personnel. One of those two personnel who remain outside may be engaged in other activities, as long as this does not jeopardize fire fighter safety and the individual's ability to participate in any rescue operation.

Structural Firefighting

One of the basic responsibilities of any fire department is fighting structure fires **FIGURE 14-2**. The equipment in the form of pumpers, hose, and nozzles is designed with this primarily in mind. Previously this was the type of emergency work that fire fighters performed the most. This is no longer true with the move of the fire department into medical aid delivery, but it is still one of the most important jobs performed. Most departments spend the bulk of their time and training budget preparing for this one aspect of emergency service. A fire department that can keep its structure losses to a minimum is performing an important part of its function very well. Every once in a while, there

FIGURE 14-2 Fire fighter advancing a hose line to attack a structure fire.
Courtesy of Kern County Fire Department

is bound to be a fire that exceeds the control capabilities of the fire department and its resources. When this happens, the public is sure to take a close look at the fire department and its budget and ask whether the money is being used effectively.

Firefighting operations at structure fires take one of two modes: offensive or defensive. In the offensive mode, fire fighters enter the burning structure and attack the seat of the fire. In the defensive mode, the water is applied through windows or into other openings to control the fire. A key point is that hose streams should not be directed through ventilation openings when resources are inside the structure. In some cases, the initial attack is made through a window to darken down the fire, and then fire fighters enter and complete the extinguishment. If fire fighters are able to enter the structure upon arrival, a primary search is made for victims.

SAFETY TIP

Hose streams should not be directed through ventilation openings when resources are inside the structure.

A firefighting operation at a structure must be a coordinated attack. The first-arriving unit sizes up the fire and decides which methods will be the most effective in bringing the fire under control. Hose lines are

pulled and charged, the power is cut off at the electric meter, ventilation is performed, and the fire fighters attack the fire **FIGURE 14-3**. If ventilation is performed before the lines are ready, the fire will intensify and may exceed the ability of the personnel at the scene to control it. Depending on the size and complexity of the fire, any number of operations are required to bring it under control. If the fire is already through the roof, ventilation may be unnecessary.

Mitigating Hazards

When fire fighters respond to structure fires, several questions are important, regardless of whether the structure is a trash bin, shed, home, business, or warehouse. One of these questions is, "What are the contents?" In the past, people were not as likely to have an assortment of household chemicals stored in structures, and firefighting was somewhat safer **FIGURE 14-4**. To limit exposure to these items, fire fighters have better protective equipment. Because of improvements to protective equipment, fire fighters sometimes think that they are invincible—which is certainly not the case. An SCBA will protect you from inhaling chemicals but will not protect you from skin contact. In a hot-structure firefighting operation, you will be sweating profusely, and all of your pores will be open. If a chemical that can harm you through skin contact comes into contact with an unprotected area, you will suffer an exposure. Many of these chemicals do not show immediate effects. After repeated

FIGURE 14-3 Fire fighters performing roof ventilation.
Courtesy of Kern County Fire Department

SAFETY TIP

An SCBA will protect you from inhaling chemicals but will not protect you from skin contact. Always wear the appropriate PPE for the situation.

FIGURE 14-4 Commonly encountered hazardous household chemicals.
© Jones & Bartlett Learning

exposures over a 30-year career, you may find yourself suffering from cancer.

Leather gloves and boots will protect you from some hazards and not from others. They are made of animal skin which, like your own, is porous. If your gloves or boots become contaminated with a harmful chemical, they cannot be decontaminated. Every time you wear them and start to sweat, you are going to be exposed again. If you get a chemical on your turnout gear or uniform and take it home for washing, you can expose your whole family. A dose of a harmful chemical that will not show any ill effect on you may be lethal to small children.

Other building contents can pose a hazard. In a kitchen fire, canned goods can explode and splatter you with their boiling contents. During gasoline shortages, people were known to store trash cans full of gasoline in their homes. Some room contents, in the form of furniture, pose a hazard. Polyurethane foam can burn as hot and produce as much smoke as gasoline. Fires have occurred where the use of flammable liquids was suspected because of heavy black smoke. After extinguishment, it was determined that the smoke was caused by burning foam, not gasoline. In areas where hunting is popular, many homes have a supply of ammunition on hand. The list of hazards in a common structure fire can go on almost indefinitely.

Interior structural components pose dangers as well. Suspended ceilings have crashed down and trapped fire fighters in their framing, concealed wiring, and ductwork. A ceiling fan is heavy enough to do some damage to you if you are hit. Tall bookcases and high-piled stock can be knocked down by hose streams and land on fire fighters. One of the worst interior structural hazards is stairs. Their use may present no problem at the beginning of the fire, but when the situation deteriorates, they can act as a

chimney for smoke and flame and cut off your escape, trapping you above the fire. In heavy smoke and darkness, it is easy to fall down the stairs if you are not paying attention to where you are going. At other times the fire can weaken the stairs that you went up. As you come back down, they collapse, dropping you into a raging inferno. Curtains hanging over a window that you are using as an emergency escape route can become wrapped around you and explode into flame in a flashover situation.

Structural collapse is one of the fire fighter's worst nightmares **FIGURE 14-5**. A floor or roof above you can collapse and land on you, pinning you down until you run out of air or are burned to death. The floor or roof you are standing on can fail, dropping you into the fire below. Many types of modern buildings use truss roof construction; older construction utilized the bowstring truss type of roof construction. Both of these types can fail dramatically when exposed to fire. Parapet walls and overhangs have a history of falling onto fire fighters. Even whole walls have fallen, burying fire fighters and equipment in the rubble.

As you walk around a burning structure, do not walk upright in front of windows. If you happen to be in front of a window at the time the fire backdrafts, you will get burned. Hose lines inside the structure are quite capable of blowing the glass out of a window and cutting you badly. The old story about walking under ladders being bad luck is especially true in firefighting. If someone above you is breaking glass or drops an axe, you are likely to get hit.

A type of structure fire that is becoming more common is the clandestine drug lab (meth lab) that has caught fire. The people preparing the drugs usually set up the operation, start the process, and then often leave for the period of time it takes to cook. They may have even booby-trapped the area. In any event, they are not going to be standing out front to warn you about the dangers of what is going on inside when

FIGURE 14-5 Structure collapse.
Courtesy of Kern County Fire Department

you arrive. Should you arrive at this type of incident, isolate and deny entry; an action that can be taken is to confine the fire to the smallest possible area while still making a defensive attack. Do not enter the structure yourself and do not touch anything. Law enforcement has access to specially trained persons who handle this type of incident.

Not all of the hazards at a structure fire are on the inside. Heavy roof loads in the form of air conditioning units and other equipment have a tendency to come crashing through when the structure is weakened by fire **FIGURE 14-6**. Chimneys pose the same hazard. A television antenna can act as a spear if it comes through the roof and pins you to the floor.

What can you do to protect yourself? Always wear your full PPE and remain alert for anything out of the ordinary. If a fire resists normal extinguishment methods, it may mean that there is some type of unusual fuel involved. Flame and smoke colors are an indicator. Normal structure fires burn with a reddish orange to yellow flame and black smoke. If the flames are dark orange, blue, green, or some other color, there is something going on you should be extra careful of. Reddish or other unusual color smoke also indicates that something is out of the ordinary.

> ### SAFETY TIP
>
> Not all of the hazards at a structure fire are on the inside. Always be aware of the entire incident scene.

Be suspicious. If you arrive at a structure and the windows are blown out or the doors are lying on the ground and appear to have been blown off, there is something

FIGURE 14-6 Partial roof collapse at a residential structure fire.
Courtesy of Kern County Fire Department

unusual going on. If you are on your way to the fire and you see someone running down the street away from the fire with a gas can in his hand, that is a definite clue.

Leave yourself a second way out. Before entering a structure, try to locate an alternate opening. Make sure there is another way out if the fire gets out of control and cuts off the path through which you entered. When working on a roof, above a fire, there should always be enough ladders placed for all of the personnel to make a quick escape. Never fewer than two ladders should be placed. If fire fighters are going upstairs, ladders should be placed at windows so they can escape without using the stairs. Jumping to the ground from a second or third floor window or from the roof is not good practice and may very well end your career, if not your life.

Do not freelance. Stay with your company and your officer. Your officer has much more firefighting experience than you do and is more likely to recognize a bad situation developing before you do. By staying together, fire fighters can help each other out. If one should become trapped under some rubble or drop a leg through the floor or roof, the others can lend a hand. At the least they can direct the hose stream while you disentangle yourself. Someone should always know where you have gone. If you are missing, it is better that someone recognizes that fact right away and not after returning to the station.

High-Rise Buildings

High-rise firefighting presents a whole list of additional hazards to the structural fire fighter. Some fire fighters have gone as far as to call these structures "elevated crematoriums." The floor areas of high-rises are often rented to different occupants. Each individual floor can have its own unique layout according to the tenant's wishes. This makes it easy to become disoriented. A practice that sometimes works is to take a quick tour of the floor below the fire floor to get an idea of the layout before entering the fire floor. The interior walls on each floor are often nothing more than partitions ending at the suspended ceiling. Above the ceiling the fire can run freely across the hidden attic space above your head. You cannot see the fire in this space, and it may come out between you and your escape route. The best way to access a high-rise fire is up the stairways. When the fire is on the 50th floor, this will take a long time to accomplish, and after you carry your gear to the fire floor, you may be too fatigued to be of much good. At this point the elevators sound like a good option. The trouble is, if they open onto the fire floor, you cannot be sure what you will be confronted with when the door opens.

The fire department hose connections are in the stair-wells. You may not be able to make it from the elevator lobby to the stairwell because of fire and heat. It is easy to get confused in heavy smoke. When you cannot make it from the elevator to the hose connection, you have no water with which to attack the fire to try to improve your situation.

One alternative is to take the elevator to the floor below the fire and to use the stairs from there. This requires being absolutely sure which floor the fire is on. The elevators can be used if they are the type that open at a sky lobby below the fire. This way you will not end up with the door opening onto an inferno. There are systems available on some elevators that allow fire fighters to override their automatic functions. If this system is installed and the fire fighters are familiar with how it works, it is another option **FIGURE 14-7**.

In high-rise fire situations, one of the exterior hazards is falling glass **FIGURE 14-8**. Most of today's high-rise buildings have glass exteriors. When there is a fire, these glass panels have a tendency to fall out or get knocked out by occupants or by fire fighters effecting

FIGURE 14-8 Fire in high-rise building.
© George Widman/AP Images

FIGURE 14-7 Fire department elevator control.
© Johnny Habell/ShutterStock

ventilation. A large piece of glass falling from far above the street is capable of severing hose lines and doing severe damage to fire fighters. Pieces of glass have the ability to sail quite far as they fall. No one should be standing out in the open within 200 ft of the fire building if possible. When underground access to the structure can be found, that is often the best way to enter.

Electrical Hazards

A hazard almost always present in structure fires is electricity. One of the first tasks at a structure fire is to cut off the power to the structure. You should never try to pull the electrical meter. It can explode in your face, showering you with glass. Even though someone has cut the switch or main breaker to the structure, it is no guarantee that the power is off. This is especially true in multifamily, commercial, and industrial occupancies. If the occupants are stealing electricity, they may have wired around the meter. If you must walk through a darkened area, it is best to put your arms up with your palms toward you. This prevents you from reflexively grabbing an energized electrical wire if you contact one. Great care must be taken in watching where apparatus is parked because overhead wires to burning structures often come disconnected and can fall across the apparatus. If this happens, jump clear and do not touch the apparatus. Any time you see a wire, consider it a live wire. Just because a wire is not arcing and jumping does not mean that it is deactivated. Wires can also burn off and fall across metal fences, such as chain-link, energizing them. There are numerous electrical safety videos available on You-Tube; look for one put out by an energy company.

One of the cardinal rules of raising ladders of any type is to look up first. Make sure that you are not raising your ladder into electrical wires. A wooden ladder will not conduct electricity as well as an aluminum one, but they can both kill you in the right situation. Your rubber boots contain carbon and are not electrical insulators. Pike poles are not to be used for moving electrical wires. They may be resistive when new, but as they get older and build up carbon on the handles, their resistance is lowered. Do not attempt to cut wires with bolt cutters. The only people who are fully trained and carry the required equipment to deal with electrical wires and equipment are the people from the power company.

Pets

A hazard most people do not stop to consider at structure fires is pets. It can be very exciting to open a door or gate and see a large pit bull launching itself at you. Some people keep exotic pets, such as rattlesnakes, in their homes. Livestock can also pose a hazard to fire fighters in certain situations. Entering a fenced area to extinguish a grass fire and meeting up with an enraged bull is a bad position to be in. In some situations your PPE may help to prevent injury. Your best defense is always a cautious approach.

Fire Department Operations at Sprinklered Occupancies

The three principal causes of unsatisfactory sprinkler performance include a closed valve in the water supply line, delivery of inadequate water supply to the system, and occupancy changes that render the installed sprinkler system unsuitable. These can be prevented through effective preincident planning and the implementation of testing and maintenance programs.

Departments should establish standard operating procedures for operations at sprinklered or standpiped occupancies. One of the first actions to be taken is to ensure the supply to the system is boosted through the use of a pumper attached to the fire department connection. All of the system's valves should be checked to ensure that they are in the fully open position, unless they are marked closed for maintenance. This is not the case for residential systems. They are designed to handle only normal pressures provided by the water main system. Once flow from the system is ensured,

hose lines can be advanced to the seat of the fire. Do not turn off the system to enhance visibility before the fire is confirmed to be under control. When a part of the system must be turned off for overhaul or salvage operations, try to turn off only the portion of the system in the affected area. The system should be placed back into service immediately before leaving the scene. If the system cannot be placed back into service, the property owner or representative should be notified.

After any fire-caused sprinkler system activation, an investigation as to cause should be conducted. Also include system effectiveness in the reporting. A full list of the components of this type of investigation are contained in *NFPA 13E: Recommended Practice for Fire Department Operations in Properties Protected by Sprinkler and Standpipe Systems.*

Electrical Distribution Equipment and Installations

Areas that require careful preplanning and consideration are electrical substations and vaults **FIGURE 14-9**. They are a hazard from a safety standpoint because of the high potential for getting electrocuted. It is also possible that indiscriminate use of water or dry chemical extinguishers could cause much more damage than the fire would. A more common fire is the power pole fire. Water can conduct electricity. When extinguishing electrical equipment, it is better to use a fog pattern or short bursts of water to accomplish the task. This lessens the chances of getting electrocuted.

Transformers used to contain PCB oil, which is carcinogenic. Power companies have, for the most part, phased out this oil. There may still be some out there, however, and you should proceed with great caution around electrical equipment.

FIGURE 14-9 Results of fire in electrical equipment.
Courtesy of Steve Pendergrass

As a general rule, fire fighters should not enter or extinguish fires in electrical distribution equipment or installations unless guided by electric company employees. These personnel are trained and know where it is safe to operate and which firefighting technique should be applied. They know which switches to throw to shut down the power. Remember from Chapter 4, *Chemistry and Physics of Fire* that a Class C fire reverts to Class A, B, D, or K once the power is shut down.

New challenges for fire fighters include solar panels and electric vehicles. Solar panels are generating electricity when the sun shines on them and they cannot just be turned off. Some owners may also have a large battery attached to the solar system to store electricity when the sun is not shining on the panels. Electric vehicles have large, powerful batteries and special care must be taken with these as well. Procedures for fighting Class C fires and avoiding shock hazards must be taken whenever electricity is involved.

> **Tip**
>
> As a general rule, fire fighters should not enter or extinguish fires in electrical distribution equipment unless guided by electric company employees.

Wildland Firefighting

In wildland firefighting operations, the basic methods of extinguishment are to apply water or fire retardant to the fire edge or to create a fire break or control line around the perimeter **FIGURE 14-10**. This is done with a variety of methods. In grasslands, pumpers can be used to make a direct attack on the fire edge. Crews using hand tools can also create a scratch line that breaks the fuel's continuity and stops the fire spread. Dozers are used in the same manner. In heavier fuels, because of radiated heat and flame lengths, a direct attack is not possible. In these fuels, crews get well ahead of the fire and make an indirect attack, creating fire breaks with hand crews and dozers. Natural barriers, like lakes and roads, are also used as part of the fire break. When Class A foam is available, it can be used to protect exposures by applying it as a wetting agent to raise the surface fuel moisture to where it will not burn.

A method often used to create a safe "black line" is to backfire. The backfire removes the fuel between the control line and the head of the advancing fire. The backfire burns back toward the main fire and away from the control line **FIGURE 14-11**. This allows a relatively narrow control line to be used. The line is effectively widened by the removal of the fuel between the advancing fire and the control line. When burning is done to remove fuel along the flanks or to remove unburned islands left in the fire perimeter, it is called "firing out." These firefighting methods should never be attempted by untrained personnel. When performed improperly, they can cause the fire to jump control lines.

The safety rules developed for wildland firefighting can be applied to all types of firefighting. The rules cover the main points that should be observed in any fire situation. The basic safety rules of operating at a wildland fire are the *10 Standard Firefighting Orders* and the *18 Situations That Shout Watch Out*.

> **SAFETY TIP**
>
> The safety rules developed for wildland firefighting can be applied to all types of firefighting.

FIGURE 14-10 Fire fighters attacking wildland fire with hoseline (A) and hand tools (B).

FIGURE 14-11 Engine in the safety zone as fire burns past during backfiring operation.
© Jones & Bartlett Learning

The *10 Standard Firefighting Orders* are (National Wildfire Coordinating Group 2018):*

1. Keep informed on fire weather conditions and forecasts.
2. Know what your fire is doing at all times.
3. Base all actions on current and expected behavior of the fire.
4. Identify escape routes and safety zones, and make them known.
5. Post lookouts when there is possible danger.
6. Be alert. Keep calm. Think clearly. Act decisively.
7. Maintain prompt communications with your forces, your supervisor, and adjoining forces.
8. Give clear instructions and be sure they are understood.
9. Maintain control of your forces at all times.
10. Fight fire aggressively, but provide for safety first.

The *18 Situations That Shout Watch Out* are (National Wildfire Coordinating Group 2018):*

1. The fire is not scouted and sized up.
2. You are in country not seen in daylight.
3. Safety zones and escape routes are not identified.
4. You are unfamiliar with the weather and local factors influencing fire behavior.
5. You are uninformed on strategy, tactics, and hazards.
6. Instructions and assignments are not clear.
7. There is no communications link with crew members/supervisor.
8. You are constructing a line without a safe anchor point.
9. You are constructing a fire line downhill with fire below.
10. You are attempting a frontal assault on the fire.
11. There is an unburned fuel between you and the fire.
12. You cannot see the main fire and are not in contact with anyone who can.
13. You are on a hillside where rolling material can ignite fuel below.
14. The weather is getting hotter and drier.
15. The wind increases and/or changes direction.
16. You are getting frequent spot fires across the line.
17. Terrain and fuels make escape to safety zones difficult.
18. You feel like taking a nap near the fire line.

All of these listed orders and situations were developed because someone was seriously injured or killed. In recent incidents where wildland fire fighters were injured or killed, violations of these orders and situations were found to be contributing factors. The four most common causative factors involved in tragedy and near-miss wildland fires are: (1) small fires or deceptively quiet sectors of large fires; (2) light fuels, such as grass or brush; (3) an unexpected shift in the wind direction or speed; and (4) fires running uphill.

LCES

Because it is very hard to memorize the *10 Standard Firefighting Orders* and all of the *18 Situations That Shout Watch Out*, two more condensed safety messages have been adopted. They cannot totally take the place of those described previously, but they serve as a general reminder.

The first of these messages is LCES (Lookouts, Communications, Escape routes, and Safety zones) (National Wildfire Coordinating Group 2018).

- **Lookouts** should always be posted to keep an eye on the fire and the weather. These personnel are not to become involved in actual firefighting. Their job is to keep an eye on the fire and to let the crew know if the fire starts to increase in intensity or make a run at the crew's position. In areas where hazard trees are a problem, this may require one person watching for every two working.

- **Communications** must be maintained between the lookouts and the crews working, between the crews and their leader, between the leader and the adjoining crews, and between the crew leaders and the command personnel.

- **Escape routes** should always be planned and communicated to all of the crew members in

---

* National Wildfire Coordinating Group. 2018. *NFES 1077 Incident Response Pocket Guide.* Boise, ID: National Interagency Fire Center.

case of a change in fire behavior. They must provide access to a safety zone. As in other types of firefighting, two escape routes are better than one, in case one gets cut off.

- **Safety zones** should be placed as often as necessary. They must be large enough to accommodate all of the crews in the area in case of a burn-over. The safety zones need to be spaced closely enough that the crew has time to travel to them if the fire dictates that they must seek refuge. They must be clearly indicated and accessible.

F LCES Δ

When looking at the application of LCES, a modified version includes what Brad Mayhew (a former hotshot fire fighter with the U.S. Forest Service) terms *F LCES Δ* (Mayhew 2009b). The *F* stands for *fire behavior*, which is to prompt you to think of the worst case scenario. In other words, what is the worst this incident can become? The World Trade Center incident was handled as a structure fire with rescue in two high-rise buildings. It then became much worse. Nothing as extreme as their total collapse had happened before—and it was not foreseen. Other fire fighters face somewhat similar situations. They respond to an incident that is fairly commonplace and then it escalates into something much worse. This could be a trash bin fire that then turns out to contain hazardous materials. To protect ourselves from the extraordinary happening during the ordinary, we need to be sure that we are trained, equipped, and led appropriately. This includes looking out for ourselves as well as others.

At the scene of a structure fire, the lookout is the IC or his or her designee. This person remains outside the structure to observe any changes to the structure or the fire intensity. The communications is the link between the exterior and the interior, either by radio or other voice communication. The escape route is the path of egress off the roof or out of the structure. The safety zone is the safe area outside the structure, clear of smoke and falling debris. These elements of safety (LCES) should be in place prior to engaging at any incident, regardless of type.

The Greek letter delta (Δ) signifies change. Fire occurs in an ever-changing environment. Over the life of the fire, combustibles are consumed. If the fire is in a structure, the structure may be weakened. Out in the hills, the fire moves over varying terrain and fuel types. Fire fighters need to pay attention to what is changing at the time and what may change in the future. The personnel assigned to the incident are changing as well. They are becoming fatigued, possibly bored, or even overconfident. Maybe they are becoming overly stressed and entering the fight-or-flight mind-set. All of this must be reviewed constantly in context of the LCES that is in place. Is the LCES that was put into place at the onset of the attack still valid? If not, why not, and what can be done to bring it into compliance? It all goes back to situational awareness on the part of everyone assigned to the incident. They all need to be focused, alert, and vocal if something does not look right.

Remember that information is what drives decision making. The information must be gathered and considered and appropriate action taken to respond to it. The last step in any decision-making process is to reevaluate. Once the response to the change occurs, it must be reevaluated again, continuously. Fire fighters must all remember to do so at every incident, every time.

Look Up, Look Down, Look Around

The second of the condensed safety messages, "look up, look down, look around," is as follows (National Wildfire Coordinating Group 2018):

Look up before you start to work, and every so often while working, get your eyes off the ground and look up. See what the wind, smoke, and fire are doing. Is the fire moving toward you or away from you? Are there aircraft starting to work overhead, close enough to be a danger to you? Look uphill from your location. Is there heavy equipment operating farther up the hill that may roll rocks or logs down on top of your position?

Look down. Watch your footing. At night it is easy to become blinded from the light of the fire and walk into holes or fall from drop-offs. Logs and pipes in tall grass can easily trip you. When carrying a razor-sharp Pulaski, this could be disastrous. Look downhill and be sure that the fire has not hooked around or been ignited by rolling material below your position. The moon can illuminate the area in which you are working quite well. The problem is that moonlight does not provide you with good depth perception. It is plenty light enough to see where you are going, but easy to walk right into a ditch.

Look around. Be aware of what is going on around you. You may be working too close to someone operating a chain saw or swinging a chopping tool. Be aware of what the fire is doing. Keep track of the escape routes and safety zones. Do not become separated from your crew.

These safety messages can be applied to every type of firefighting, not just wildland. Their main focus is that you keep informed on what is happening and remain aware of how you are going to seek safe refuge. Fire fighters tend to fall victim to the "candle and moth" syndrome and are drawn into seeing only the fire and nothing else. Do not become so intent on extinguishing the fire that you do not consider where it is going and what damage it has caused to trees and structures, either of which may fall on you if you are not careful.

Wildland–Urban Interface/ Intermix Firefighting

Wildland–urban interface firefighting is becoming more common. More and more people have escaped the cities by building their homes in the foothills and mountainous areas. This has given rise to situations where the fire fighters cannot afford to just back off a couple of ridges, create a fire break, and wait for the fire to get there. Now fire fighters must place themselves and their equipment in the path of the advancing fire **FIGURE 14-12**. In this type of firefighting, the main objective is to protect as many structures as possible. This is primarily done by placing a pumper at each structure. It would seem that a pumper with a 1000 gpm pump and 500 gallons of water could take on most any fire. This is untrue and is proved so on a regular basis, with the result being injured fire fighters and a destroyed pumper. In areas of dry brush and trees, flame lengths can easily exceed 50 ft. Fire fighters do not have the capability to stop this type of fire in a direct, frontal assault.

This firefighting method is not meant to be a suicide mission. There are several things to consider when assessing your ability to protect a structure and provide for your own safety. One of the first things to do is to determine your water needs versus the availability of water at the scene. Perform a structure triage to determine whether the structure is truly defendable based on its construction type, roof covering, proximity to the fuel, and topography. Make a fair evaluation of what is to be protected and the probability of a successful outcome. Last but not least, determine whether it is truly safe to take the action you are considering.

When the decision is made to attempt to save the structure from the advancing fire, the following safety considerations should be followed.

Have a planned and understood escape route that is constantly reevaluated to be sure that it is available. Park your engine on the side of the structure away from the advancing fire to protect it from radiant heat. Keep in communication with your crew and surrounding crews. Plan for the necessity of being overrun, should it happen, and use the cab of the engine or the structure you are protecting as a safe refuge. Keep at least 100 gallons in the engine water tank for your own protection. Headlights should be kept on, windows rolled up, and outside radio speakers turned up so you can hear a warning of advancing fire. Coil a charged 1½-in. line at the engine for engine protection. Try to stay as mobile as possible for a rapid escape. Keep structure protection lines short. Use water only as necessary. Never compromise crew safety to protect property.

In these safety messages, great emphasis is placed on maintaining communications. Without good communications among resource groups, a coordinated attack cannot be implemented. As a factor directly relating to safety, communications are important to keep everyone informed as to what the incident is doing and what action to take when things go wrong.

Petroleum Firefighting

In areas with oil production or refining, oil firefighting is a skill that needs to be developed **FIGURE 14-13**. The primary objectives in oil firefighting are to extinguish the fire and control the source of any leaks. Fire props

FIGURE 14-12 Engine strike team at a wildland fire.
Courtesy of Kern County Fire Department

FIGURE 14-13 Fire fighters making upwind approach to burning petroleum piping to shut off fuel supply at valve.
Courtesy of University of Nevada

at oil firefighting schools are designed so that fire fighters advance hose lines toward the burning product and push it away from them. Once at the burning flange or pipe, the valve is closed to shut off the flow. In plumbing fires, this is sometimes the technique used to control fires. Often a refinery employee can shut down the leak from a location remote from the fire, and the fire department's job is to confine the fire and protect exposures until the fuel remaining in the pipe burns itself out.

In refinery fires, many different products can be encountered. All of them have different burning characteristics. On the receiving side, the product is crude oil. It is very viscous; it is then refined into lighter products, such as diesel fuel and gasoline. The lightest product is hydrogen gas. It burns so cleanly that the flame is often invisible. It would be possible for a fire fighter to walk into the jet of burning hydrogen gas without seeing it. You should project a hose stream out in front or even use a regular broom, held out in front of you to search out the flame. Generally, fire fighters should be accompanied by refinery employees any time they enter the refinery for firefighting purposes. It is dangerous to shut off valves or attempt other mitigation efforts without the proper guidance and expertise.

When the fire occurs in a storage tank, different methods are used. In storage tank fires, only the surface is burning. This may appear rather easy to control because the surface area is contained by the walls of the tank. The problem is that just pouring water into the tank at a high rate will not extinguish the fire.

Much of the water will be turned to steam by the heat of the fire, and oil floats on water, leading to three main problems in this type of firefighting; these are boilover, slopover, and frothover (FireEngineeringUniversity.com 2013):*

1. *Boilover* occurs when water is trapped under the surface of the oil. Such a layer of water is called a lens, because of its shape. As the oil surface burns, a heat wave travels downward through the oil at a rate of 12–48 inches per hour (Shell Oil Co. 1991). The temperature of the heat wave is higher than the boiling point of the trapped water. When this heat wave hits a lens of water, it turns the water to steam. The water, when turned to vapor, expands at a rate of approximately 1700 to 1. The expansion of the water to steam causes the oil to erupt upward and may cause oil to be ejected from the tank. If you were too close to the wall of the tank, you could be in the path of the falling, burning, or at least extremely hot oil.

2. *Slopover* occurs when fire streams are applied to the burning liquid surface at such an angle that hot and/or burning oil is forced over the edge of the tank. The oil then flows down the side of the tank, and you are now confronted with a fire on the ground as well as the one in the tank.

3. *Frothover* occurs when fire streams are directed at such an angle that they plunge under the surface of the burning oil. This would be most likely to occur if the water were applied from elevated stream devices, such as ladder trucks. As the water enters the surface, it is turned to steam, much like the boilover, and the surface is turned to burning oil froth. This froth can easily overflow the walls of the tank and spread as a ground fire.

In tanks with a cone roof and vapor space above the contents, subsurface foam injection plumbing is often provided **FIGURE 14-14**. The foam-generating equipment is hooked to a manifold on the tank, and foam is pumped in until the fire is extinguished. Other tanks may have a system of piping that extends up the outside of the tank. The foam-generating equipment is then attached to this plumbing and the foam flows onto the surface of the product, extinguishing the fire. When neither of these is provided, the foam must be applied by aerial apparatus or from the ground through large-bore nozzles that can project a stream over the wall of the tank. In some cases, a nozzle is mounted to fixed plumbing for tank protection **FIGURE 14-15**. Throwing a ladder on the outside of the tank to gain access to the surface is extremely dangerous and definitely not recommended. If other tanks are endangered by the radiated heat of the fire, they may be cooled with water spray. The water must be

FIGURE 14-14 Weak seam failure in cone roof crude oil tank caused by fire.
Courtesy of Kern County Fire Department

FIGURE 14-15 Fixed monitor nozzle used for petroleum storage tank fire protection.
© Jones & Bartlett Learning

FIGURE 14-16 Floating roof tanks.
© Jones & Bartlett Learning

applied directly on the tank shell. Setting up a **water curtain** to absorb radiated heat in the air is ineffective.

Most modern storage tanks are of the floating roof type. The roof floats directly on the product, which reduces the vapor space inside the tank. The most common fire is in the seal area between the floating lid and the tank shell **FIGURE 14-16**. When large amounts of water are pumped onto the lid, it can cause it to sink, increasing the burning surface area greatly. The two methods most commonly used on a seal fire are attack lines using foam or dry chemical extinguishers. If the tank is equipped with a plumbing system for foam application, it should be designed with a dam to keep the foam applied over the area of the seal.

Gasoline Spills

The mass release of gasoline, or any other flammable liquid, requires that the vapors be controlled, if it can be done in a safe manner. Class B foam is the agent of choice for this operation, though it is not effective on all Class B materials. When Class B foam is properly applied, it can seal the liquid surface so that the flammable vapors are contained and ignition of the vapors is prevented. When Class B foam is used, it must be periodically reapplied because it will break down after a time and become ineffective in controlling the vapors.

Any time Class B foam is used on a petroleum spill, fire fighters should not enter the area of the spill. Walking through the foam blanket breaks the seal and releases the flammable vapors. If the vapors find an ignition source, you will be surrounded by flames. In this situation, the natural reaction is to run. This just disrupts the foam blanket further and makes the situation worse. If you must walk through the foam, shuffle your feet slowly to allow the foam to reseal as you move; do not pick your feet up as you walk. If you must pick up your feet to go over an obstacle, lift them up and put them down slowly. Also, try to keep hand lines out of the foam blanket; the best course is to stand back and lob the foam to where it needs to go. This helps prevent breaking the foam blanket.

SAFETY TIP

Any time Class B foam is used on a flammable liquid spill, fire fighters should not enter the area of the spill. Walking through the foam breaks the seal and releases the flammable vapors.

Tip

Any time Class A or B foam is mentioned in this Chapter, Encapsulator Agents (Chapter 12) may also be used.

Liquefied Petroleum Gas

A commonly used by-product of oil refining is liquefied petroleum gas (LPG). The storage containers for these products, including butane and propane, are made of a steel shell that is welded together. They are easily recognizable by their hemispherical ends. They may be of many different sizes and are either mounted on the ground, carried on trains, or mounted on trucks and trailers in the larger sizes. Some other places these containers are encountered are tanks on backyard barbecues, forklifts, travel trailers, motor homes, and propane-powered vehicles. The smaller versions are sometimes encountered in garage fires.

The tanks are built to withstand high pressures, above that of the **working pressure** of the tank. They are equipped with a **relief valve** on the top that should keep them from ever reaching the tank's **burst pressure**.

This works well under normal conditions. The problem is that not all conditions remain normal. In earthquakes, tornadoes, and floods, the tanks become dislodged from their mountings. In vehicle accidents, with truck- or train-mounted tanks, the tank can end up in any position. It may have its structural integrity compromised from the tank skidding along the ground or striking another solid object. In accident situations the relief valve may end up on the bottom because the tank is upside down.

When one of these containers is involved in a fire, the recommendation is that large volumes of water be applied at each point of flame contact. The volume is determined by the size of fire and number of containers involved. When the fire is burning from the relief valve on top of the tank, there may not be any direct flame contact, but there will be a problem caused by the radiated heat. As the container is heated, the rate of gas release increases. This only compounds the problem as the fire is intensified and the container is then heated at a faster rate. Steel starts to weaken as it is heated (Fire 2009). When the pressure in the container increases to the point where it overcomes the strength of the weakened steel, the container shell will separate, releasing the contents **FIGURE 14-17**.

When a release of the contents happens, it is called a boiling liquid expanding vapor explosion (BLEVE), sometimes referred to as a "blast leveling everything very effectively." When a flammable gas is involved, the release creates a vapor cloud that ignites from the fire, causing the rupture. The resultant cloud of flaming material and heat and shock waves have tremendous destructive potential. Anyone in the immediate area has very little chance of survival.

When fighting fires involving containers with a flammable BLEVE potential, water should be applied with remotely supplied master stream appliances. Many facilities are set up with large nozzles for this

purpose. On the roadway these are not going to be readily available. When personnel set up appliances, they should get them set up and get out of the area as quickly as possible. Container pieces have been known to fly as far as one-half mile from the explosion site.

A BLEVE can also occur in a nonflammable liquid. Any container that has liquid contents and will withstand a pressure rise inside before it bursts can BLEVE. When a can of creamed corn explodes in a kitchen fire, the same forces are at work. The results are just not as spectacular. An example of this type of explosion on a small scale is to make popcorn. As the water inside the kernel is heated to the burst pressure of the container, the kernel pops open.

SAFETY TIP

Container pieces have been known to fly as far as one-half mile from the explosion site in a BLEVE. Stay well away from any container with BLEVE potential. If you must make an attack, put unmanned streams in place and withdraw all personnel to a safe distance.

Not all flammable gas releases are caused by or result in fires. LPG can be released because of mechanical failure of plumbing or human error. The vapor density is heavier than that of air, and the LPG vapors will tend to pool in low areas. The course of action in these situations is to control ignition sources and disperse the vapors with water fog.

Natural Gas

A common flammable gas, natural gas, poses little hazard when released into the open, because it is lighter than air and will disperse itself. However, it has the potential for collecting in structures when leaks occur in interior-mounted gas meters and plumbing. This has been the cause of many explosions in structures, sometimes destroying more than one building and causing structural collapse. There have been numerous instances where digging work was being done with either backhoe tractors, plows, or shovels and gas lines were cut. The resultant leakage of gas should be handled by the gas company. The danger of static electricity causing ignition should not be ignored, and an attack line should be pulled and charged just in case. When a backhoe or other equipment ruptures the line, it should be shut down immediately and not restarted until the source of the gas is turned off and any remaining gas is dispersed.

FIGURE 14-17 Gas cylinders: distorted (left) and ruptured because of heat from fire (right).

A very real hazard to fire fighters is the person bent on suicide who releases natural gas or LPG into a structure. If the neighbors smell the gas and the fire department responds before the person has expired, the suicidal person may decide to take you with him or her by igniting the gas as you approach. Any gas smell incident should be treated as a true emergency with life-threatening potential. The worst scenario for fire fighters is the cloud of flammable gas that is not yet ignited. You may find yourself in the cloud without realizing it. If at that time the gas finds its ignition source, you may become engulfed in a fireball.

Hazardous Materials Incidents

Hazardous materials incidents are becoming more common, partly because of greater awareness and partly because of increased use of these materials **FIGURE 14-18**. Approximately 2000 new chemical combinations are introduced every year. Those that have no commercial value are not created in any great quantity, but those with commercial value are shipped in every mode of transportation and stored at all kinds of facilities (SBCCOM 1999). If you were to inventory the garage in almost any home in the country, you would find some type of material that could be classified as hazardous.

The days of charging in and taking immediate action at all incidents are long gone. In every situation, you must be alert for the presence of materials that pose more than the ordinary hazard. This subject was already discussed to some extent in the portion of this chapter on structure fires. This section deals with incidents that are recognized as involving hazardous materials either before arrival or upon arrival.

The federal government has legislated, under the Superfund Amendments and Reauthorization Act

FIGURE 14-18 Fire fighters operating at a hazardous materials incident.
Courtesy of Kern County Fire Department

of 1986, that all employers and employees follow the requirements of OSHA 29, CFR Part 1910.20. Part 1910.20 requires training to at least the first responder level for employees responding to take action at incidents that involve hazardous materials. Annual refresher training is also required. This law also requires the establishment of an ICS on any incident involving hazardous materials. In states without a state-level OSHA-type department, Section 126 requires the Environmental Protection Agency to issue an identical set of regulations to cover state and local government employees.

When responding to a hazardous materials incident, the primary consideration is to make a precautionary approach. Slow down and think about what you are going to do before you take action. Regular PPE is not designed to protect you from hazardous materials. Many chemicals can penetrate your turnout gear and attack you. The effects may not be immediately evident and may accumulate over many years. Even SCBA is not enough protection when the chemical involved is a skin absorption hazard. Leather gloves, as previously mentioned, are no protection from chemicals. They just soak up the chemical and hold it next to your skin as you sweat and your pores open up.

All approaches to suspected hazardous materials incidents should be made from upwind, uphill, and upstream. If the wind is blowing in an uphill direction, however, stay out of the plume. This may require a detour from the most direct route in some situations. The approach should be made with as few personnel as possible. Fire fighters should work in pairs. If it takes only two fire fighters to make the approach and one to get close enough to read the label, then only two should make entry. Once at the scene, any apparatus should be parked facing out. An engine can make a much faster escape from a deteriorating situation with its forward gears than it can in reverse.

The initial responsibility of the fire department is to isolate, identify, and deny entry. The purpose of this operation is to ensure that persons who have not been exposed do not become exposed. **Perimeters** are set up to control the access of personnel to the scene. The innermost perimeter is the exclusionary or hot zone, and only properly equipped and trained personnel should be allowed into this area. Proper recognition and identification must take place before such actions are taken. Any offensive actions take place only after proper risk–benefit analysis has taken place. Should hot zone entry take place, only the minimum number of responders necessary should be allowed into the hot zone. If your department has a hazardous materials team, this is where their additional training, expertise, and equipment are used. Around this perimeter, a secondary zone, called the contamination reduction or warm zone, is set up. Anyone who passes from the hot zone to the warm zone is to be properly decontaminated. This is often done by trained engine company personnel. Any nonessential personnel are to stay in the support or cold zone. Victims should be brought out of the hot zone and properly decontaminated before they are turned over to emergency medical services (EMS) personnel. The EMS personnel should not be allowed to enter the hot zone to rescue them unless they have the proper training and PPE **FIGURE 14-19**.

By identifying the material involved, a decision can be made as to the level of hazard and how it should be handled. The truck driver who says, "I've handled this stuff for years" is not a reliable source of hazard information. Reference sources are available to research the hazards of many but not all substances. The problem is that in many incidents there may be two or more chemicals mixed together. This can create a combination that is not clearly defined. When in doubt, go with the worst hazard of each of the chemicals involved and act accordingly.

When the material is positively identified, information can be gathered from the U.S. Department of Transportation's *Emergency Response Guidebook*, CHEMTREC, the manufacturer, and computer databases. Many chemicals have more than one hazard. For example, many liquid pesticides are flammable as well as toxic. The *Emergency Response Guidebook* tends to describe the hazards of the materials it lists from a transportation viewpoint. It is not the definitive answer in all situations. Before any action is taken, at least three sources should be referenced as to the material's hazards. When the material has been positively identified and the proper protective clothing is determined and available, three basic actions are taken: berming (sometimes called diking), diverting, and controlling. Berming is done to contain the flow of a material from the area of the leak. A berm can be constructed from dirt or other readily available materials **FIGURE 14-20**. Sawdust is a poor choice because it is combustible and some chemicals can react with it, starting a fire and compounding the problems you already have. Before berming with any material other than dry sand or dirt, a reference source should be consulted.

FIGURE 14-20 Fire fighter berming a spill to contain liquid material.
Courtesy of Kern County Fire Department

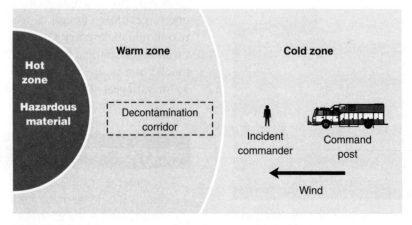

FIGURE 14-19 Zones set up at hazardous materials incident.
© Jones & Bartlett Learning

Diverting is done to direct the flow of a hazardous material that cannot be contained in a berm because of amount or availability of berming materials. It may also need to be done because approaching the material closely enough to berm it would pose an unnecessary safety risk. Diverting can consist of covering a storm drain grate with plastic and a ring of dirt to keep the material out of the sewer. Often in releases of large amounts of material, diverting is all that the first-in company can accomplish until more help arrives at the scene.

Controlling is done to stop or reduce the flow rate of the leak. If deemed safe, a leaking 55-gallon drum can be rolled so the hole is above the liquid level. In other situations, a plastic bucket can be placed where it will catch a dripping material. Spilled powder materials can be covered with a piece of plastic sheeting to keep them from being spread by the wind or draft from passing vehicles. Releases from plumbing can often be controlled by finding the valve and turning it off. All of these procedures are fairly simple, commonsense techniques that require no special equipment or training. Many specially designed leak control devices should be used only by trained personnel. Controlling usually requires a close approach to the source of the leak and is not advised until the material and its hazards are identified.

Once the incident is under control and the source of the leak is stopped, the cleanup phase begins. Many fire departments do not do cleanup. This is left to private contractors that specialize in this type of operation. Cleanup often requires specialized equipment and permits for storage and transportation of hazardous materials. The health department usually determines how clean the area needs to be and has the final say on how clean is clean.

Weapons of Mass Destruction

As we have seen from the attacks on the Murrah Federal Building (Oklahoma City, Oklahoma, 1995), the Pentagon (Arlington County, Virginia, 2001), and the World Trade Center (New York, New York, 2001), and in other parts of the world, acts of terrorism with large loss of life are a very real threat. As with other incidents, fire fighters are often the first responders. In the World Trade Center incident alone, 343 fire fighters lost their lives.

The purpose of weapons of mass destruction (WMD) attacks is to take human lives and/or to cause panic and disruption. These attacks may be chemical, biological, radioactive, nuclear, and/or explosive (CBRNE) in nature.

Some of the conditions found in a terrorist CBRNE incident, but not necessarily at a hazardous materials incident, are:

- *Crime scene.* Requires you to preserve as much evidence as possible.

- *Major interaction with local, state, and federal agencies.* These agencies might include the Federal Bureau of Investigation (FBI), the Bureau of Alcohol, Tobacco, Firearms and Explosives, and the Environmental Protection Agency.

- *Scene communication overload.* Radio traffic will be intense and the cellular phone systems may become overloaded with citizens, responders, and the news media.

- *Chaos.* People will be trying to escape the area and the number of responding agencies will be many more than for an ordinary hazardous materials incident. An effective incident command structure must therefore be implemented early on in the incident.

- *Overwhelming of resources.* The number of victims and destruction/contamination of the area may rapidly overwhelm the first responders' ability to handle, or even assess, the scope of the incident.

- *Secondary devices designed to kill responders.* A secondary device may be a bomb planted to go off after emergency personnel have arrived at the scene of a bombing or other attack.

- *Preincident indicators.* There may be a threat phoned or mailed, or an agency, such as the FBI, may be aware of a threat prior to the incident.

- *Deliberate attack.* Terrorist attacks are designed to cause as many casualties as possible and are done deliberately, usually in crowded places.

- *Super toxic material.* A material, such as sarin (200 times more toxic than chlorine) may be used to increase the number of casualties.

- *Identification of material used.* In hazardous materials incidents, there are usually some indicators as to what material is involved. In a terrorist attack, the only visible indicators may be the symptoms displayed by the victims.

- *Mass casualties with many fatalities.* The purpose of a terrorist attack is to draw attention. A high death/casualty toll is certain to gain media attention.
- *Psychological effects.* Psychological effects will go far beyond those directly involved and add casualties to the number directly affected by the incident.
- *Mass decontamination.* In an attack utilizing a toxic agent, contamination must be confined to as small an area as possible. Any persons, whether victims or responders, and equipment must be decontaminated before leaving the scene.
- *Unusual risk to emergency responders and civilians.* Regular PPE may not be sufficient to protect you from the effects of toxic materials, and secondary devices may be present. Remember, the purpose of a terrorist attack is to draw attention by means of numerous casualties.

Emergency Medical Service Operations

The fire department of today has become much more involved in providing EMS and in doing so has moved into a new area with its own problems. Operations in the area of emergency medicine for most fire departments revolve around first aid for injuries, basic life support, and extrication of victims from vehicle collisions. Fire fighters also respond to victims of assaults, including gunshot wounds, knife wounds, and rape. Medical emergencies can include diabetic problems, overdoses, and emergency childbirth. All of these scenarios pose their own types of hazards.

One of the primary problems involved in providing emergency medical assistance is avoiding exposure to bloodborne and airborne pathogens. The federal government has enacted legislation that addresses this problem. The legislation is part 29 CFR 1910.1030, *Occupational Exposure to Bloodborne Pathogens.* This regulation identifies emergency response personnel as Category 1 employees. Category 1 employees are at the greatest risk of occupational exposure to communicable diseases. These bloodborne communicable diseases include, but are not limited to, human immunodeficiency virus and hepatitis. The National Fire Protection Association (NFPA) addresses this issue in the NFPA 1500 series. *NFPA 1581: Standard on Fire*

Department Infection Control Program defines minimum requirements and criteria. The standard lists required program components and includes recommendations on fire department facilities, personnel, PPE, and procedures for cleaning, disinfecting, and disposal.

> ## SAFETY TIP
>
> One of the primary concerns involved in providing emergency medical assistance is avoiding exposure to bloodborne and airborne pathogens. Appropriate PPE for a medical incident may include gowns, properly rated gloves, and face and eye protection.

When responding to emergency medical incidents, personnel should be provided with special PPE. The PPE required provides a liquid-proof barrier between the fire fighter and the victim to prevent any of the victim's body fluids from coming into contact with the fire fighter's skin or mucous membranes. The two terms used to describe this PPE system are *universal precautions,* used by the American Red Cross and Centers for Disease Control and Prevention, and *body substance isolation,* from the National Fire Academy course, Infection Control for Emergency Response Personnel. Both of these terms refer to the wearing of appropriate gowns or garments, properly rated gloves, and face and eye protection. All of these are available as disposable items; this way they can be bagged and disposed of properly without having to be touched again. The rule of thumb when it comes to wearing medical PPE is, "If it's wet and it isn't yours, don't get it on you."

Regular turnout gear is equipped with a vapor barrier inside the flame-resistant outer shell and will protect you fairly well. The problem is that once you get blood from a victim all over your turnouts, where are you going to clean them? To take them home is extremely dangerous to your family and yourself and may be against the law. They should instead be placed in a plastic bag and professionally laundered to remove any contaminants. This is great if you have two sets of turnouts, but most people do not. There are going to be times when working in your turnouts is unavoidable. An appropriate gown or garment is no protection from the dangers of getting burned if gasoline ignites when extricating a victim from a wreck. In situations where there is no danger from fire, an appropriate gown or garment should be worn. Properly rated gloves can be worn under your leather gloves to prevent the properly rated gloves from becoming cut on glass and sharp objects and to protect your hands from liquids soaking through.

When involved in situations that require you to care for more than one victim, it is required that you change gloves between victims. When the incident is concluded and you are ready to pick up and return to the station, it is a good idea to bag up any gloves and other materials that have become contaminated in a biohazard container **FIGURE 14-21**. Always be sure that contaminants are not spread to the handles of the oxygen administration equipment or medical aid kit. If someone on the crew has touched a victim's body fluids and then opened the medical aid kit with the same gloves on, the handles of the kit will be contaminated. If you remove your gloves and place the kit back into the compartment on the apparatus, you have contaminated the compartment handle and yourself. In this same way, contamination can be spread to the steering wheel, radios, and other equipment. Always be sure to check the bottom of your boots for contamination as well.

Be cautious when working around paramedics. They will likely be starting intravenous (IV) lines. After the IV is inserted, the needle must be placed in a sharps container. They should never be dropped on the ground or stuck into the mattress on the ambulance cot. If needles are dropped on the ground and you kneel down on one, you are likely to get stuck. It is your responsibility to provide for your own safety and to keep an eye on where the needles are placed after use.

Some other precautions you can take are to stay in good health: Do not report to work sick; eat a good diet; wash your hands well and often, especially after incidents; and approach every victim, regardless of age or gender, as if he or she has a deadly infection. This does not mean that you should not treat a victim; it means you should treat him or her with caution.

FIGURE 14-21 Biohazard container for infectious waste.
© Jones & Bartlett Learning

Highway Incident Safety

A major concern of fire fighters and other responders to roadside incidents is the threat of passing vehicles. Every year fire fighters and other personnel are hit by passing vehicles while working at the scene of roadside incidents. These can easily be fatal to the responder. This has become a national concern, and many states are now passing and enforcing traffic laws and posting signs that direct drivers to either slow down or, whenever possible, to change lanes to avoid roadside workers.

The rules for working roadside incidents of any type must focus on safety: first on the safety of responders and then on the safety of those involved in the incident. One effective way of creating a relatively safe area on a highway is to position a large piece of apparatus between oncoming traffic and the incident, angled into the highway to divert vehicles on the highway from the incident lane or area. Once the apparatus is placed, ensure that the brakes are set and the wheels are chocked so the apparatus stays securely in its location. Turn on all emergency lights and set flares behind the apparatus (provided there is no spilled fuel) to give approaching vehicles plenty of visual warning of an incident ahead and to ensure motorists have ample opportunity to move into the far lane to avoid the incident scene. Keep in mind that highway traffic typically moves at 60 mph or faster, and that the higher the speed, the longer it will take the vehicle to slow or stop. Therefore, ensure that lights and flares are visible at least a few hundred feet before the incident scene.

These precautions help protect responders and victims by making the area as safe as possible, but continued diligence is also necessary. Do not become so focused on the incident that you lose situational awareness and become a victim yourself. For example, it may be that hose or equipment are needed from the apparatus during the incident mitigation. Whenever possible, these items should be retrieved from the incident side of the apparatus, rather than the open highway side. Always keep an eye on potential routes away from the scene should the dangers escalate. When backing away from a highway incident, always use a backer; you do not want to find yourself backing into a lane of traffic, law enforcement vehicle, or ambulance that arrived after you did.

The first element of highway incident traffic safety is to remain aware that vehicles are going to be passing on the roadway until they are stopped completely, which will usually occur only with the arrival of law enforcement. Even when law enforcement arrives, it will take them some time to get into position and stop the traffic. At least one member of the responding crew should be assigned to monitor traffic and sound a warning if someone enters the incident work area. Responders must minimize the amount of time they spend working in the dangerous area of the roadside. Use LCES and have an escape route and safety zone.

In addition, always don all appropriate protective equipment. This includes fire resistive clothing and vest (as shown in Figure 11-3) as specified in 23 CFR Rule 634, *Fire fighter High-Visibility Safety Apparel* and *NFPA 1901: Standard for Automotive Fire Apparatus.*

An excellent source of information concerning traffic and roadway safety is the *Manual of Uniform Traffic Control Devices* published by the Department of Transportation, and the *Emergency Vehicle Safety Initiative FA-272* published by the National Highway Traffic Safety Administration.

Vehicle Accidents

When working vehicle accidents, it is important always to consider the dangers of spilled fuel. The apparatus should be parked uphill of the accident so spilled fuel cannot run down underneath it. If you must work in an area with spilled fuel to rescue a victim, the spill should be covered with class B foam. A fire fighter should be designated to stay with the attack line; if he or she gets involved in helping you or gets bored and sets down the nozzle, that may be when the fire starts.

A new set of hazards has become prevalent with the introduction of new technology in automobiles. Hybrid and electric vehicles have batteries that are high voltage and are either lithium-ion or nickel-metal-hydride and contain chemical hazards as well. There are also high-voltage cables running through the vehicle. Many new cars have keyless ignitions. With keyless ignition and an electric motor, it may be difficult to tell if the car is shut off. If someone depresses the accelerator, the car may move when rescuers are doing their jobs. High-strength steel that is

much tougher to cut or bend when performing extrications is now being used.

When working with power tools to extricate accident victims, caution must be exercised to protect yourself and the victims. Cars and trucks are equipped with multiple air bags that may deploy if disturbed during operations, striking victims and/or rescuers. Power tools should not be used when there is spilled fuel in the immediate area. A car roof can be cut off with hacksaws when necessary. It is slower than a rescue tool but is not an ignition hazard. When using the **power shears**, all loose trim metal should be removed first. A power rescue tool is capable of exerting in excess of 120,000 pounds of force (Hurst Performance 2019). If a door is popped off incorrectly, it could easily hit someone or fall and cut off your toes. The main point is not to use this equipment unless you and the others operating it are trained in its use.

Vehicle Fires

Vehicle fires can be described as structure fires with wheels. They can have all of, if not more than, the hazards of a structure fire. A fire in a truck tractor-trailer rig can have numerous hazards from both the large amounts of diesel fuel they carry and the cargo itself. Buses are more commonly fueled with liquefied natural gas, a BLEVE hazard. Smaller vehicles are now presenting differing hazards with the use of alternative fuels such as LPG, as are hybrid vehicles with their large acid-containing batteries and hydrogen fuel cells in the near future.

Automobile fires should be approached from the side or from a 45-degree angle, never straight in front of a bumper. A five-mile-an-hour bumper is mounted on two shock absorbers, and if they explode, the bumper can be propelled forward with great force. It would hit you just under your knees. Exploding tires can be a shrapnel hazard as well. Full PPE, including SCBA, should always be worn. There are many materials used in today's cars that are toxic if inhaled when burning. Cars do not generally blow up in a fire as they do on television and in the movies, but the seams of the gas tank can let go, spilling forth a large amount of burning fuel **FIGURE 14-22**.

FIGURE 14-22 Fire fighters operating at a vehicle fire. Both of these fire fighters should be wearing their SCBA face masks.
Courtesy of Kern County Fire Department

Aircraft Firefighting

Aircraft are encountered by fire fighters in numerous situations, either as an emergency incident or as a tool. We will first look at aircraft incidents.

When an aircraft crashes or catches fire, it can present many different hazards. It is carrying a certain amount of fuel and it has batteries, tires, flammable metals, and possibly oxygen cylinders. If the incident involves military aircraft, there is always the possibility of munitions on board and scattered about a crash scene. When an aircraft crashes, the fuel is often atomized and can explode into flames. This can cause a spectacular fire with the need for rescue of the occupants.

The first priority in aircraft firefighting is to create a path, which allows the apparatus and firefighting personnel to approach and the victims who can rescue themselves to escape. Next, entry is made into the body of the aircraft to rescue anyone alive inside and to complete extinguishment of the fire. The last operation is to complete overhaul of the fire scene.

Over the years there have been crashes of aircraft into structures. This scenario is further complicated by the presence of a structure fire and associated rescue problems. The aircraft are often just completing takeoff and carrying a large amount of fuel. When the crash is a distance from a large airport, specialized crash fire apparatus may not be available and the fire must be controlled by regular apparatus.

EMS and Firefighting with Aircraft

Aircraft are used as a tool by fire fighters as well. The type of aircraft fire fighters most commonly encounter at incidents are helicopters (rotary wing aircraft) used for transportation of victims on emergency medical incidents and for tactical and logistical needs on other incident types. The rules for working around helicopters are the same no matter what the type of incident (National Wildfire Coordinating Group 2018):*

- Approach and depart helicopters from the side or front in a crouching position, in view of the pilot.
- Approach and depart the helicopter from the downhill side to avoid the main rotor.
- Approach and depart the helicopter in the pilot's field of vision; do not go anywhere near the **tail rotor**.
- Use a chin strap or secure your head gear (hard hat) when working under the **main rotor**.
- Carry tools horizontally, beneath waist level to avoid contact with the main rotor.
- Fasten your seat belt when you enter the helicopter and refasten it when you leave. A seat belt dangling out of the door can cause major damage to the thin aluminum skin of a helicopter.
- Use the door latches as instructed. Use caution around windows, antennas, and any moving parts.
- When entering or exiting the helicopter, step on the **skid**. If you place your foot next to the skid and the weight of the helicopter changes because of loading, the skid may run over your foot.
- Any time you ride in a helicopter in a wildland fire situation, you are required to wear full PPE.
- Do not throw articles from the helicopter because they may end up in the main or tail rotors.

When setting up a landing zone for a helicopter, there are some very important precautions to consider. Secure all loose articles in the area, such as boxes or other items that may become airborne. Perform dust abatement by wetting down the area, if possible. If the helicopter is to land on a roadway, make sure that vehicle and pedestrian traffic is stopped. When a main or tail rotor hits a vehicle, the helicopter is out of service and unable to fly. If a rotor hits a person, it will most likely kill the person and disable the helicopter. Wear eye and hearing protection when working around helicopters. When a helicopter lands or takes off, it kicks up a tremendous amount of wind and debris. Be sure to roll up the windows on any vehicles in the area. Provide for plenty of clearance for the helicopter to land **FIGURE 14-23**. The recommended safety circle for a helicopter depends on its size.

---

* National Wildfire Coordinating Group. 2018. *NFES 1077 Incident Response Pocket Guide.* Boise, ID: National Interagency Fire Center.

FIGURE 14-23 Helicopter with external mounted hoist being used to medevac an injured person.
Courtesy of Kern County Fire Department

FIGURE 14-24 Improvised wind indicator using shovel and flagging tape.
© Jones & Bartlett Learning

A large firefighting helicopter should have a safety circle of 110 ft in diameter, and a smaller helicopter, such as the type commonly used for EMS, requires a 90-ft safety circle. It is better to keep unnecessary personnel well out of the diameter of the safety circle as a precautionary measure.

When the helicopter is coming in to land, and you have established radio communications with the pilot, there are several updates the pilot should receive. Identify and notify the pilot of any elevated hazards in the area, such as power lines, fences, and light poles. Either set up a wind indicator, with flagging tape on a pike pole or shovel, for example, or throw several shovels of dirt in the air, or stand with your back to the wind with your arms out in front of you **FIGURE 14-24**. Notify everyone at the scene that the helicopter is about to land so they can protect their eyes from debris.

One of the main points to remember when working around any aircraft is that contact with any moving part on an aircraft is often fatal. Also, all of their parts are extremely expensive. In most situations, the aircraft carries its own crew that will assist you in operating in the vicinity of the aircraft. You should not touch anything until you are briefed on its use, nor should you approach the aircraft until given permission by a crew member or the pilot.

On wildland incidents, helicopters are used for dropping water and retardant on the fire. The larger helicopters (helitankers) can carry up to 2500 gallons of water and create a very strong rotor downwash.

When the helicopter comes in to make its drop, get out of the area or lie on the ground and cover up. Try to be uphill of the drop so you do not get washed down the hill or hit by rocks loosened by the water. A large helicopter can cause the tops of dead trees to break loose and become falling hazards. The downwash from the main rotor will also cause a change in fire activity as the flames are fanned.

Another activity is logistical support. Helicopters are used to sling load materials to crews on the fire line. When sling loads are delivered, fire fighters should avoid placing themselves in the approach and departure path of the helicopter and should stay clear of the landing zone. If the helicopter develops trouble in flight, one of the first things the pilot does is release the load to reduce weight.

Airplanes are used to drop fire retardant at wildland fires (fixed-wing aircraft). The material dropped is either a chemical compound mixed with water or water alone. Some of these aircraft are capable of delivering up to 3000 gallons at one time. The DC-10 air tanker can carry up to 11,600 gallons (10 Tanker Air Carrier 2019). If you consider that 3000 gallons, at over 9 pounds a gallon, is coming down from around 200 ft above ground level at 130 mph, the dangers become evident. The turbulence caused by the plane's passage can knock the tops out of trees and fan the fire. If you are in the area where a drop is to be made, it is best to clear the area prior to the drop. If this is not possible, lie on the ground on your stomach, facing the aircraft, and place your hands on top of your head. If you are hit by a retardant drop you will be covered by a sticky, slippery coating that has a strong ammonia odor. It is not really harmful unless you get it in your eyes, but it is uncomfortable. Be aware that the retardant is slippery on dry grass and rocks and will make hand tools hard to hold onto. This material is slightly corrosive and should be washed off vehicles as soon as possible (National Wildfire Coordinating Group 2018).

Technical Rescue

Technical rescues are those incidents that employ the use of tools and skills that exceed those of the fire department's normal duties, such as firefighting, EMS, and rescue. Technical rescues can be performed in types of situations including rope rescue (both high and low angle), swiftwater rescue, confined space, caves and mines, pipes, wells, storage tanks, sink holes, trenches, and building collapse—basically anywhere a person can be trapped, by their own actions or an accident. All of these rescue types require specialized equipment and training. There are so many types of specialized (technical) rescue scenarios that may occur that they are well beyond the scope of this text to pursue.

All too often, rescuers without the proper training and equipment become victims themselves in these types of incidents. One of the problems fire departments face is that they are expected to do something to mitigate the situation when they arrive at the scene. This may or may not be possible within the training, experience, and equipment of the responders. Sometimes all that can be done is to isolate and deny entry. This is extremely difficult when family members are present at the scene. They are beseeching you to do something and there is nothing you can do until the proper resources arrive. This brings us back to the concept of a precautionary approach. Is there really a rescue to be performed—in other words, a live person to be saved—or is it a body recovery? Fire fighters cannot fall into the trap of believing that doing something is better than doing nothing when the doing something endangers rescuers, and they become victims.

Decision Making

Any discussion of safety in the emergency services has to include decision making. Decisions were involved in all the places and situations in which fire fighters and other responders were killed or injured, including during training, responding to incidents, while performing at incidents, and when returning from incidents. Without a review and understanding of how we make decisions, especially under stress, we are missing one of the key elements of keeping ourselves and those around us safe from harm. As has been stated, all fire fighters deserve a safe trip home.

Decision making is being evaluated and discussed across the fire service, on both the structural firefighting side and the wildland firefighting side. Many of the elements of decision making can be directly cross-applied or adapted from one side to the other. In some cases, fire agencies have both responsibilities. All of those involved in emergency response have lessons they can learn from each other when it comes to responding to emergencies, which tend to be dynamic, high-stress situations.

For a new fire fighter, it is going to require some tact should he or she choose to approach his or her supervisor on these items. Not everyone has bought into the more recently emphasized concepts in emergency-response decision making. Some are unaware and some do not necessarily agree. So step lightly when discussing this with others.

Proper decision making leads to safer, more efficient operations by fire fighters at incidents. The first decision that must be made correctly is that safety is the first priority, not that safety is the first priority until it gets in the way of getting the job done. All fire fighters must think first and act second, not the other way around.

Crew Resource Management

In adopting the concept of **crew resource management (CRM)**, the fire service has followed a model developed in the aviation industry. In the early 1980s, the aviation industry recognized that human error was the prevailing cause in aviation disasters (IAFC 2003).

Additional industries adopted the concept of CRM in the 1980s and 1990s. The resultant increase in safety and reduction in injuries is an indicator of how effective CRM can be when utilized properly. The U.S. Coast Guard reported a 74 percent reduction in its injury rate since the adoption of CRM. U.S. air disasters showed a marked reduction from approximately 20 per year to 1 or 2 per year (IAFC 2003).

CRM is based on the following five elements:

1. *Communication.* CRM requires that people speak directly and respectfully and communicate responsibly. A model of this is the one presented in the National Wildfire Coordinating Group's *Incident Response Pocket Guide*. It states under communications responsibilities that fire fighters must brief others as needed, debrief their own actions, communicate hazards to others, acknowledge messages, and ask if they don't know. It also states that all leaders of fire fighters have the responsibility to provide complete briefings that include a clearly stated

leader's intent. This includes (National Wildfire Coordinating Group 2018):*

- *Task.* What is to be done
- *Purpose.* Why it is to be done
- *End state.* How it should look when done

Implied in this is that *all* personnel at the scene should have an idea of what is going on, what they are supposed to be doing, and what is to be accomplished.

2. *Situational awareness.* This can be described as maintaining attentiveness to an event. Situations in the emergency services are particularly dynamic and require the full attention of everyone to recognize hazards and changing elements at the scene. The *Incident Response Pocket Guide* lists the elements of situational awareness as:

- Objectives
- Communication
- Who's in charge
- Previous fire (incident) behavior
- Weather forecast
- Local factors

3. *Decision making.* Decision making must be based on information. This includes information from training, from experience, and gathered at the scene. The decision making is focused on risk–benefit analysis. We must look at what we are doing and decide whether it is worth it. No fire fighter should die to save property, though every year some do. Too much information overloads the decision maker and too little leads to poor decision making. Somewhere in the middle is the right amount of information selectively focused on the right elements of the scene to make sound decisions. We must also recognize that there are some things we know, some things that we do not know, and some things that we do not know that we do not know. This is where experience enters the equation. A person with quality experience can focus on the important elements, pay less attention to the less important elements, and not forget that there are factors in a dynamic situation that may not be readily perceived. This reinforces the need to maintain situational awareness.

4. *Teamwork.* Teamwork is critical. Every member of the team must know his or her job and know where he or she fits in the organization. This could be a company level organization or a major operation with hundreds of personnel. This refers back to the incident command concepts of *unity of command* and *chain of command.* Failure to perform as a team at emergency incidents can lead to a reduction in operational effectiveness, excessive property loss, injury, and death.

5. *Barriers.* Barriers can affect any or all of the previous factors and are discussed further in the next section.

The 2 and 7 Tool

Brad Mayhew, author of *F LCES Δ*, discussed previously in this chapter, is conducting ongoing research into how fire fighters make decisions. His **2 and 7 tool** illustrates two errors and seven barriers common to poor decision making. Though he is from a wildland fire background, his tool can be applied to other types of emergency situations as well (Mayhew 2009a).

The two errors:

1. Underestimating hazards and using inadequate safety measures.
2. Failing to notice changing conditions and adjust tactics accordingly.

If you were to apply these two errors to any number of case studies of fire fighter near-misses and fatalities in all risk situations, you could see that they often apply. Fire fighters commit too deeply to the situation and overestimate their ability to escape. Even a RIC cannot get you out if you are in too far or the situation deteriorates to the point that they cannot locate you or get to you. Driving too fast on the way to the incident is an example of error number 1, because you are unable to stop quickly enough, or avoid the hazard, due to your excessive speed (Utah Department of Transportation 2019). Another example is placing only one or two ladders to the roof for one engine or truck company. As the incident progresses and the company is joined by other fire fighters on the roof, there would be too many fire fighters on the roof to make a quick escape down the number of ladders available, should it become necessary for them to do so. In a wildland fire, the example would be getting too far from the safety zone, creating too long an escape time, and then the fire flares up, cutting off the escape. Or, similar to the ladder example, more personnel are assigned to the division than can safely fit in the safety zone. This sort of thing just seems to creep up on fire fighters because they are often overly focused on getting the job done and not on what else is happening (lack of situational awareness).

The seven barriers are:

1. *Inexperience.* As previously noted, experience is valuable. That is, if it is quality experience in

---

* National Wildfire Coordinating Group. 2018. *NFES 1077 Incident Response Pocket Guide.* Boise, ID: National Interagency Fire Center.

performing safely, not learning cowboy tactics and thinking they are OK just because you got away with it before.

2. *Getting too comfortable.* Thinking nothing bad will happen. Or as in barrier 1, nothing bad has happened yet, so what is the likelihood that it will this time? State legislatures would not have to pass laws against talking and texting on cell phones while driving if people were attentive and not overly comfortable using phones while driving, thinking nothing bad will happen while texting when they are supposed to be focused on their driving.

3. *Distraction from primary duty.* You can focus fully on only one thing at a time. Outside factors at the incident scene can divert your attention from what you should be focusing on. A part of the structure collapsing down the street diverts your attention and the piece you are working under comes down on top of you. The driver/operator's primary duty is to get you there safely, yet we still suffer from fatal responding vehicle rollovers because of excessive speed and accidents.

4. *Priorities out of order.* The first priority in any incident is life safety, including your own. Fire fighters have been killed trying to perform "rescues" that were obviously body recoveries. Some fight fires for the thrill, danger, and adrenaline rush. These are the ones who will get you hurt or killed.

5. *Social influences.* One of the strongest of these barriers is peer pressure. In the fire service, it is common that new fire fighters have to prove themselves. This occurs at the station and on the fireground. This does not mean that we should all get into the mode of groupthink where we stop thinking for ourselves or overextend just to prove something to others.

6. *Stress reaction.* As stress builds, the rational thinking part of the brain tends to shut down. We focus on a plan of action and stick with it. This is the do-or-die mind-set and leads to tunnel vision, which reduces situational awareness. This becomes evident when the interior attack crew is ordered out of the building by the IC because the roof is about to fall in and they reply, "We almost got it, Chief, just a few more minutes." They are not in position to see the big picture and need to do as ordered and not argue.

7. *Physical impairment.* Physical impairment can come in many forms—carbon monoxide poisoning, lack of sleep, drinking alcohol the night before coming on duty, drugs, and too many incidents in too short a time. These all lead to a general diminishment in awareness and decision-making ability.

We must all guard ourselves and those around us from making the two errors and falling victim to the seven barriers if we are to be safe and effective fire fighters. This applies to every incident, every time, no matter how simple or commonplace.

Wrap-Up

CHAPTER SUMMARY

- A well-trained fire fighter should be able to deal with most types of emergency situations.
- The fire department is not the only agency that shows up at an incident that has a say in determining the objectives, strategy, and tactics to be used.
- In 2004, the National Fallen Firefighter Foundation examined the causes of line-of-duty fire fighter deaths and put forth the *16 Firefighter Life Safety Initiatives* to be followed by all fire fighters.
- The assignment of RICs at working incidents provides the ability to initiate a rescue effort immediately to locate, rescue, or assist fire fighters who are in trouble at the scene of an incident.
- OSHA has created a regulation, commonly referred to as "two in, two out," in which personnel who are operating in atmospheres that require SCBA must use the two in, two out procedure.
- Firefighting operations at structure fires take one of two modes: offensive or defensive.
  - In the offensive mode, fire fighters enter the burning structure and attack the seat of the fire.
  - In the defensive mode, the water is applied through windows or into other openings to control the fire.

- Improvements to protective equipment have led some fire fighters to think that they are invincible—which is certainly not the case.

- High-rise firefighting presents a whole list of additional hazards to the structural fire fighter. Some fire fighters have gone as far as to call these structures "elevated crematoriums."

- A hazard almost always present in structure fires is electricity. One of the first tasks at a structure fire is to cut off the power to the structure.

- A hazard most people do not stop to consider at structure fires is pets. Your best defense is always a cautious approach.

- The three principal causes of unsatisfactory sprinkler performance include a closed valve in the water supply line, delivery of inadequate water supply to the system, and occupancy changes that render the installed sprinkler system unsuitable.

- When extinguishing electrical equipment, it is better to use a fog pattern or short bursts of water to accomplish the task. This lessens the chances of getting electrocuted.

- In wildland firefighting operations, the basic methods of extinguishment are to apply water or fire retardant to the fire edge or to create a fire break or control line around the perimeter.

- Brief safety messages include the following:
  - *LCES*: Lookout, communications, escape routes, and safety zones.
  - *F LCES* Δ: Fire behavior (worst case scenario) lookout, communications, escape routes, safety zones, and change.
  - *Look up, look down, look around.*

- Wildland–urban interface firefighting is becoming more common. More and more people have escaped the cities by building their homes in the foothills and mountainous areas, changing the way wildland fires are fought.

- The primary objectives in oil firefighting are to extinguish the fire and control the source of any leaks.

- The mass release of gasoline, or any other flammable liquid, requires that the vapors be controlled, if it can be done in a safe manner.

- When a LPG container is involved in a fire, the recommendation is that large volumes of water be applied at each point of flame contact. The volume is determined by the size of fire and number of containers involved.

- A common flammable gas, natural gas poses little hazard when released into the open, because it is lighter than air and will disperse itself. However, it has the potential for collecting in structures when leaks occur in interior-mounted gas meters and plumbing.

- Hazardous materials incidents are becoming more common, partly because of greater awareness and partly because of increased use of these materials.

- WMD attacks may be CBRNE in nature.

- Operations in the area of emergency medicine for most fire departments revolve around first aid for injuries, basic life support, and extrication of victims from vehicle collisions.

- The rules for working roadside incidents of any type are summed up in the three Es of highway survival, which are:
  - Eye on traffic
  - Exposure
  - Escape route

- At vehicle accidents, the apparatus should be parked uphill of the accident so spilled fuel cannot run down underneath it.

- Vehicle fires can be described as structure fires with wheels. They can have all of, if not more than, the hazards of a structure fire.

- When an aircraft crashes or catches fire, it can present many different hazards. It is carrying a certain amount of fuel and it has batteries, tires, flammable metals, and possibly oxygen cylinders.

- Aircraft are used as a tool by fire fighters as well. The type of aircraft fire fighters most commonly encounter at incidents are helicopters used for transportation of victims on emergency medical incidents and for tactical and logistical needs on other incident types.
- Without a review and understanding of how we make decisions, especially under stress, we are missing one of the key elements of keeping ourselves and those around us safe from harm.
- CRM is based on the following five elements:
  1. Communication
  2. Situational awareness
  3. Decision making
  4. Teamwork
  5. Barriers
- Mayhew's 2 and 7 Tool illustrates two errors and seven barriers common to poor decision making.
  - Two errors: underestimating hazards and using inadequate safety measures; failing to notice changing conditions and adjust tactics accordingly
  - Seven barriers: inexperience, getting too comfortable, distraction from primary duty, priorities out of order, social influences, stress reaction, and physical impairment

KEY TERMS

2 and 7 Tool The two errors and seven barriers common to poor decision making.

Arcing Spark created when electrical contact is made.

Boiling liquid expanding vapor explosion (BLEVE) An explosion that occurs when a tank containing a volatile liquid at the bottom of the tank and a flammable gas at the top of the tank is heated to the point where the tank ruptures.

Burst pressure The pounds per square inch (psi) of pressure at which a container will fail.

Cone roof A type of petroleum product storage tank construction with a vapor space over the product. The lid is connected to the tank with a weak seam that will rupture before the tank wall seams.

Contaminated Coated with a harmful substance.

Crew resource management (CRM) A behavioral modification training system developed by the aviation industry to reduce its accident rate. It is based on the assumption that human error is the primary cause of fire-ground fatalities and injuries, and by using this training the fire service can reduce the number of negative outcomes.

Decontaminated Physical removal of contaminants from people or equipment.

Delta (Δ) Change in factors affecting the incident, including personnel fatigue, time of day, structural weakening, and so forth.

Dry chemical extinguishers Fire extinguisher using a chemically active powder.

Escape route A preplanned and understood route to a safety zone.

Flanks The sides of an advancing wildland fire.

Hazard trees Trees that have burned out at the base or are liable to drop large limbs. Dead trees are often called snags; these are classified as hazard trees as well.

Immediately dangerous to life and health (IDLH) Atmosphere that is capable of causing death, irreversible adverse health effects, or the impairment of an individual's ability to escape.

Main rotor Horizontal blades that create lift for a helicopter.

Mitigation Reducing the hazard, making less severe.

Mucous membranes Inside of the nose, mouth, and covering of the eye.

Parapet walls Walls that extend above the roof line.

Pathogens Disease-causing agents.

PCB oil Oil containing polychlorinated biphenyl, a compound that can cause cancer.

Perimeters Boundaries for controlled access (hazardous materials). The fire's edge (wildland).

Plume (1) The path of the material released from a container in a hazardous materials incident. (2) The smoke column from a fire.

Power shears Cutting attachment for a rescue tool.

Primary search A rapid search of all involved and exposed areas affected by the fire that can be entered to verify a removal and/or safety of all occupants. Should this not be possible, a secondary search is conducted as soon as it is safe to do so.

Relief valve Device used to release unwanted pressure.

Safety zones A place where fire fighters can be safe from the incident's hazards.

Scratch line A quickly created wildland fire control line, constructed using hand tools.

Skid The long tubular shaped feet that helicopters sit on when on the ground.

Sky lobby A lobby on a high floor level in a high-rise building. Elevators leave from this area to service the upper floors.

Subsurface foam injection Plumbing installed on a tank to allow for the introduction of foam under the surface of the contained liquid.

Tail rotor Vertical propeller used for steering control that is installed on the tail of a helicopter.

Unburned islands Areas of unburned fuel within a fire perimeter.

Water curtain A screen of water spray set up to protect exposures.

Working pressure The pounds per square inch of pressure that a tank is designed to contain.

CASE STUDY*

On December 23, 2011, a 42-year-old male career fire fighter died during firefighting operations on the second floor of a three-story apartment building. The victim was assigned to Engine 5 (E5) with a lieutenant and driver/pump operator. E5 was the first-due engine company at this fire. The IC ordered E5 to take a 1¾-in. hose line and attack the fire in a second-floor apartment. The lieutenant stretched the line to the landing of the second floor but did not realize there were two apartments on that floor. Because of heavy smoke conditions, he went to apartment 4 instead of the fire apartment (apartment 3). Apartment 4 was locked, so he went to get the ladder company, which was operating on the third floor. At this time, the lieutenant lost contact with the victim. The IC went to the second floor landing, contacted the lieutenant from E5, advised him the fire was in apartment 3, and the door was open. The lieutenant then entered the fire apartment, attempted to knock down the fire, and the apartment flashed. The lieutenant, with his helmet on fire, was pulled out of the apartment by members of Engine 3 and Ladder 1. At this time, the location of the victim was unknown. The lieutenant returned to the fire apartment with a thermal imaging camera, but the image was featureless because of the amount of heat and fire in the apartment. Several fire fighters stated they heard a personal alert safety system alarm sounding but were unable to determine the location. The officer of the fourth-due engine company entered the fire apartment, located the victim, and removed him with the help of two other fire fighters. Despite receiving cardiopulmonary resuscitation and advanced life

support outside the structure, in the ambulance, and in the local hospital's emergency department, the victim died.

1. When the attacking company was split up, what happened to crew integrity?

   A. It was lost.
   B. It was not a factor.
   C. It was still in place.
   D. It was maintained by radio.

2. The fire operation was missing which key company assigned to rescuing lost fire fighters?

   A. FF rescue squad
   B. RIT/RIC
   C. Evacuation group
   D. FF salvage

3. The fire attack was delayed because of initially going into the wrong apartment; this resulted in the fire developing into which stage before the attack was made?

   A. Incipient stage
   B. Free burning stage
   C. Fully developed stage
   D. Decay stage

4. When the initial attack company was split up at the beginning of the fire attack, which concept of fire fighter safety was violated?

   A. Mass attack
   B. Provide ventilation
   C. Provide adequate water supply
   D. The buddy system

---

* NIOSH. 2013. "Career Firefighter Dies during Fire-Fighting Operations at a Multi-family Residential Structure Fire—Massachusetts." #F2011-31. Centers for Disease Prevention and Control. Retrieved from https://www.cdc.gov/niosh/fire/reports/face201131.html

REVIEW QUESTIONS

1. List at least three interior hazards encountered at structure fires.

2. List at least three exterior hazards encountered at structure fires.

3. What is wrong with freelancing at emergency scenes?

4. Why should the IC set up a RIC at the scene?

5. Why shouldn't you enter electrical substations without electrical company personnel?

6. List the *10 Standard Firefighting Orders*.

7. List the *18 Situations That Shout Watch Out*.

8. What are the four components of LCES?

9. List several of the safety considerations when providing structure protection.

10. What is meant by the abbreviations WMD and CBRNE?

11. What is meant by universal precautions in regard to EMS incidents?

12. Why should an attack line be pulled on all vehicle rescue situations?

13. What are the three priorities in an aircraft fire-fighting incident?

14. What action should you take if an air tanker is about to make a drop on your position?

15. What are the safety rules for approaching a helicopter?

16. What are the two common errors and seven barriers to decision making as described in the 2 and 7 Tool?

DISCUSSION QUESTIONS

1. Why is the buddy system important when operating in IDLH atmospheres?

2. If a WMD event were to happen in your community, is the fire department prepared to deal with it? Which other local agencies would be involved?

3. Why is a good safety attitude as necessary as safety training in staying safe in emergency and nonemergency situations?

4. Explain, using examples, the meaning of the following statement: To fight fire one must be aggressive; to survive it one must be cautious.

5. Why is it sometimes a better option to take no action?

6. In some states emergency lights are not to be left on when working roadside incidents if all vehicles and victims are safely out of the roadway. What are your thoughts on this practice?

7. Why is decision making a skill that needs to be highly developed by fire fighters?

REFERENCES AND ADDITIONAL RESOURCES

10 Tanker Air Carrier. 2019. *10 Tanker Aerial Firefighting.* https://www.10tanker.com/index.html

Automobile Club of Southern California. 2003. *ERS Tow and Service Instruction.* Los Angeles, CA: Automobile Club of Southern California.

Department of Transportation. 2009. *Manual of Uniform Traffic Control Devices.* Washington, DC: Department of Transportation.

Fire, Frank L. 2009. *The Common Sense Approach to Hazardous Materials.* New York, NY: Fire Engineering.

FireEngineeringUniversity.com. 2013. "Slopover, Frothover, and Boilover." FireEngineeringUniversity. https://www.fireengineeringuniversity.com/courses/22/HTML/section_6.htm

Gabbert, Bill. 2013. "Weather Conditions during the Tragedy at Yarnell Hill, and Where Do We Go from Here." *Wildfire Today,* July 4. https://wildfiretoday.com/2013/07/04/weather-conditions-during-the-tragedy-at-yarnell-hill-and-where-do-we-go-from-here/

Hurst Performance. 2019. *Hurst Jaws of Life.* Shelby, NC: Edraulic. www.jawsoflife.com/rescue-products/edraulic

International Association of Fire Chiefs (IAFC). 2003. *Crew Resource Management.* Fairfax, VA: International Association of Fire Chiefs.

Martinez, Michael. 2013. "Expert: 'Almost Impossible' for Firefighters to Evade Surprise Attacks." *CNN,* April 12. https://www.cnn.com/2013/04/11/us/Fire fighters-violent-scenarios/index.html?iref=allsearch

Mayhew, Brad. 2009a. "The 2 & 7 Tool." Fireline Factors. www.firelinefactors.com

Mayhew, Brad. 2009b. "The Intent of LCES." Fireline Factors. www.firelinefactors.com

National Fallen Firefighters Foundation (NFFF). 2019. "Everyone Goes Home—16 Firefighter Life Safety Initiatives." Everyone Goes Home. www.everyonegoeshome.com

National Fire Protection Association. 2008. *Fire Protection Handbook.* 20th ed. Quincy, MA: National Fire Protection Association.

National Fire Protection Association. *NFPA 13E: Recommended Practice for Fire Department Operations in Properties Protected by Sprinkler and Standpipe Systems.* Quincy, MA: National Fire Protection Association.

National Fire Protection Association. *NFPA 1500: Standard on Fire Department Occupational Safety, Health and Wellness Program.* Quincy, MA: National Fire Protection Association.

National Fire Protection Association. *NFPA 1581: Standard on Fire Department Infection Control Program.* Quincy, MA: National Fire Protection Association.

National Fire Protection Association. *NFPA 1901: Standard for Automotive Fire Apparatus.* Quincy, MA: National Fire Protection Association.

National Highway Traffic Safety Administration (NHTSA). 2019. *Emergency Vehicle Safety Initiative FA-272.* Washington, DC: National Highway Traffic Safety Administration.

National Institute for Occupational Safety and Health (NIOSH). 2013. "Career Fire Fighter Dies during Fire-Fighting Operations at a Multi-Family Residential Structure Fire—Massachusetts." Centers for Disease Prevention and Control. #F2011-31. https://www.cdc.gov/niosh/fire/reports/face201131.html

National Wildfire Coordinating Group. 2014. *Wildland Fire Incident Management Field Guide.* Boise, ID: National Wildfire Coordinating Group.

National Wildfire Coordinating Group. 2018. *NFES 1077 Incident Response Pocket Guide.* Boise, ID: National Interagency Fire Center.

Occupational Safety and Health Administration (OSHA). 1986. *Standard 29 CFR 1910.120: Hazardous Waste Operations and Emergency Response.* Washington, DC: OSHA.

Shell Oil. (1991). *Oil Fire Training Course Manual.* Reno, NV: Shell Oil Company. p. 13.

Soldier and Biological Chemical Command (SBCCOM). 1999. *Domestic Preparedness Training Program.* Aberdeen Proving Ground, MD: U.S. Army Edgewood Research, Development, and Engineering Center.

Utah Department of Transportation (UDOT). 2019. Trucks Need More Time to Stop. https://www.udot.utah.gov/trucksmart/motorist-home/stopping-distances/

Appendix

FESHE Correlation Guide

Principles of Emergency Services (C0273) Course Outcomes	Introduction to Fire Protection and Emergency Services, Sixth Edition Chapter Correlation
1. Illustrate the history of the fire service	
a. Illustrate and explain the history	Chapter 3
b. Evaluate the culture of the fire service	Chapters 1, 3, and 11
c. Analyze the basic components of fire as a chemical chain reaction, as well as the major phases of fire	Chapter 4
d. Examine the main factors that influence fire spread and fire behavior	Chapters 4 and 14
2. Compare and contrast the components and development of the fire and emergency services	
a. List and describe the major organizations that provide emergency response service, and illustrate how they interrelate	Chapter 5
b. Explain the scope, purpose, and organizational structure of fire and emergency services	Chapters 3, 7, 8, 9, and 13
c. Differentiate between fire service training and education	Chapters 1 and 9
d. Explain the value of higher education to the professionalization of the fire service	Chapter 1
e. Define the role of national, state, and local support organizations in fire and emergency services	Chapter 5
f. Describe the common types of fire and emergency service facilities, equipment, and apparatus	Chapter 6
g. Compare and contrast effective management concepts for various emergency situations	Chapter 7, 13, and 14

3. Analyze careers in fire and emergency services	
a. Identify fire protection and emergency-service careers in both the public and private sector	Chapters 2, 7, and 8
b. Explain the primary responsibilities of fire prevention personnel, including code enforcement, public information, and public and private protection systems	Chapters 10, 11, and 12
c. Develop the components of career preparation and goal setting	Chapter 1
d. Demonstrate the importance of wellness and fitness as it relates to emergency services	Chapters 1 and 14

Glossary

2 and 7 Tool The two errors and seven barriers. Common to poor decision making.

Absolute zero The lowest temperature that is theoretically possible. The temperature at which all molecular motion ceases. Absolute zero is expressed as −459.67°F, −273.15°C, 0K, and 0°R.

Acceptable risk A risk that is considered to be of low enough severity or frequency that it is considered acceptable.

Accountability officer The person at an incident scene responsible for tracking the location of personnel operating at the incident. This may be the incident commander or a designee.

Accreditation A process in which certification of competency, authority, or credibility is presented.

Administrative procedures Written procedures for performing staff-related functions, such as reports and other paperwork.

All clear The short descriptive phrase indicating that a primary search of the structure for victims has been completed.

Alley lights Lights mounted in a light bar that shine to the side of the vehicle; commonly used for spotting addresses on structures at night.

Ambient temperature The temperature surrounding an object; air temperature.

Ambulance gurney The wheeled cot that patients are placed on prior to transport in an ambulance.

Aquifer The underground layer of water-bearing permeable rock or unconsolidated materials (sand, gravel, etc.).

Arcing Spark created when electrical contact is made.

Articulated boom Elevating device consisting of a boom that is hinged in the middle.

Atmospheric pressure The pressure of the atmosphere exerted on any point, which is 14.7 psi at sea level.

Automatic aid Under this system, departments assist each other without regard to jurisdictional boundaries. It is often used in areas where there are county islands within city limits or in interagency areas where a fire starting in one agency's area is a direct threat to another's jurisdiction.

Automatic vehicle location (AVL) A means for automatically determining and transmitting the geographic location of a vehicle.

Average daily consumption The amount of water used daily by water system customers; computed by dividing the total water used by the number of customers over a period of a year by 365. Expressed in gallons or liters.

Backdraft A type of explosion caused by the sudden influx of air into a mixture of gases, which have been heated above the ignition temperature of at least one of them.

Backfires Fires lit in front of an advancing fire to remove fuel and widen control lines.

Baffles Partitions placed in tanks that prevent the water from sloshing and making the vehicle unstable when turning corners.

Banding A civil service selection process tool in which candidates who score between certain points on the scale (such as between 90 and 100 percent) are grouped, and the employer may choose candidates for employment from within the group.

Base The location from which primary logistics and administrative functions are coordinated and administered. The ICP may be co-located with the base.

Black fire A situation where heavy, dense, black smoke is being forcefully emitted by a fire.

Boiling liquid expanding vapor explosion (BLEVE) An explosion that occurs when a tank containing a volatile liquid at the bottom of the tank and a flammable gas at the top of the tank is heated to the point where the tank ruptures.

Bonnet The top of a hydrant.

Buddy breathing A technique in which two people share the same SCBA air supply to avoid breathing smoke or toxic fumes.

Burning out Lighting a fire to remove fuel along the flanks of a fire; also used to remove unburned islands that remain as the fire advances.

Burst pressure The pounds per square inch (psi) of pressure at which a container will fail.

Bury The piping that extends from the water main to the hydrant.

Call-back A recall of personnel to on-duty status, usually because of an emergency situation.

Camp The location where resources may be kept to support incident operations if a base is not accessible to all resources or the incident is of a large enough scale as to require extended transportation times from the base to the tactical work assignments.

Cascade systems Systems of large compressed gas cylinders connected to a manifold.

Cause and origin The source of heat that started the fire and exactly where it started.

Cavitate To form small vapor bubbles in the interior of a pump.

Certification A document that specifies that a student has successfully completed the prerequisite education and training to perform a job function, such as Fire Fighter I or emergency medical technician.

Certify A formal, written document issued and undersigned by an official authority to recognize an individual or a group possessing certain qualifications or meeting certain standards.

Chain of command An orderly line of authority within the ranks of the organization, with lower levels subordinate to, and connected to, higher levels. The organization of management at the incident starts with the IC and develops downward.

Civil liability The accountability of an individual under civil law.

Codes A law that can be established by legislative action, but is most commonly created by an administrative agency or a local entity.

Combustible construction The use of unprotected wood and wood by-products in building construction.

Combustible gas indicators A device that measures the percentage of lower explosive limit concentration of gas in the atmosphere. This device must be used by trained personnel for proper interpretation of the readings.

Command presence The ability to maintain composure in situations that are stressful.

Company officer The first-line supervisor in the fire department. Depending on the jurisdiction involved, this position may be identified by various titles, such as captain, lieutenant, sergeant, station manager, module leader, or unit manager.

Composition roofing Tar paper and shingles or tar paper covered with roofing asphalt.

Cone roof A type of petroleum product storage tank construction with a vapor space over the product. The lid is connected to the tank with a weak seam that will rupture before the tank wall seams.

Conflagration Very large fire that defies control efforts and causes extensive damage over a large area.

Confined space An area a person can enter to do work, but which has limited means of entry or exit and is not designed for continuous occupancy by a person.

Consensus standards Standards that are developed through the consensus process. Usually representatives from government and industry meet to determine the language of the standard. Input is sought and then meetings are held to determine the final language used in the standard. This process is often used in the creation of codes, such as the NFPA's National Fire Codes.

Contaminated Coated with a harmful substance.

Continuity The manner in which a fuel is spread across an area. Horizontal continuity is expressed as either uniform or patchy.

Control lines An area where fuel has been removed, water or other extinguishing agent has been applied, or natural barriers exist to stop a wildland fire from spreading.

Corrosive Able to destroy and damage other substances with which it comes into contact.

Cost recovery System on the part of public agencies to recover expenditures incurred in mitigating an incident.

Cover To move resources into a fire station when the regular crew is assigned to an incident. In some departments this is called a *move up*.

Coverage levels The number of gallons applied per 100 square feet. Light fuels, such as grass would require a coverage level of 1 and heavier fuels (i.e., timber and brush) would require higher coverage levels.

Credibility A state established between persons based on expertise (subject matter knowledge) and relationship (the ability to get along with others). It is especially important between instructors and students.

Crew resource management (CRM) A behavioral modification training system developed by the aviation industry to reduce its accident rate. It is based on the assumption that human error is the primary cause of fire-ground fatalities and injuries, and by using this training the fire service can reduce the number of negative outcomes.

Crime An unlawful act as defined in the criminal codes.

Critical incident stress debriefing (CISD) A discussion in which personnel are encouraged to express their feelings after responding to and operating at particularly stressful incidents that result in high loss of life, loss of life by a co-worker, or other significant conditions. These are conducted after an incident to assist personnel to better deal with their emotions.

Critical incident stress management (CISM) An intervention protocol developed specifically for dealing with traumatic events. It is a formal, highly structured, and professionally recognized process for helping those involved in a critical incident to share their experiences, vent emotions, learn about stress reactions and symptoms and given referral for further help, if required. It is not psychotherapy. It is a confidential, voluntary, and educative process, sometimes called "psychological first aid." This is performed through a peer process utilizing Critical Incident Stress Debriefings.

Critical thinking The process of examining, analyzing, questioning, and challenging situations, issues, and information of all kinds.

Crown fires Fires in the tops of trees. These fires move very rapidly and defy control efforts.

Curriculum A particular course of study.

Dead-end mains Water mains that are not gridded into the system. Water flows into them from only one way.

Decontaminated Physically removed of contaminants from people or equipment.

Delegation of authority The action by which a commander assigns part of his or her authority commensurate with the assigned task to a subordinate commander. While ultimate responsibility cannot be relinquished, delegation of authority carries with it the imposition of a measure of responsibility.

Delta (Δ) Change in factors affecting the incident, including personnel fatigue, time of day, structural weakening, and so forth.

Demographics The statistical characteristics (e.g., age, race, gender, income) of the population of an area.

Division of labor The assignment of work to those most qualified to carry it out or the division of a complex task into several less complex tasks.

Drafting pit An open-topped underground tank that is used for drafting operations and pump testing.

Drills Tasks and jobs being practiced to improve performance.

Driver/operator The position responsible for operating the pumping or aerial apparatus assigned to the fire department. Depending on the jurisdiction involved, this position may be identified by various titles, such as engineer, chauffeur, or truck operator.

Dry chemical extinguisher Fire extinguisher using a chemically active powder.

Duff Leaves, pine needles, and other dead forest material.

Education Memorization of specific pieces of information and development of an understanding of concepts or philosophies.

Eligible list A certified list of persons who have successfully completed the testing process.

Emergency medical dispatch A system in which dispatchers are trained to give medical advice to the persons at the incident, such as CPR instructions, until emergency help arrives.

Emergency medical technician (EMT) A specified level of medical training that usually consists of approximately 100 hours of classroom and practical training and the completion of a national registry examination.

Endothermic reaction Reaction that absorbs heat.

Energy The capacity for doing work.

Escape route A preplanned and understood route to a safety zone.

Exception principle A method or plan of supervision under which only significant deviations from normally expected results or conditions are brought to the attention of a supervisor for consideration and decision.

Exothermic reaction Reaction that results in the release of energy in the form of heat.

Explorers A program of the Boy Scouts of America for persons 15–21 years of age. The Explorers work in conjunction with a professional organization such as the fire or police department to learn the operation and job requirements. Females are now allowed in the Boy Scouts and its programs.

Evaporation The changing of liquid to a vapor.

Evolution Combination of skills to perform a task. Example: performing all the skills required to don structural PPE and advance a 1¾-in. hose line from an engine up a ladder and into a second-story window.

Extrication The act of removing trapped victims. This term is usually used in reference to vehicle accidents.

Felony A serious crime, such as murder, arson, or rape, for which the punishment is either imprisonment in a state prison for more than 1 year or death.

First-alarm complement The equipment normally dispatched when a fire is first reported.

Fire department connection Fittings connected to the fire protection system used by the fire department to boost the pressure and/or add water to the system.

Fire flow The total volume of water required to control the fire incident expressed in gallons per minute.

Fire-resistive construction Construction that has been designed to resist the effects of heat from fire.

Fire stream The flow of water projected from a fire nozzle (may also be called a hose stream as it is coming from the nozzle attached to the end of a fire hose).

Flameover A condition occurring in a structure fire where the fire extends across the ceiling consuming heated gases. Same as rollover.

Flanks The sides of an advancing wildland fire.

Flashover A condition during a fire in a room when the contents are heated to their ignition temperature and flames break out over the entire area almost simultaneously.

Foam The finished product of water combined with certain agents that aid in the water's ability to extinguish fires.

Foam eductor An in-line device that draws foam concentrate from a container into the hose stream.

Forest litter The components of duff, including tree limbs.

Freelance The act of performing operations without a coordinated effort or the knowledge of one's superior officer.

Free radical An atom or group of atoms (molecule) that is unstable and must combine with other atoms to achieve stability.

Fuel Anything that will burn.

Fuels management A program where naturally growing fuels, such as brush, are reduced to lessen fire intensity or to open up areas for wildlife and cattle.

Fugitive pigment A coloring agent added to fire retardant that is dropped from aircraft.

Fully encapsulated suit A suit that includes total body protection. When worn with gloves, it gives head-to-toe protection from certain chemicals. The interior is sealed from the outside air.

Fusees Road flares, sometimes called railroad flares.

Generalist A person with general knowledge of varying levels of depth in many subject areas.

Geographic information system (GIS) A geographic information system (GIS) is a system designed to capture, store, manipulate, analyze, manage, and present all types of geographical data.

Good Samaritan laws Laws stating that a person who voluntarily assists an injured person is not chargeable with responsibility for any errors or omissions in the care provided.

Ground strikes When lightning bolts reach objects on the ground, often starting fires in trees when there is no rain received with the storm.

Ground sweep nozzles Nozzles mounted underneath apparatus to sweep fire from under the vehicle.

Halogenated agents Fire-extinguishing agents containing the elements from Group 7 on the periodic table of the elements (halogens).

Harassment Coercive or repeated, unsolicited, and unwelcome verbal comments, gestures, or physical contacts, including retaliation for confrontation or reporting harassment.

Hazards Things within the environment that can cause harm to people or equipment.

Hazard trees Trees that have burned out at the base or are liable to drop large limbs. Dead trees are often called snags; these are classified as hazard trees as well.

Helibase The location from which helicopter-centered air operations are conducted.

Helispot A natural or improved takeoff and landing area intended for temporary or occasional helicopter use.

Helitack Personnel whose primary means of transportation to fires is by helicopter. They also assist in helicopter operations when the helicopter is being used for water drops or for crew and equipment transportation.

High-angle rescue Rescue utilizing ropes and other equipment. Examples are removing persons from smokestacks, wind turbines, or water towers.

High-rise building A multi-storied building that is over 75 ft in height; commonly encountered as an office building, hotel, or condominiums.

Hose lay The method of laying out hose at a fire scene.

Hydraulics The required pressure to be applied to water to overcome the effects of pressure loss because of friction in piping and fire hose.

Hydrant hookups Attaching the suction hose from the pumper to the hydrant.

Ignition temperature The minimum temperature to which a substance must be raised before it will ignite. The piloted ignition temperature is usually much lower than the auto-ignition temperature. Piloted ignition may be provided by a spark or flame or by raising the general temperature.

Immediately dangerous to life and health (IDLH) Atmosphere that is capable of causing death, irreversible adverse health effects, or the impairment of an individual's ability to escape.

Incendiary Deliberately set.

Incendiary device A device used to light a fire. This can be as simple as a lit cigarette folded into a matchbook or more complicated, involving chemical mixtures and a timer or cell phone to activate.

Incident An occurrence, either caused by humans or natural phenomena, that requires response actions to prevent or minimize loss of life or damage to property and/or the environment.

Inert A substance that will not react with other substances.

Infrared sensing devices Devices that can detect heat energy through smoke and clouds. Used for aerial mapping of fire edges and locating hot spots.

Interoperability The ability of different departments to communicate on common radio frequencies at incidents. This includes assisting fire departments and other departments, such as public works and law enforcement.

Inventory The act of accounting for all of the equipment and tools assigned to a piece of apparatus.

Inverter An electrical device that converts 12-volt current to 110 volt. Used to operate lights and tools from a vehicle's charging system.

Jake brake Common name for the Jacobs Engine Brake and other brakes of that type. Used on diesel motors.

Latent heat of vaporization The amount of heat a material must absorb when it changes from a liquid to a vapor or gas.

Lateral transfer A change of jobs from one fire department to another without moving up or down in rank.

Liaison officer A contact person for outside agencies.

Light bar Roof-mounted unit containing emergency warning lights.

Local area network A system linking the computer terminals of a department on a local scale—for example, at the different desks at headquarters. A wide area network links the computers of several different geographic locations, such as between headquarters and the fire stations.

Logging slash The remnants of logging operations, including limbs trimmed from downed trees and broken tree trunks.

Main rotor Horizontal blades that create lift for a helicopter.

Malfeasance Dishonest, intentionally illegal, or immoral action.

Manipulative training Training in the operation of tools and equipment.

Master stream appliance Large-bore nozzle equipped with a base. Not designed for hand-held use.

Maximum daily consumption The highest amount of water used in a 1-day period by the customers of a water system; computed by finding the highest amount of water used in 1 day out of 365 days. Expressed in gallons or liters.

McLeod A tool with a scraping blade on one side of the head and a rake on the other; used for fighting wildland fires.

Mechanical aptitude The ability to figure out the operation and construction of equipment from drawings.

Mentor A person who guides and directs another person toward a goal.

Mixable Capable of mixing without separation.

Misdemeanor A crime punishable by up to 1 year in a county jail or by a fine usually not to exceed $1000, or both.

Misfeasance Mistaken, careless, or inadvertent action that results in a violation of law.

Mitigation Reducing the hazard, making less severe.

Mixture A substance made up of two or more substances physically mixed together.

Mobile data computer (MDC) A computer mounted in the apparatus and connected to an antenna to provide and receive CAD information.

Model curriculum A series of courses meeting standardized criteria, including titles, descriptions, outcomes, and outlines.

Moisture content A description of the amount of moisture contained in a natural fuel, such as brush, grass, or other natural fiber. It is usually expressed as a percentage by weight.

Molecules Combined groups of atoms. Molecules composed of two or more different kinds of atoms are called compounds.

Motor block heaters Electrical devices that keep oil in the motor warm, make for easier starting, and help prevent damage when the motor is started in cold weather.

Mucous membranes Inside of the nose, mouth, and covering of the eye.

Mutual aid System in which departments draw up an agreement that they will assist each other upon request.

National response plan The plan that delineates the all-discipline, all-hazards response and responsibilities of all federal agencies for the management of domestic incidents in the United States.

Nonfeasance A failure to act when action is required.

Nonmixable Not capable of mixing; will separate.

Objectives Steps to be taken to achieve goals.

Occupancy The use or intended use of a building.

On duty The time fire fighters spend performing their jobs.

One-hour fire-rated separation A fire-rated assembly that should resist breakthrough for a period of 1 hour. An example of this type of construction is the use of 5/8-in.-thick fire-rated gypsum (15.875 mm) wallboard or a combination of wallboard and plaster. All of the electrical boxes must be metal and not plastic. Any penetrations through the assembly must be properly protected to prevent the spread of fire.

Open screw and yoke valve (OS and Y) A valve with a hand wheel that exposes a threaded rod when in the open position. The hand wheel looks much like a steering wheel and can be locked with a chain and padlock so it cannot turn.

Operational period The period of time scheduled for execution of a given set of tactical actions as specified in the IAP. It may be as long as 24 hours for a wildland incident or as short as an hour for a hazardous materials incident.

Operational procedures Written procedures for performing operational functions.

Oral interview panel An interview technique in which the interviewers ask questions and evaluate the answers given by job candidates. They assign a score to the candidate's responses for ranking purposes during the selection process.

Overhaul The operation performed to ensure that all embers are extinguished after a fire is controlled.

Oxidation The chemical combination of any substance with an oxidizer.

Oxidizer A substance that gains electrons in a chemical reaction.

Paramedic A person with an advanced level of medical training. Paramedics can perform invasive procedures on the patient, such as starting intravenous lines.

Parapet walls Walls that extend above the roof line.

Pathogens Disease-causing agents.

PCB oil Oil containing polychlorinated biphenyl, a compound that can cause cancer.

Peak hourly consumption The highest amount of water used in 1 hour of 1 day; determined by finding the highest use per hour in a 24-hour period, and expressed in gallons or liters.

Performance-based certification A training program in which the student must meet prerequisites of education and experience to take a training course. Once the course is completed, the student then completes a task book for the position, requiring classroom and/or incident experience. Once the task book is completed, the person receives certification for the new position.

Perimeter control Controlling the edges of a wildland fire.

Perimeters Boundaries for controlled access (hazardous materials). The fire's edge (wildland).

Perjury False statements in a sworn document or testimony.

Petrochemical Related to oil refining and production facilities.

Placards Signs or notices for display in a public place.

Plume (1) The path of the material released from a container in a hazardous materials incident. (2) The smoke column from a fire.

Polar solvents Liquids that will mix readily with water because water is a polar substance. The common polar solvents are alcohols, aldehydes, esters, ketones, and organic acids.

Policies General guidelines of how things will be done.

Positive pressure mode SCBA regulator function that keeps positive pressure in the mask face piece at all times.

Post indicator valve (PI) A valve with an indicator body that sticks up out of the ground. The body has a small window that says either "shut" or "open" depending on the position of the valve.

Post-Traumatic Stress Disorder (PTSD) A disorder that develops in some people who have experienced a shocking, scary, or dangerous event.

Power shears Cutting attachment for a rescue tool.

Prescribed burning Planned application of fire under specified conditions in a predetermined area to achieve management objectives; includes removal or modification of fuels, clearing paths through brush, and killing unwanted plant growth.

Primary search A rapid search of all involved and exposed areas that are affected by the fire but can be entered, to verify removal and/or safety of all occupants. Should this not be possible, a secondary search is conducted as soon as it is safe to do so.

Probationary fire fighter A person hired by the fire department who has not been granted permanent status.

Pulaski A tool for fighting wildland fires with an axe on one side of the head and a grub hoe on the other.

Pump and roll A tactic used in grass fires utilizing pumpers that can drive while the pump is operating. Hose lines are connected to the apparatus, and water is sprayed to extinguish the fire edge.

Pyrolysis The chemical decomposition of matter through the action of heat.

Quality assurance (QA) Checking work to establish a consistent standard of care provided to patients at incidents by trained medical personnel.

Rappel Descend by means of a rope.

Recurring exposure to trauma (RET) The effects of responding numerous times in one's career to traumatic events and their effect on the responder.

Reforestation The planting of seedling trees in areas destroyed by fire or logging.

Regulations Rules designed to implement a statute based on an agency's interpretation of that statute.

Rehab Short for rehabilitation. A time in which fire fighters rest, cool off, and rehydrate.

Rehabilitated (1) To rehabilitate personnel means that they rest, cool off, and replenish body fluids. (2) To rehabilitate a fire line means to construct water bars to direct water runoff and prevent erosion. Under the federal "Minimum Impact Suppression Tactics" policy, rehabilitation may mean the erasure of fire lines as much as possible. In other words, to cause as little damage as possible controlling the fire as fire is a part of the natural environment.

Rekindle To reignite after a fire was thought to be extinguished. This commonly happens in attics, basements, and walls of structure fires and in logs on wildland fires. This is usually due to incomplete overhaul/mop-up.

Relief valve Device used to release unwanted pressure.

Repeater A device that receives radio transmissions, boosts the signal, and retransmits the signal. It is used in areas where topography or tall buildings interrupt clear communications.

Reserve/cadet programs Organized programs sponsored by paid fire departments that provide training in return for personnel volunteering their time.

Resource designators An identification system using numbers and letters to identify resources by agency and type.

Résumé A listing of a person's areas of experience and education.

Retardant A material spread on fuels that inhibits their burning.

Retard chamber A small tank attached to sprinkler systems that allows pressure surges to dissipate their energy before they enter the system and set off the water flow alarm.

Retroreflective A surface, material, or device (retroreflector) that reflects light or other radiation back to its source; reflective.

Risks The chance that humans take in relationship to the hazard(s).

Risk management Any activity that involves the evaluation or comparison of risks and the development of approaches that change the probability or the consequences of a harmful action.

Rollover A condition occurring in a structure fire where the fire extends across the ceiling consuming heated gases. Same as flameover.

Rule One of a set of explicit or understood regulations or principles governing conduct within a particular activity or sphere.

Safety section A body of law that sets the retirement benefit rate for certain professions, primarily those that are high hazard and deal with public safety, namely fire and police.

Safety zone A place where fire fighters can be safe from the incident's hazards.

Salvage A firefighting procedure for protecting building contents from damage due to water or falling debris.

Salvage covers Tarps carried on fire apparatus used to cover building contents to prevent damage from water and falling debris. They may also be used to create water chutes to remove water from a building or to build a temporary sump when combined with ladders.

Scope of employment The complete range of activities an employee might reasonably be expected to perform while carrying out the business of the employer.

Scratch line A quickly created wildland fire control line, constructed using hand tools.

Skids The long tubular shaped feet that helicopters sit on when on the ground.

Sky lobby A lobby on a high floor level in a high-rise building. Elevators leave from this area to service the upper floors.

Sling load Material transported by being placed in a net suspended underneath a helicopter.

Smoke jumpers Highly trained personnel who parachute in to suppress fires in remote areas.

Span of control The number of subordinates a manager can directly control. The number varies with the complexity of the operation and the skill of the subordinates.

Specialists People with extensive training in one or more areas of operations or information.

Specific heat The ratio between the amount of heat necessary to raise the temperature of a substance compared to the amount of heat required to raise the same weight of water the same number of degrees. Water has a specific heat value of 1.

Spot fires In heavier fuels, flying fire brands can land outside the fire perimeter and start new fires.

Staging areas Temporary locations at an incident where personnel and equipment are kept while waiting for tactical assignments.

Standard A document, the main text of which contains only mandatory provisions and is in a form suitable for mandatory reference by another standard or code or adoption into law.

Standard operating guidelines (SOG) Another term for Standard Operating Procedure (SOP).

Standard operating procedures (SOP) A particular way of accomplishing something or acting in a specified situation.

Standpipe system Plumbing system installed in multistory buildings for fire department use with outlets on each floor for attaching fire hose.

Static water source Pond, lake, or tank used to supply fire engines.

Statutory laws Laws adopted by Congress (federal statutes) and those that have been passed by state legislatures (state statutes).

Strategy The method used to coordinate the tactical operations of units to achieve the desired incident objectives.

Stress drills Drills conducted under realistic conditions to develop and test the fire fighters' ability to perform in stressful situations.

Structure protection Protecting structures in danger of being consumed by an advancing wildland fire.

Subsurface foam injection Plumbing installed on a tank to allow for the introduction of foam under the surface of the contained liquid.

Tactics Actions taken to achieve strategies.

Tactical support A vehicle equipped to provide the needs of fire fighters at the emergency scene. See *rehab*.

Tail rotor Vertical propeller used for steering control that is installed on the tail of a helicopter.

Tasks Specific things that must be done to bring a job to completion.

Task book Log used to verify competency in particular skills. A trainer must certify that the skill was performed in a satisfactory manner in a field and/or classroom environment.

Thrust block A mass of concrete poured on the outside of an angle fitting and extending back to native soil. The purpose is to prevent surges in flow through a pipe from flexing the fitting and wiggling it in the ground, which would, over time, form a larger and larger underground space, possibly allowing the pipe fitting to pull apart.

Tort A civil wrong leading to a legal claim for damages.

Toxic A substance that is poisonous and possibly lethal.

Toxicology The science of materials that are poisonous to living things, especially humans and animals.

Trail drop This is when the aircraft drops a specified coverage level for a long distance. Versus a salvo drop when a large amount of retardant or water is dropped on a small area, such as a patch of dead trees. Another type of drop is the segmented drop when the tanker drops in one area, without releasing its full load, then flies on and drops in other areas.

Training The pursuit of a particular skill.

Transfer of command The transfer of the role of incident commander from one person to another.

Turret nozzles Roof- or bumper-mounted nozzles remotely controlled from inside the cab.

Unburned islands Areas of unburned fuel within a fire perimeter.

Unity of command The organizational principle in which every individual is accountable to only one designated supervisor to whom he or she reports at the scene of an incident.

Upper division College-level courses that are applicable to a degree program for a bachelor's degree or higher.

Urban interface The area where built-up areas of homes and businesses have little separation from the natural-growing wildland area.

Urban intermix Buildings interspersed in wildland areas; homes built in the woods.

Vacuum trucks Tank trucks equipped with a pump that evacuates the air from inside the tank, causing it to draw a vacuum. Used for picking up liquids from spills or tanks, commonly used at crude oil production facilities.

Validate To make sure that the items included in the testing process are actual requirements of the job.

Vertical arrangement The manner in which a fuel is arranged vertically above ground, divided into ground fuels, surface fuels, and aerial fuels.

Veterans' points Points added to a person's final score on a competitive examination process; given to persons who have satisfactorily performed military service.

Voice over A presentation technique, often used in television, where the audience sees video and hears the voice of the narrator but does not see the narrator.

Volunteer firefighting Performing firefighting services without pay. In some areas a variation of this is the Paid Call Fire Fighter program. Under this program, fire fighters are paid a specified sum when they respond to incidents or attend training.

Water curtain A screen of water spray set up to protect exposures.

Water hammer A pressure surge or wave caused by water in motion when it is stopped suddenly.

Water mains A pipe that carries water in a water system.

Watersheds Complex geographic, geologic, and vegetative components that control runoff of rainwater and support varied ecosystems.

Water table The underground depth at which point the ground is totally saturated by water.

Wildland Open land in its natural state.

Wildland fire An unplanned fire burning in vegetation.

Worker's compensation Money paid to persons who have been injured in the course of their employment and are unable to work either temporarily or permanently.

Working pressure The pounds per square inch of pressure that a tank is designed to contain.

Index

Note: Page numbers followed by *f* or *t* indicate material in figures or tables, respectively.